SCIENTIFIC AMERICAN | Resource Library

Life Sciences 7

SCIENTIFIC AMERICAN

Resource Library

READINGS IN THE

Life Sciences   VOLUME 7

OFFPRINTS 1074–1120

W. H. FREEMAN AND COMPANY   San Francisco

Each of the articles in this volume is available as a separate Offprint. For a complete listing of Offprints available in the Life Sciences, Chemistry, Physics, Technology, Psychology, the Social Sciences, the History and Philosophy of Science, and the Earth Sciences, write to W. H. Freeman and Company, 660 Market Street, San Francisco, California 94104 or to W. H. Freeman and Company, Ltd., Warner House, Folkstone, Kent.

Printed in the United States of America.
Library of Congress Catalog Card Number: 71-87739
Standard Book Number: 7167 0989-9

Organization of the  Resource Library

Subject Series:

The *SCIENTIFIC AMERICAN Resource Library* is a multi-volume compilation of more than 700 articles selected from the magazine. These are organized under five subject classifications.

Readings in Earth Sciences (2 volumes)
Readings in Life Sciences (7 volumes)
Readings in Physical Sciences and Technology (3 volumes)
Readings in Psychology (2 volumes)
Readings in Social Sciences (1 volume)

Numbering System:

Each article is numbered and the articles in the volumes are arranged in numerical order. The numbers assigned to each series are:

Earth Sciences, 801–999.
Life Sciences, 1–199 and 1001–1999.
Physical Sciences and Technology, 201–399.
Psychology, 401–599.
Social Sciences, 601–799.

Topic Index:

This index classifies the Readings in Life Sciences by topic. Note that articles from other subject series that are relevant to the topic are also listed.

Author Index:

The authors of the articles in all five subject series are given. The numbers after the authors' names are the article numbers, not page numbers.

Scientific American Offprints:

Every article in these volumes is published separately in the SCIENTIFIC AMERICAN Offprint Series and may be purchased in any quantity. Order by article number and title.
The number of each offprint corresponds to the number on each article in the Resource Library.
A catalog of SCIENTIFIC AMERICAN offprints may be obtained from the publisher.

Additions to the Series:

New titles are added to the SCIENTIFIC AMERICAN Offprint Series every month and when enough new articles on a subject are available, a new bound volume will be added to the *SCIENTIFIC AMERICAN Resource Library.*

W. H. Freeman and Company
660 Market Street, San Francisco, California 94104
Warner House, Folkestone, Kent, England

Contents

VOLUME **2**

Topic Index
Author Index

VOLUME **5**

Topic Index
Author Index

VOLUME **6**

Topic Index
Author Index

VOLUME **7**

Topic Index
Author Index

Topic Index

(This index includes not only the articles in Readings in the Life Sciences but also relevant articles from other subject series.)

THE CELL AND ITS ENERGETICS

Author Index

(The authors of the articles in all five subject series are given here.)

SCIENTIFIC AMERICAN | Resource Library

Life Sciences   7

SCIENTIFIC
AMERICAN   May 1967, Vol. 216, No. 5, pp. 80-94

OFFPRINT 1074

GENE STRUCTURE AND PROTEIN STRUCTURE

by Charles Yanofsky

A linear correspondence between these two chainlike
molecules was postulated more than a dozen years ago.
Here is how the correspondence was finally demonstrated.

The present molecular theory of genetics, known irreverently as "the
central dogma," is now 14 years
old. Implicit in the theory from the out-
set was the notion that genetic information is coded in linear sequence in molecules of deoxyribonucleic acid (DNA)
and that the sequence directly determines the linear sequence of amino acid
units in molecules of protein. In other
words, one expected the two molecules
to be colinear. The problem was to prove
that they were.

Over the same 14 years, as a consequence of an international effort, most
of the predictions of the central dogma
have been verified one by one. The results were recently summarized in these
pages by F. H. C. Crick, who together
with James D. Watson proposed the
helical, two-strand structure for DNA on

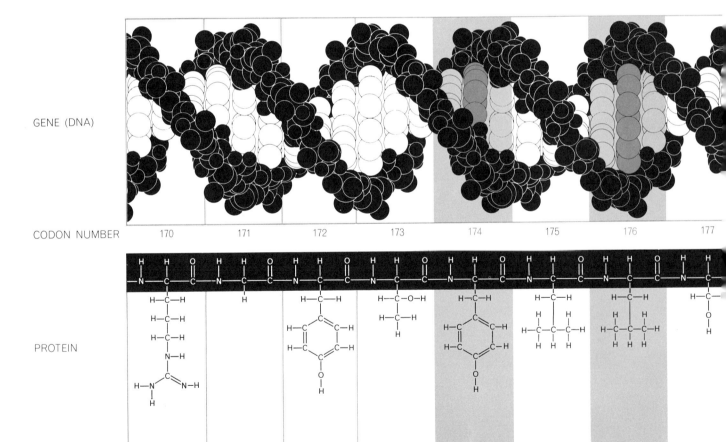

GENE (DNA)

| CODON NUMBER | 170 | 171 | 172 | 173 | 174 | 175 | 176 | 177 |

PROTEIN

| AMINO ACID | ARG | GLY | TYR | THR | TYR→CYS | LEU | LEU→ARG | SER |
| | 170 | 171 | 172 | 173 | 174 | 175 | 176 | 177 |

STRUCTURES OF GENE AND PROTEIN have been shown to
bear a direct linear correspondence by the author and his colleagues at Stanford University. They demonstrated that a particular
sequence of coding units (codons) in the genetic molecule deoxyribonucleic acid, or DNA (top), specifies a corresponding sequence
of amino acid units in the structure of a protein molecule (bottom).

In the DNA molecule depicted here the black spheres represent repeating units of deoxyribose sugar and phosphate, which
form the helical backbones of the two-strand molecule. The white
spheres connecting the two strands represent complementary pairs
of the four kinds of base that provide the "letters" in which the
genetic message is written. A sequence of three bases attached to

which the central dogma is based [see "The Genetic Code: III," by F. H. C. Crick; SCIENTIFIC AMERICAN Offprint 1052]. Here I shall describe in somewhat more detail how our studies at Stanford University demonstrated the colinearity of genetic structure (as embodied in DNA) and protein structure.

Let me begin with a brief review. The molecular subunits that provide the "letters" of the code alphabet in DNA are the four nitrogenous bases adenine (A), guanine (G), cytosine (C) and thymine (T). If the four letters were taken in pairs, they would provide only 16 different code words—too few to specify the 20 different amino acids commonly found in protein molecules. If they are taken in triplets, however, the four letters can provide 64 different code words, which would seem too many for the efficient specification of the 20 amino acids. Accordingly it was conceivable that the cell might employ fewer than the 64 possible triplets. We now know that na-

ture not only has selected the triplet code but also makes use of most (if not all) of the 64 triplets, which are called codons. Each amino acid but two (tryptophan and methionine) are specified by at least two different codons, and a few amino acids are specified by as many as six codons. It is becoming clear that the living cell exploits this redundancy in subtle ways. Of the 64 codons, 61 have been shown to specify one or another of the 20 amino acids. The remaining three can act as "chain terminators," which signal the end of a genetic message.

A genetic message is defined as the amount of information in one gene; it is the information needed to specify the complete amino acid sequence in one polypeptide chain. This relation, which underlies the central dogma, is sometimes expressed as the one-gene-one-enzyme hypothesis. It was first clearly enunciated by George W. Beadle and Edward L. Tatum, as a result of their studies with the red bread mold *Neurospora crassa* around 1940. In some cases

a single polypeptide chain constitutes a complete protein molecule, which often acts as an enzyme, or biological catalyst. Frequently, however, two or more polypeptide chains must join together in order to form an active protein. For example, tryptophan synthetase, the enzyme we used in our colinearity studies, consists of four polypeptide chains: two alpha chains and two beta chains.

How might one establish the colinearity of codons in DNA and amino acid units in a polypeptide chain? The most direct approach would be to separate the two strands of DNA obtained from some organism and determine the base sequence of that portion of a strand which is presumed to be colinear with the amino acid sequence of a particular protein. If the amino acid sequence of the protein were not already known, it too would have to be established. One could then write the two sequences in adjacent columns and see if the same codon (or its synonym) always appeared adjacent to a particular amino acid. If it

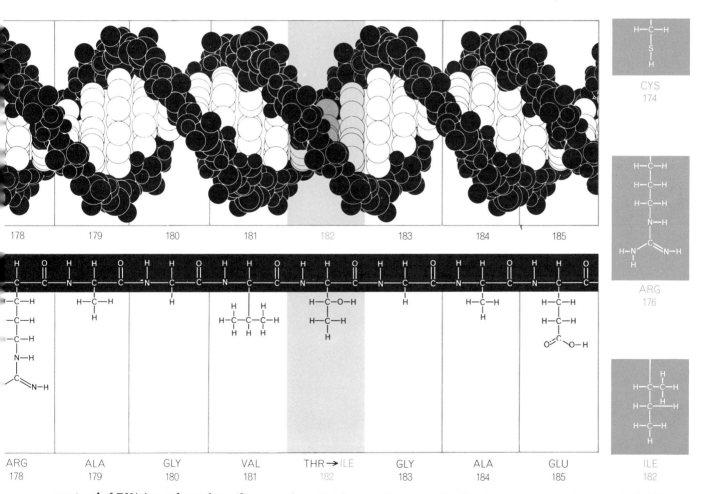

one strand of DNA is a codon and specifies one amino acid. The amino acid sequence illustrated here is the region from position 170 through 185 in the A protein of the enzyme tryptophan synthetase produced by the bacterium *Escherichia coli*. It was found that mutations in the A gene of *E. coli* altered the amino acids at three places (174, 176 and 182) in this region of the A protein. (A key to the amino acid abbreviations can be found on page 2447.) Three amino acids that replace the three normal ones as a result of mutation are shown at the extreme right. Each replacement is produced by a mutation at one site (*dark color*) in the DNA of the A gene. In all, the author and his associates correlated mutations at eight sites in the A gene with alterations in the A protein.

did, a colinear relation would be established. Unfortunately this direct approach cannot be taken because so far it has not been possible to isolate and identify individual genes. Even if one could isolate a single gene that specified a polypeptide made up of 150 amino acids (and not many polypeptides are that

small), one would have to determine the sequence of units in a DNA strand consisting of some 450 bases.

It was necessary, therefore, to consider a more feasible way of attacking the problem. An approach that immediately suggests itself to a geneticist is to construct a genetic map, which is a

representation of the information contained in the gene, and see if the map can be related to protein structure. A genetic map is constructed solely on the basis of information obtained by crossing individual organisms that differ in two or more hereditary respects (a refinement of the technique originally

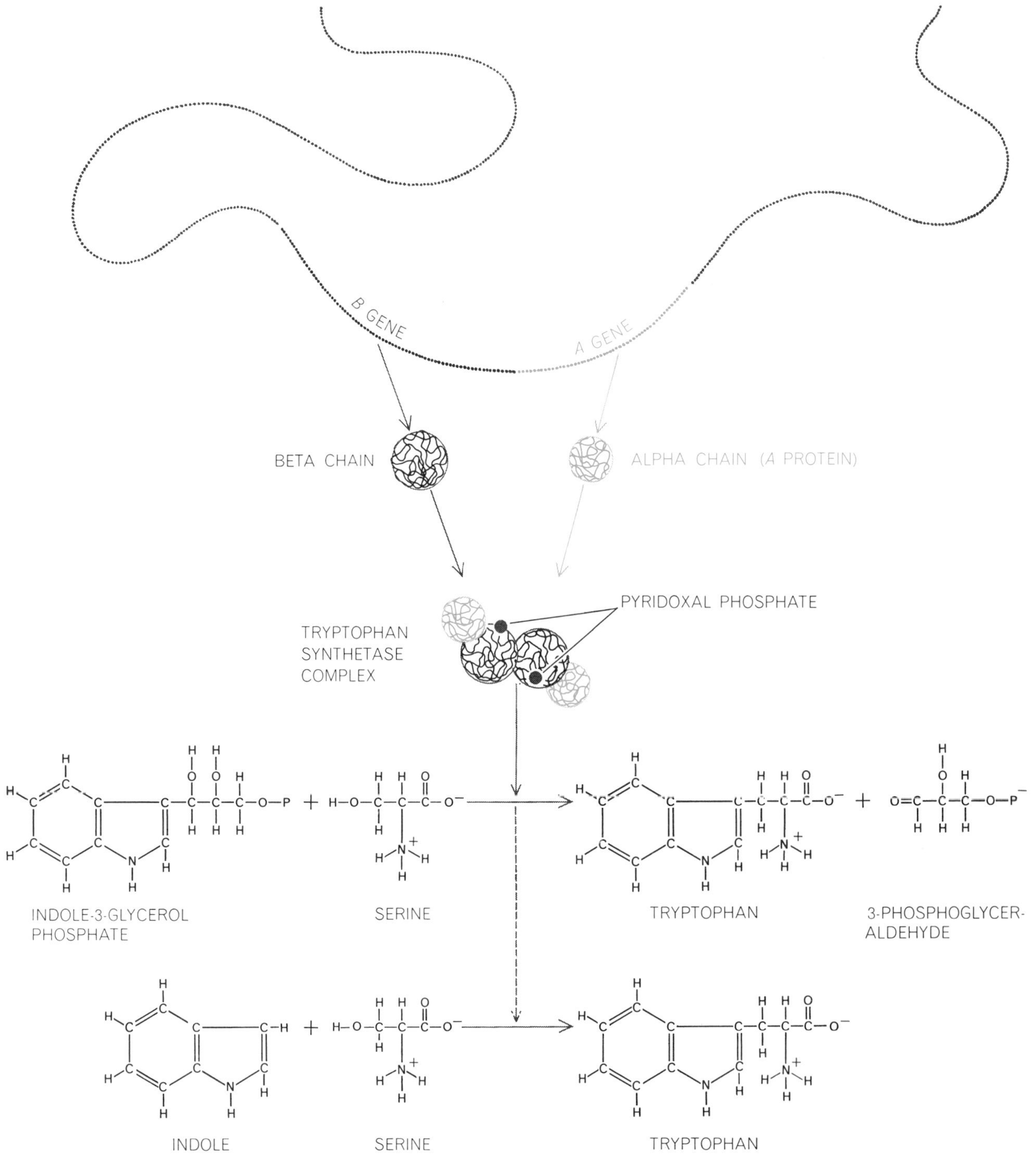

INDOLE-3-GLYCEROL PHOSPHATE

SERINE

TRYPTOPHAN

3-PHOSPHOGLYCER-ALDEHYDE

INDOLE

SERINE

TRYPTOPHAN

GENETIC CONTROL OF CELL'S CHEMISTRY is exemplified by the two genes in *E. coli* that carry the instructions for making the enzyme tryptophan synthetase. The enzyme is actually a complex of four polypeptide chains: two alpha chains and two beta chains. The alpha chain is the *A* protein in which changes produced by mutations in the *A* gene have provided the evidence for gene-pro-

tein colinearity. One class of *A*-protein mutants retains the ability to associate with beta chains but the complex is no longer able to catalyze the normal biochemical reaction: the conversion of indole-3-glycerol phosphate and serine to tryptophan and 3-phosphoglyceraldehyde. But the complex can still catalyze a simpler nonphysiological reaction: the conversion of indole and serine to tryptophan.

used by Gregor Mendel to demonstrate how characteristics are inherited).

By using bacteria and bacterial viruses in such studies one can catalogue the results of crosses involving millions of individual organisms and thereby deduce the actual distances separating the sites of mutational changes in a single gene. The distances are inferred from the frequency with which parent organisms, each with at least one mutation in the same gene, give rise to offspring in which neither mutation is present. As a result of the recombination of genetic material the offspring can inherit a gene that is assembled from the mutation-free portions of each parental gene. If the mutational markers lie far apart on the parental genes, recombination will frequently produce mutation-free progeny. If the markers are close together, mutation-free progeny will be rare [see bottom illustration on next page].

In his elegant studies with the "rII" region of the chromosome of the bacterial virus designated T4, Seymour Benzer, then at Purdue University, showed that the number of genetically distinguishable mutation sites on the map of the gene approaches the estimated number of base pairs in the DNA molecule corresponding to that gene. (Mutations involve pairs of bases because the bases in each of the two entwined strands of the DNA molecule are paired with and are complementary to the bases in the other strand. If a mutation alters one base in the DNA molecule, its partner is eventually changed too during DNA replication.) Benzer also showed that the only type of genetic map consistent with his data is a map on which the sites altered by mutation are arranged linearly. Subsequently A. D. Kaiser and David Hogness of Stanford University demonstrated with another bacterial virus that there is a linear correspondence between the sites on a genetic map and the altered regions of a DNA molecule isolated from the virus. Thus there is direct experimental evidence indicating that the genetic map is a valid representation of DNA structure and that the map can be employed as a substitute for information about base sequence.

This, then, provided the basis of our approach. We would pick a suitable organism and isolate a large number of mutant individuals with mutations in the same gene. From recombination studies we would make a fine-structure genetic map relating the sites of the mutations. In addition we would have to be able to isolate the protein specified by that gene and determine its amino acid

sequence. Finally we would have to analyze the protein produced by each mutant (assuming a protein were still produced) in order to find the position of the amino acid change brought about in its amino acid sequence by the mutation. If gene structure and protein structure were colinear, the positions at which amino acid changes occur in the protein should be in the same order as the positions of the corresponding mutationally altered sites on the genetic map. Although this approach to the question of colinearity would require a great deal of work and much luck, it was logical and experimentally feasible. Several research groups besides our own set out to find a suitable system for a study of this kind.

The essential requirement of a suitable system was that a genetically

ALA	ALANINE	GLY	GLYCINE	PRO	PROLINE
ARG	ARGININE	HIS	HISTIDINE	SER	SERINE
ASN	ASPARAGINE	ILE	ISOLEUCINE	THR	THREONINE
ASP	ASPARTIC ACID	LEU	LEUCINE	TRP	TRYPTOPHAN
CYS	CYSTEINE	LYS	LYSINE	TYR	TYROSINE
GLN	GLUTAMINE	MET	METHIONINE	VAL	VALINE
GLU	GLUTAMIC ACID	PHE	PHENYLALANINE		

AMINO ACID ABBREVIATIONS identify the 20 amino acids commonly found in all proteins. Each amino acid is specified by a triplet codon in the DNA molecule (see below).

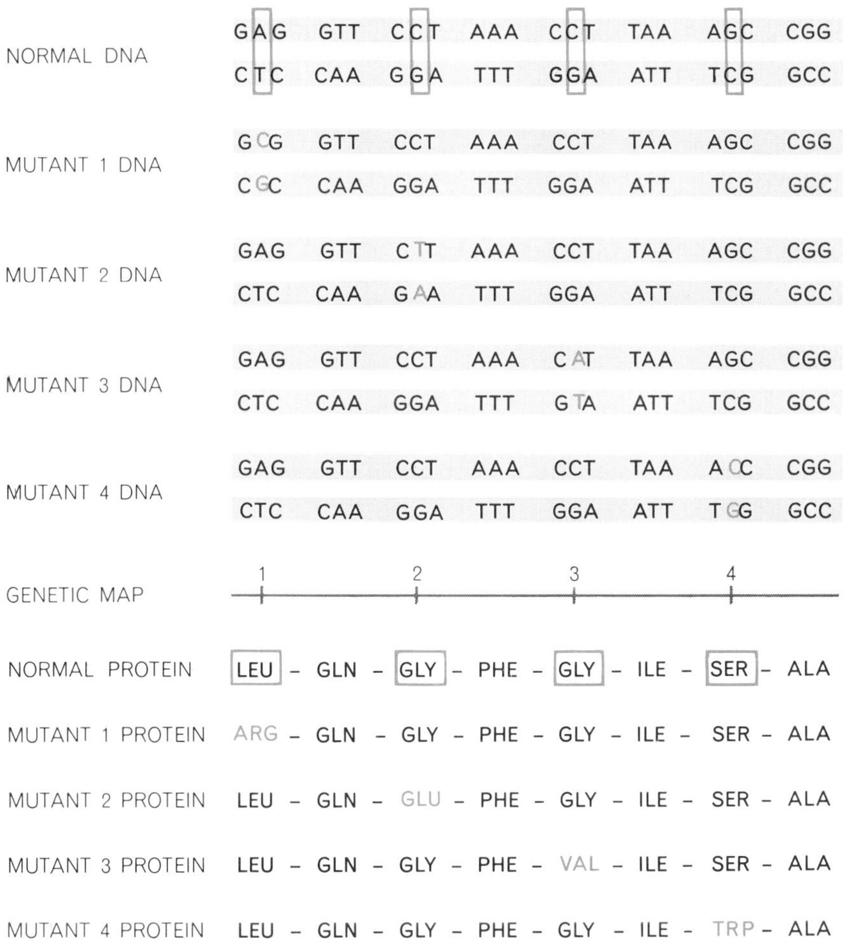

NORMAL DNA	GAG	GTT	CCT	AAA	CCT	TAA	AGC	CGG
	CTC	CAA	GGA	TTT	GGA	ATT	TCG	GCC
MUTANT 1 DNA	GCG	GTT	CCT	AAA	CCT	TAA	AGC	CGG
	CGC	CAA	GGA	TTT	GGA	ATT	TCG	GCC
MUTANT 2 DNA	GAG	GTT	CTT	AAA	CCT	TAA	AGC	CGG
	CTC	CAA	GAA	TTT	GGA	ATT	TCG	GCC
MUTANT 3 DNA	GAG	GTT	CCT	AAA	CAT	TAA	AGC	CGG
	CTC	CAA	GGA	TTT	GTA	ATT	TCG	GCC
MUTANT 4 DNA	GAG	GTT	CCT	AAA	CCT	TAA	ACC	CGG
	CTC	CAA	GGA	TTT	GGA	ATT	TGG	GCC

GENETIC MAP   1   2   3   4

NORMAL PROTEIN   LEU – GLN – GLY – PHE – GLY – ILE – SER – ALA

MUTANT 1 PROTEIN   ARG – GLN – GLY – PHE – GLY – ILE – SER – ALA

MUTANT 2 PROTEIN   LEU – GLN – GLU – PHE – GLY – ILE – SER – ALA

MUTANT 3 PROTEIN   LEU – GLN – GLY – PHE – VAL – ILE – SER – ALA

MUTANT 4 PROTEIN   LEU – GLN – GLY – PHE – GLY – ILE – TRP – ALA

GENETIC MUTATIONS can result from the alteration of a single base in a DNA codon. The letters stand for the four bases: adenine (A), thymine (T), guanine (G) and cytosine (C). Since the DNA molecule consists of two complementary strands, a base change in one strand involves a complementary change in the second strand. In the four mutant DNA sequences shown here (top) a pair of bases (color) is different from that in the normal sequence. By genetic studies one can map the sequence and approximate spacing of the four mutations (middle). By chemical studies of the proteins produced by the normal and mutant DNA sequences (bottom) one can establish the corresponding amino acid changes.

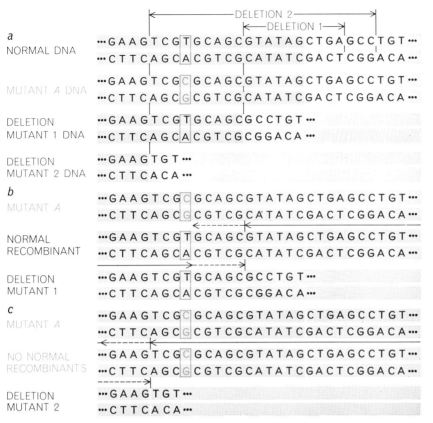

a
NORMAL DNA

···G A A G T C G T G C A G C G T A T A G C T G A G C C T G T ···
···C T T C A G C A C G T C G C A T A T C G A C T C G G A C A ···

MUTANT *A* DNA

···G A A G T C G C G C A G C G T A T A G C T G A G C C T G T ···
···C T T C A G C G C G T C G C A T A T C G A C T C G G A C A ···

DELETION
MUTANT 1 DNA

···G A A G T C G T G C A G C G C C T G T ···
···C T T C A G C A C G T C G C G G A C A ···

DELETION
MUTANT 2 DNA

···G A A G T G T ···
···C T T C A C A ···

b
MUTANT *A*

···G A A G T C G C G C A G C G T A T A G C T G A G C C T G T ···
···C T T C A G C G C G T C G C A T A T C G A C T C G G A C A ···

NORMAL
RECOMBINANT

···G A A G T C G T G C A G C G T A T A G C T G A G C C T G T ···
···C T T C A G C A C G T C G C A T A T C G A C T C G G A C A ···

DELETION
MUTANT 1

···G A A G T C G T G C A G C G C C T G T ···
···C T T C A G C A C G T C G C G G A C A ···

c
MUTANT *A*

···G A A G T C G C G C A G C G T A T A G C T G A G C C T G T ···
···C T T C A G C G C G T C G C A T A T C G A C T C G G A C A ···

NO NORMAL
RECOMBINANTS

···G A A G T C G C G C A G C G T A T A G C T G A G C C T G T ···
···C T T C A G C G C G T C G C A T A T C G A C T C G G A C A ···

DELETION
MUTANT 2

···G A A G T G T ···
···C T T C A C A ···

"DELETION" MUTANTS provide one approach to making a genetic map. Here (*a*) normal DNA and mutant *A* differ by only one base pair (*C–G* has replaced *T–A*) in a certain portion of the *A* gene (*colored area*). In deletion mutant 1 a sequence of 10 base pairs, including six pairs from the *A* gene, has been spontaneously deleted. In deletion mutant 2, 22 base pairs, including 15 pairs from the *A* gene, have been deleted. By crossing mutant *A* with the two different deletion mutants in separate experiments (*b, c*), one can tell whether the mutated site (*C–G*) in the *A* gene falls inside or outside the deleted regions. A normal-type recombinant will appear (*b*) only if the altered base pair falls outside the deleted region.

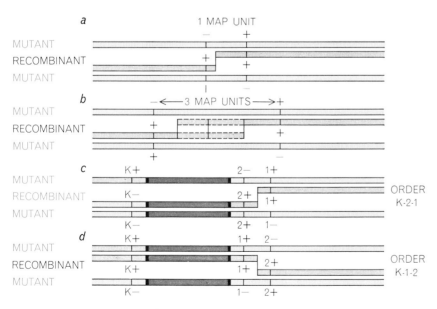

OTHER MAPPING METHODS involve determination of recombination frequency (*a, b*) and the distribution of outside markers (*c, d*). The site of a mutational alteration is indicated by "–," the corresponding unaltered site by "+." If the altered sites are widely spaced (*b*), normal recombinants will appear more often than if the altered sites are close together (*a*). In the second method the mutants are linked to another gene that is either normal (*K⁺*) or mutated (*K⁻*). Recombinant strains that contain 1⁺ and 2⁺ will carry the *K⁻* gene if the correct order is *K–2–1*. They will carry the *K⁺* gene if the order is *K–1–2*.

mappable gene should specify a protein whose amino acid sequence could be determined. Since no such system was known we had to gamble on a choice of our own. Fortunately we were studying at the time how the bacterium *Escherichia coli* synthesizes the amino acid tryptophan. Irving Crawford and I observed that the enzyme that catalyzed the last step in tryptophan synthesis could be readily separated into two different protein species, or subunits, one of which could be clearly isolated from the thousands of other proteins synthesized by *E. coli*. This protein, called the tryptophan synthetase A protein, had a molecular weight indicating that it had slightly fewer than 300 amino acid units. Furthermore, we already knew how to force *E. coli* to produce comparatively large amounts of the protein—up to 2 percent of the total cell protein—and we also had a collection of mutants in which the activity of the tryptophan synthetase A protein was lacking. Finally, the bacterial strain we were using was one for which genetic procedures for preparing fine-structure maps had already been developed. Thus we could hope to map the A gene that presumably controlled the structure of the A protein.

To accomplish the mapping we needed a set of bacterial mutants with mutational alterations at many different sites on the A gene. If we could determine the amino acid change in the A protein of each of these mutants, and discover its position in the linear sequence of amino acids in the protein, we could test the concept of colinearity. Here again we were fortunate in the nature of the complex of subunits represented by tryptophan synthetase.

The normal complex consists of two A-protein subunits (the alpha chains) and one subunit consisting of two beta chains. Within the bacterial cell the complex acts as an enzyme to catalyze the reaction of indole-3-glycerol phosphate and serine to produce tryptophan and 3-phosphoglyceraldehyde [*see illustration on page 2446*]. If the A protein undergoes certain kinds of mutations, it can still form a complex with the beta chains, but the complex loses the ability to catalyze the reaction. It retains the ability, however, to catalyze a simpler reaction when it is tested outside the cell: it will convert indole and serine to tryptophan. There are still other kinds of A-gene mutants that evidently lack the ability to form an A protein that can combine with beta chains; thus these strains are not able to catalyze even the simpler reaction. The first class of mutants—those that produce an A protein that is still able to combine with beta chains and exhibit catalytic activity when they are tested outside the cell—proved to be the most important for our study.

A fine-structure map of the A gene was constructed on the basis of genetic crosses performed by the process called transduction. This employs a particular bacterial virus known as transducing phage *P1kc*. When this virus multiplies in a bacterium, it occasionally incorporates a segment of the bacterial DNA within its own coat of protein. When the virus progeny infect other bacteria, genetic material of the donor bacteria is introduced into some of the recipient cells. A fraction of the recip-

ients survive the infection. In these survivors segments of the bacterium's own genetic material pair with like segments of the "foreign" genetic material and recombination between the two takes place. As a result the offspring of an infected bacterium can contain characteristics inherited from its remote parent as well as from its immediate one.

In order to establish the order of mutationally altered sites in the A gene we have relied partly on a set of mutant bacteria in which one end of a deleted segment of DNA lies within the A gene. In each of these "deletion" mutants a segment of the genetic material of the bacterium was deleted spontaneously.

Thus each deletion mutant in the set retains a different segment of the A gene. This set of mutants can now be crossed with any other mutant in which the A gene is altered at only a single site. Recombination can give rise to a normal gene only if the altered site does not fall within the region of the A gene that is missing in the deletion mutant [*see top illustration on opposite page*]. By crossing many A-protein mutants with the set of deletion mutants one can establish the linear order of many of the mutated sites in the A gene. The ordering is limited only by the number of deletion mutants at one's disposal.

A second method, which more closely

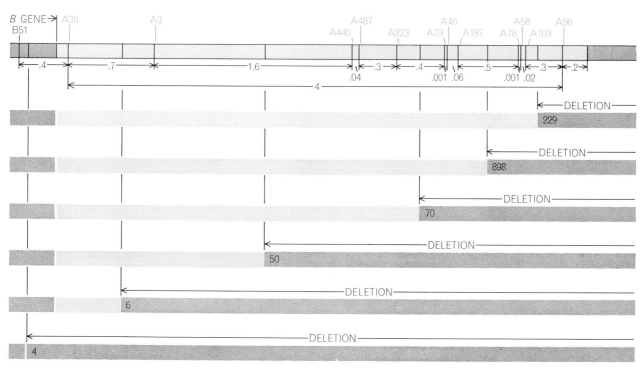

MAP OF A GENE shows the location of mutationally altered sites, drawn to scale, as determined by the three genetic-mapping methods illustrated on the opposite page. The total length of the A gene is slightly over four map units (probably 4.2). Below map are six

deletion mutants that made it possible to assign each of the 12 A-gene mutants to one of six regions within the gene. The more sensitive mapping methods were employed to establish the order of mutations and the distance between mutation sites within each region.

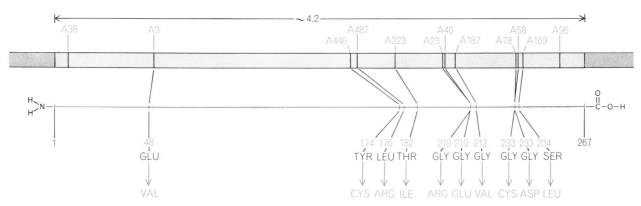

COLINEARITY OF GENE AND PROTEIN can be inferred by comparing the A-gene map (*top*) with the various amino acid

changes in the A protein (*bottom*), both drawn to scale. The amino acid changes associated with 10 of the 12 mutations are also shown.

MET – GLN – ARG – TYR – GLU – SER – LEU – PHE – ALA – GLN – LEU – LYS – GLU – ARG – LYS – GLU – GLY – ALA – PHE – VAL –
1   20

PRO – PHE – VAL – THR – LEU – GLY – ASP – PRO – GLY – ILE – GLU – GLN – SER – LEU – LYS – ILE – ASP – THR – LEU – ILE –
21   40

A3

GLU – ALA – GLY – ALA – ASP – ALA – LEU – GLU – LEU – GLY – ILE – PRO – PHE – SER – ASP – PRO – LEU – ALA – ASP – GLY –
41   60
                                                VAL

PRO – THR – ILE – GLN – ASN – ALA – THR – LEU – ARG – ALA – PHE – ALA – ALA – GLY – VAL – THR – PRO – ALA – GLN – CYS –
61   80

PHE – GLU – MET – LEU – ALA – LEU – ILE – ARG – GLN – LYS – HIS – PRO – THR – ILE – PRO – ILE – GLY – LEU – LEU – MET –
71   100

TYR – ALA – ASN – LEU – VAL – PHE – ASN – LYS – GLY – ILE – ASP – GLU – PHE – TYR – ALA – GLN – CYS – GLU – LYS – VAL –
101   120

GLY – VAL – ASP – SER – VAL – LEU – VAL – ALA – ASP – VAL – PRO – VAL – GLN – GLU – SER – ALA – PRO – PHE – ARG – GLN –
121   140

ALA – ALA – LEU – ARG – HIS – ASN – VAL – ALA – PRO – ILE – PHE – ILE – CYS – PRO – PRO – ASP – ALA – ASP – ASP – ASP –
141   160

                                                                A446   A487

LEU – LEU – ARG – GLN – ILE – ALA – SER – TYR – GLY – ARG – GLY – TYR – THR – TYR – LEU – LEU – SER – ARG – ALA – GLY –
161   CYS   ARG   180

A223

VAL – THR – GLY – ALA – GLU – ASN – ARG – ALA – ALA – LEU – PRO – LEU – ASN – HIS – LEU – VAL – ALA – LYS – LEU – LYS –
181   ILE   200

                                    A23 A46   A187

GLU – TYR – ASN – ALA – ALA – PRO – PRO – LEU – GLN – GLY – PHE – GLY – ILE – SER – ALA – PRO – ASP – GLN – VAL – LYS –
201   ARG GLU   VAL   220

                                                                A78 A58   A169

ALA – ALA – ILE – ASP – ALA – GLY – ALA – ALA – GLY – ALA – ILE – SER – GLY – SER – ALA – ILE – VAL – LYS – ILE – ILE –
221   CYS ASP LEU   240

GLU – GLN – HIS – ASN – ILE – GLU – PRO – GLU – LYS – MET – LEU – ALA – ALA – LEU – LYS – VAL – PHE – VAL – GLN – PRO –
241   260

MET – LYS – ALA – ALA – THR – ARG – SER
261   267

AMINO ACID SEQUENCE OF *A* PROTEIN is shown side by side with a ribbon representing the DNA of the *A* gene. It can be seen that 10 different mutations in the gene produced alterations in the amino acids at only eight different places in the *A* protein. The explanation is that at two of them, 210 and 233, there were a total of four alterations. Thus at No. 210 the mutation designated A23 changed glycine to arginine, whereas mutation A46 changed glycine to glutamic acid. At No. 233 glycine was changed to cysteine by one mutation (A78) and to aspartic acid by another mutation (A58). On the genetic map A23 and A46, like A78 and A58, are very close.

resembles traditional genetic procedures, relies on recombination frequencies to establish the order of the mutationally altered sites in the *A* gene with respect to one another. By this method one can assign relative distances—map distances—to the regions between altered sites. The method is often of little use, however, when the distances are very close.

In such cases we have used a third method that involves a mutationally altered gene, or genetic marker, close to the *A* gene. This marker produces a recognizable genetic trait unrelated to the *A* protein. What this does, in effect, is provide a reading direction so that one can tell whether two closely spaced mutants, say No. 58 and No. 78, lie in the order 58–78, reading from the left on the map, or vice versa [*see bottom illustration on page 2448*].

With these procedures we were able to construct a genetic map relating the altered sites in a group of mutants responsible for altered *A* proteins that could themselves be isolated for study. Some of the sites were very close together, whereas others were far apart [*see upper illustration on page 2449*]. The next step was to determine the nature of amino acid changes in each of the mutationally altered proteins.

It was expected that each mutant of the *A* protein would have a localized change, probably involving only one amino acid. Before we could hope to identify such a specific change we would have to know the sequence of amino acids in the unmutated *A* protein. This was determined by John R. Guest, Gabriel R. Drapeau, Bruce C. Carlton and me, by means of a well-established procedure. The procedure involves breaking the protein molecule into many short fragments by digesting it with a suitable enzyme. Since any particular protein rarely has repeating sequences of amino acids, each digested fragment is likely to be unique. Moreover, the fragments are short enough—typically between two and two dozen amino acids in length—so that careful further treatments can release one amino acid at a time for analysis. In this way one can identify all the amino acids in all the fragments, but the sequential order of the fragments is still unknown. This can be established by digesting the complete protein molecule with a different enzyme that cleaves it into a uniquely different set of fragments. These are again analyzed in detail. With two fully analyzed sets of fragments in hand, it is not difficult to

SEGMENT OF PROTEIN	MUTANT										NOR-MAL
	H11	C140	B17	B272	H32	B278	C137	H36	A489	C208	
I	+	+	+	+	+	+	+	+	+	+	+
II	−	+	+	+	+	+	+	+	+	+	+
III	−	−	+	+	+	+	+	+	+	+	+
IV	−	−	−	+	+	+	+	+	+	+	+
V	−	−	−	−	+	+	+	+	+	+	+
VI	−	−	−	−	−	+	+	+	+	+	+
VII	−	−	−	−	−	+	+	+	+	+	+
VIII	−	−	−	−	−	−	−	+	+	+	+
IX	−	−	−	−	−	−	−	−	+	+	+
X	−	−	−	−	−	−	−	−	−	+	+
XI	−	−	−	−	−	−	−	−	−	−	+

GENETIC MAP   H11 C140 B17 B272 H32   B278 C137 H36   A489 C208

INDEPENDENT EVIDENCE FOR COLINEARITY of gene and protein structure has been obtained from studies of the protein that forms the head of the bacterial virus T4D. Sydney Brenner and his co-workers at the University of Cambridge have found that mutations in the gene for the head protein alter the length of head-protein fragments. In the table "+" indicates that a given segment of the head protein is produced by a particular mutant; "−" indicates that the segment is not produced. When the genetic map was plotted, it was found that the farther to the right a mutation appears, the longer the fragment of head protein.

find short sequences of amino acids that are grouped together in the fragment of one set but that are divided between two fragments in the other. This provides the clue for putting the two sets of fragments in order. In this way we ultimately determined the identity and location of each of the 267 amino acids in the unmutated *A* protein of tryptophan synthetase.

Simultaneously my colleagues and I were examining the mutants of the *A* protein to identify the specific sites of mutational changes. For this work we used a procedure first developed by Vernon M. Ingram, now at the Massachusetts Institute of Technology, in his studies of naturally occurring abnormal forms of human hemoglobin. This procedure also uses an enzyme (trypsin) to break the protein chain into peptides, or polypeptide fragments. If the peptides are placed on filter paper wetted with certain solvents, they will migrate across

the paper at different rates; if an electric potential is applied across the paper, the peptides will be dispersed even more, depending on whether they are negatively charged, positively charged or uncharged under controlled conditions of acidity. The former separation process is chromatography; the latter, electrophoresis. When they are employed in combination, they produce a unique "fingerprint" for each set of peptides obtained by digesting the *A* protein from a particular mutant bacterium. The positions of the peptides are located by spraying the filter paper with a solution of ninhydrin and heating it for a few minutes at about 70 degrees centigrade. Each peptide reacts to yield a characteristic shade of yellow, gray or blue.

When the fingerprints of mutationally altered *A* proteins were compared with the fingerprint of the unmutated protein, they were found to be remarkably similar. In each case, however, there was a difference. The mutant fingerprint usually lacked one peptide spot that appears in the nonmutant fingerprint and exhibited a spot that the nonmutant fingerprint lacks. The two peptides would presumably be related to each other with the exception of the change resulting from the mutational event. One can isolate each of the peptides and compare their amino acid composition. Guest, Drapeau, Carlton and I, together with D. R. Helinski and U. Henning, identified the amino acid substitutions in

each of a variety of altered A proteins.

The final step was to compare the locations of these changes in the A protein with the genetic map of the mutationally altered sites. There could be no doubt that the amino acid sequence of the A protein and the map of the A gene are in fact colinear [*see lower illustration on page 2449*].

One can also see that the distances between mutational sites on the map of the A gene correspond quite closely to the distances separating the corresponding

amino acid changes in the A protein. In two instances two separate mutational changes, so close as to be almost at the same point on the genetic map, led to changes of the same amino acid in the unmutated protein. This is to be expected if a codon of three bases in DNA is required to specify a single amino acid in a protein. Evidently the most closely spaced mutational sites in our genetic map represent alterations in two bases within a single codon.

Thus our studies have shown that each

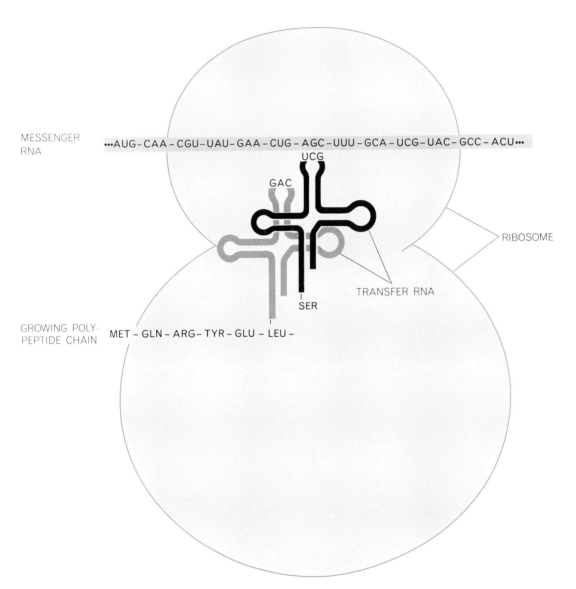

DNA

...ATG – CAA – CGT – TAT – GAA – CTG – AGC – TTT – GCA – TCG – TAC – GCC – ACT – GTT – TCT – ATT – GCA...

...TAC – GTT – GCA – ATA – CTT – GAC – TCG – AAA – CGT – AGC – ATG – CGG – TGA – CAA – AGA – TAA – CGT...

MESSENGER RNA

...AUG – CAA – CGU – UAU – GAA – CUG – AGC – UUU – GCA – UCG – UAC – GCC – ACU...

UCG

GAC

RIBOSOME

TRANSFER RNA

SER

GROWING POLY-PEPTIDE CHAIN   MET – GLN – ARG – TYR – GLU – LEU –

SCHEME OF PROTEIN SYNTHESIS, according to the current view, involves the following steps. Genetic information is transcribed from double-strand DNA into single-strand messenger ribonucleic acid (RNA), which becomes associated with a ribosome. Amino acids are delivered to the ribosome by molecules of transfer RNA, which embody codons complementary to the codons in messenger RNA. The next to the last molecule of transfer RNA to arrive (*color*) holds the growing polypeptide chain while the arriving molecule of transfer RNA (*black*) delivers the amino acid that is to be added to the chain next (serine in this example). The completed polypeptide chain, either alone or in association with other chains, is the protein whose specification was originally embodied in DNA.

```
      AGY                                    AGX  AGY                                                 AGY
    CGZ  GGZ  UAX  ACZ  [UAX] XUZ  [CUZ]    UCZ  CGZ  GCZ  GGZ  GUZ [ACW] GGZ  GCZ  GAY  AAX  CGZ  GCZ  GCZ  XUZ
  – ARG – GLY – TYR – THR – [TYR] – LEU – [LEU] – SER – ARG – ALA – GLY – VAL – [THR] – GLY – ALA – GLU – ASN – ARG – ALA – ALA – LEU –
    170                                                                                                               190


    CCZ  XUZ  AAX  CAX  XUZ  GUZ  GCZ  AAY  XUZ  AAY  GAY  UAX  AAX  GCZ  GCZ  CCZ  CCZ  XUZ  CAY  [GGA]
    PRO – LEU – ASN – HIS – LEU – VAL – ALA – LYS – LEU – LYS – GLU – TYR – ASN – ALA – ALA – PRO – PRO – LEU – GLN – [GLY] –
    191                                                                                                          210

            AGX
    UUX  [GGZ] AUW  UCZ  GCZ  CCZ  GAX  CAY  GUZ  AAY  GCZ  GCZ  AUW  GAX  GCZ  GGZ  GCZ  GCZ  GGZ  GCZ
    PHE – [GLY] – ILE – SER – ALA – PRO – ASP – GLN – VAL – LYS – ALA – ALA – ILE – ASP – ALA – GLY – ALA – ALA – GLY – ALA –
    211                                                                                                          230

        AGX
    AUW  UCZ  [GGX] [UCZ] GCZ  AUW  GUZ  AAY  AUW  AUW  GAY  CAY  CAX  AAX  AUW  GAY  CCZ  GAY  AAY  AUG
    ILE – SER – [GLY] – [SER] – ALA – ILE – VAL – LYS – ILE – ILE – GLU – GLN – HIS – ASN – ILE – GLU – PRO – GLU – LYS – MET –
    231                                                                                                          250
```

W = U, C or A   X = U or C   Y = A or G   Z = U, C, A or G

PROBABLE CODONS IN MESSENGER RNA that determines the sequence of amino acids in the *A* protein are shown for 81 of the protein's 267 amino acid units. The region includes seven of the eight mutationally altered positions (*colored boxes*) in the *A* protein. The codons were selected from those assigned to the amino acids by Marshall Nirenberg and his associates at the National Institutes of Health and by H. Gobind Khorana and his associates at the University of Wisconsin. Codons for the remaining 186 amino acids in the *A* protein can be supplied similarly. In most cases the last base in the codon cannot be specified because there are usually several synonymous codons for each amino acid. With a few exceptions the synonyms differ from each other only in the third position.

unique sequence of bases in DNA—a sequence constituting a gene—is ultimately translated into a corresponding unique linear sequence of amino acids— a sequence constituting a polypeptide chain. Such chains, either by themselves or in conjunction with other chains, fold into the three-dimensional structures we recognize as protein molecules. In the great majority of cases these proteins act as biological catalysts and are therefore classed as enzymes.

The colinear relation between a genetic map and the corresponding protein has also been convincingly demonstrated by Sydney Brenner and his co-workers at the University of Cambridge. The protein they studied was not an enzyme but a protein that forms the head of the bacterial virus T4. One class of mutants of this virus produces fragments of the head protein that are related to one another in a curious way: much of their amino acid sequence appears to be identical, but the fragments are of various lengths. Brenner and his group found that when the chemically similar regions in fragments produced by many mutants were matched, the fragments could be arranged in order of increasing length. When they made a genetic map of the mutants that produced these fragments, they found that the mutationally altered sites on the genetic map were in the same order as the termination points in the protein fragments. Thus the length of the fragment of the head protein produced by a mutant increased as the site of mutation was displaced farther from one end of the genetic map [*see illustration on page 2451*].

The details of how the living cell translates information coded in gene structure into protein structure are now reasonably well known. The base sequence of one strand of DNA is transcribed into a single-strand molecule of messenger ribonucleic acid (RNA), in which each base is complementary to one in DNA. Each strand of messenger RNA corresponds to relatively few genes; hence there are a great many different messenger molecules in each cell. These messengers become associated with the small cellular bodies called ribosomes, which are the actual site of protein synthesis [*see illustration on page 2452*]. In the ribosome the bases on messenger RNA are read in groups of three and translated into the appropriate amino acid, which is attached to the growing polypeptide chain. The messenger also contains in code a precise starting point and stopping point for each polypeptide.

From the studies of Marshall Nirenberg and his colleagues at the National Institutes of Health and of H. Gobind Khorana and his group at the University of Wisconsin the RNA codons corresponding to each of the amino acids are known. By using their genetic code dictionary we can indicate approximately two-thirds of the bases in the messenger RNA that specifies the structure of the *A*-protein molecule. The remaining third cannot be filled in because synonyms in the code make it impossible, in most cases, to know which of two or more bases is the actual base in the third position of a given codon [*see illustration*

above]. This ambiguity is removed, however, in two cases where the amino acid change directed by a mutation narrows down the assignment of probable codons. Thus at amino acid position 48 in the *A*-protein molecule, where a mutation changes the amino acid glutamic acid to valine, one can deduce from the many known changes at this position that of the two possible codons for glutamic acid, GAA and GAG, GAG is the correct one. In other words, GAG (specifying glutamic acid) is changed to GUG (specifying valine). The other position for which the codon assignment can be made definite in this way is No. 210. This position is affected by two different mutations: the amino acid glycine is replaced by arginine in one case and by glutamic acid in the other. Here one can infer from the observed amino acid changes that of the four possible codons for glycine, only one—GGA—can yield by a single base change either arginine (AGA) or glutamic acid (GAA).

Knowledge of the bases in the messenger RNA for the *A* protein can be translated, of course, into knowledge of the base pairs in the *A* gene, since each base pair in DNA corresponds to one of the bases in the RNA messenger. When the ambiguity in the third position of most of the codons is resolved, and when we can distinguish between two quite different sets of codons for arginine, leucine and serine, we shall be able to write down the complete base sequence of the *A* gene—the base sequence that specifies the sequence of the 267 amino acids in the *A* protein of the enzyme tryptophan synthetase.

The Author

CHARLES YANOFSKY is professor of biology at Stanford University. After his graduation from the City College of the City of New York in 1948 he did graduate work in the department of microbiology at Yale University, receiving a Ph.D. there in 1951. He remained at Yale until 1954, when he joined the faculty of the Western Reserve University School of Medicine. Four years later he went to the department of biological sciences at Stanford. Yanofsky has received several awards for his work in molecular biology. He was elected to the American Academy of Arts and Sciences in 1964 and to the National Academy of Sciences last year.

Bibliography

CO-LINEARITY OF β-GALACTOSIDASE WITH ITS GENE BY IMMUNOLOGICAL DETECTION OF INCOMPLETE POLYPEPTIDE CHAINS. Audree V. Fowler and Irving Zabin in *Science*, Vol. 154, No. 3752, pages 1027–1029; November 25, 1966.

CO-LINEARITY OF THE GENE WITH THE POLYPEPTIDE CHAIN. A. S. Sarabhai, A. O. W. Stretton, S. Brenner and A. Bolle in *Nature*, Vol. 201, No. 4914, pages 13–17; January 4, 1964.

THE COMPLETE AMINO ACID SEQUENCE OF THE TRYPTOPHAN SYNTHETASE A PROTEIN (α SUBUNIT) AND ITS CO-LINEAR RELATIONSHIP WITH THE GENETIC MAP OF THE A GENE. Charles Yanofsky, Gabriel R. Drapeau, John R. Guest and Bruce C. Carlton in *Proceedings of the National Academy of Sciences*, Vol. 57, No. 2, pages 296–298; February, 1967.

MUTATIONALLY INDUCED AMINO ACID SUBSTITUTIONS IN A TRYPTIC PEPTIDE OF THE TRYPTOPHAN SYNTHETASE A PROTEIN. John R. Guest and Charles Yanofsky in *Journal of Biological Chemistry*, Vol. 240, No. 2, pages 679–689; February, 1965.

ON THE COLINEARITY OF GENE STRUCTURE AND PROTEIN STRUCTURE. C. Yanofsky, B. C. Carlton, J. R. Guest, D. R. Helinski and U. Henning in *Proceedings of the National Academy of Sciences*, Vol. 51, No. 2, pages 266–272; February, 1964.

SCIENTIFIC
AMERICAN June 1967, Vol. 216, No. 6, pp. 64-76   OFFPRINT 1075

MOLECULAR ISOMERS IN VISION

by Ruth Hubbard and Allen Kropf

Certain organic compounds can exist in two or more forms that have
the same chemical composition but different molecular architecture.
One of them is the basis for vision throughout the animal kingdom.

Molecular biology, which today is
so often associated with very
large molecules such as the nucleic acids and proteins, actually em-
braces the entire effort to describe the
structure and function of living organisms in molecular terms. We are coming
to see how the manifold activities of the
living cell depend on interactions among
molecules of thousands of different sizes
and shapes, and we can speculate on how
evolutionary processes have selected
each molecule for its particular functional properties. The significance of precise molecular architecture has become
a central theme of molecular biology.

One of the more recent observations
is that biological molecules are not
static structures but, in a number of well-
established cases, change shape in response to outside influences. As an example, the molecule of hemoglobin has
one shape when it is carrying oxygen
from the lungs to cells elsewhere in the
body and a slightly different shape when
it is returning to the lungs without oxygen [see "The Hemoglobin Molecule,"
by M. F. Perutz; SCIENTIFIC AMERICAN
Offprint 196]. A somewhat similar
changeability in the molecule of lysozyme, which breaks down the walls of
certain bacterial cells, was described in
these pages last November by David C.
Phillips of the University of Oxford. In
this article we shall describe some of the
simplest changes in shape that can take
place in much smaller organic molecules
and show how change of this type provides the basis of vision throughout the
animal kingdom.

A "Childish Fantasy"

The notion that molecules of the
same atomic composition might have
different spatial arrangements is less

than 100 years old. It dates back to a
paper titled "Sur les formules de structure dans l'espace," written in 1874
by Jacobus Henricus van't Hoff, then
an obscure chemist at the Veterinary
College of Utrecht. At that time it was
still respectable to doubt the existence
of atoms; to speak of the three-dimensional arrangement of atoms in molecules
was a speculative leap of great audacity.
Van't Hoff's paper provoked Hermann
Kolbe, one of the most eminent organic
chemists of his day, to publish a wither-
ing denunciation.

"Not long ago," Kolbe wrote in 1877,
"I expressed the view that the lack
of general education and of thorough
training in chemistry of quite a few pro-
fessors of chemistry was one of the causes of the deterioration of chemical research in Germany. . . . Will anyone to
whom my worries may seem exaggerated please read, if he can, a recent
memoir by a Herr van't Hoff on 'The Arrangements of Atoms in Space,' a document crammed to the hilt with the out-
pourings of a childish fantasy. This Dr.
J. H. van't Hoff, employed by the Veterinary College at Utrecht, has, so it
seems, no taste for accurate chemical
research. He finds it more convenient to
mount his Pegasus (evidently taken
from the stables of the Veterinary College) and to announce how, on his dar-
ing flight to Mount Parnassus, he saw
the atoms arranged in space."

Van't Hoff's "childish fantasy" was
put forth independently by the French
chemist Jules Achille le Bel and was
soon championed by a number of leading
chemists. In spite of Kolbe's opinion,
evidence in support of the three-dimensional configuration of molecules rapidly
accumulated. In 1900 van't Hoff was
named the first recipient of the Nobel
prize in chemistry.

Even before van't Hoff's paper of
1874 chemists had begun using the concept of the valence bond, commonly
represented by a line connecting two
atoms. It was not unnatural, therefore, to
associate the valence bond concept with
the idea that atoms were arranged precisely in space. The simplest hydrocarbon, methane (CH_4), would then be represented as a regular tetrahedron with a
hydrogen atom at each vertex joined by
a single valence bond to a carbon atom
at the center of the structure [see illustration on page 2457].

The valence bond remained an elusive concept, however, until G. N. Lewis postulated in 1916 that a common
type of bond—the covalent bond—was
formed when two atoms shared two
electrons. "When two atoms of hydrogen
join to form the diatomic molecule," he
wrote, "each furnishes one electron of
the pair which constitutes the bond. Representing each valence electron by a dot,
we may therefore write as the graphical
formula of hydrogen H : H." He visualized this bond to be "that 'hook and
eye,' which is part of the creed of the organic chemist." To explain why electrons
should tend to pair in this manner, Lewis
could offer nothing beyond an intuitive
principle that he called "the rule of two."

The rule of two entered the physicist's description of the atom when
Wolfgang Pauli put forward the exclusion principle in 1923. This states that
electrons in atoms and molecules are
found in "orbitals" that can accommo-
date at most two electrons. Since electrons can be regarded as minuscule spinning negative charges, and thus as tiny
electromagnets, the two electrons in each
orbital must be spinning in opposite directions.

Let us return, however, to some of
the chemical observations that gave rise

to van't Hoff's ideas of stereochemistry late in the 19th century. Chemists were confronted by a series of puzzling observations best exemplified by two simple compounds: maleic acid and fumaric acid [see illustrations on page 2458]: two distinct, chemically pure substances, each with four atoms of carbon, four of hydrogen and four of oxygen ($C_4H_4O_4$). It was known, moreover, that the connections between atoms in the two molecules were exactly the same and that the two central carbon atoms in each molecule were connected by a double bond. Yet the two compounds were indisputably different. Whereas crystals of maleic acid melted at 128 degrees centigrade, crystals of fumaric acid did not melt until heated to about 290 degrees C. Furthermore, maleic acid was

about 100 times more soluble in water and 10 times stronger as an acid than fumaric acid. When maleic acid was heated in a vacuum, it gave off water vapor and became a new substance, maleic anhydride, which readily recombined with water and reverted to maleic acid. Fumaric acid underwent no such reaction. On the other hand, if either compound was heated in the presence of hydrogen, it was transformed into the identical compound, succinic acid ($C_4H_6O_4$), which contains two more hydrogen atoms per molecule than maleic or fumaric acid.

It was known that the four carbon atoms in maleic and fumaric acids form a chain. The only way to explain the differences between the compounds is to assume that the two halves of a molecule

that are connected by a double bond are not free to rotate with respect to each other. Thus the form of the $C_4H_4O_4$ molecule in which the two terminal COOH groups lie on the same side of the double bond (maleic acid) is not identical with the form in which the COOH groups lie on opposite sides (fumaric acid). Molecules that assume distinct shapes in this way are called geometrical or cis-trans isomers of one another. "Cis" is from the Latin meaning on the same side; "trans" means on opposite sides. Therefore maleic acid is the cis isomer of the $C_4H_4O_4$ molecule and fumaric acid is the trans isomer. When the double carbon-carbon bond of either isomer is reduced to a single bond by the addition of two more hydrogen atoms, the two halves of the molecule are free

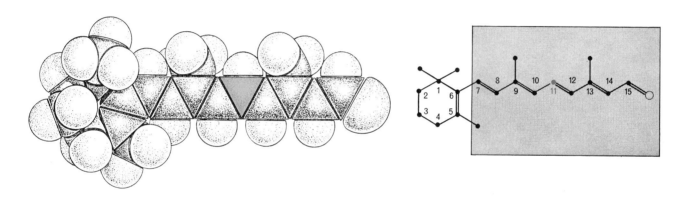

ALL-*TRANS* RETINAL

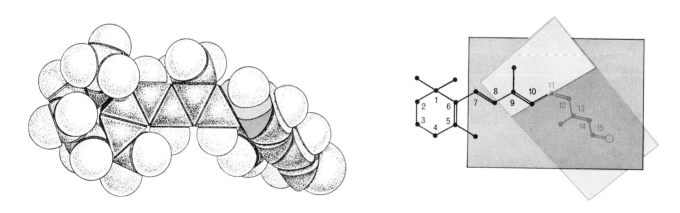

11-*CIS* RETINAL

FUNDAMENTAL MOLECULE OF VISION is retinal ($C_{20}H_{28}O$), also known as retinene, which combines with proteins called opsins to form visual pigments. Because the nine-member carbon chain in retinal contains an alternating sequence of single and double bonds, it can assume a variety of bent forms. Each distinct form is termed an isomer. Two isomers of retinal are depicted here. In the models (*left*) carbon atoms are dark, except carbon No. 11, which is shown in color; hydrogen atoms are light. The large atom attached to carbon No. 15 is oxygen. In the structural formulas (*right*) hydrogen atoms are omitted. The parts of each isomer that lie in a plane are marked by background panels. When tightly bound to opsin, retinal is in the bent and twisted form known as 11-*cis*. When struck by light, it straightens out into the all-*trans* configuration. This simple photochemical event provides the basis for vision.

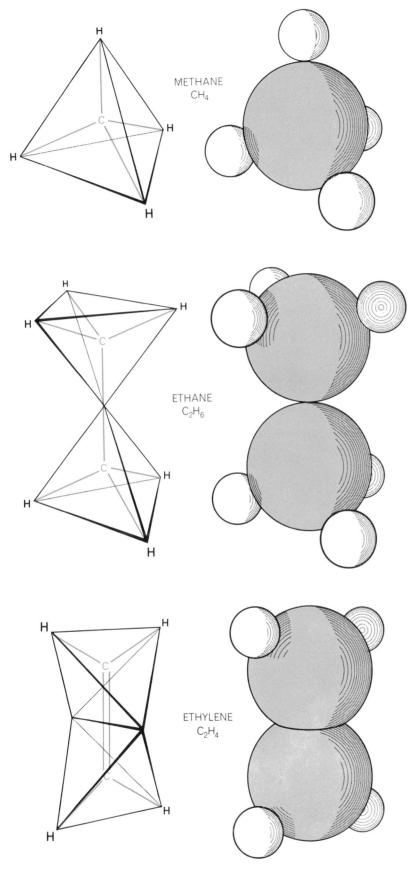

METHANE
CH₄

ETHANE
C₂H₆

ETHYLENE
C₂H₄

SIMPLEST HYDROCARBON MOLECULES are methane, ethane and ethylene. In methane the carbon atom lies at the center of a tetrahedron that has a hydrogen atom at each apex. The models at right show relative diameters of carbon and hydrogen. Ethane can be visualized as two tetrahedrons joined apex to apex. Ethylene, the simplest hydrocarbon that has a carbon-carbon double bond, can be visualized as two tetrahedrons joined edge to edge. The C=C bond in ethylene is about 15 percent shorter than the C—C bond in ethane.

to rotate with respect to each other and a single compound results: succinic acid.

Van't Hoff proposed that when two carbon atoms are joined by a single bond they can be regarded as the centers of two tetrahedrons that meet apex to apex, thus allowing the two bodies to rotate freely. To represent a double bond, he visualized the two tetrahedrons as being joined edge to edge so that they were no longer free to rotate. Apart from minor modifications his proposals have stood up extremely well.

Electrons in Orbitals

Van't Hoff's explanation, of course, was a purely formal one and provided no real insight into *why* a double bond prevents the parts of a molecule it joins from rotating with respect to each other. This was not understood for another 50 years, when the development of wave mechanics by Erwin Schrödinger set the stage for one of the most productive periods in theoretical chemistry. With Schrödinger's wave equation to guide them, chemists and physicists could compute the orbitals around atoms where pairs of electrons could be found. The valence bonds, which chemists had been drawing as lines for almost a century, now took on physical reality in the form of pairs of electrons confined to orbitals that were generally located in the regions where the valence lines had been drawn.

The first molecule to be analyzed successfully by the new wave mechanics was hydrogen (H_2). Walter Heitler and Fritz London applied Schrödinger's prescription and obtained the first profound insight into the nature of chemical bonding. Their results define the region in space most likely to be frequented by the pair of electrons associated with the two hydrogen atoms in the hydrogen molecule. The region resembles a peanut, each end of which contains a proton, or hydrogen nucleus [*see top illustration on page 2459*].

The phrase "most likely to be frequented" must be used because, as Max Born convincingly argued in the late 1920's, the best one can do in the new era of quantum mechanics is to calculate the probability of finding electrons in certain regions; all hope of placing them in fixed orbits must be abandoned. The new methods were quickly applied to many kinds of molecule, including some with double bonds.

One of the most fruitful methods for describing doubly bonded molecules—the molecular-orbital method—was devised by Robert S. Mulliken of the University of Chicago, who last year received the Nobel prize in chemistry. In Mulliken's concept a double bond can be visualized as three peanut-shaped regions [*see bottom illustration on next page*]. The central peanut, the "sigma" orbital, encloses the nuclei of the two adjacent atoms, as in the hydrogen molecule. The other two peanuts, which jointly form the "pi" orbital, lie along each side of the sigma orbital. The implication of this model is that in forming the sigma orbital the two electrons occupy a common volume, whereas in forming the pi orbital both electrons tend to occupy the two separate volumes simultaneously.

One can say that the sigma bond connects the atoms like an axle that joins two wheels but leaves them free to rotate separately. The pi bond ties the

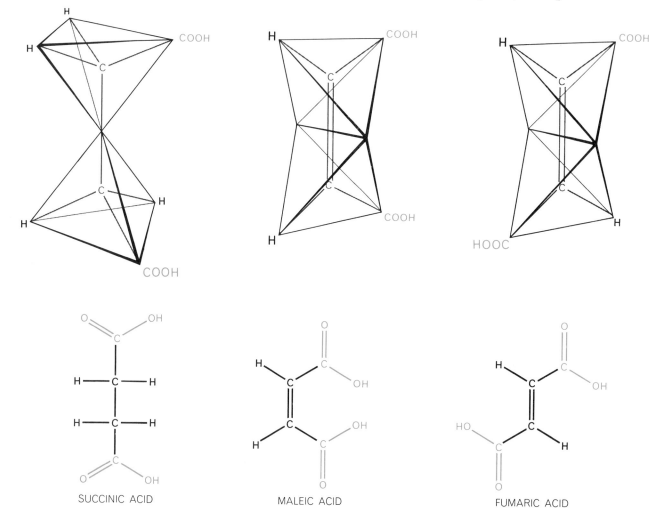

SUCCINIC ACID   MALEIC ACID   FUMARIC ACID

MOLECULAR PUZZLE was presented to chemists of the 19th century by maleic acid and fumaric acid, which have the same formula, $C_4H_4O_4$, and can be converted to succinic acid by the addition of two hydrogen atoms. Nevertheless, maleic and fumaric acids have very different properties. In 1874 Jacobus Henricus van't Hoff suggested that the central pair of carbon atoms in the three acids could be visualized as occupying the center of tetrahedrons that were joined edge to edge in the case of maleic and fumaric acids and apex to apex in the case of succinic acid. Thus the spatial relations of the two carboxyl (COOH) groups would be rigidly fixed in maleic and fumaric acids but not in succinic acid, because in the latter molecule the tetrahedrons would be free to rotate.

MALEIC ACID   MALEIC ANHYDRIDE + WATER   FUMARIC ACID

ONE CONSEQUENCE OF ISOMERISM is that maleic acid readily loses a molecule of water when heated, yielding maleic anhydride. Fumaric acid does not undergo this reaction because its carboxyl groups are held apart at opposite ends of the molecule.

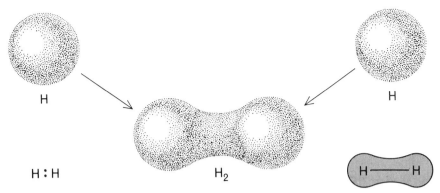

HYDROGEN MOLECULE is formed when two atoms of hydrogen (*H*) are joined by a chemical bond. The bond is created by the pairing of two electrons, one from each atom, which must have opposite magnetic properties if the atoms are to attract each other. The position of the electrons as they orbit around the hydrogen nuclei cannot be precisely known but can be represented by an "orbital," a fuzzy region in which the electrons spend most of their time. Known as a sigma orbital, it can be stylized as at lower right. The formula for the hydrogen molecule can be written as at lower left; the dots indicate electrons.

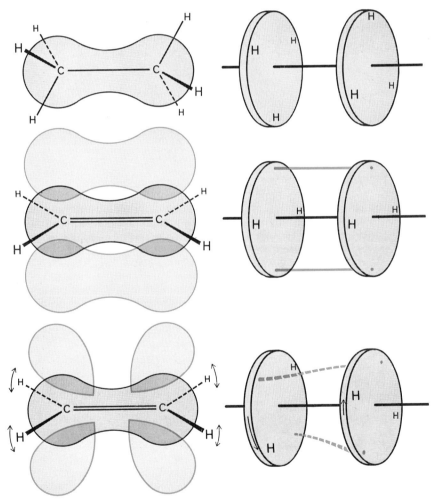

MOLECULAR ORBITALS help to explain why molecules held together by single bonds differ from molecules with double bonds. For example, the two carbon atoms in ethane are joined by two electrons in a sigma orbital, similar to the orbital in the hydrogen molecule. The two ends of the molecule, like wheels joined by a simple axle, are able to rotate. The two carbon atoms in ethylene (*middle*) are joined by two additional electrons in a "pi" orbital (*color*), as well as by two electrons in a sigma orbital. The four hydrogen atoms in ethylene are held in a plane perpendicular to the plane of the orbitals. The effect is as if two wheels were held together by two rigid rods in addition to an axle. When ethylene is in an "excited" state (*bottom*), one of the pi electrons occupies the four-lobed orbital. This lessens the rigidity of the double bond and gives it more of the character of a single bond.

two wheels together so that they must rotate as a unit. It also forces the two halves of the molecule to lie in the same plane, exactly as if two tetrahedrons were cemented edge to edge. In this way the molecular-orbital description of bonding provides a quantum-mechanical explanation of *cis-trans* isomerism.

The single bond joining two atoms, such as the carbon atoms of the two methyl groups in ethane (CH_3-CH_3), is a sigma bond, which leaves the groups attached to the two carbons free to rotate with respect to each other. Actually the two methyl groups in ethane are known to have a preferred configuration, so that they are not completely freewheeling. Nonetheless, at ordinary temperatures enough energy is available to make 360-degree rotations so frequent that derivatives of ethane (in which one hydrogen in each methyl group is replaced by a different kind of atom) do not form *cis-trans* isomers. There are exceptions, however, if the groups of atoms that replace hydrogen are so bulky that they collide and prevent rotation. In general, therefore, *cis-trans* isomerism is confined to molecules incorporating double bonds.

Electrons Delocalized

So much for molecules that have one double bond. What is the situation when a molecule has two or more double bonds? Specifically, what stereochemical behavior can be expected when single and double bonds alternate to form what is called a conjugated system?

The simplest conjugated system is found in 1,3-butadiene, a major ingredient in the manufacture of synthetic rubber, which can be written C_4H_6 or $CH_2=CH-CH=CH_2$. The designation "1,3" indicates that the double bonds originate at the first and third carbon atoms. Some of the properties of the biologically more interesting conjugated molecules are exhibited by butadiene. From the foregoing discussion one might expect that the second and third carbon atoms in the molecule would be free to rotate around the sigma bond connecting them. In actuality the rotation is not free: all the atoms in butadiene tend to lie in a plane.

It can also be shown that the energy content of each double bond in butadiene differs significantly from the energy content of the one double bond in the closely related compound 1-butene ($CH_2=CH-CH_2-CH_3$). The energy released in changing the two double bonds

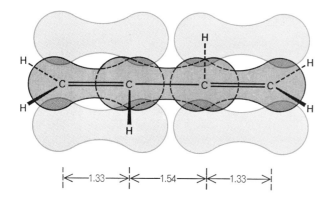

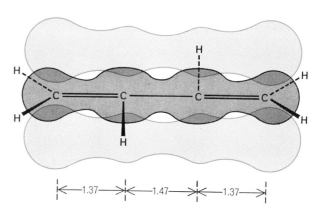

|←—1.33—→|←——1.54——→|←—1.33—→|

|←—1.37—→|←——1.47——→|←—1.37—→|

DELOCALIZED ORBITALS are found in "conjugated" systems: molecules in which single and double bonds alternate. The simplest conjugated molecule is 1,3-butadiene (C_4H_6). If its pi orbitals (*color*) were simply confined to the double bonds, as in ethylene, its orbital structure and carbon-carbon distances (in angstrom units) would be as shown at the left. Even in the lowest energy state, however, the pi electrons tend to spread across the entire molecule (*right*). As a result double bonds are lengthened and the single bond is shortened, making each type of bond more like the other. As a consequence the entire molecule is planar, or flat.

2 [structural formula 1-BUTENE] + 2H₂ ⟶ 2 [structural formula BUTANE] + 58.98 KILOCALORIES PER TWO MOLES

1-BUTENE   HYDROGEN   BUTANE

[structural formula 1,3-BUTADIENE] + 2H₂ ⟶ [structural formula BUTANE] + 55.36 KILOCALORIES PER MOLE

1,3-BUTADIENE   HYDROGEN   BUTANE   3.62 KILOCALORIES DIFFERENCE

EVIDENCE FOR BOND MODIFICATION in a conjugated molecule can be obtained by measuring the energy released when double bonds are converted to single bonds by adding hydrogen. Hydrogenation of the double bond in 1-butene, which is not a conjugated molecule, yields 29.49 kilocalories for every mole of reactant. A mole is a weight in grams equal to the molecular weight of a substance: 56 for butene and 54 for butadiene. Hydrogenation of two moles of butene, hence the hydrogenation of twice as many double bonds, would therefore yield 58.98 kilocalories. Hydrogenation of the same number of double bonds in butadiene (present in a single mole) yields only 55.36 kilocalories. The difference is 3.62 kilocalories per mole for the two bonds, or 1.81 kilocalories for each double bond. The lesser energy in the butadiene double bonds indicates that they are more stable than the double bond in 1-butene.

of butadiene into the single bonds of butane (CH_3–CH_2–CH_2–CH_3) is about 55,400 calories per mole of butane formed. (A mole is a weight in grams equal to the molecular weight of the molecule: 58 for butane, 54 for butadiene.) The energy released in converting 1-butene, which has only one double bond, into butane is about 29,500 calories per mole. When expressed in terms of equivalent numbers of double bonds hydrogenated, the latter reaction yields some 1,800 calories more than the former [see *lower illustration above*]. The greater energy release means that the double bond in 1-butene is more reactive than either of those in 1,3-butadiene.

The added stability of the bonds in 1,3-butadiene was not unexpected; the same kind of result had been obtained for benzene, whose famous ring structure is formed by six carbon atoms connected alternately by single and double bonds. One can picture the extra energy of stabilization as arising from the tendency of electrons in the pi orbitals to leak out and become delocalized. Indeed, the phenomenon is called delocalization. The pi orbitals spread over larger portions of conjugated molecules than one might have thought, so that the properties of delocalized systems can no longer be described in terms of the properties of the double and single bonds as they are usually drawn. In order to represent the pi orbitals of 1,3-butadiene more accurately one must stretch them across all four carbon atoms of the molecule [see *upper illustration above*]. The stretching helps to explain why butadiene is not completely free to rotate around the central single bond: the bond has some

of the characteristics of a double bond.

The altered character of butadiene's central carbon-carbon bond has been confirmed by X-ray-diffraction studies of butadiene. Whereas the usual carbon-carbon bond lengths are about 1.54 angstrom units for a single bond and 1.33 angstroms for a double bond, the length of the central single bond in 1,3-butadiene is only 1.47 angstroms. (An angstrom is 10^{-8} centimeter.) Linus Pauling, who did much to clarify the nature of the chemical bond, has estimated that the observed shortening of the central carbon-carbon bond of butadiene implies that it has about 15 percent of the double-bond character. One consequence of this is that the molecular configuration of butadiene tends to remain planar, or flat.

The tendency toward planarity in

DIBENZYL

TRANS STILBENE

CIS STILBENE

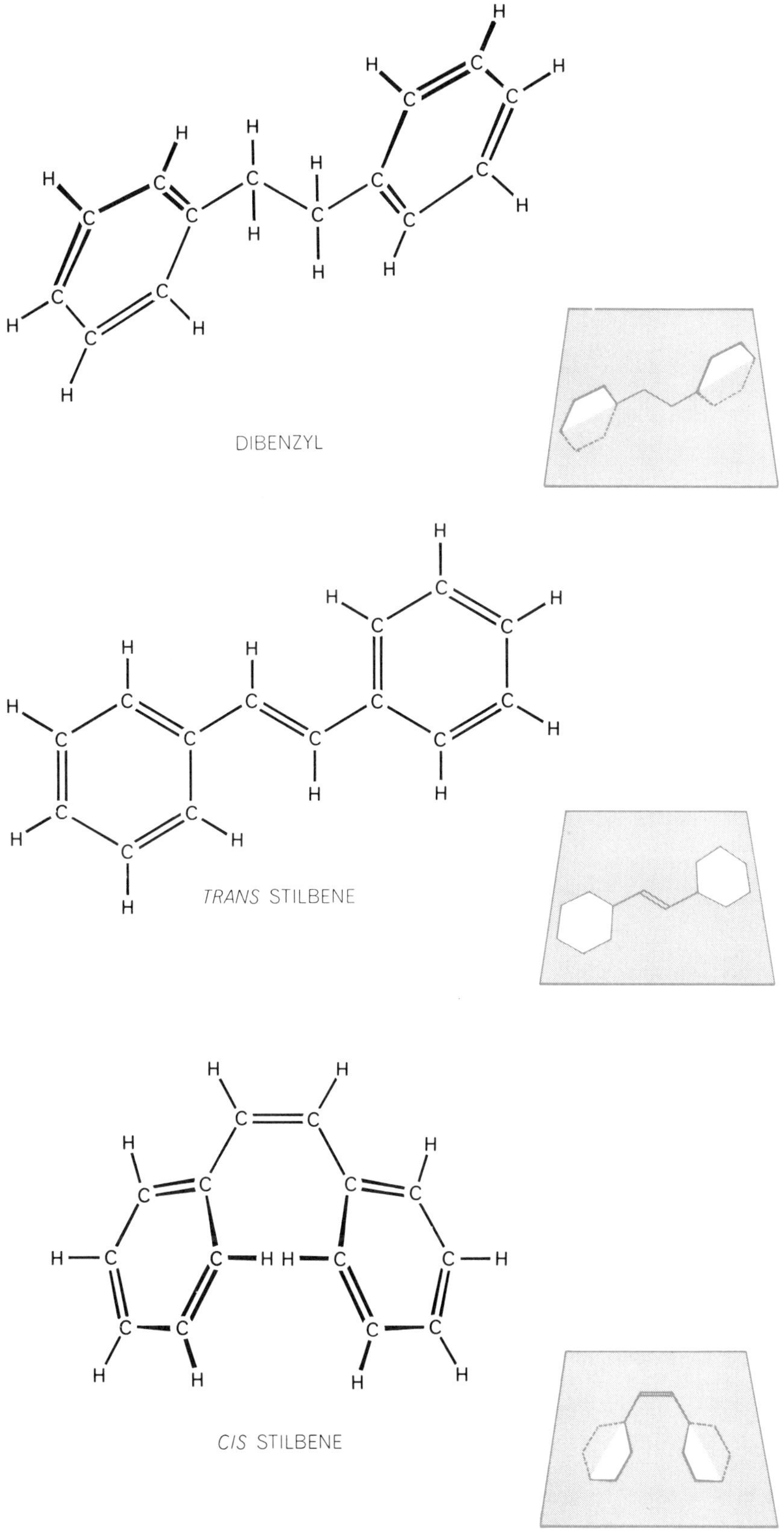

PREFERENCE FOR FLATNESS in conjugated systems is exhibited by molecules of dibenzyl and two isomers of stilbene. The latter have a double bond in the carbon-carbon bridge linking the two benzene rings, whereas dibenzyl has a single bond. In dibenzyl the two rings are practically at right angles to the plane of the bridge. In *trans* stilbene all the atoms lie essentially in a plane. In *cis* stilbene, as can be demonstrated with molecular models, the two rings interfere with each other and thus cannot lie flat. The twisting of the rings has been established by X-ray studies of a related compound, *cis* azobenzene, in which the two rings are joined through a doubly bonded nitrogen (N=N) bridge.

conjugated systems was clearly demonstrated by the Scottish X-ray crystallographer J. M. Robertson and his colleagues in the mid-1930's. They compared the configurations of dibenzyl and *trans* stilbene, both of which contain two benzene rings joined by two carbon atoms [*see illustration at left*]. The difference between the two molecules is that in dibenzyl the two carbons are joined by a single bond, whereas in stilbene they are joined by a double bond. Robertson showed that the rings in dibenzyl are essentially at right angles to the connecting carbon-carbon bridge. In the *trans* form of stilbene the rings and the bridge lie in a plane, and the single bonds that join the rings to the two carbons in the bridge are foreshortened from the normal single-bond length to about 1.44 angstroms. In the *cis* form of stilbene the two rings cannot lie in a plane because they bump into each other.

Light-sensitive Molecules

We turn now to *cis-trans* isomerism in the family of molecules we have worked with most directly, the carotenoids and their near relatives, retinal (also known as retinene) and vitamin A. These molecules are built up from units of isoprene, which is like 1,3-butadiene in every respect but one: at the second carbon of isoprene a methyl group (CH_3) replaces the hydrogen atom present in butadiene. Natural rubber is polyisoprene, a long conjugated chain of carbon atoms with a methyl group attached to every fourth carbon.

The compound known as beta-carotene, which is responsible for the color of carrots, consists of an 18-carbon conjugated chain terminated at both ends by a six-member carbon ring, each of which adds another double bond to the conjugated system. The molecule has 40 carbon atoms in all and is presumably assembled from eight isoprene units [*see illustration on page 2462*].

Until about 15 years ago the *cis-trans* isomers of the carotenoids entered biology in only one rather trivial way: in determining the color of tomatoes. Laszlo T. Zechmeister and his collaborators at the California Institute of Technology found in the early 1940's that normal red tomatoes contain the carotenoid lycopene in the all-*trans* configuration. (Lycopene differs from beta-carotene only in that the six carbon atoms at each end of the molecule do not close to form rings.) The yellow mutant known as the tangerine tomato contains a yellow *cis* isomer of lycopene called prolycopene. Zechmeister, who contributed more than anyone else to our present understanding of carotenoid chemistry, liked to demonstrate how a yellow solution of prolycopene, extracted from tangerine tomatoes, could be converted into a brilliant orange solution of all-*trans* lycopene simply by adding a trace of iodine in the presence of a strong light.

The discovery that the *cis-trans* isomerism of a carotenoid plays a crucial role in biology was made in the laboratory of George Wald of Harvard University, where it was found that the *cis-trans* isomerism of retinal is intrinsic to the way in which visual pigments react to light. The discovery of these pigments is usually attributed to the German physiologist Franz Boll.

The Chemistry of Vision

In 1877, the same year that Kolbe was ridiculing van't Hoff's work on stereochemistry, Boll noted that a frog's retina, when removed from the eye, was initially bright red but bleached as he watched it, becoming first yellow and finally colorless. Subsequently Boll observed that in a live frog the red color of the retina could be bleached by a strong light and would slowly return if the animal was put in a dark chamber. Recognizing that the bleachable substance must somehow be connected with the frog's ability to perceive light, Boll named it "erythropsin" or "Sehrot" (visual red). Before long Willy Kühne of Heidelberg found the red pigment in the retinal rod cells of many animals and renamed it "rhodopsin" or "Sehpurpur" (visual purple), which it has been called ever since. Kühne also named the yellow product of bleaching "Sehgelb" (visual yellow) and the white product "Sehweiss" (visual white).

The chemistry of the rhodopsin system remained largely descriptive until 1933, when Wald, then a postdoctoral fellow working in Otto Warburg's laboratory in Berlin and Paul Karrer's laboratory in Zurich, demonstrated that the eye contains vitamin A. Wald showed that the vitamin appears when rhodopsin is bleached by light—the physiological process known as light adaptation—and disappears when rhodopsin is resynthesized during dark adaptation [see "Night Blindness," by John E. Dowling; SCIENTIFIC AMERICAN Offprint 1053]. He found rhodopsin consists of a colorless protein (later named opsin) that carries as its chromophore, or color bearer, an unknown yel-

ALL-*TRANS* CAROTENE

ALL-*TRANS* LYCOPENE

TWO NATURAL CAROTENOIDS are examples of highly conjugated systems. Like other carotenoids, they are built up from units of isoprene (C_5H_8), also known as 2-methyl butadiene. In these diagrams hydrogen atoms are omitted so that the carbon skeletons can be seen more clearly. Both molecules contain 40 carbon atoms and are symmetrical around the central carbon-carbon double bond, numbered 15–15'. Beta-carotene gives carrots their characteristic orange color. *Trans* lycopene is responsible for the red color of tomatoes.

low carotenoid that he called retinene. Wald went on to show that the bleaching of rhodopsin to visual yellow corresponds to the liberation of retinene from its attachment to opsin, and that the fading of visual yellow to visual white represents the conversion of retinene to vitamin A. During dark adaptation rhodopsin is resynthesized from these precursors.

The chemical relation between retinene and vitamin A was elucidated in 1944 by R. A. Morton of the University of Liverpool. He showed that retinene is formed when vitamin A, an alcohol, is converted to an aldehyde, a change that involves the removal of two atoms of hydrogen from the terminal carbon atom of the molecule. As a result of Morton's finding the name retinene was recently changed to retinal.

In 1952 one of us (Hubbard), then a graduate student in Wald's laboratory, demonstrated that only the 11-*cis* isomer of retinal can serve as the chromophore of rhodopsin. This has since been confirmed for all the visual pigments whose chromophores have been examined. These pigments found in both the rod and cone cells of the eye contain various opsins, which combine either with retinal (strictly speaking retinal₁) or with a slightly modified form of retinal known as retinal₂. One other isomer of retinal, the 9-*cis* isomer, also combines with opsins to form light-sensitive pigments, but they are readily distinguishable from the visual pigments in their properties and have never been found to occur naturally. They have been called isopigments.

In 1959 we showed that the only thing light does in vision is to change the shape of the retinal chromophore by isomerizing it from the 11-*cis* to the all-*trans* configuration [*see illustration on page 2456*]. Everything else—further chemical changes, nerve excitation, perception of light, behavioral responses—are consequences of this single photochemical act.

The change in the shape of the chromophore alters its relation to opsin and ushers in a sequence of changes in the mutual interactions of the chromophore and opsin, which is observed as a sequence of color changes. In vertebrates the all-*trans* isomer of retinal and opsin are incompatible and come apart. In some invertebrates, such as the squid, the octopus and the lobster, a metastable state is reached in which the all-*trans* chromophore remains bound to opsin.

Until the structure of opsin is established there is no way to know just how 11-*cis* retinal is bound to the opsin molecule. In the 1950's F. D. Collins, G. A. J. Pitt and others in Morton's laboratory showed that in cattle rhodopsin the aldehyde (C=O) group of 11-*cis* retinal forms what is called a Schiff's base with an amino (NH₂) group in the opsin molecule. Recently Deric Bownds in Wald's laboratory has found that the amino group belongs to lysine, one of the amino acid units in the opsin molecule, and has identified the amino acids in its immediate vicinity. There is little doubt that 11-*cis* retinal also has secondary points of attachment to opsin; otherwise it would be hard to explain why only the 11-*cis* isomer serves as the chromophore in visual pigments. Light changes the shape of the chromophore and thus alters its spatial relation to opsin. This leads, in turn, to changes in the shape of the

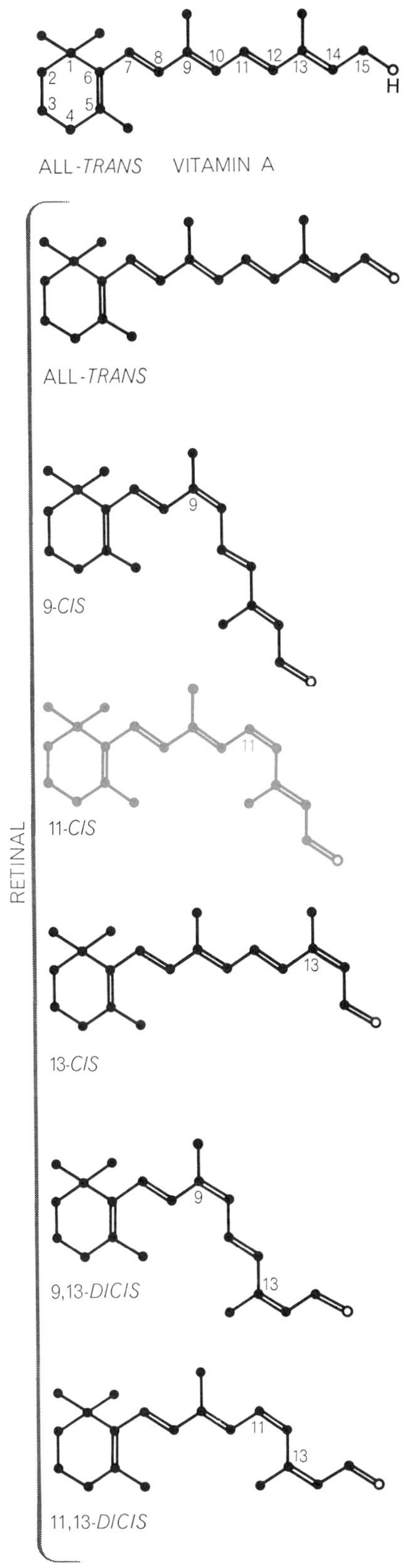

ALL-*TRANS*   VITAMIN A

RETINAL

ALL-*TRANS*

9-*CIS*

11-*CIS*

13-*CIS*

9,13-*DICIS*

11,13-*DICIS*

SIX ISOMERS OF RETINAL are represented in skeleton form below the structure of all-*trans* vitamin A. Hydrogen atoms are omitted, except for the H in the hydroxyl group of vitamin A. If that H and one other on the final carbon are removed, all-*trans* retinal results. This isomer and 11-*cis* retinal, which combines with opsin to form rhodopsin, are the isomers involved in vision.

opsin molecule [*see lower illustration on page 2462*]. The details of these however, are still obscure.

How Molecules Twist

Let us examine somewhat more closely the various isomers of retinal. The six known isomers are illustrated at the left: the all-*trans* isomer and five *cis* isomers of one kind or another. Experiments with models, together with other evidence, show that four of the six isomers are essentially planar. The two that are not are the 11-*cis* isomer and the 11,13-*dicis* isomer. In these isomers there is considerable steric hindrance, or intramolecular crowding, between the hydrogen atom on carbon No. 10 (C_{10}) and the methyl group attached to C_{13}. Thus the double bond that joins C_{11} and C_{12} cannot be rotated by 180 degrees from the *trans* to a planar *cis* configuration. In the 11-*cis* isomers the tail of the molecule from C_{11} through C_{15} is therefore twisted out of the plane formed by the rest of the molecule.

This twisted geometry introduces two configurations, called enantiomers, that are mirror images of each other; if the molecule could be viewed from the ring end, one form would be twisted to the left and the other to the right. It is possible that opsin may combine selectively with only one enantiomer.

As Pauling had predicted in the 1930's, the steric hindrance that necessitates the twist in the 11-*cis* isomer makes it less stable than the all-*trans* or the 9-*cis* and 13-*cis* forms. We have recently found, for example, that the 11-*cis* form contains about 1,500 calories more "free energy" per mole than the *trans* form. One has to put in about 25,000 calories per mole, however, to rotate the molecule from one form to the other. This amount of energy, which is much more than a molecule is likely to acquire through chance collisions with its neighbors, is known as the activation energy: the energy required to surmount the barrier that separates the *cis* and *trans* states.

This raises an important point. How can two parts of a molecule be rotated around a double bond? The interconversion of *cis* and *trans* isomeric forms to another requires gross departures from flatness. How can this be accomplished?

Here we must introduce the concept of the excited state. One can think of molecules as existing in two kinds of state: a "ground," or stable, state of relatively low internal energy and various less stable states of higher energy—

the excited states. Molecules are raised from the ground state into one or another excited state by a sudden influx of energy, which can be in the form of heat or light. They return to the ground state by giving up their excess energy, usually as heat but occasionally as light, as in fluorescence or phosphorescence.

The orbital diagrams we have described apply to molecules in the ground state. When molecules are in an excited state, their electrons have more energy and therefore occupy different orbitals. Quantum-mechanical calculations show that an excited pi electron divides its time between the two ends of a double bond [*see bottom illustration on page 2459*]. The effect is to make the double bond in an excited molecule more like a single bond and less like a double bond. In a conjugated molecule, in which pi electrons are already delocalized, the changes in bond character are not uniform throughout the conjugated system but depend on the nature of the excitation and the structure of the molecule that is excited.

When one tries to isomerize carotenoids in the laboratory, it is usually helpful to add catalysts such as bromine or iodine. (The reader will recall that Zechmeister used iodine in his demonstrations.) Heat and light favor the existence of excited states. Bromine and iodine probably function by dissociating into atomic bromine and iodine, a process that is also favored by light. A bromine or iodine atom adds fleetingly to the double bond and converts it into a single bond, which is then momentarily free to rotate until the bromine or iodine atom has departed. The actual lifetime of the singly bonded form can be very brief indeed: the time required for one rotation around a carbon-carbon single bond is only about 10^{-12} second.

The Sensitivity of Eyes

It may seem remarkable that all animal visual systems so far studied depend on the photoisomerization of retinal for light detection. Three main branches of the animal kingdom—the mollusks, the arthropods and the vertebrates—have evolved types of eyes that differ profoundly in their anatomy. It seems that various anatomical (that is, optical) arrangements will do; apparently the photochemistry, once it had evolved, was universally accepted. Presumably the visual pigments of all animals must within narrow limits be equally sensitive to light, otherwise the more light-sensitive animals would eventually replace those whose eyes were less sensitive.

How sensitive to light is the animal retina? In a series of experiments conducted about 1940 Selig Hecht and his collaborators at Columbia University showed that the dark-adapted human eye will detect a very brief flash of light when only five quanta of light are absorbed by five rod cells. From this Hecht concluded that a single quantum is enough to trigger the discharge of a dark-adapted rod cell in the retina.

It is therefore essential that the quantum efficiency of the initial photochemical event be close to unity. In other words, virtually every quantum of light absorbed by a molecule of rhodopsin must isomerize the 11-*cis* chromophore to the all-*trans* configuration. It was shown many years ago by the British workers H. J. A. Dartnall, C. F. Goodeve and R. J. Lythgoe that an absorbed quantum has about a 60 percent chance of bleaching frog rhodopsin. One of us (Kropf) has found a similar quantum efficiency for the isomerization of the 11-*cis* retinal chromophore of cattle rhodopsin. Our work also shows that 11-*cis* retinal is more photosensitive than either the 9-*cis* or the all-*trans* isomers when they are attached as chromophores to opsin, and this may be the reason why the geometrically hindered and therefore comparatively unstable 11-*cis* isomer has evolved into the chromophore of the visual pigments.

We have also recently measured the quantum efficiency of the photoisomerization of retinal and several closely related carotenoids in solution. Retinal

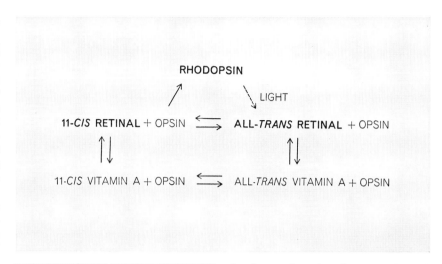

PHOTOCHEMICAL EVENTS IN VISION involve the protein opsin and isomers of retinal and its derivative, vitamin A. Opsin joined to 11-*cis* retinal forms rhodopsin. When struck by light, the 11-*cis* chromophore is converted to an all-*trans* configuration and subsequently all-*trans* retinal becomes detached from opsin. With the addition of two hydrogen atoms, all-*trans* retinal is converted to all-*trans* vitamin A. Within the eye this isomer must be converted to 11-*cis* vitamin A, thence to 11-*cis* retinal, which recombines with opsin to form rhodopsin.

turns out to be considerably more photosensitive than any of them and nearly as photosensitive as rhodopsin.

Although all animal eyes seem to employ 11-*cis* retinal as their light-sensitive agent, there are slight variations in the opsins that combine with retinal, just as there are variations in other proteins, such as hemoglobin, from species to species. Within the next few years we may learn the complete amino acid sequence of one of the opsins, and thereafter we should be able to compare such sequences for two or more species. It may be many years, however, before X-ray crystallographers have established the complete three-dimensional structure of an opsin molecule and are able to describe the site that binds it to retinal. One can conjecture that the binding site will be quite similar in the various opsins, even those from animals of different phyla, but there may be surprises in store. Whatever the precise details, it is clear that evolution has produced a remarkably efficient system for translating the absorption of light into the language of biochemistry—a language whose vocabulary and syntax are built on the various ways proteins interact with one another and with smaller molecules in their environment.

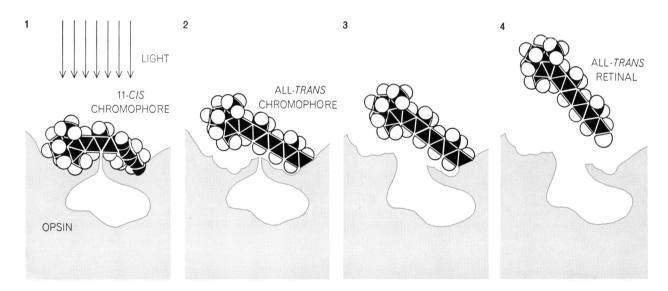

MOLECULAR EVENTS IN VISION can be inferred from the known changes in the configuration of 11-*cis* retinal after the absorption of light. In these schematic diagrams the twisted isomer is shown attached to its binding site in the much larger protein molecule of opsin (*1*). After absorbing light the 11-*cis* chromophore straightens into the all-*trans* isomer (*2*). Presumably a change in the shape of opsin (*3*) facilitates the release of all-*trans* retinal (*4*). The configuration of the binding site in opsin is not yet known.

The Authors

RUTH HUBBARD and ALLEN KROPF are respectively resident associate in biology at Harvard University and associate professor of chemistry at Amherst College. Miss Hubbard, who was born in Vienna, took a Ph.D. in biology at Radcliffe College in 1950 and since then, except for a year as a Guggenheim fellow in Copenhagen, has worked with George Wald (to whom she was married in 1958) in his laboratory at Harvard. Kropf, who was graduated from Queens College in 1951 and obtained a Ph.D. at the University of Utah in 1954, became interested in the chemistry of vision after hearing Wald lecture on the subject at the University of Utah. From 1956 to 1958 Kropf worked with Wald at Harvard, going to Amherst in 1958.

Bibliography

THE CHEMISTRY OF VISUAL PHOTO-RECEPTION. Ruth Hubbard, Deric Bownds and Tôru Yoshizawa in *Cold Spring Harbor Symposia on Quantitative Biology: Vol. XXX,* Cold Spring Harbor Biological Laboratory, 1965.

CIS-TRANS ISOMERS OF RETINENE IN VISUAL PROCESSES. G. A. J. Pitt and R. A. Morton in *Steric Aspects of the Chemistry and Biochemistry of Natural Products,* edited by J. K. Grant and W. Klyne. Cambridge University Press, 1960.

MOLECULAR ASPECTS OF VISUAL EXCITATION. Ruth Hubbard and Allen Kropf in *Annals of the New York Academy of Sciences,* Vol. 81, Art. 2, pages 388–398; August 28, 1959.

VALENCE. Charles A. Coulson. Oxford University Press, 1961.

SPECIAL NOTE TO TEACHERS: Each article in this volume, plus more than 660 others, is available as a separate, self-bound SCIENTIFIC AMERICAN Offprint. Offprints may be ordered in any combination and in any quantity. Teachers who want to adopt articles for their courses, therefore, can ensure that each student has his own set. Students' sets are collated by the publisher before shipment.

BUTTERFLY EGGS (*top*) stand upright on a leaf of clover, the egg-laying site selected by the gravid female. Clover is the food plant preferred by this species: *Colias philodice*, the clouded sulphur. After hatching (*bottom*), growing clouded sulphur larvae feed on the plant preselected for them by the parent. When they metamorphose, they too will seek out clover as an egg-laying site.

SCIENTIFIC
AMERICAN   June 1967, Vol. 216, No. 6, pp. 104-113   OFFPRINT 1076

BUTTERFLIES AND PLANTS

by Paul R. Ehrlich and Peter H. Raven

The hungry larvae of butterflies are selective in choosing the plants they eat. This reflects the fact that the evolution of both plants and the animals that feed on them is a counterpoint of attack and defense.

Anyone who has been close to nature or has wandered about in the nonurban areas of the earth is aware that animal life sometimes raises havoc with plant life. Familiar examples are the sudden defoliation of forests by hordes of caterpillars or swarms of locusts and the less abrupt but nonetheless thorough denudation of large areas by grazing animals. A visitor to the Wankie National Park of Rhodesia can see a particularly spectacular scene of herbivore devastation. There herds of elephants have thinned the forest over hundreds of square miles and left a litter of fallen trees as if a hurricane had passed through.

Raids such as these are rare, and the fertile regions of the earth manage to remain rather green. This leads most people, including many biologists, to underestimate the importance of the perennial onslaught of animals on plants. Detailed studies of the matter in recent years have shown that herbivores are a major factor in determining the evolution and distribution of plants, and the plants in turn play an important part in shaping the behavior and evolution of herbivores.

The influence of herbivores on plants is usually far from obvious, even when it is most profound. In Australia huge areas in Queensland used to be infested with the spiny prickly-pear cactus, which covered thousands of square miles of the area and made it unusable for grazing herds. Today the plant is rare in these areas. It was all but wiped out by the introduction of a cactus moth from South America, which interestingly enough is now hardly in evidence. When one searches scattered remaining clumps of the cactus, one usually fails to find any sign of the insect. The plant survives only as a fugitive species; as soon as a clump of the cactus is discovered by the moth it is devoured, and the population of moths that has flourished on it then dies away. A similar situation is found in the Fiji Islands. There a plant pest of the genus *Clidemia* was largely destroyed by a species of thrips brought in from tropical America, and the parasitic insect, as well as the plant, has now become rare in Fiji.

The interplay of plant and animal populations takes many forms—some direct, some indirect, some obvious, some obscure. In California the live oak is disappearing from many areas because cattle graze on the young seedlings. In Australia a native pine that was decimated by rabbits has made a dramatic comeback since the rabbit population was brought under control by the myxomatosis virus. Australia also furnishes a striking example of how the evolution of a plant can be influenced by the presence or absence of certain animals. The plant involved is the well-known acacia. In Africa and tropical America, where grazing mammals abound, the acacia species are protected by thorns that are often fearsomely developed. Until recently there were comparatively few grazing mammals in Australia, and most of the acacia plants there are thornless, apparently having lost these weapons of their relatives on other continents.

By far the most important terrestrial herbivores are, of course, the insects. They have evolved remarkably efficient organs for eating plants: a great variety of mouthparts with which to pierce, suck or chew plant material. They eat leaves from the outside and the inside, bore through stems and roots and devour flowers, fruits and seeds. In view of the abundance, variety and appetites of the insects, one may well wonder how it is that any plants are left on the earth. The answer, of course, is that the plants have not taken the onslaught of the herbivores lying down. Some of their defenses are quite obvious: the sharp spines of the cactus, the sharp-toothed leaves of the holly plant, the toxins of poison ivy and the oleander leaf, the odors and pungent tastes of spices. The effectiveness of these weapons against animal predators has been demonstrated by laboratory experiments. For example, it has been shown that certain leaf-edge-eating caterpillars normally do not feed on holly leaves but will devour the leaves when the sharp points are cut away.

The plant world's main line of defense consists in chemical weapons. Very widespread among the plants are certain chemicals that apparently perform no physiological function for the plants themselves but do act as potent insecticides or insect repellents. Among these are alkaloids, quinones, essential oils, glycosides, flavonoids and raphides (crystals of calcium oxalate). Long before man learned to synthesize insecticides he found that an extract from chrysanthemums, pyrethrin, which is harmless to mammals, is a powerful killer of insects.

Particularly interesting are the alkaloids, a heterogeneous group of nitrogenous compounds found mainly in flowering plants. They include nicotine, caffeine, quinine, marijuana, opium and peyote. Considering the hallucinogenic properties of the last three drugs, it is amusing to speculate that the plants bearing them may practice "chemopsychological warfare" against their enemies! Does an insect that has fed on a fungus containing lysergic acid diethylamide (LSD) mistake a spider for its mate? Does a zebra that has eaten a

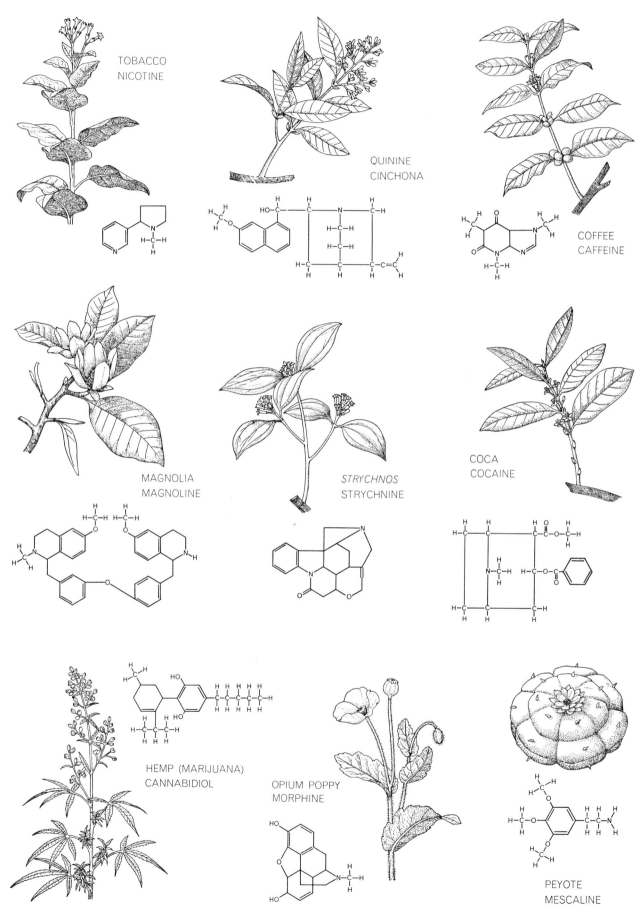

ALKALOIDS give the plants that contain them protection from predators; nine such plants are illustrated. The authors note that plant alkaloids can disturb a herbivore's physiology and that hallucinogenic alkaloids may be "chemopsychological" weapons.

plant rich in alkaloids become so intoxicated that it loses its fear of lions? At all events, there is good reason to believe eating plant alkaloids produces a profound disturbance of animals' physiology.

Of all the herbivores, the group whose eating habits have been studied most intensively is the butterflies—that is to say, butterflies in the larval, or caterpillar, stage, which constitutes the major part of a butterfly's lifetime. Around the world upward of 15,000 species of butterflies, divided taxonomically into five families, have been identified. The five families are the Nymphalidae (four-footed butterflies), the Lycaenidae (blues, metalmarks and others), the Pieridae (whites and yellows), the Papilionidae (including the swallowtails, the huge bird-wings of the Tropics and their relatives) and the Libytheidae (a tiny family of snout butterflies). The Nymphalidae and Lycaenidae account for most (three-fourths) of the known genera and species.

A caterpillar is a formidable eating machine: by the time it metamorphoses into a butterfly it has consumed up to 20 times its dry weight in plant material. The numerous species vary greatly in their choice of food. Some are highly selective, feeding only on a single plant family; others are much more catholic in their tastes, but none feeds on all plants indiscriminately. Let us examine the food preferences of various groups and then consider the evolutionary consequences.

One group that is far-ranging in its taste for plants is the Nymphalinae, a subfamily of the Nymphalidae that comprises at least 2,500 species and is widespread around the world. The plants that members of this group feed on include one or more genera of the figwort, sunflower, maple, pigweed, barberry, beech, borage, honeysuckle, stonecrop, oak, heather, mallow, melastome, myrtle, olive, buttercup, rose, willow and saxifrage families. Another group that eats a wide variety of plants is the Lycaeninae, a subfamily of the Lycaenidae that consists of thousands of species of usually tiny but often beautifully colored butterflies. The Lycaeninae in general are catholic in their tastes, and among their many food plants are members of the pineapple, borage, pea, buckwheat, rose, heather, mistletoe, mint, buckthorn, chickweed, goosefoot, morning glory, gentian, oxalis, pittosporum and zygophyllum families.

What determines the caterpillars'

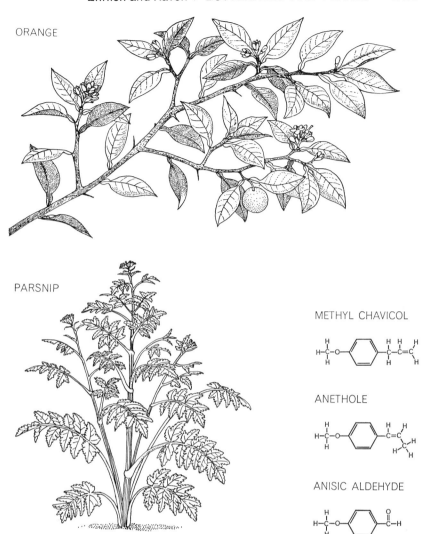

ORANGE

PARSNIP

METHYL CHAVICOL

ANETHOLE

ANISIC ALDEHYDE

PLANTS OF TWO FAMILIES, citrus (*top*) and parsley (*bottom*), produce the same three essential oils attractive to the larvae of black swallowtail butterflies. The chemical kinship between these plant families suggests a closer ancestral tie than had been suspected.

food preferences? We learn a great deal about this subject by examining the diets of those butterfly species that are particularly selective in their choice of plants. One large group of swallowtails, for example, confines its diet mainly to plants of the Dutchman's-pipe family. Another feeds only on the "woody Ranales," a group of primitive angiosperms that includes the magnolias, the laurels and many tropical and subtropical plants. A third group of swallowtails is partial to plants of the citrus and parsley families; the striped caterpillars of these butterflies, which extrude two bright orange scent horns when they are disturbed, are familiar to gardeners, who often see them feeding on parsley, dill, fennel and celery plants. The caterpillars of the white butterfly group (a subfamily of the Pieridae) feed primarily on caper plants in the Tropics and on plants of the mustard family in temperate regions. Similarly, the monarch butterfly and its relatives (a subfamily of the Nymphalidae) confine their diet primarily to plants of the milkweed and dogbane families.

Analysis of the plant selections by the butterfly groups has made it clear that their choices have a chemical basis, just as parasitic fungi choose hosts that meet their chemical needs. Vincent G. Dethier, then at Johns Hopkins University, noted some years ago that plants of the citrus and parsley families, although apparently unrelated, have in common certain essential oils (such as methyl chavicol, anethole and anisic aldehyde) that presumably account for their attractiveness to the group of swallowtails that feeds on them. Dethier found that caterpillars of the black swallowtail would even attempt to feed on filter paper soaked in these substances. The

FIVE BUTTERFLIES protected by their unpalatability are illustrated with their preferred plants. They are *Thyridia themisto* and one of the nightshades (*a*), *Battus philenor* and Dutchman's-pipe (*b*), *Danaus plexippus* and milkweed (*c*), *Heliconius charitonius* and passion flower (*d*) and *Pardopsis punctatissima* and a representative of the violet family (*e*).

same caterpillars could also be induced to feed on plants of the sunflower family (for example goldenrod and cosmos), which contain these oils but are not normally eaten by the caterpillars in nature.

The chemical finding, incidentally, raises an interesting question about the evolutionary relationship of plants. The sunflower, citrus and parsley families have been considered to be very different from one another, but their common possession of the same group of substances suggests that there may be a chemical kinship after all, at least between the citrus and the parsley family. Chemistry may therefore become a basis for reconsideration of the present classification system for plants.

In the case of the cabbage white butterfly larva the attractive chemical has been shown to be mustard oil. The pungent mustard oils are characteristic of plants of the caper and mustard families (the latter family includes many familiar food plants, such as cabbages, Brussels sprouts, horseradish, radishes and watercress). The whites' larvae also feed occasionally on plants of other families that contain mustard oils, including the garden nasturtium. The Dutch botanist E. Verschaeffelt found early in this century that these larvae would eat flour, starch or even filter paper if it was smeared with juice from mustard plants. More recently the Canadian biologist A. J. Thorsteinson showed that the larvae would eat the leaves of plants on which they normally do not feed when the plants were treated with mustard oil glucosides.

In contrast to the attractive plants, there are plant families on which butterfly larvae do not feed (although other insects may). One of these is the coffee family. Although this family, with some 10,000 species, is probably the fourth largest family of flowering plants in the world and is found mainly in the Tropics, as the butterflies themselves are, butterfly larvae rarely, if ever, feed on these plants. A plausible explanation is that plants of the coffee family are rich in alkaloids. Quinine is one example. Other plant families that butterflies generally avoid eating are the cucurbits (rich in bitter terpenes), the grape family (containing raphides) and the spiny cactus family.

One of the most interesting findings is that butterflies that are distasteful to predators (and that are identified by conspicuous coloring) are generally narrow specialists in their choice of food. They tend to select plants on which other butterfly groups do not feed, notably plants that are rich in alkaloids. It seems highly probable that their use of these plants for food has a double basis: it provides them with a feeding niche in which they have relatively little competition, and it may supply them with the substances, or precursors of substances, that make them unpalatable to predators. The distasteful groups of butterflies apparently have evolved changes in physiology that render them immune to the toxic or repellent plant substances and thus enable them to turn the plants' chemical defenses to their own advantage. Curiously, the butterfly species that mimic the coloring of the distasteful ones are in general more catholic in their feeding habits; evidently their warning coloration alone is sufficient to protect them.

The fact that some butterflies' diets are indeed responsible for their unpalatability has been demonstrated recently by Lincoln P. Brower of Amherst College and his co-workers. They worked with the monarch butterfly, whose larvae normally feed on plants of the milkweed family. Such plants are rich in cardiac glycosides, powerful poisons that are used in minute quantities to treat heart disease in man. When adult butterflies of this species are offered to hand-reared birds (with no previous experience with butterflies), the butterflies are tasted and then promptly rejected, as are further offerings of either the monarch or its close mimic, the viceroy. Recently Brower succeeded in spite of great difficulties in rearing a generation of monarch butterflies on cabbage and found that the resulting adults were perfectly acceptable to the birds, although they were refused by birds that had had previous experience with milkweed-fed monarchs.

The concept of warfare between the plants and the butterflies leads to much enlightenment on the details of evolutionary development on both sides. On the plants' side, we can liken their problem to that of the farmer, who is obliged to defend his crops from attack by a variety of organisms. The plants must deploy their limited resources to protect themselves as best they can. They may confine their growing season to part of the year (limiting their availability to predators); they may be equipped with certain mechanical or chemical defenses; some develop a nutrient-poor sap or nutritional imbalances that make them an inefficient or inadequate source of food. The herbivorous insects, for their part, reply with specializations to cope with the special defenses, as a hunter uses a high-powered rifle to hit deer or bear, a shotgun to hit birds or a hook to catch fish. No butterfly larva (or other herbivore) possesses the varieties of physical equipment that would allow it to feed on all plants; in order to feed at all it must specialize to some degree. Some of the specializations are extremely narrow; certain sap-sucking insects, for example, have developed filtering mechanisms that trap the food elements in nutrient-poor sap, and some of the caterpillars possess detoxifying systems that enable them to feed on plants containing toxic substances.

By such devices herbivores of one kind or another have managed to breach the chemical defenses of nearly every group of plants. We have already noted several examples. The mustard oils of the mustard and caper plant families,

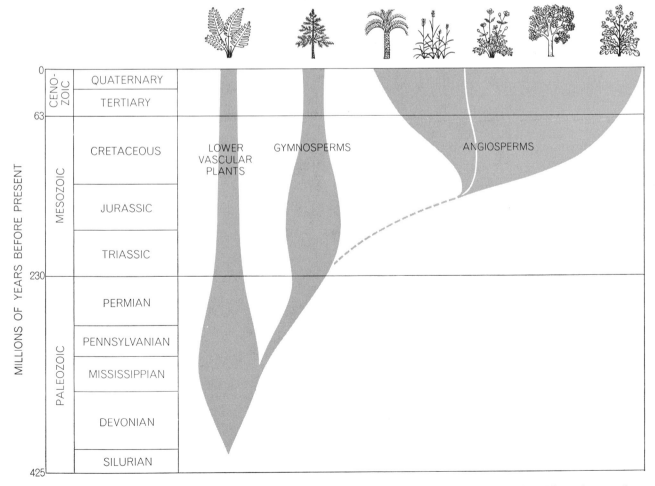

RECORD OF EVOLUTION within the plant kingdom shows that among the vascular plants the gymnosperms (*center*) declined as the angiosperms (*right*) became abundant. The authors attribute this to the acquisition of chemical defenses by the angiosperms.

for instance, serve to make these plants unpalatable to most herbivores, but the white butterflies and certain other insects have become so adapted to this defense mechanism that the mustard oils actually are a feeding stimulus for them. O. L. Chambliss and C. M. Jones, then at Purdue University, showed that a bitter, toxic substance in fruits of the squash family that repels honeybees and yellow jackets is attractive to the spotted cucumber beetle. Incidentally, this substance has been bred out of the cultivated watermelon, as any picnicker who has had to wave yellow jackets away from the watermelon can testify. By selecting against this bitter taste man has destroyed one of the natural protective mechanisms of the plant and must contend with a much wider variety of predators on it than the watermelon had to in the wild.

An important aspect of the insects' chemical adaptability is the recent finding that insects that feed on toxic plants are often immune to man-made insecticides. They evidently possess a generalized detoxifying mechanism. H. T. Gordon of the University of California at Berkeley has pointed out that this is commonly true of insects that are in the habit of feeding on a wide variety of plants. He suggests that through evolutionary selection such insects have evolved a high tolerance to biochemical stresses.

What can we deduce, in the light of the present mutual interrelations of butterflies and plants, about the evolutionary history of the insects and flowering plants? We have little information about their ancient history to guide us, but a few general points seem reasonably clear.

First, we can surmise that the great success of the angiosperm plants (plants with enclosed seeds), which now dominate the plant world since most of the more primitive gymnosperm lines have disappeared, is probably due in large measure to the angiosperms' early acquisition of chemical defenses. One important group of protective secondary plant substances, the alkaloids, is found almost exclusively in this class of plants and is well represented in those groups of angiosperms that are considered most primitive. Whereas other plants were poorly equipped for chemical warfare, the angiosperms were able to diversify behind a biochemical shield that gave them considerable protection from herbivores.

As the flowering plants diversified, the insect world also underwent a tremendous diversification with them. The intimate present relation between butterflies and plants leaves no doubt that the two groups evolved together, each

MODEL MIMETIC FORM NONMIMETIC FORM

UNPALATABLE BUTTERFLIES, whose disagreeable taste originates with the plants they ate as larvae, are often boldly marked and predators soon learn to avoid them. The three "models," so called because unrelated species mimic them, are the monarch, *Danaus* (*a*), another Danaine, *D. chrysippus* (*d*) and a third Danaine, *Amauris* (*f*). Their imitators are the viceroy, *Limenitis* (*b*), one form of *Papilio dardanus* (*e*) and another form of *P. dardanus* (*g*). Mimicry is not a genus-wide phenomenon: *L. astyanax* (*c*), a relative of the viceroy, is nonmimetic. So is a third form of *P. dardanus* (*h*), whose cousins (*e, g*) mimic two of the Danaine models.

influencing the development of the other. In all probability the butterflies, which doubtless descended from the primarily nocturnal moths, owe their success largely to the decisive step of taking to daytime feeding. By virtue of their choice of food plants all butterflies are somewhat distasteful, and Charles L. Remington of Yale University has suggested that this is primarily what enabled them to establish themselves and flourish in the world of daylight. The butterflies and their larvae did not, of course, overwhelm the plant world; on the contrary, in company with the other herbivores they helped to accelerate the evolution of the plants into a great variety of new and more resistant forms.

From what little we know about the relationships between other herbivore groups and their associated plants, we can assume that the butterfly-plant association is typical of most herbivore-plant pairings. This information gives us an excellent starting point for understanding the phenomenon that we might call "communal evolution," or coevolution. It can help, for example, to account for the great diversity of plant and insect species in the Tropics compared with the much smaller number of species in the temperate zones. The abundance of plant-eating insects in the Tropics, interacting with the plants, unquestionably has been an important factor, perhaps the most important one, in promoting the species diversity of both plants and animals in those regions. Indeed, the interaction of plants and herbivores may be the primary mechanism responsible for generating the diversity of living forms in most of the earth's environ-

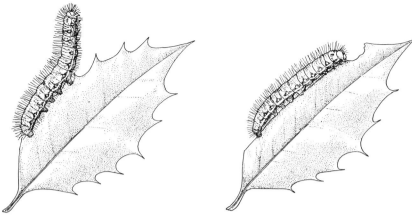

TOOTHED EDGE of the holly (*left*) normally protects it from leaf-edge eaters, such as the tent caterpillar. After the leaf's teeth are trimmed (*right*) the insect readily devours it.

ments.

Since the welfare, and even the survival, of mankind depend so heavily on the food supply and on finding ways to deal with insects without dangerous contamination of the environment with insecticides, great benefits might be derived from more intensive study of plant-herbivore associations. With detailed knowledge of these associations, plants can be bred for resistance to insects. Crop plants might be endowed with bred-in repellents, and strains of plants containing strong attractants for pests might be planted next to the crops to divert the insects and facilitate their destruction. New methods of eliminating insects without danger to man might be developed. Carroll M. Williams of Harvard University and his co-workers have discovered, for example, that substances analogous to the juvenile hormone of some insects are present in tissues of the American balsam fir. Since the juvenile

hormone acts to delay metamorphosis in insects, plants bred for such substances might be used to interfere with insect development. It is even possible that insects could be fought with tumor-inducing substances: at least one plant alkaloid, nicotine, is known to be a powerful carcinogen in vertebrates.

Such methods, together with techniques of biological control of insects already in use and under development, could greatly reduce the present reliance on hazardous insecticides. The insects have shown that they cannot be conquered permanently by the brand of chemical warfare we have been using up to now. After all, they had become battle-hardened from fighting the insecticide warfare of the plants for more than 100 million years. By learning from the plants and sharpening their natural weapons we should be able to find effective ways of poisoning our insect competitors without poisoning ourselves.

The Authors

PAUL R. EHRLICH and PETER H. RAVEN are respectively professor and associate professor of biological sciences at Stanford University. Ehrlich, who was graduated from the University of Pennsylvania in 1953 and received a doctorate in entomology from the University of Kansas in 1957, has been at Stanford since 1959. His primary interest is in the structure, dynamics and genetics of natural populations of animals. Raven, who was graduated from the University of California at Berkeley in 1957 and obtained a Ph.D. in botany from the University of California at Los Angeles in 1960, went to Stanford in 1962. His principal concerns are the biosystematics and evolution of the higher plants and their pollination systems.

Bibliography

BIRDS, BUTTERFLIES, AND PLANT POISONS: A STUDY IN ECOLOGICAL CHEMISTRY. Lincoln Pierson Brower and Jane Van Zandt Brower in *Zoologica*, Vol. 49, No. 3, pages 137–159; 1964.

BUTTERFLIES AND PLANTS: A STUDY IN COEVOLUTION. Paul R. Ehrlich and Peter H. Raven in *Evolution*, Vol. 18, No. 4, pages 586–608; January 28, 1965.

COEVOLUTION OF MUTUALISM BETWEEN ANTS AND ACACIAS IN CENTRAL AMERICA. Daniel H. Janzen in *Evolution*, Vol. 20, No. 3, pages 249–275; 1966.

SCIENTIFIC
AMERICAN   June 1967, Vol. 216, No. 6, pp. 115-122   OFFPRINT **1077**

MEMORY AND PROTEIN SYNTHESIS

by Bernard W. Agranoff

If a goldfish is trained to perform a simple task and shortly
thereafter a substance that blocks the manufacture of protein
is injected into its skull, it forgets what it has been taught.

What is the mechanism of memory? The question has not yet been answered, but the kind of evidence needed to answer it has slowly been accumulating. One important fact that has emerged is that there are two types of memory: short-term and long-term. To put it another way, the process of learning is different from the process of memory-storage; what is learned must somehow be fixed or consolidated before it can be remembered. For example, people who have received shock treatment in the course of psychiatric care report that they cannot remember experiences they had immediately before the treatment. It is as though the shock treatment had disrupted the process of consolidating their memory of the experiences.

In our laboratory at the University of Michigan we have demonstrated that there is a connection between the consolidation of memory and the manufacture of protein in the brain. Our experimental animal is the common goldfish (*Carassius auratus*). Basically what we do is train a large number of goldfish to perform a simple task and at various times before, during and after the training inject into their skulls a substance that interferes with the synthesis of protein. Then we observe the effect of the injections on the goldfish's performance.

Why seek a connection between memory and protein synthesis? For one thing, enzymes are proteins, and enzymes catalyze all the chemical reactions of life. It would seem reasonable to expect that memory, like all other life processes, is dependent on enzyme-catalyzed reactions. What is perhaps more to the point, the manufacture of new enzymes is characteristic of long-term changes in living organisms, such as growth and the differentiation of cells in the embryo. And long-term memory is by definition a long-term change.

The investigation of a connection between memory and protein synthesis is made possible by the profound advances in knowledge of protein synthesis that have come in the past 10 years. A molecule of protein is made from 20 differ-

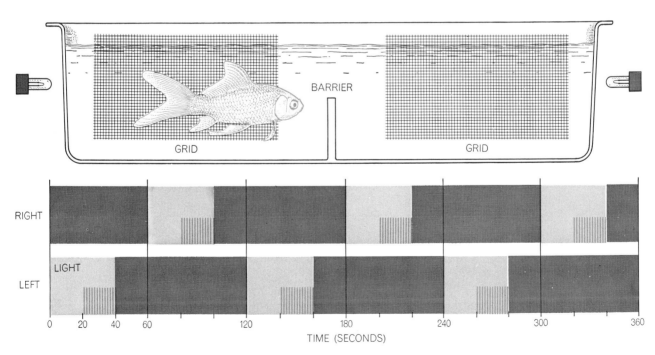

TRAINING TANK the author used was designed so that goldfish learned to swim from the light end to the dark end. A learning trial began with the illumination of the left end of the tank (*chart at bottom*), followed after a pause by mild electric shocks (*colored vertical lines*) from grids at that end. At first a fish would swim over the central barrier in response to shock; then increasingly the fish came to respond to light cue alone as sequence of light, shock and darkness was alternately repeated at each end of the tank.

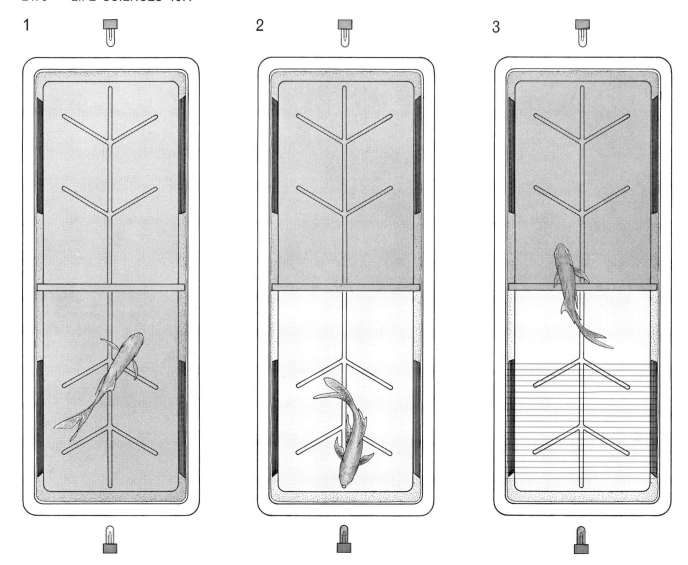

GOLDFISH LEARN in successive trials to solve the problem the shuttle box presents. Following 20 seconds of darkness (1) the end of the box where the fish is swimming is lighted for an equal period of time (2). The fish fails to respond, swimming over the barrier

ent kinds of amino acid molecule, strung together in a polypeptide chain. The stringing is done in the small bodies in the living cell called ribosomes. Each amino acid molecule is brought to the ribosome by a molecule of transfer RNA, a form of ribonucleic acid. The instructions according to which the amino acids are linked in a specific sequence are brought to the ribosome by another form of ribonucleic acid: messenger RNA. These instructions have been transcribed by the messenger RNA from deoxyribonucleic acid (DNA), the cell's central library of information.

With this much knowledge of protein synthesis one can begin to think of examining the process by interfering with it in selective ways. Such interference can be accomplished with antibiotics. Whereas some substances that interfere with the machinery of the cell, such as cyanide, are quite general in their ef-

fects, antibiotics can be highly selective. Indeed, some of them block only one step in cellular metabolism. As an example, the antibiotic puromycin simply stops the growth of the polypeptide chain in the ribosome. This it does by virtue of the fact that its molecule resembles one end of the transfer RNA molecule with an amino acid attached to it. Accordingly the puromycin molecule is joined to the growing end of the polypeptide chain and blocks its further growth. The truncated chain is released attached to the puromycin molecule.

Numerous workers have had the idea of using agents such as puromycin to block protein synthesis in animals and then observing the effects on the animals' behavior. Among them have been C. Wesley Dingman II and M. B. Sporn of the National Institutes of Health, who injected 8-azaguanine into rats; T. J. Chamberlain, G. H. Rothschild and

Ralph W. Gerard of the University of Michigan, who administered the same substance to rats, and Josefa B. Flexner, Louis B. Flexner and Eliot Stellar of the University of Pennsylvania, who injected puromycin into mice. Such experiments encouraged us to try our hand with the goldfish.

We chose the goldfish for our experiments because it is readily available and can be accommodated in the laboratory in large numbers. Moreover, a simple and automatically controlled training task for goldfish had already been developed by M. E. Bitterman of Bryn Mawr College. One might wonder if a fish has such a thing as long-term memory; in the opinion of numerous psychologists and anglers there can be no doubt of it.

Our training apparatus is called a shuttle box. It is an oblong plastic tank

4   5   6

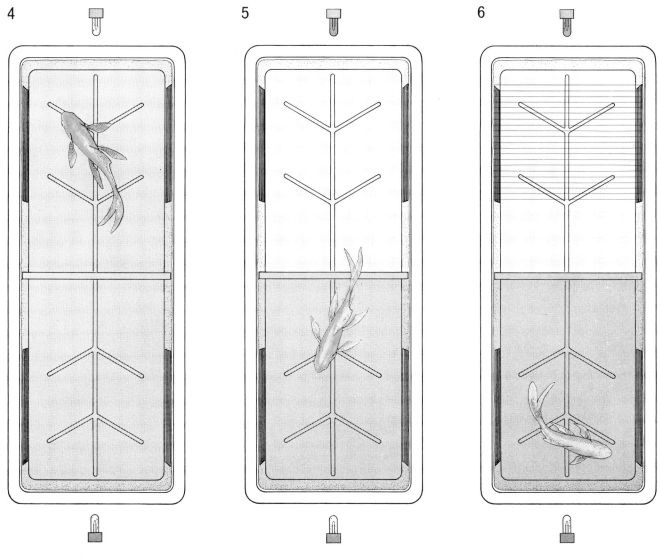

only after the shock period heralded by light has begun (3). When the same events are repeated at the other end of the box (4, 5 and 6), the fish shown here succeeds in crossing the barrier during the 20 seconds of light that precede the period of intermittent shock.

divided into two compartments by a barrier that comes to within an inch of the water surface [*see illustration on page 2475*]. At each end of the box is a light that can be turned on and off. On opposite sides of each compartment are grids by means of which the fish can be given a mild electric shock through the water.

The task to be learned by the fish is that when it is in one compartment and the light goes on at that end of the box, it should swim over the barrier into the other compartment. In our initial experiments we left the fish in the dark for five minutes and then gave it five one-minute trials. Each trial consisted in (1) turning on the light at the fish's end of the box, (2) 20 seconds later intermittently turning on the shocking grids and (3) 20 seconds after that turning off both the shocking grids and the light. If the fish crossed the barrier into the other

compartment during the first 20 seconds, it *avoided* the shock; if it crossed the barrier during the second 20 seconds, it *escaped* the shock.

An untrained goldfish almost always escaped the shock, that is, it swam across the barrier only when the shock began. Whether the fish escaped the shock or avoided it, it crossed the barrier into the other compartment. Then, after 20 seconds of darkness, the light at that end was turned on to start the second trial. Thus the fish shuttled back and forth with each trial. If a fish failed to either avoid or escape, it missed the next trial. Such missed trials were rare and generally came only at the beginning of training.

In these experiments the goldfish went through five consecutive cycles of five minutes of darkness followed by five training trials; accordingly they received a total of 20 trials in 40 minutes. They

were then placed in individual "home" tanks—plastic tanks that are slightly smaller than the shuttle boxes—and kept there for three days. On the third day they were returned to the shuttle box, where they were given 10 more trials in 20 minutes.

The fish readily learned to move from one compartment to the other when the light went on, thereby avoiding the shock. Untrained fish avoided the shock in about 20 percent of the first 10 trials and continued to improve with further trials. If they were allowed to perform the task day after day, the curve of learning flattened out at about 80 percent correct responses.

What was even more significant for our experiments was what happened when we changed the interval between the first cycle of trials and the second, that is, between the 20th and the 21st of the 30 trials. If the second cycle was

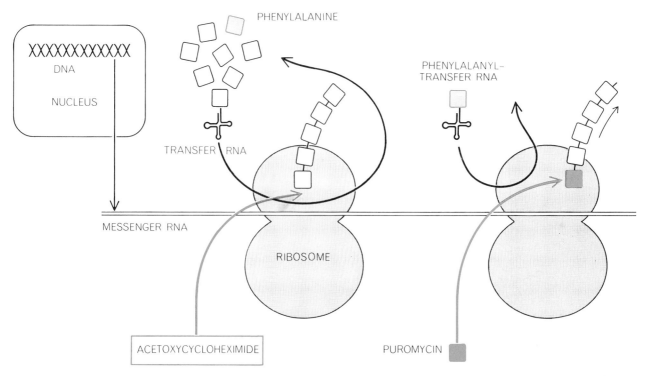

PROTEIN-BLOCKING AGENTS can interrupt the formation of molecules at the ribosome, where the amino acid units of protein are linked according to instructions embodied in messenger ribonucleic acid (mRNA). One agent, acetoxycycloheximide, interferes with the bonding mechanism that links amino acids brought to the ribosome by transfer RNA (tRNA). Puromycin, another agent, resembles the combination of tRNA and the amino acid phenylalanine. Thus it is taken into chain and prematurely halts its growth.

MOLECULAR DIAGRAMS show the resemblance between puromycin and the combination phenylalanyl-tRNA. In both cases the portion of the molecule below the broken line is incorporated into a growing protein molecule, joining at the free amino group (1). But in puromycin the CONH group (2), unlike the corresponding group (COO) of phenylalanyl, will not accept another amino acid and the chain is broken. Acetoxycycloheximide does not resemble amino acid but slows rate at which the chain forms.

begun a full month after the first, the fish performed as well as they did on the third day. If the second cycle was begun on the day after the first, the fish performed equally well, as one would expect. In short, the fish had perfect memory of their training.

We found that we could predict the training scores of groups of fish on the third day on the basis of their scores on the first day. This made it easier for us to determine the effect of antibiotics on the fish's memory: we could compare the training scores of fish receiving antibiotics with the predicted scores. Since we conducted these initial experiments we have made several improvements in our procedure. We now record the escapes and avoidances automatically with photodetectors, and we have arranged matters so that a fish does not miss a trial if it fails to escape. We have altered the trial sequence and the time interval between the turning on of the light and the turning on of the shocking grid. The results obtained with these improved procedures are essentially the same as our earlier ones.

The principal antibiotic we use in our experiments is puromycin, whose effect on protein synthesis was described earlier. We inject the drug directly into the skull of the goldfish with a hypodermic syringe. A thin needle easily penetrates the skull; 10 microliters of solution is then injected over the fish's brain (not into it). In an early series of experiments we injected 170 micrograms of puromycin in that amount of solution at various stages in our training procedures.

We found that if the puromycin was injected immediately after training, memory of the training was obliterated. If the same amount of the drug was injected an hour after training, on the other hand, memory was unaffected. Injection 30 minutes after training produced an intermediate effect. Reducing the amount of puromycin caused a smaller loss of memory.

After the injection the fish seemed to swim normally. We were therefore encouraged to test whether or not puromycin interferes with the changes that occur in the brain as the fish is being trained. This we did by injecting the fish before their initial training. We found that they learned the task at a normal rate, that is, their improvement during the first 20 trials was normal. Fish tested three days later, however, showed a profound loss of memory. This indicated to us that puromycin did not block the short-term memory demonstrated during

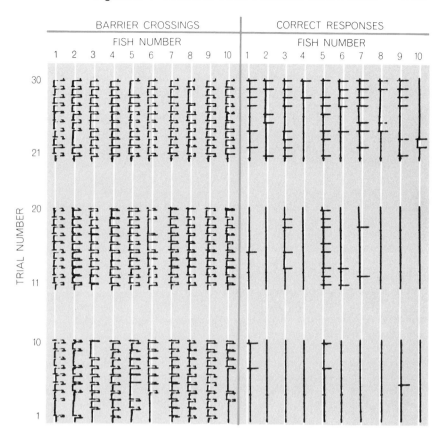

TRACE FROM RECORDER shows the performances of 10 goldfish in 30 trials. Each horizontal row represents a trial, beginning at the bottom with trial 1. A blip (*left side*) indicates that a fish either escaped or avoided the shock; a dash in the same row (*right*) signifies an avoidance, that is, a correct response for the trial. These fish learned at the normal rate.

learning but did interfere with the consolidation of long-term memory. And since an injection an hour after training has no effect on long-term memory, whereas an injection immediately after training obliterates it, it appears that consolidation can take place within an hour.

One observation puzzled us. The animals had received their initial training during a 40-minute period, 20 minutes of which was spent in the dark. Puromycin could erase all memory of this training; none of the memory was consolidated. Yet the experiment in which we injected puromycin 30 minutes after training had shown that more than half of the memory was consolidated during that period. How was it that no memory at all was consolidated at least toward the end of the 40-minute training period? To be sure, the fish that had been injected 30 minutes after the training period had been removed from the shuttle boxes and placed in their home tanks. But what was different about the time spent in the shuttle box and the time spent in the home tank that memory could be consolidated in the home tank

but could not be in the shuttle box?

Roger E. Davis of our laboratory undertook further experiments to clarify the phenomenon. He found that fish that were allowed to remain in the shuttle box for several hours after training and were then returned to their home tank showed no loss of memory when they were tested four days later. On the other hand, fish that were allowed to remain in the shuttle box for the same length of time and were then injected with puromycin and returned to their home tank had a marked memory loss! In other words, the fish in the first group did not consolidate memory of their training until after they had been placed in their home tank. It appears that simply being in the shuttle box prevents the fixation of memory. Subsequent studies have led us to the idea that memory fixation is blocked when the organism is in an environment associated with a high level of stimulation. This effect indicates that the formation of memory is environment-dependent, just as the consolidation of memory is time-dependent.

We conclude from all these experiments that long-term memory of training in the goldfish is formed by a puromycin-sensitive step that begins after training and requires that the animal be removed from the training environment. The initial acquisition of information by the fish is puromycin-insensitive and is a qualitatively different process. But what does the action of puromycin on memory formation have to do with its known biochemical effect: the inhibition of protein synthesis?

We undertook to establish that puromycin blocks protein synthesis in the goldfish brain under the conditions of our experiments. This we did in the following manner. First we injected puromycin into the skull of the fish. Next we injected into the abdominal cavity of the fish leucine that had been labeled with tritium, or radioactive hydrogen. Now, leucine is an amino acid, and if labeled leucine is injected into a goldfish's abdominal cavity, it will be incorporated into whatever protein is being synthesized throughout the goldfish's body. By measuring the amount of labeled leucine incorporated into protein after, say, 30 minutes, one can determine the rate of protein synthesis during that time.

We compared the amount of labeled leucine incorporated into protein in goldfish that had received an injection of puromycin with the amount incorporated in fish that had received either no

injection or an injection of inactive salt solution. We found that protein synthesis in the brain of fish that had been injected with puromycin was deeply inhibited. The effects of different doses of puromycin and the length of time it took the drug to act did not, however, closely correspond to what we had observed in our experiments involving the behavioral performance of the goldfish. In retrospect this result is not surprising. Various experiments, including our own, had shown that the rate of memory consolidation can be altered by changes in the conditions of training. Moreover, the rate of leucine incorporation can be affected by complex physiological factors.

Another way to check whether or not puromycin exerts its effects on memory by inhibiting protein synthesis would be to perform the memory experiments with a second drug known to inhibit such synthesis. Then if puromycin blocks long-term memory by some other mechanism, the second drug would have no effect on memory. It would be even better if the second drug did not resemble puromycin in molecular structure, so that its effect on protein synthesis would not be the same as puromycin's. Such a drug exists in acetoxycycloheximide. Where puromycin blocks the growth of the polypeptide chain by taking the place of an amino acid, acetoxycycloheximide simply slows down the rate at

which the amino acids are linked together. We found that a small amount of this drug (.1 microgram, or one 1,700th the weight of the amount of puromycin we had been using) produced a measurable memory deficit in goldfish. Moreover, it commensurately inhibited the synthesis of protein in the goldfish brain.

These experiments suggest that protein synthesis is required for the consolidation of memory, but they are not conclusive. Louis Flexner and his colleagues have found that puromycin can interfere with memory in mice. On the other hand, they find that acetoxycycloheximide has no such effect. They conclude that protein is required for the expression of memory but that experience acts not on protein synthesis directly but on messenger RNA. The conditions of their experiments and the fact that they are working with a different animal do not allow any ready comparison with our experiments.

Our studies of the goldfish have led us to view learning and memory as a form of biological development. One may think of the brain of an animal as being completely "wired" by heredity; all possible pathways are present, but not all are "soldered." It may be that in short-term memory, pathways are selected rapidly but impermanently. In that case protein synthesis would not be required, which may explain why puromycin has no effect on short-term memory. If the consolidation of memory calls for more permanent connections among pathways, it seems reasonable that protein synthesis would be involved. The formation of such connections, of course, would be blocked by puromycin and acetoxycycloheximide.

Another possibility is that the drugs block not the formation of permanent pathways but the transmission of a signal to fix what has just been learned. There is some evidence for this notion in what happens to people who suffer damage to certain parts of the brain (the mammillary bodies and the hippocampus). They retain older memories and are capable of new learning, but they cannot form new long-term memories. Experiments with animals also provide some evidence for a "fix" signal. We are currently doing experiments in the hope of determining which of these hypotheses best fits the effects of puromycin and acetoxycycloheximide on memory in the goldfish.

Quite apart from our own work, it has been suggested by others that it is possible to transfer patterns of behavior

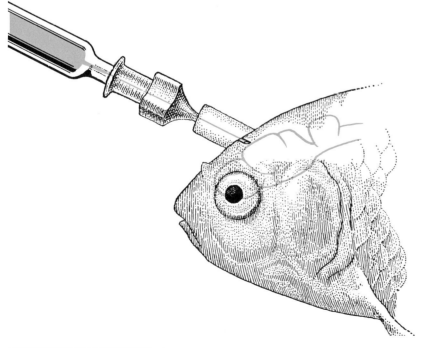

ANTIBIOTIC WAS INJECTED through the thin skull of a goldfish and over rather than into the brain. The antibiotic was puromycin, which inhibits protein synthesis. Following its injection the fish were able to swim normally. They could then be tested for memory loss.

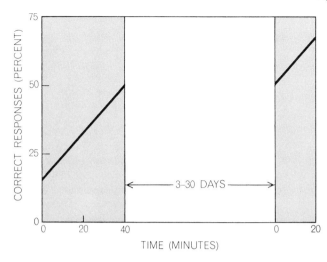

NORMAL LEARNING RATE of goldfish in 30 shuttle-box trials is shown by the black curve. Whether the last 10 trials were given three days after the first 20 (the regular procedure) or as much as a month later, fish demonstrated the same rate of improvement.

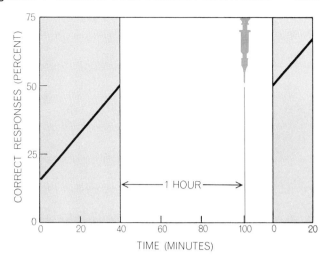

INJECTION WITH PUROMYCIN one hour after completion of 20 learning trials did not disrupt memory. Goldfish given the antibiotic at this point scored as well as those in the control group in the sequence of 10 trials that followed three days afterward.

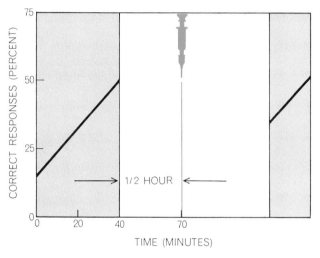

INJECTION HALF AN HOUR AFTER the first 20 trials cut the level of correct responses to half the level without such injection.

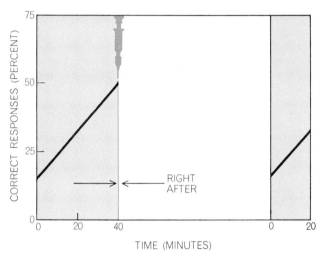

INJECTION IMMEDIATELY AFTER the first 20 trials erased all memory of training. The fish scored at the untrained level.

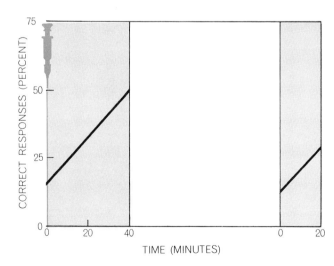

INJECTION PRIOR TO TRAINING did not affect the rate at which goldfish learned to solve the shuttle-box problem. But puromycin given at this point did suppress the formation of long-term memory, as shown by the drop in the scores three days afterward.

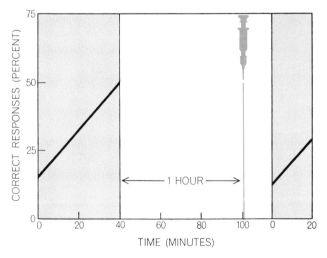

ENVIRONMENTAL FACTOR in the formation of lasting memory was seen when fish remained in training (instead of "home") tanks during the fixation period. Under these conditions fixation did not occur. Puromycin given at end of period still erased memory.

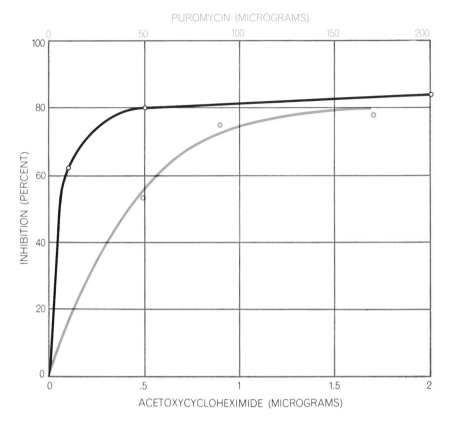

PUROMYCIN (MICROGRAMS)

ACETOXYCYCLOHEXIMIDE (MICROGRAMS)

SLOWED PROTEIN SYNTHESIS in the brain of goldfish is induced both by acetoxycyclo-heximide (*black line*) and the antibiotic puromycin (*colored line*), agents that block the fixation of memory. The author tested the effect on goldfish of various quantities of the two drugs; acetoxycycloheximide was found several hundred times more potent than puromycin.

at the speed required for learning.

It might also be that learning and memory involve the formation of short segments of RNA or protein that somehow label an individual brain cell. Richard Santen of our laboratory has calculated (on the basis of DNA content) the number of cells in the brain of a rat: it comes to 500 million. With this figure one can calculate further that a polypeptide chain of seven amino acids, arranged in every possible sequence, could provide each cell in the rat's brain with two unique markers.

The concept that each nerve cell has its own chemical marker is supported by experiments on the regeneration of the optic nerve performed by Roger W. Sperry of the California Institute of Technology. If the optic nerve of a frog is cut and the two ends of the nerve are put back together rotated 180 degrees with respect to each other, the severed fibers of the nerve link up with the same fiber as before. This of course suggests that each fiber has a unique marker that in the course of regeneration enables it to recognize its mate.

Is it possible, then, that a cell "turned on" by the learning process manufactures a chemical marker? And could such a process give rise to a substance that, when it is injected into another animal, finds its way to the exact location where it can effectuate memory? Thus far the evidence put forward in support of such ideas has not been impressive. In this exciting period of discovery in brain research clear-cut experiments are more important than theories. Certain long-term memories held by investigators in this area may be more of a hindrance than a help in exploring all its possibilities.

from one animal to another (even to an animal of a different species) by injecting RNA or protein from the brain of a trained animal into the brain (or even the abdominal cavity) of an untrained one. If such transfers of behavior patterns can actually be accomplished, they imply that memory resides in molecules of RNA or protein. Nothing we have learned with the goldfish argues for or against the possibility that a behavior pattern is stored in such a molecule.

It can be observed, however, that there is no precedent in biology for such storage. What could be required would be a kind of somatic mutation: a change in the cell's store of information that would give rise to a protein with a new sequence of amino acids. It seems unlikely that such a process could operate

The Author

BERNARD W. AGRANOFF is coordinator of biological science in the Mental Health Research Institute at the University of Michigan. He holds a medical degree, which he received at Wayne State University in 1950, but he has worked mainly as a research biochemist. From 1954 to 1960 he was a research biochemist at the National Institute of Neurological Diseases and Blindness; he went to Michigan in 1960. For the past three years he has been engaged in research on biochemical aspects of the formation of memory in the goldfish.

Bibliography

ANTIMETABOLITES AFFECTING PROTEINS OR NUCLEIC ACID SYNTHESIS: PHLEOMYCIN, AN INHIBITOR OF DNA POLYMERASE. Arturo Falaschi and Arthur Kornberg in *Federation Proceedings*, Vol. 23, No. 5, Part I, pages 940–989; September–October, 1964.

CHEMICAL STUDIES ON MEMORY FIXATION IN GOLDFISH. Bernard W. Agranoff, Roger E. Davis and John J. Brink in *Brain Research*, Vol. 1, No. 3, pages 303–309; March–April, 1966.

MEMORY IN MICE AS AFFECTED BY INTRACEREBRAL PUROMYCIN. Josefa B. Flexner, Louis B. Flexner and Eliot Stellar in *Science*, Vol. 141, No. 3575, pages 57–59; July 5, 1963.

SCIENTIFIC
AMERICAN   July 1967, Vol. 217, No. 1, pp. 13–17   OFFPRINT 1078

THIRD-GENERATION PESTICIDES

by Carroll M. Williams

The first generation is exemplified by arsenate of lead; the second, by DDT. Now insect hormones promise to provide insecticides that are not only more specific but also proof against the evolution of resistance.

Man's efforts to control harmful insects with pesticides have encountered two intractable difficulties. The first is that the pesticides developed up to now have been too broad in their effect. They have been toxic not only to the pests at which they were aimed but also to other insects. Moreover, by persisting in the environment— and sometimes even increasing in concentration as they are passed along the food chain—they have presented a hazard to other organisms, including man. The second difficulty is that insects have shown a remarkable ability to develop resistance to pesticides.

Plainly the ideal approach would be to find agents that are highly specific in their effect, attacking only insects that are regarded as pests, and that remain effective because the insects cannot acquire resistance to them. Recent findings indicate that the possibility of achieving success along these lines is much more likely than it seemed a few years ago. The central idea embodied in these findings is that a harmful species of insect can be attacked with its own hormones.

Insects, according to the latest estimates, comprise about three million species—far more than all other animal and plant species combined. The number of individual insects alive at any one time is thought to be about a billion billion (10^{18}). Of this vast multitude 99.9 percent are from the human point of view either innocuous or downright helpful. A few are indispensable; one need think only of the role of bees in pollination.

The troublemakers are the other .1 percent, amounting to about 3,000 species. They are the agricultural pests and the vectors of human and animal disease. Those that transmit human disease are the most troublesome; they have joined with the bacteria, viruses and protozoa in what has sometimes seemed like a grand conspiracy to exterminate man, or at least to keep him in a state of perpetual ill health.

The fact that the human species is still here is an abiding mystery. Presumably the answer lies in changes in the genetic makeup of man. The example of sickle-cell anemia is instructive. The presence of sickle-shaped red blood cells in a person's blood can give rise to a serious form of anemia, but it also confers resistance to malaria. The sickle-cell trait (which does not necessarily lead to sickle-cell anemia) is appreciably more common in Negroes than in members of other populations. Investigations have suggested that the sickle cell is a genetic mutation that occurred long ago in malarial regions of Africa. Apparently attrition by malaria-carrying mosquitoes provoked countermeasures deep within the genes of primitive men.

The evolution of a genetic defense, however, takes many generations and entails many deaths. It was only in comparatively recent times that man found an alternative answer by learning to combat the insects with chemistry. He did so by inventing what can be called the first-generation pesticides: kerosene to coat the ponds, arsenate of lead to poison the pests that chew, nicotine and rotenone for the pests that suck.

Only 25 years ago did man devise the far more potent weapon that was the first of the second-generation pesticides. The weapon was dichlorodiphenyltrichloroethane, or DDT. It descended on the noxious insects like an avenging angel. On contact with it mosquitoes, flies, beetles—almost all the insects—were stricken with what might be called the "DDT's." They went into a tailspin, buzzed around upside down for an hour or so and then dropped dead.

The age-old battle with the insects appeared to have been won. We had the stuff to do them in—or so we thought. A few wise men warned that we were living in a fool's paradise and that the insects would soon become resistant to DDT, just as the bacteria had managed to develop a resistance to the challenge of sulfanilamide. That is just what happened. Within a few years the mosquitoes, lice, houseflies and other noxious insects were taking DDT in their stride. Soon they were metabolizing it, then they became addicted to it and were therefore in a position to try harder.

Fortunately the breach was plugged by the chemical industry, which had come to realize that killing insects was —in more ways than one—a formula for

2484 LIFE SCIENCES 1078

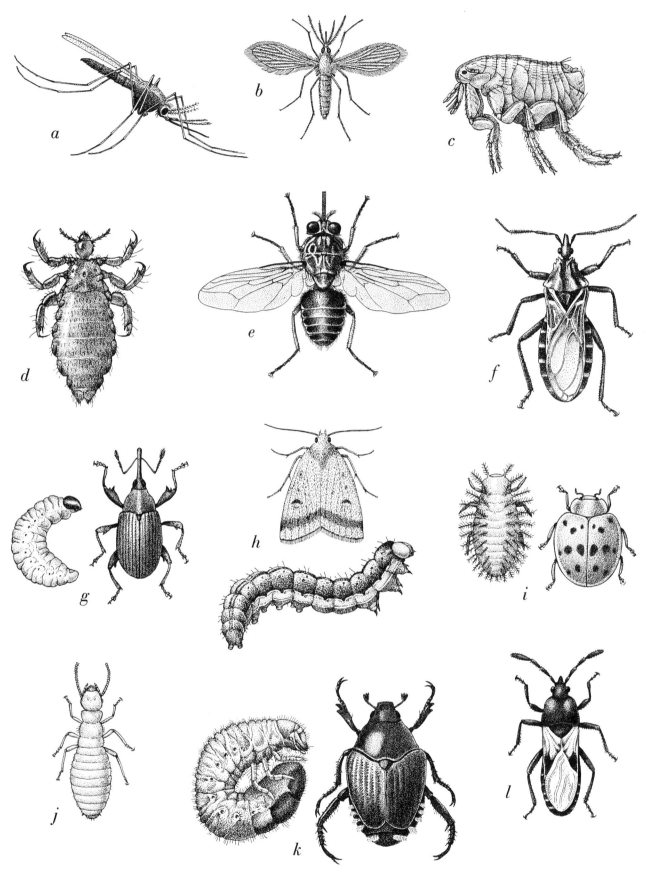

INSECT PESTS that might be controlled by third-generation pesticides include some 3,000 species, of which 12 important examples
are shown here. Six (a–f) transmit diseases to human beings; the
other six are agricultural pests. The disease-carriers, together with
the major disease each transmits, are (a) the *Anopheles* mosquito,
malaria; (b) the sand fly, leishmaniasis; (c) the rat flea, plague;
(d) the body louse, typhus; (e) the tsetse fly, sleeping sickness,
and (f) the kissing bug, Chagas' disease. The agricultural pests,
four of which are depicted in both larval and adult form, are (g)
the boll weevil; (h) the corn earworm; (i) the Mexican bean beetle; (j) the termite; (k) the Japanese beetle, and (l) the chinch
bug. The species in the illustration are not drawn to the same scale.

getting along in the world. Organic chemists began a race with the insects. In most cases it was not a very long race, because the insects soon evolved an insensitivity to whatever the chemists had produced. The chemists, redoubling their efforts, synthesized a steady stream of second-generation pesticides. By 1966 the sales of such pesticides had risen to a level of $500 million a year in the U.S. alone.

Coincident with the steady rise in the output of pesticides has come a growing realization that their blunderbuss toxicity can be dangerous. The problem has attracted widespread public attention since the late Rachel Carson fervently described in *The Silent Spring* some actual and potential consequences of this toxicity. Although the attention thus aroused has resulted in a few attempts to exercise care in the application of pesticides, the problem cannot really be solved with the substances now in use.

The rapid evolution of resistance to pesticides is perhaps more critical. For example, the world's most serious disease in terms of the number of people afflicted continues to be malaria, which is transmitted by the *Anopheles* mosquito—an insect that has become completely resistant to DDT. (Meanwhile the protozoon that actually causes the disease is itself evolving strains resistant to antimalaria drugs.)

A second instance has been presented recently in Vietnam by an outbreak of plague, the dreaded disease that is conveyed from rat to man by fleas. In this case the fleas have become resistant to pesticides. Other resistant insects that are agricultural pests continue to take a heavy toll of the world's dwindling food supply from the moment the seed is planted until long after the crop is harvested. Here again we are confronted by an emergency situation that the old technology can scarcely handle.

The new approach that promises a way out of these difficulties has emerged during the past decade from basic studies of insect physiology. The prime candidate for developing third-generation pesticides is the juvenile hormone that all insects secrete at certain stages in their lives. It is one of the three internal secretions used by insects to regulate growth and metamorphosis from larva to pupa to adult. In the living insect the juvenile hormone is synthesized by the corpora allata, two tiny glands in the head. The corpora allata are also responsible for regulating the flow of the hormone into the blood.

At certain stages the hormone must be

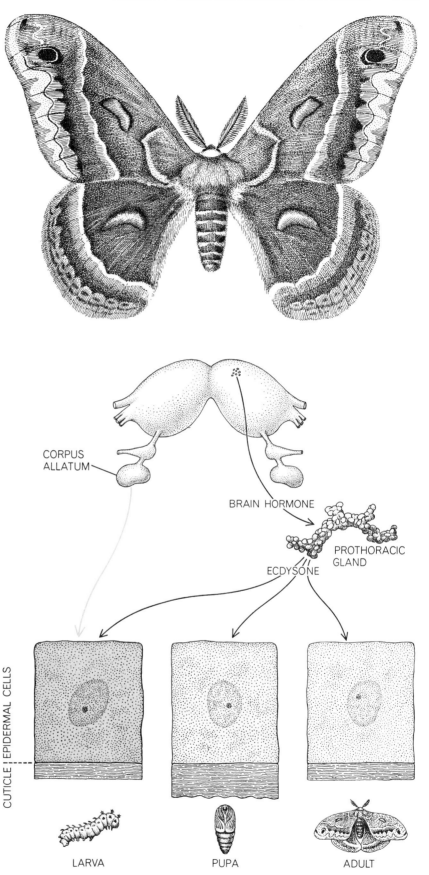

HORMONAL ACTIVITY in a Cecropia moth is outlined. Juvenile hormone (*color*) comes from the corpora allata, two small glands in the head; a second substance, brain hormone, stimulates the prothoracic glands to secrete ecdysone, which initiates the molts through which a larva passes. Juvenile hormone controls the larval forms and at later stages must be in low concentration or absent; if applied then, it deranges insect's normal development. The illustration is partly based on one by Howard A. Schneiderman and Lawrence I. Gilbert.

CHEMICAL STRUCTURES of the Cecropia juvenile hormone (*left*), isolated this year by Herbert Röller and his colleagues at the University of Wisconsin, and of a synthetic analogue (*right*) made in 1965 by W. S. Bowers and others in the U.S. Department of Agriculture show close similarity. Carbon atoms, joined to one or two hydrogen atoms, occupy each angle in the backbone of the molecules; letters show the structure at terminals and branches.

JUVENILE HORMONE ACTIVITY has been found in various substances not secreted by insects. One (*left*) is a material synthesized by M. Romanuk and his associates in Czechoslovakia. The other (*right*), isolated and identified by Bowers and his colleagues, is the "paper factor" found in the balsam fir. The paper factor has a strong juvenile hormone effect on only one family of insects, exemplified by the European bug *Pyrrhocoris apterus*.

secreted; at certain other stages it must be absent or the insect will develop abnormally [*see illustration on preceding page*]. For example, an immature larva has an absolute requirement for juvenile hormone if it is to progress through the usual larval stages. Then, in order for a mature larva to metamorphose into a sexually mature adult, the flow of hormone must stop. Still later, after the adult is fully formed, juvenile hormone must again be secreted.

The role of juvenile hormone in larval development has been established for several years. Recent studies at Harvard University by Lynn M. Riddiford and the Czechoslovakian biologist Karel Sláma have resulted in a surprising additional finding. It is that juvenile hormone must be absent from insect eggs for the eggs to undergo normal embryonic development.

The periods when the hormone must be absent are the Achilles' heel of insects. If the eggs or the insects come into contact with the hormone at these times, the hormone readily enters them and provokes a lethal derangement of further development. The result is that the eggs fail to hatch or the immature insects die without reproducing.

Juvenile hormone is an insect invention that, according to present knowledge, has no effect on other forms of life. Therefore the promise is that third-generation pesticides can zero in on in-

sects to the exclusion of other plants and animals. (Even for the insects juvenile hormone is not a toxic material in the usual sense of the word. Instead of killing, it derails the normal mechanisms of development and causes the insects to kill themselves.) A further advantage is self-evident: insects will not find it easy to evolve a resistance or an insensitivity to their own hormone without automatically committing suicide.

The potentialities of juvenile hormone as an insecticide were recognized 12 years ago in experiments performed on the first active preparation of the hormone: a golden oil extracted with ether from male Cecropia moths. Strange to say, the male Cecropia and the male of its close relative the Cynthia moth remain to this day the only insects from which one can extract the hormone. Therefore tens of thousands of the moths have been required for the experimental work with juvenile hormone; the need has been met by a small but thriving industry that rears the silkworms.

No one expected Cecropia moths to supply the tons of hormone that would be required for use as an insecticide. Obviously the hormone would have to be synthesized. That could not be done, however, until the hormone had been isolated from the golden oil and identified.

Within the past few months the difficult goals of isolating and identifying the hormone have at last been attained by a team of workers headed by Herbert Röller of the University of Wisconsin. The juvenile hormone has the empirical formula $C_{18}H_{36}O_2$, corresponding to a molecular weight of 284. It proves to be the methyl ester of the epoxide of a previously unknown fatty-acid derivative [*see upper illustration on this page*]. The apparent simplicity of the molecule is deceptive. It has two double bonds and an oxirane ring (the small triangle at lower left in the molecular diagram), and it can exist in 16 different molecular configurations. Only one of these can be the authentic hormone. With two ethyl groups ($CH_2 \cdot CH_3$) attached to carbons No. 7 and 11, the synthesis of the hormone from any known terpenoid is impossible.

The pure hormone is extraordinarily active. Tests the Wisconsin investigators have carried out with mealworms suggest that one gram of the hormone would result in the death of about a billion of these insects.

A few years before Röller and his colleagues worked out the structure of the authentic hormone, investigators at several laboratories had synthesized a number of substances with impressive juvenile hormone activity. The most potent of the materials appears to be a crude mixture that John H. Law, now at the University of Chicago, prepared by a simple one-step process in which hydrogen chloride gas was bubbled through an alcoholic solution of farnesenic acid. Without any purification this mixture was 1,000 times more active than crude Cecropia oil and fully effective in killing all kinds of insects.

One of the six active components of Law's mixture has recently been identified and synthesized by a group of workers headed by M. Romaňuk of the Czechoslovak Academy of Sciences. Romaňuk and his associates estimate that from 10 to 100 grams of the material would clear all the insects from 2½ acres. Law's original mixture is of course even more potent, and so there is much interest in its other five components.

Another interesting development that preceded the isolation and identification of true juvenile hormone involved a team of investigators under W. S. Bowers of the U.S. Department of Agriculture's laboratory at Beltsville, Md. Bowers and his colleagues prepared an analogue of juvenile hormone that, as can be seen in the accompanying illustration [*top of opposite page*], differed by only two carbon atoms from the authentic Cecropia hormone (whose structure was then, of course, unknown). In terms of the dosage required it appears that the Beltsville compound is about 2 percent as active as Law's mixture and about .02 percent as active as the pure Cecropia hormone.

All the materials I have mentioned are selective in the sense of killing only insects. They leave unsolved, however, the problem of discriminating between the .1 percent of insects that qualify as pests and the 99.9 percent that are helpful or innocuous. Therefore any reckless use of the materials on a large scale could constitute an ecological disaster of the first rank.

The real need is for third-generation pesticides that are tailor-made to attack only certain predetermined pests. Can such pesticides be devised? Recent work that Sláma and I have carried out at Harvard suggests that this objective is by no means unattainable. The possibility arose rather fortuitously after Sláma arrived from Czechoslovakia, bringing with him some specimens of the European bug *Pyrrhocoris apterus*— a species that had been reared in his laboratory in Prague for 10 years.

To our considerable mystification the bugs invariably died without reaching sexual maturity when we attempted to rear them at Harvard. Instead of metamorphosing into normal adults they continued to grow as larvae or molted into adult-like forms retaining many larval characteristics. It was evident that the bugs had access to some unknown source of juvenile hormone.

Eventually we traced the source to the paper toweling that had been placed in the rearing jars. Then we discovered that almost any paper of American origin—including the paper on which *Scientific American* is printed—had the same effect. Paper of European or Japanese manufacture had no effect on the bugs. On further investigation we found that the juvenile hormone activity originated in the balsam fir, which is the principal source of pulp for paper in Canada and the northern U.S. The tree synthesizes what we named the "paper factor," and this substance accompanies the pulp all the way to the printed page.

Thanks again to Bowers and his associates at Beltsville, the active material of the paper factor has been isolated and characterized [*see lower illustration on opposite page*]. It proves to be the methyl ester of a certain unsaturated fatty-acid derivative. The factor's kinship with the other juvenile hormone analogues is evident from the illustrations.

Here, then, is an extractable juvenile hormone analogue with selective action against only one kind of insect. As it happens, the family Pyrrhocoridae includes some of the most destructive pests of the cotton plant. Why the balsam fir should have evolved a substance against only one family of insects is unexplained. The most intriguing possibility is that the paper factor is a biochemical memento of the juvenile hormone of a former natural enemy of the tree—a pyrrhocorid predator that, for obvious reasons, is either extinct or has learned to avoid the balsam fir.

In any event, the fact that the tree synthesizes the substance argues strongly that the juvenile hormone of other species of insects can be mimicked, and perhaps has been by trees or plants on which the insects preyed. Evidently during the 250 million years of insect evolution the detailed chemistry of juvenile hormone has evolved and diversified. The process would of necessity have gone hand in hand with a retuning of the hormonal receptor mechanisms in the cells and tissues of the insect, so that the use as pesticides of any analogues that are discovered seems certain to be effective.

The evergreen trees are an ancient lot. They were here before the insects; they are pollinated by the wind and thus, unlike many other plants, do not depend on the insects for anything. The paper factor is only one of thousands of terpenoid materials these trees synthesize for no apparent reason. What about the rest?

It seems altogether likely that many of these materials will also turn out to be analogues of the juvenile hormones of specific insect pests. Obviously this is the place to look for a whole battery of third-generation pesticides. Then man may be able to emulate the evergreen trees in their incredibly sophisticated self-defense against the insects.

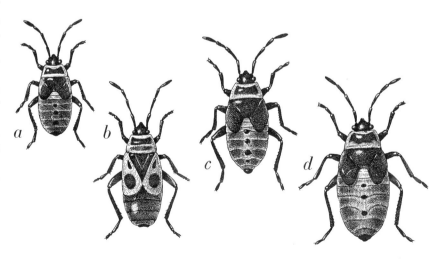

EFFECT OF PAPER FACTOR on *Pyrrhocoris apterus* is depicted. A larva of the fifth and normally final stage (*a*) turns into a winged adult (*b*). Contact with the paper factor causes the insect to turn into a sixth-stage larva (*c*) and sometimes into a giant seventh-stage larva (*d*). The abnormal larvae usually cannot shed their skin and die before reaching maturity.

The Author

CARROLL M. WILLIAMS is Bussey Professor of Biology at Harvard University. After his graduation from the University of Richmond in 1937 he went to Harvard and successively obtained master's and doctor's degrees in biology and, in 1946, an M.D. He joined the Harvard faculty in 1946 and became full professor in 1953. From 1959 to 1961 he was chairman of the biology department. He has been a member of the National Academy of Sciences since 1961. Williams' studies of insects have won a number of awards; two months ago he received the George Ledlie Prize of $1,500, which is given every two years to the member of the Harvard faculty who has made "the most valuable contribution to science, or in any way for the benefit of mankind."

Bibliography

THE EFFECTS OF JUVENILE HORMONE ANALOGUES ON THE EMBRYONIC DEVELOPMENT OF SILKWORMS. Lynn M. Riddiford and Carroll M. Williams in *Proceedings of the National Academy of Sciences,* Vol. 57, No. 3, pages 595–601; March, 1967.

THE HORMONAL REGULATION OF GROWTH AND REPRODUCTION IN INSECTS. V. B. Wigglesworth in *Advances in Insect Physiology: Vol. II,* edited by J. W. L. Bement, J. E. Treherne and V. B. Wigglesworth. Academic Press Inc., 1964.

SYNTHESIS OF A MATERIAL WITH HIGH JUVENILE HORMONE ACTIVITY. John H. Law, Ching Yuan and Carroll M. Williams in *Proceedings of the National Academy of Sciences,* Vol. 55, No. 3, pages 576–578; March, 1966.

SPECIAL NOTE TO TEACHERS: Each article in this volume, plus more than 660 others, is available as a separate, self-bound SCIENTIFIC AMERICAN Offprint. Offprints may be ordered in any combination and in any quantity. Teachers who want to adopt articles for their courses, therefore, can ensure that each student has his own set. Students' sets are collated by the publisher before shipment.

SCIENTIFIC
AMERICAN   July 1967, Vol. 217, No. 1, pp. 60–74   OFFPRINT 1079

BUILDING A BACTERIAL VIRUS

by William B. Wood and R. S. Edgar

T4 viruses with mutations in certain genes produce unassembled viral components. These particles are combined in the test tube in an effort to learn how the genes of a virus specify its shape.

Slice an orange in half, squeeze the juice into a pitcher and then drop in the rind. It comes as no surprise that the orange does not reconstitute itself. If, on the other hand, the components of the virus that causes the mosaic disease of tobacco are gently dissociated and then brought together under the proper conditions, they do reassociate, forming complete, infectious virus particles. The tobacco mosaic virus consists of a single strand of ribonucleic acid with several thousand identical protein subunits assembled around it in a tubular casing. The orange, of course, is a large and complex structure composed of a variety of cell types incorporating many different kinds of proteins and other materials. Yet both orange and virus are examples of biological architecture that must arise as a consequence of the action of genes.

Molecular biologists have now provided a fairly complete picture of how

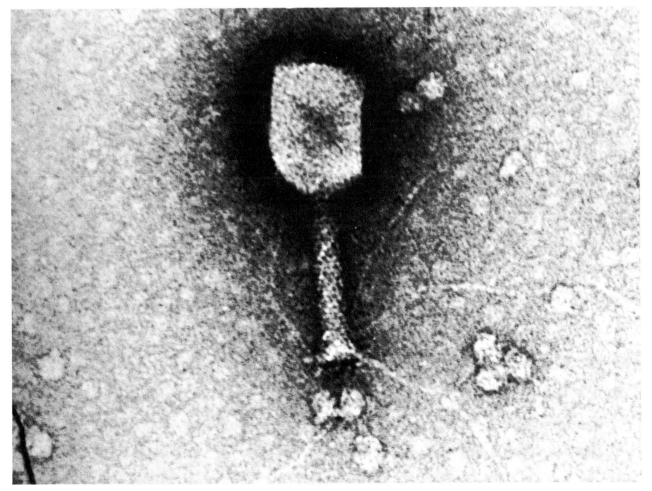

COMPLETE T4 PARTICLE was built by assembling component parts in the test tube. The virus is enlarged about 300,000 diameters in this electron micrograph made, like the ones on the next page, by Jonathan King of the California Institute of Technology.

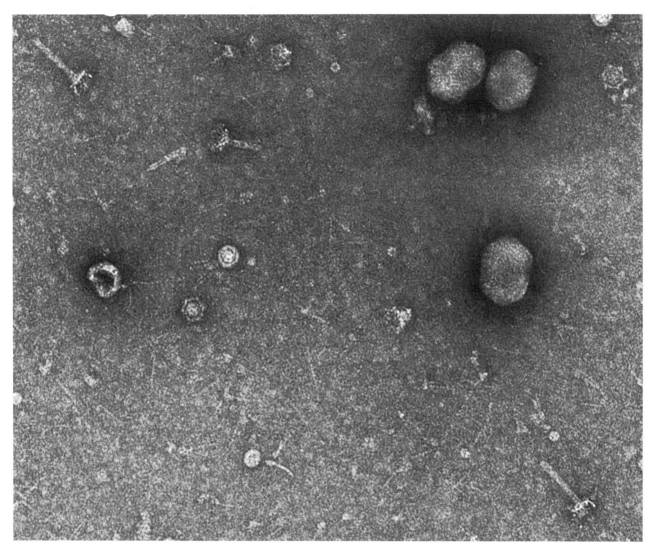

UNASSEMBLED PARTS of the T4 virus are present in this extract. It was prepared by infecting colon bacilli with a mutant virus defective in gene No. 18, which specifies the synthesis of the sheath (*see upper illustration on page 2491*). The result is the accumulation of all major components except the sheath: heads, free tail fibers and "naked" tails consisting of cores and end plates.

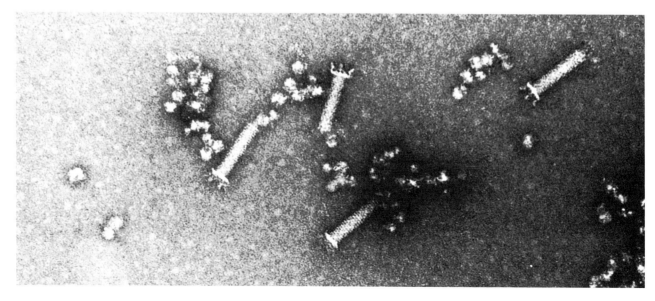

COMPLETE TAILS, enclosed in sheaths, were produced by a different mutant, defective in a gene involved in head formation. The tails were separated from the resulting extract (along with some spherical bacterial ribosomes) by being spun in a centrifuge. If the tails are added to the extract (*top photograph*), they combine with the heads and free fibers in it to form infectious virus.

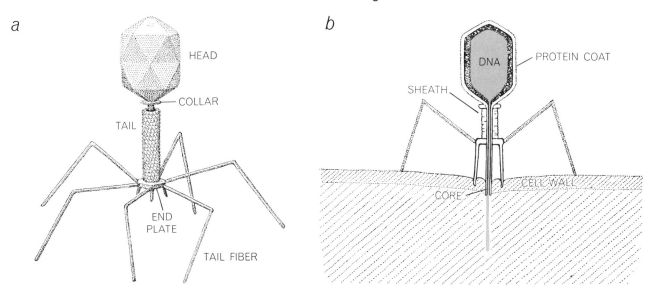

a

HEAD

COLLAR

TAIL

END
PLATE

TAIL FIBER

b

DNA

PROTEIN COAT

SHEATH

CORE

CELL WALL

T4 BACTERIAL VIRUS is an assembly of protein components (*a*). The head is a protein membrane, shaped like a kind of prolate icosahedron with 30 facets and filled with deoxyribonucleic acid (DNA). It is attached by a neck to a tail consisting of a hol-low core surrounded by a contractile sheath and based on a spiked end plate to which six fibers are attached. The spikes and fibers affix the virus to a bacterial cell wall (*b*). The sheath contracts, driving the core through the wall, and viral DNA enters the cell.

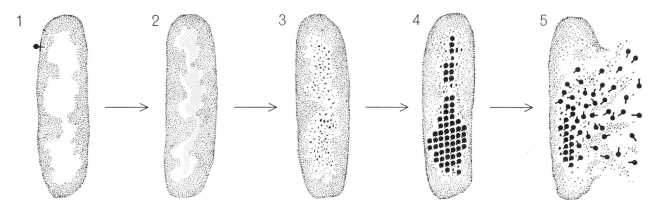

1   2   3   4   5

VIRAL INFECTION begins when viral DNA (*color*) enters a bacterium (*1*). Bacterial DNA is disrupted and viral DNA replicated (*2*). Synthesis of viral structural proteins (*3*) and their assembly into virus (*4*) continues until the cell bursts, releasing particles (*5*).

genes carry out their primary function: the specification of protein structure. The segment of nucleic acid (DNA or RNA) that constitutes a single gene specifies the chain of amino acids that comprises a protein molecule. Interactions among the amino acids cause the chain to fold into a unique configuration appropriate to the enzymatic or structural role for which it is destined. In this way the information in one gene determines the three-dimensional structure of a single protein molecule.

Where does the information come from to direct the next step: the assembly of many kinds of protein molecules into more complex structures? To build the relatively simple tobacco mosaic virus no further information is required; the inherent properties of the strand of RNA and the protein subunits cause them to interact in a unique way that results in the formation of virus particles. Clearly such a self-assembly process cannot explain the morphogenesis of an orange. At some intermediate stage on the scale of biological complexity there must be a point at which self-assembly becomes inadequate to the task of directing the building process. Working with a virus that may be just beyond that point, the T4 virus that infects the colon bacillus, we have been trying to learn how genes supply the required additional information.

Although the T4 virus is only a few rungs up the biological ladder from the tobacco mosaic virus, it is considerably more complex. Its DNA, which comprises more than 100 genes (compared with five or six in the tobacco mosaic virus), is coiled tightly inside a protein membrane to form a polyhedral head. Connected to the head by a short neck is a springlike tail consisting of a contractile sheath surrounding a central core and attached to an end plate, or base, from which protrude six short spikes and six long, slender fibers.

The life cycle of the T4 virus begins with its attachment to the surface of a colon bacillus by the tail fibers and spikes on its end plate. The sheath then contracts, driving the tubular core of the tail through the wall of the bacterial cell and providing an entry through which the DNA in the head of the virus can pass into the bacterium. Once inside, the genetic material of the virus quickly takes over the machinery of the cell. The bacterial DNA is broken down, production of bacterial protein stops and within less than a minute the cell has begun to manufacture viral proteins under the control of the injected virus genes. Among the first proteins to be made are the enzymes needed for viral DNA replication, which begins five minutes after infection. Three minutes later a second set of genes starts to direct the synthesis of the structural proteins that will form the head components and the tail components, and the process of viral morpho-

genesis begins. The first completed virus particle materializes 13 minutes after infection. Synthesis of both the DNA and the protein components continues for 12 more minutes until about 200 virus particles have accumulated within the cell. At this point a viral enzyme, lysozyme, attacks the cell wall from the inside to break open the bacterium and liberate the new viruses for a subsequent round of infection.

Additional insight into this process has come from studying strains of T4

carrying mutations—molecular defects that arise randomly and infrequently in the viral DNA during the course of its replication [see "The Genetics of a Bacterial Virus," by R. S. Edgar and R. H. Epstein; SCIENTIFIC AMERICAN Offprint 1004]. When a mutation is present, the protein specified by the mutant gene is synthesized in an altered form. This new protein is often nonfunctional, in which case the development of the virus stops at the point where the protein is required. Normally such a mutation has little experimental use, since the virus in

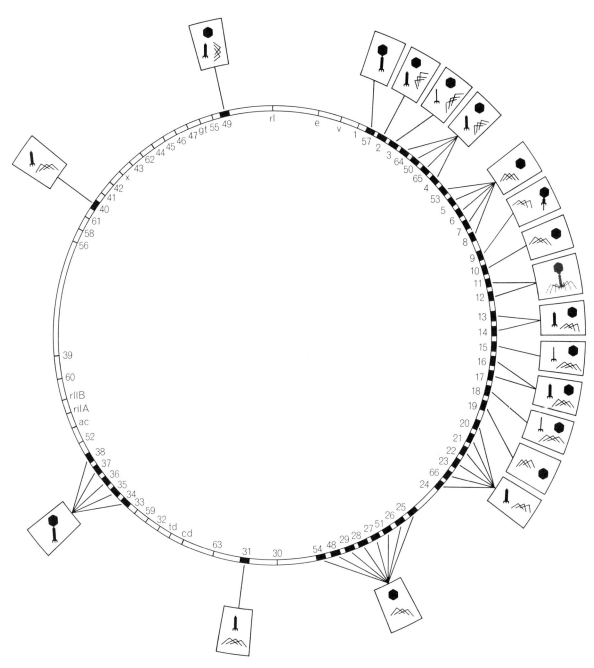

GENETIC MAP of the T4 virus shows the relative positions of more than 75 genes so far identified on the basis of mutations. The solid black segments of the circle indicate genes with morphogenetic functions. The boxed diagrams show which viral components are seen in micrographs of extracts of cells infected by mutants defective in each morphogenetic gene. A defect in gene No. 11 or 12 produces a complete but fragile particle. Heads, all tail parts, sheaths or fibers are the missing components in other extracts.

which it arises is dead and hence cannot be recovered for study. Edgar and Epstein, however, found mutations that are only "conditionally lethal": the mutant protein is produced in either a functional or a nonfunctional form, depending on the conditions of growth chosen by the experimenter. Under "permissive" conditions reproduction is normal, so that the mutants can be cultured and crossed for genetic studies. Under "restrictive" conditions, however, viral development comes to a halt at the step where the protein is needed, and by determining the point at which development is blocked the investigator can infer the normal function of the mutated gene. In this way a number of conditionally lethal mutations have been assigned to different genes, have been genetically mapped and have been tested for their effects on viral development under restrictive conditions [see illustration on page 2492].

In the case of genes that control the later stages of the life cycle, involving the assembly of virus particles, mutations lead to the accumulation of unassembled viral components. These can be identified with the electron microscope. By noting which structures are absent as a result of mutation in a particular gene, we learn about that morphogenetic gene's normal function. For example, genes designated No. 23, No. 27 and No. 34 respectively appear to control steps in the formation of the head, the tail and the tail fibers; these are the structures that are missing from the corresponding mutant-infected cells.

A blockage in the formation of one of these components does not seem to affect the assembly of the other two, which accumulate in the cell as seemingly normal and complete structures. This information alone provides some insight into the assembly process. The virus is apparently not built up the way a sock is knitted—by a process starting at one end and adding subunits sequentially until the other end is reached. Instead, construction seems to follow an assembly-line process, with three major branches that lead independently to the formation of heads, tails and tail fibers. The finished components are combined in subsequent steps to form the virus particle.

A second striking aspect of the genetic map is the large number of genes controlling the morphogenetic process. More than 40 have already been discovered, and a number probably remain to be identified. If all these genes specify proteins that are component parts of the virus, then the virus is considerably more complex than it appears to be. Alternatively, however, some gene products

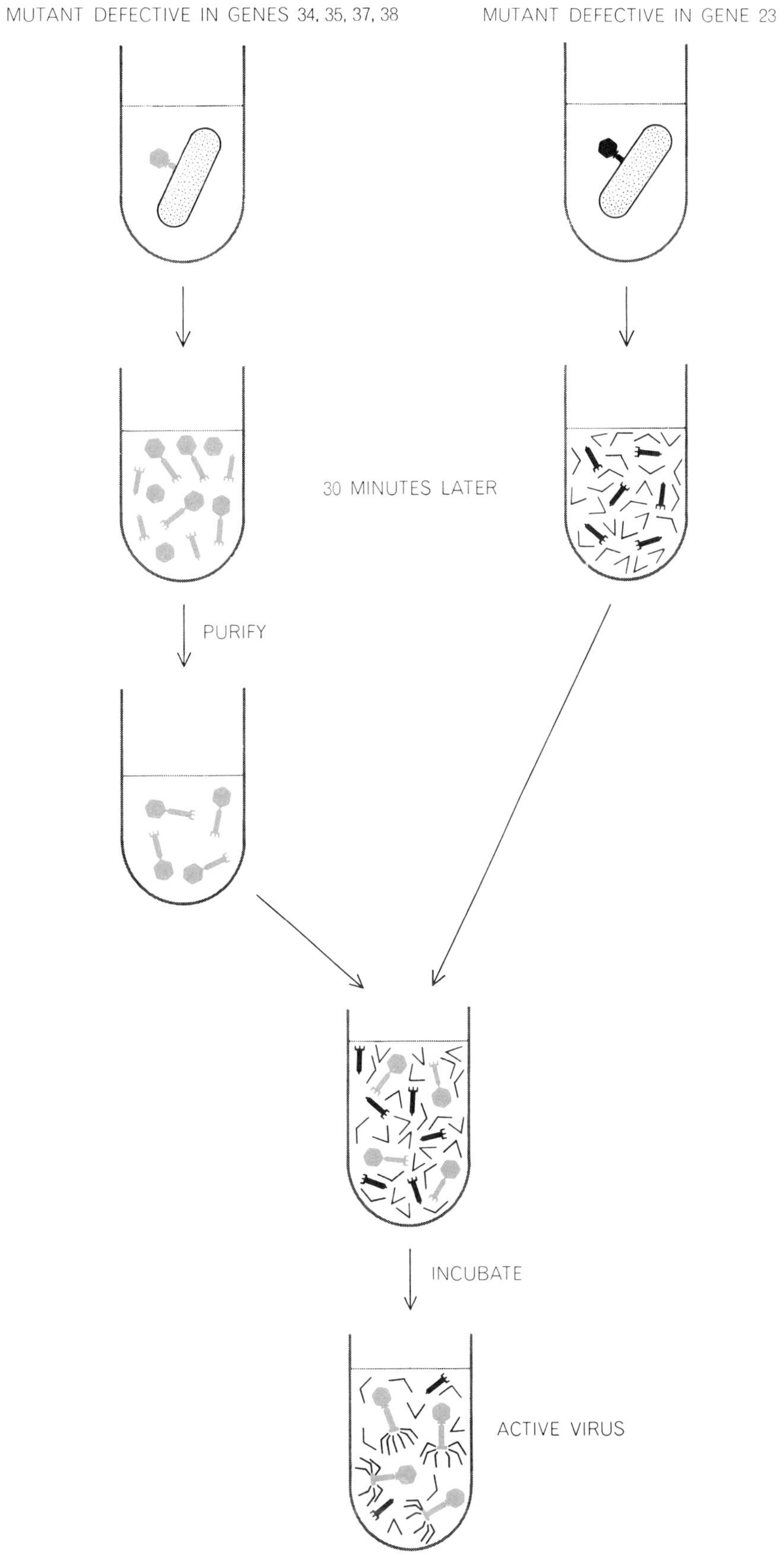

MUTANT DEFECTIVE IN GENES 34, 35, 37, 38   MUTANT DEFECTIVE IN GENE 23

30 MINUTES LATER

PURIFY

INCUBATE

ACTIVE VIRUS

TAIL FIBERS are attached to fiberless particles in the experiment diagrammed here. Cells are infected with a virus (color) bearing defective tail-fiber genes. The progeny particles, lacking fibers, are isolated with a centrifuge. A virus with a head-gene mutation (black) infects a second bacterial culture, providing an extract containing free tails and fibers. When the two preparations are mixed and incubated at 30 degrees centigrade, the fiberless particles are converted to infectious virus particles by the attachment of the free fibers.

may play directive roles in the assembly process without contributing materially to the virus itself. Studies of seven genes controlling formation of the virus's head support this possibility [see "The Genetic Control of the Shape of a Virus," by Edouard Kellenberger; SCIENTIFIC AMERICAN Offprint 1058].

In order to determine the specific functions of the many gene products involved in morphogenesis, it seemed necessary to seek a way to study individual assembly steps under controlled conditions outside the cell. One of us (Edgar) is a geneticist by training, the other (Wood) a biochemist. The geneticist is inclined to let reproductive processes take their normal course and then, by analyzing the progeny, to deduce the molecular events that must have occurred within the organism. The biochemist is eager to break the organism open and search among the remains for more direct clues to what is going on inside. For our current task a synthesis of these two approaches has proved to be most fruitful. Since it seemed inconceivable that the T4 virus could be built from scratch like the tobacco mosaic virus, starting with nucleic acid and individual protein molecules, we decided to let cells infected with mutants serve as sources of preformed viral components. Then we would break open the cells and, by determining how the free parts could be assembled into complete infectious virus, learn the sequence of steps in assembly, the role of each gene product and perhaps its precise mode of action.

Our first experiment was an attempt to attach tail fibers to the otherwise complete virus particle—a reaction we suspected was the terminal step in morphogenesis. Cells infected with a virus bearing mutations in several tail fiber genes (No. 34, 35, 37 and 38) were broken open, and the resulting particles —complete except for fibers and noninfectious—were isolated by being spun in a high-speed centrifuge. Other cells, infected with a gene No. 23 mutant that was defective in head formation, were similarly disrupted to make an extract containing free fibers and tails but no heads. When a sample of the particles was incubated with the extract, the level of infectious virus in the mixture increased rapidly to 1,000 times its initial value. Electron micrographs of samples taken from the mixture at various times showed that the particles were indeed acquiring tail fibers as the reaction proceeded.

In that first experiment the production of infectious virus required only one kind of assembly reaction—the attachment of completed fibers to completed particles. We went on to test more demanding mixtures of defective cell extracts. For example, with a mutant blocked in head formation and another one blocked in tail formation we prepared two extracts, one containing tails and free tail fibers but no heads and another containing heads and free tail fibers but no tails. When a mixture of these two extracts also gave rise to a large number of infectious viruses, we concluded that at least two reactions must have occurred: the attachment of heads to tails and the attachment of fibers to the resulting particles.

MUTANT DEFECTIVE IN GENE 27   MUTANT DEFECTIVE IN GENE 23

30 MINUTES LATER

INCUBATE

ACTIVE VIRUS

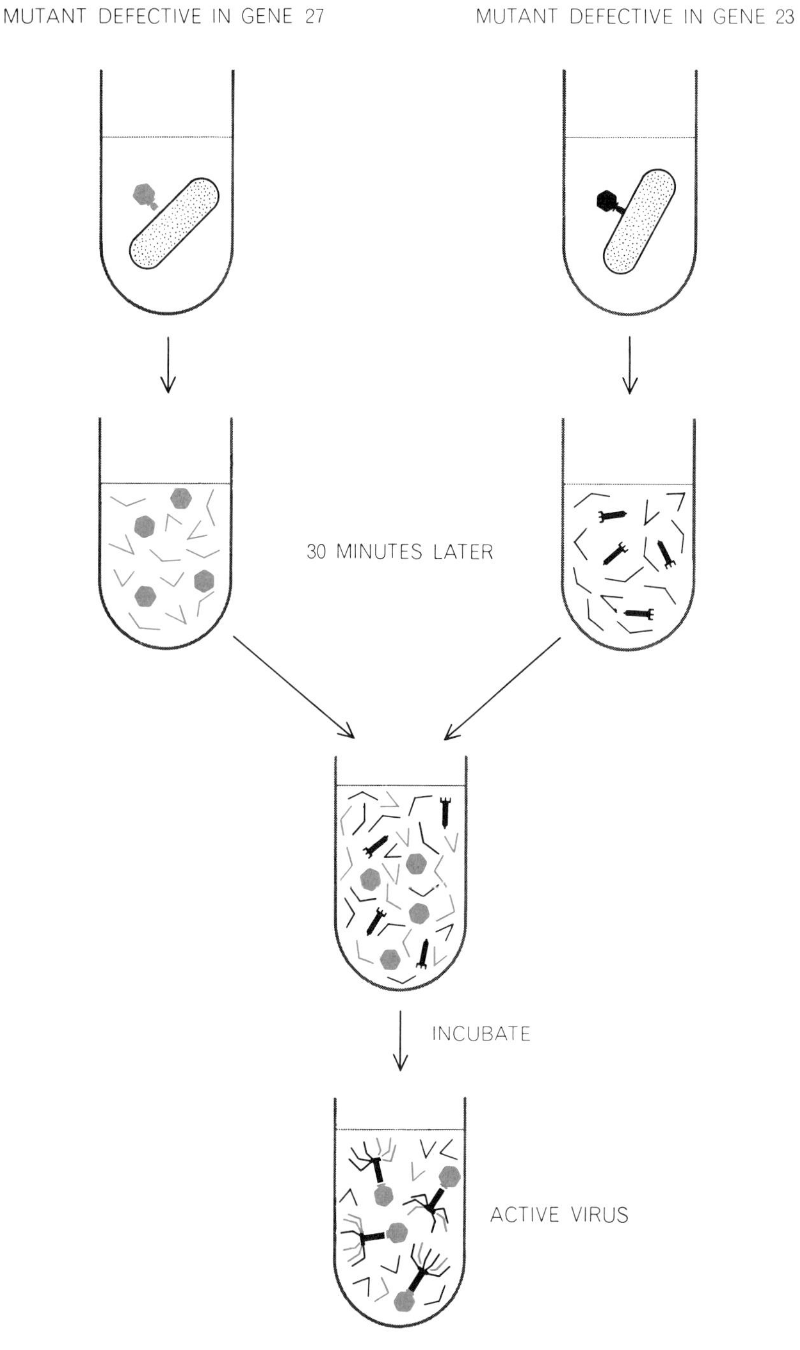

TWO ASSEMBLY REACTIONS occur in this experiment: union of heads and tails and attachment of fibers. One virus (color), with a defective tail gene, produces heads and fibers. Another (black), with a mutation in a head gene, produces tails and fibers. When the two extracts are mixed and incubated, the parts assemble to produce infectious virus.

By infecting bacilli with mutants bearing defects in different genes concerned with assembly, we prepared 40 different extracts containing viral components but no infectious virus. When we tested the extracts by mixing pairs of them in many of the appropriate combinations, some mixtures produced active virus and others showed no detectable activity. The production of infective virus implied that the two extracts were complementing each other in the test tube, that each was supplying a component that was missing or defective in the other and that could be assembled into complete, active virus under our experimental conditions. Lack of activity,

on the other hand, suggested that both extracts were deficient in the same viral component—a component being defined as a subassembly unit that functions in our experimental system. By analyzing the pattern of positive and negative results we could find out how many functional components we were dealing with.

It developed that there are at least 13 such components. That is, analysis of our pair combinations produced 13 complementation groups, the members of which did not complement one another but did complement any member of any other group. Two of these groups were

quite large [see illustration below]. Since one gene produces one protein and since each extract has a different defective gene product, a mixture of any two extracts should include all the proteins required for building the virus. The fact that members of these large groups do not complement one another must mean that our experimental system is not as efficient as an infected cell; whatever the gene products that are missing in each of these extracts do, they cannot do it in the test tube.

The idea that a complementation group consisted of extracts deficient in

EXTRACT GROUP	MUTANT GENES	COMPONENTS PRESENT	INFERRED DEFECT
I	5, 6, 7, 8, 10, 25, 26, 27, 28, 29, 48, 51, 53		TAIL
II	20, 21, 22, 23, 24, 31	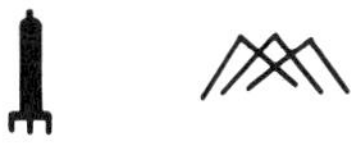	HEAD (FORMATION)
	2, 4, 16, 17, 49, 50, 64, 65		HEAD (COMPLETION)
III	54		TAIL CORE
IV	13, 14		?
V	15		
VI	18		
VII	9		?
VIII	11	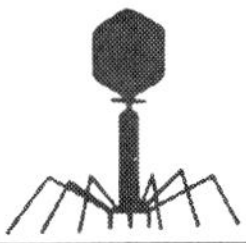	?
IX	12		
X	37, 38		TAIL FIBERS
XI	36		
XII	35		
XIII	34		

COMPLEMENTATION TESTS defined 13 groups of defective extracts, as described in the text. Mixing any two extracts in a single group fails to produce infectious virus in the test tube, but mixing any two members of different groups yields infectious virus. Apparently each group represents the genes concerned with the synthesis of a component that is functional under experimental conditions. The precise nature of the defect in some extracts, and hence the function of the missing gene product, could not be identified on the basis of the structures recognized in electron micrographs and remained to be determined by additional experiments.

the same functional component could be checked against the earlier electron micrograph results. Micrographs of the 12 defective extracts of Group I, for example, all show virus heads and tail fibers but no tails. Each of these extracts must therefore be deficient in a gene product that has to do with a stage of tail formation that cannot be carried out in our extracts. The second large complementation group appeared at first to be anomalous in terms of electron micrography: some extracts contained only tails and tail fibers, whereas others contained heads as well. Tests against extracts known to contain active tails revealed, however, that these heads—although they looked whole—could not combine to produce active virus in the test tube. In other words, heads, like tails, must be nearly completed within an infected cell before they become active for comple-

mentation. The early stages of head formation are still inaccessible to study in mixed extracts.

The remaining defective extracts gave rise to active virus in almost all possible pair combinations, segregating into another 11 complementation groups. With a total of 13 groups, there must be at least 12 assembly steps that can occur in mixtures of extracts. The defects recognizable in micrographs suggest what some of these steps must be: the completion and union of heads and tails, the assembly of tail fibers and the attachment of fibers to head-tail particles. These, then, are the steps that can be studied further in our present experimental system. We have in effect a virus-building kit, some of whose more intricate parts have been preassembled at the cellular factory.

Our next experiments were designed

to determine the normal sequence of assembly reactions and further characterize those whose nature remained ambiguous. Examples of the latter were the steps controlled by genes No. 13, 14, 15 and 18. Defects in the corresponding gene products resulted in the accumulation of free heads and tails, suggesting that they are somehow involved in head-tail union. It was unclear, however, whether these gene products are required for the attachment process itself or for completion of the head or the tail before attachment. We could distinguish the alternatives by complementation tests using complete heads and tails. These we isolated from the appropriate extracts in the centrifuge, taking advantage of their large size in relation to the other materials present. On the basis of the evidence for the independent assembly of heads and tails, we assumed that

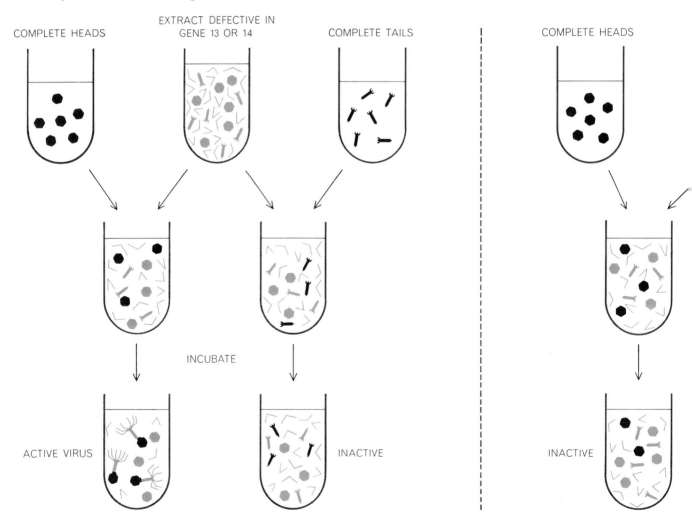

ASSEMBLY DEFECTS of mutants (*color*) that seem to produce complete heads, tails and fibers are identified, using isolated complete heads and tails (*black*) as test reagents. When complete heads are added to some extracts to be tested (*left*), infectious virus is produced, but the addition of complete tails is ineffective. This indicates that the tails made by these mutants must be functional,

the heads we isolated from a tail-defective extract would be complete, as would tails isolated from a head-defective extract.

The results of the tests were unambiguous. The addition of isolated heads to extracts lacking the products of gene No. 13 or 14 resulted in virus production, whereas the addition of tails did not. We could therefore conclude that the components missing from these extracts normally affect the head structure, and that genes No. 13 and 14 control head completion rather than tail completion or head-tail union. The remaining two of the four extracts gave the opposite result; these were active with added tails but not with added heads, indicating that genes No. 15 and 18 are involved in the completion of the tail. All four of these steps must precede the attachment of heads to tails, since defects in any of

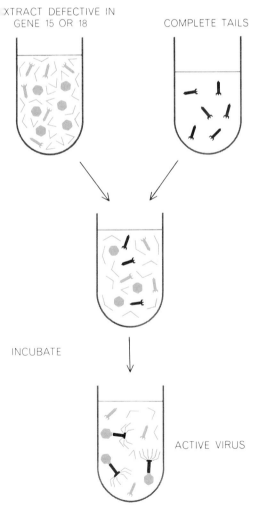

XTRACT DEFECTIVE IN GENE 15 OR 18

COMPLETE TAILS

INCUBATE

ACTIVE VIRUS

implying that the heads must be defective. In the case of other mutants (right), such tests indicate that the tails must be defective.

the corresponding genes block head-tail union.

By manipulating extracts blocked at other stages we worked out the remaining steps in the assembly process with the help of Jonathan King and Jeffrey Flatgaard. The various reactions were characterized and their sequence determined by many experiments similar to those described above. In addition, more detailed electron micrographs of defective components helped to clarify the nature of some individual steps. For example, knowing that genes No. 15 and 18 were concerned with tail completion, we went on to find just what each one did. Electron micrographs showed that in the absence of the No. 18 product no contractile sheaths were made. If No. 18 was functional but No. 15 was defective, the sheath units were assembled on the core but were unstable and could fall away. The addition of the product of gene No. 15 (and of No. 3 also, as it turned out) supplied a kind of "button" at the upper end of the core and thus apparently stabilized the sheath.

The results to date of this line of investigation can be summarized in the form of a morphogenetic pathway [*see illustration on page 2498*]. As we had thought, it consists of three principal independent branches that lead respectively to the formation of the head, the tail and the tail fibers.

The earliest stages of head morphogenesis are controlled by six genes. These genes direct the formation of a precursor that is identifiable as a head in electron micrographs but is not yet functional in extract-complementation experiments. Eight more gene products must act on this precursor to produce a head structure that is active in complementation experiments. This active structure undergoes the terminal step in head formation (the only one so far demonstrated in the test tube): conversion to the complete head that is able to unite with the tail. The nature of this conversion, which is controlled by genes No. 13 and 14, remains unclear. A likely possibility would be that these genes control the formation of the upper neck and collar, but evidence on this point is lacking. The attachment of head structures to tails has never been observed in extracts prepared with mutants defective in gene No. 13 or 14, or with any of the preceding class of eight genes. It therefore appears that completion of the head is a

prerequisite for the union of heads and tails.

The earliest structure so far identified in the morphogenesis of the tail is the end plate. It is apparently an intricate bit of machinery, since 15 different gene products participate in its formation. All the subsequent steps in tail formation can be demonstrated in the test tube. The core is assembled on the end plate under the control of the products of gene No. 54 and probably No. 19; the resulting structure appears as a tail without a sheath. The product of gene No. 18 is the principal structural component of the sheath, which is somehow stabilized by the products of genes No. 3 and 15. Tails without sheaths do not attach themselves to head structures, indicating that the tail as well as the head must be completed before head-tail union can occur. Moreover, unattached tail structures are never fitted with fibers, suggesting that these can be added only at a later stage of assembly.

Completed heads and tails unite spontaneously, in the absence of any additional factors, to produce a precursor particle that interacts in a still undetermined manner with the product of gene No. 9, resulting in the complete head-plus-tail particle. It is only at this point that tail fibers can become attached to the end plate.

At least five gene products participate in the formation of the tail fiber. In the first step, which has not yet been demonstrated in extracts, the products of genes No. 37 and 38 combine to form a precursor corresponding in dimensions to one segment of the finished fiber. This precursor then interacts sequentially with the products of genes No. 36, 35 and 34 to produce the complete structure. Again the completion of a major component—in this case the tail fiber—appears to be a prerequisite for its attachment, since we have never seen the short segments linked to particles.

The final step in building the virus is the attachment of completed tail fibers to the otherwise finished particle. We have studied this process in reaction mixtures consisting of purified particles and a defective extract containing complete tail fibers but no heads or tails. When we divided the extract into various fractions, we found that it supplies two components, both of which are necessary for the production of active virus. One of these of course is the tail fiber. The other is a factor whose properties suggest that

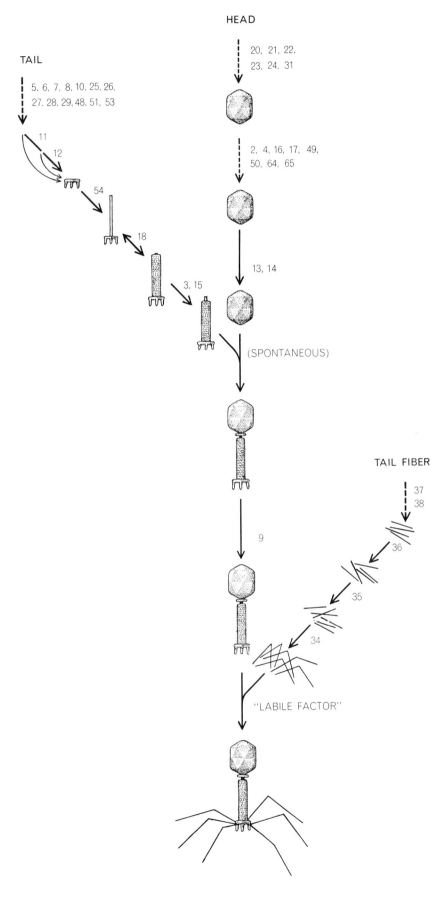

TAIL

5, 6, 7, 8, 10, 25, 26,
27, 28, 29, 48, 51, 53

11

12

54

18

3, 15

HEAD

20, 21, 22,
23, 24, 31

2, 4, 16, 17, 49,
50, 64, 65

13, 14

(SPONTANEOUS)

TAIL FIBER

37

38

36

35

9

34

"LABILE FACTOR"

MORPHOGENETIC PATHWAY has three principal branches leading independently to the formation of heads, tails and tail fibers, which then combine to form complete virus particles. The numbers refer to the gene product or products involved at each step. The solid portions of the arrows indicate the steps that have been shown to occur in extracts.

it might be an enzyme. For one thing, the rate at which fibers are attached depends on the level of this factor present in the reaction mixture, and yet the factor does not appear to be used up in the process. Moreover, the rate of attachment depends on the temperature of incubation—increasing by a factor of about two with every rise in temperature of 10 degrees centigrade. These characteristics suggest that the factor could be catalyzing the formation of bonds between the fibers and the tail end plate. At the moment we can only speculate on its possible mechanism of action, since the chemical nature of these bonds is not yet known; we call it simply a "labile factor," not an enzyme. Although no gene controlling the factor has yet been discovered, we assume that its synthesis must be directed by the virus, since it is not found in extracts of uninfected bacteria.

The T4 assembly steps so far accomplished and studied in the test tube represent only a fraction of the total number. Already, however, it is apparent that there is a high degree of sequential order in the assembly process; restrictions are somehow imposed at each step that prevent its occurrence until the preceding step has been completed. Only two exceptions to this rule have been discovered. The steps controlled by genes No. 11 and 12, which normally occur early in the tail pathway, can be bypassed when these gene products are lacking. In that case the tail is completed, attaches itself to a head and acquires tail fibers, but the result is a fragile, defective particle. The particle can, however, be converted to a normal active virus by exposure to an extract containing the missing gene products. These are the only components whose point of action in the pathway appears to be unimportant.

The problem has now reached a tantalizing stage. A partial sequence of gene-controlled assembly steps can be written, but the manner in which the corresponding gene products contribute to the process remains unclear, and the questions posed at the beginning of this article cannot yet be answered definitively. There is the suggestion that the attachment of tail fibers is catalyzed by a virus-induced enzyme. If this finding is substantiated, it would overthrow the notion that T4 morphogenesis is entirely a self-assembly process. Continued investigation of this reaction and the assembly steps that precede it can be expected to provide further insight into how genes control the building of biological structures.

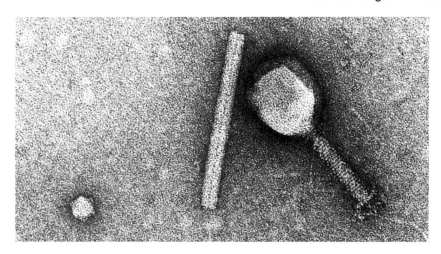

TWO SIMPLER VIRUSES are shown with the T4 in an electron micrograph made by Fred Eiserling of the University of California at Los Angeles. The icosahedral ΦX174 virus (*left*) infects the colon bacillus, as does the T4. The rod-shaped tobacco mosaic virus reassembles itself in the test tube after dissociation. The enlargement is 200,000 diameters.

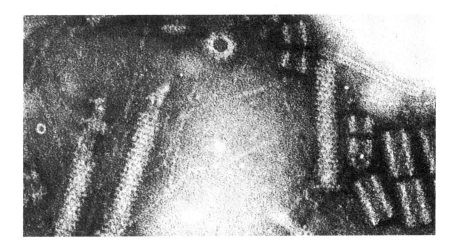

COMPLEX STRUCTURE of the T4 tail is shown in an electron micrograph made by E. Boy de la Tour of the University of Geneva. The parts were obtained by breaking down virus particles, not by synthesis, which is why fibers are attached to tails. The hollow interiors of the free core (*top right*) and pieces of sheath are delineated by dark stain that has flowed into them. There are end-on views of pieces of core (*left*) and sheath (*top center*).

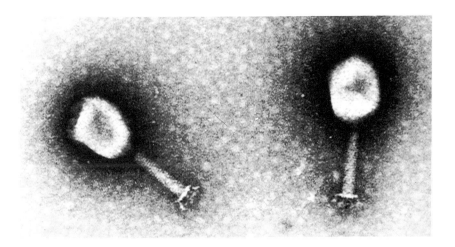

FIBERLESS PARTICLES, otherwise complete, are products of infection by a mutant defective in one of the fiber-forming genes. Heads, tails and fibers are each formed by a subassembly line (see *illustration on page 2498*). The electron micrograph was made by King.

The Authors

WILLIAM B. WOOD and R. S. ED-GAR are in the division of biology of the California Institute of Technology; Wood is assistant professor and Edgar is professor. Wood, who did his undergraduate work at Harvard College, received a doctorate in biochemistry from Stanford University in 1963 and spent a year and a half as a postdoctoral fellow in Switzerland before joining the Cal Tech faculty. Edgar, a graduate of McGill University, obtained his Ph.D. from the University of Rochester. Wood writes that they began discussing the experiments described in their article in 1963 and started work in 1965. "I suspect," he adds, "that either of us alone might never have initiated these experiments."

Bibliography

CONDITIONAL MUTATIONS IN BACTERIO-PHAGE T4. R. S. Edgar and R. H. Epstein in *Genetics Today*, edited by S. J. Geerts. Pergamon Press, 1963.

GENE ACTION IN THE CONTROL OF BACTERIOPHAGE T4 MORPHOGENESIS. W. B. Wood in *Proceedings of the Thomas Hunt Morgan Centennial Symposium*. University of Kentucky, in press.

SOME STEPS IN THE MORPHOGENESIS OF BACTERIOPHAGE T4. R. S. Edgar and I. Lielausis in *Journal of Molecular Biology*, in press.

SPECIAL NOTE TO TEACHERS: Each article in this volume, plus more than 660 others, is available as a separate, self-bound SCIENTIFIC AMERICAN Offprint. Offprints may be ordered in any combination and in any quantity. Teachers who want to adopt articles for their courses, therefore, can ensure that each student has his own set. Students' sets are collated by the publisher before shipment.

SCIENTIFIC
AMERICAN   August 1967, Vol. 217, No. 2, pp. 60-71   OFFPRINT **1080**

TETRODOTOXIN

by Frederick A. Fuhrman

It is a powerful poison that is found in two almost totally unrelated
kinds of animal: puffer fish and newts. It has been serving as a tool
in nerve physiology and may provide a model for new local anesthetics.

On September 8, 1774, His Majesty's Sloop *Resolution*, commanded by Captain James Cook, lay at anchor off the South Pacific island of New Caledonia, discovered by Cook a few days earlier. That afternoon the ship's clerk traded with a native for a fish. Captain Cook asked to have the fish prepared for a supper he was to share with the expedition's two naturalists, J. R. Forster and his son Georg. Later Cook recorded in his journal: "The opperation of describeing and drawing took up so much time till it was too late so that only the Liver and Roe was dressed of which the two Mr. Forsters and myself did but just taste. About 3 or 4 o'clock in the Morning we were siezed with an extraordinary weakness in all our limbs attended with a numbness or Sensation like to that caused by exposeing ones hands or feet to a fire after having been pinched by frost, I had almost lost the sence of feeling nor could I distinguish between light and heavy bodies, a quart pot full of Water and a feather was the same in my hand.... In [the morning] one of the Pigs which had eat the entrails was found dead."

Captain Cook's account and the Forsters' description of the fish leave no doubt that the three men had been served a puffer fish. It was fortunate that they had eaten sparingly, because many puffer fishes contain a powerful toxin that can kill a man. Curiously this toxin, which is now called tetrodotoxin, has also been found in an almost totally unrelated animal, the newt. Recently a number of investigators have become interested in tetrodotoxin's chemical nature and mode of action. It has been discovered that the toxin blocks the conduction of nerve signals, and in a highly specific manner. This action has already proved useful in the study of

nerve-signal transmission; it also suggests that tetrodotoxin may be able to serve as a model for valuable new local anesthetics.

At least 40 species of puffer fish are known to be poisonous. Most of them belong to the family Tetraodontidae, but some species in at least three other closely related families are toxic when they are eaten. In puffer fish tetrodotoxin is most highly concentrated in the ovaries and the liver; smaller amounts are found in the intestines and skin. In some species the muscle tissue is also toxic.

Many different kinds of fish and shellfish are known to be poisonous, and frequently it is difficult to identify the toxin that is responsible. Sometimes poisoning is caused by bacteria in spoiled fish; sometimes it results from one-celled organisms the fish have ingested. In the latter case a species may be safe to eat in one season of the year and poisonous in another. Mussels, for example, can be poisonous in the summer because in those months they sometimes ingest and concentrate one-celled dinoflagellates. Some of these organisms contain saxitoxin, the potent toxin that is responsible for "paralytic shellfish poisoning." By the same token a species of fish may be safe to eat if it is caught in some places and poisonous if it is caught in others. In most waters the red snapper is safe, but around certain islands in the South Pacific it can cause a form of poisoning called ciguatera. Puffer fish, on the other hand, are poisonous throughout the year (although the amount of tetrodotoxin in their viscera may vary with the season) and wherever they are caught.

Poisonous puffer fish are found in all the warm seas of the world. Their common names usually refer to their ability to inflate themselves when they

are disturbed: puffer fish, swellfish, blowfish, toado, globefish, porcupine fish, balloonfish. They are called *makimaki* in Hawaiian, *blaser* in Indonesian, *botete* in Spanish and *fugu* in Japanese. In Japan puffer poisoning was common for centuries. In his *History of Japan*, published in 1727, Engelbert Kaempfer wrote of the puffer: "He is rank'd among the poisonous Fish, and if eat whole, is said unavoidably to occasion death.... Many People die of it, for want, as they say, of thoroughly washing and cleaning it.... The Japanese won't deprive themselves of a dish so delicate in their opinion, for all they have so many Instances, of how fatal and dangerous a consequence it is to eat it."

Indeed, puffer poisoning still occurs in Japan. In 1957, 176 cases were recorded; 90 of them were fatal. An effort has been made to reduce such incidents by licensing chefs who are considered competent to prepare puffer fish, but the fascination of eating the fish complicates the problem of control. Poisoning has been known to result from drinking a mixture of hot *sake* and *fugu* testes, which is supposed to contribute to virility. The testes of some species contain tetrodotoxin and the testes of others do not, but under such circumstances the drinker is unlikely to distinguish between testes and the highly poisonous liver and roe. It is also said that a small amount of puffer liver imparts a particularly piquant flavor to certain dishes. Dangerous pleasures are not unknown in human life, but this one calls for an uncommon nicety of judgment. (There is little danger in eating *fugu sashimi*, thin slices of raw *fugu* muscle arranged in flower or bird patterns. The slices are cut from the nontoxic back meat of the fish.)

Since puffer poisoning has been com-

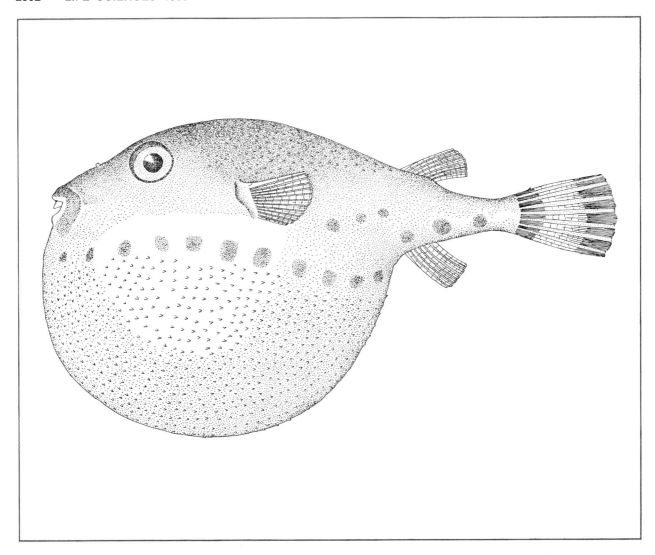

PUFFER FISH harbors the potent nerve poison tetrodotoxin. The fish's ovaries and liver are extremely poisonous, the intestines and skin less so. A true puffer, *Spheroides spengleri* of the Tetraodontidae family, is shown. Poisonous puffers are found in warm seas.

CALIFORNIA NEWT, *Taricha torosa*, also contains tetrodotoxin, principally in its skin, muscles and blood. Shown almost twice its actual size, *Taricha torosa* is the most toxic of the 11 species of Salamandridae, or true newts, that are known to contain the toxin.

paratively common in Japan, it has not escaped the attention of Japanese scientists. As early as 1883 an attempt was made to prepare a pure toxin from the fish; Yoshizumi Tahara of Tokyo named the substance tetrodotoxin, but we know today that his best preparations were only about .2 percent pure. Nevertheless, Tahara's tetrodotoxin was widely used as a drug for the relief of pain, and he was granted a U.S. patent on it in 1913. It was not until 1950 that Akira Yokoo of Okayama University and Kyosuke Tsuda of the University of Tokyo independently succeeded in crystallizing the substance in pure form.

Meanwhile a line of investigation had been started in the U.S. that had no apparent connection with the Japanese work. In 1932 Victor C. Twitty joined the department of biological sciences at Stanford University. He had come from Yale University, where he had been studying the grafting of tissues in the common eastern American salamander *Ambystoma punctatum*. In order to continue his work he needed a substitute animal, and one was easily found. In the spring the hills above the Stanford campus abound in the California newt, *Taricha torosa*. The newts come out of the forested areas to spawn in ponds and streams; spherical clusters of their eggs are found attached to submerged twigs or lying free on the bottom.

Twitty grafted eye and limb buds from embryos of the California newt into larvae of the striped salamander, *Ambystoma tigrinum*. Much to his surprise he observed that the host salamanders became paralyzed and remained so for many days. In a series of superb experiments he established that the paralysis was caused by a potent toxin found in the eggs and embryos of *Taricha torosa*. A few years later a group of biologists at Stanford succeeded in obtaining a concentrated preparation of the toxin. We now know that it was only about 1 percent pure, but when it was administered to animals, it gave rise to dramatic physiological effects. I joined the Stanford group in 1941, and somewhat later we found that the toxin acted to block the transmission of nerve impulses in frogs. We did not attempt a more intensive study. It was obvious that our best preparations of the toxin were quite impure, and we thought it unlikely that detailed examination of its effects in animals would lead to anything conclusive. Further purification of the toxin was not possible with the methods that were available at the time.

After the war powerful new methods for the separation of organic compounds, such as chromatography and electrophoresis, came into common use. By 1960 it seemed reasonable to apply some of these methods to the purification of the toxin from newt embryos. Harry S. Mosher, a Stanford chemist, agreed to collaborate with me, and with the aid of nets and a kitchen strainer we collected some 200 pounds of newt eggs. A young graduate student, M. S. Brown, undertook the task of attempting to extract pure toxin from the gelatinous mass. By the summer of 1962 he had crystallized

EGGS OF *TARICHA TOROSA*, enclosed in a translucent jelly-like substance, are deposited in spherical clusters along twigs of branches submerged in ponds and streams. Tetrodotoxin from the newt was first isolated at Stanford University from eggs found near the campus.

a substance of which each milligram was sufficiently potent to kill about 7,000 mice. We named the substance "tarichatoxin." C. Y. Kao of the Downstate Medical Center of the State University of New York joined us for the summer, and we began to investigate the pharmacological action of this new toxin.

Poisonous substances from animals fall into two broad chemical classes: proteins and nonproteins. Protein toxins are by far the most potent, but they are also extremely difficult to characterize and reconstruct in the laboratory. Nonprotein toxins are more tractable, and they also offer the possibility that they can be modified and used to treat disease. Tarichatoxin was not a protein. As its purification progressed and it became apparent that it was more toxic than any other nonprotein then known, we studied the literature on toxic substances to see if any of them resembled it. We found two: saxitoxin from shellfish and tetrodotoxin from puffer fish. We readily determined that there were certain chemical and pharmacological differences between tarichatoxin and saxitoxin, but to our astonishment every new piece of evidence indicated that tarichatoxin and tetrodotoxin were identical. Working with a sample of tetrodotoxin supplied by Tsuda, we compared the two closely. Their infrared and nuclear-magnetic-resonance spectra were the same, and they behaved identically in various chromatographic systems. Mice, rats, goldfish and other animals were killed by both toxins; newts were affected by neither of them. We concluded that puffer fish and newts, biologically different though they are, contain the same toxin. We suggested that the older name "tetrodotoxin" be retained.

Many different species of newts and salamanders have been tested in our laboratories for the presence of the toxin, but it has been found only in the 11 species of a single family: the Salamandridae, or true newts. (A different toxin, samanderin, has been isolated from specialized glands of the European fire salamander.) The most poisonous newts belong to three closely related species that range along the Pacific Coast from southern California to Alaska; one of them is *Taricha torosa*, whose eggs first yielded the newt toxin. The toxin of these species is concentrated principally in their skin, muscle and blood. It seems likely that the ability to synthesize tetrodotoxin in both newts and puffer fish has evolved as a protection against predators. Why this ability should be found in only one sub-

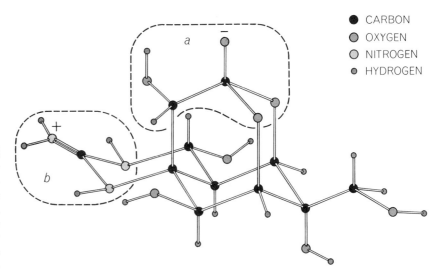

CARBON
OXYGEN
NITROGEN
HYDROGEN

STRUCTURE OF TETRODOTOXIN is unusual. The hemilactal link at *a*, which joins the molecule's two rings, was not previously known to exist. At *b* is a guanidium group, characteristic of the poisonous substance guanidine; this group may contribute to tetrodotoxin's physiological action. In acid solution the molecule carries both a positive and a negative charge. This presentation of the molecule was worked out by Harry S. Mosher of Stanford.

order of fishes and one family of amphibians, however, is an evolutionary mystery.

Tetrodotoxin is chemically unusual. The tiny colorless crystals that have been obtained from puffer fish and newt eggs are soluble in water if a trace of acid is added but decompose easily in both strongly acid and strongly alkaline solutions. The empirical formula of the molecule is $C_{11}H_{17}N_3O_8$. The problem of the structure of the molecule occupied chemists in the U.S. and Japan for many years. Finally, at a conference in Kyoto in 1964, the same structure was proposed by four different groups of chemists who had worked on the problem independently of one another. The groups were led by Tsuda, Mosher, Toshio Goto of Nagoya University and Robert B. Woodward of Harvard University. Tetrodotoxin is a weakly basic compound that in an acid solution exists as a molecule carrying both a positive and a negative charge—a "zwitterion." The molecule does not show a close resemblance to any other that is known.

The mode of action of tetrodotoxin has been investigated in several ways. The outward symptoms of tetrodotoxin poisoning are sufficiently distinct for them to be readily differentiated from other types of fish poisoning, with the possible exception of paralytic shellfish poisoning. The first symptoms, which some individuals have felt within 10 minutes of having eaten the fish, are a numbness and tingling of the lips, tongue and inner surfaces of the mouth. Weakness follows

and then a paralysis of the limb and chest muscles. Sometimes there is vomiting. The blood pressure is low and the pulse usually faster and weaker than normal. Death can occur within 30 minutes.

The most important physiological effects of the toxin can be observed directly if a small dose is injected into an animal that has been prepared so that recordings can be made of its blood pressure, rate of respiration and the response of a leg muscle when the nerve to it has been stimulated. After about 10 micrograms (millionths of a gram) of tetrodotoxin have been introduced into a vein there is a prompt drop in blood pressure and a decrease in the depth of respiration. The muscle fails to contract in spite of continued stimulation of its nerve. Administered in this way, tetrodotoxin can take effect so rapidly that an animal's respiration changes from a normal pattern to complete stoppage in the course of a single breath.

The effect of the toxin on the functioning of nerve and muscle has been particularly useful in unraveling its mode of action. The technique with which this effect has been explored is essentially a modern version of one used in 1856 by the great French physiologist Claude Bernard to study the Indian arrow poison curare. After a poison is injected into an animal electrical stimulation is applied to a motor-nerve axon, the long fiber that extends from the bodies of nerve cells in the spinal cord to the muscle cells. When an axon associated with the muscle of a limb is stimulated in this

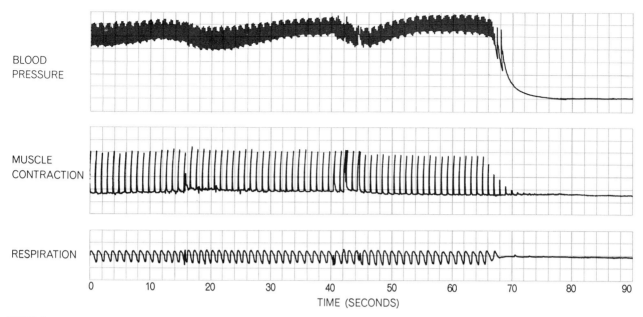

BLOOD
PRESSURE

MUSCLE
CONTRACTION

RESPIRATION

TIME (SECONDS)

INTERRUPTED RHYTHMS of blood pressure, muscle contraction and respiration occur almost immediately when a small amount of tetrodotoxin is introduced into the vein of an anesthetized animal. About 10 micrograms (millionths of a gram) of the toxin produced the changes shown in these records. Limb-muscle contractions were induced by electrical stimulation of a nerve.

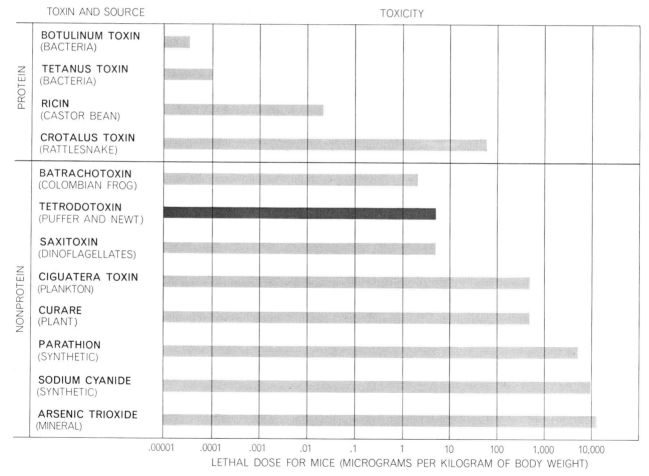

SCALE OF TOXICITY rates selected protein and nonprotein poisons and shows arsenic trioxide to be the least toxic of the group. The crotalus toxin rated here is crotactin. Batrachotoxin, the most toxic of the nonprotein compounds, is present in the skin of a South American frog. The toxin was isolated after tetrodotoxin was found in Salamandridae (true newts). Tetrodotoxin also is present in 40 species of fish, most of them Tetraodontidae (true puffers). Saxitoxin causes "paralytic shellfish poisoning"; it is present in dinoflagellates, one-celled organisms that shellfish ingest. The agent responsible for ciguatera poisoning has not been determined; it is probably caused by a toxin in plankton that are ingested by fish. Curare is an Indian arrow poison; parathion is an insecticide.

way, it causes the muscle to contract just as a normal nerve impulse does. After the injection of a toxin, however, the muscle may fail to respond to its nerve. In this event electrical stimulation is applied directly to the muscle. If the muscle then contracts, the site of action of the toxin must be either the axon or the junction between nerve and muscle.

Experiments of this kind were performed with tetrodotoxin, some of them by Fusao Ishihara of the University of Tokyo, as early as 1918. They were difficult to interpret. After tetrodotoxin was administered, a muscle stimulated through its nerve failed to contract. When the muscle was stimulated directly, it did contract at first, but after a few minutes the contractions stopped. Today we know that tetrodotoxin first affects the axons of motor nerves and then, somewhat later, it affects the muscle fibers in much the same way. This has been demonstrated in a clear-cut fashion by more elaborate experiments.

When a nerve relays an impulse from the brain to a muscle, the signal carried by the axon can be measured as a change in electric potential. A similar change occurs in a nerve that has been removed from a frog and stimulated by artificial means. An experiment can be set up so that the electric potential is measured at two points along an isolated axon, and between these points the axon can be immersed in a weak solution of tetrodotoxin. When the nerve is stimulated under these conditions, there is no change in electric potential beyond the place where the toxin is applied; conduction of the impulse has been blocked. Most of the effects suffered by those who have eaten poisonous puffer fish can be explained by this effect of tetrodotoxin.

Experiments similar to this one have been performed with nerves of the California newt. When newt axons are bathed in a solution of tetrodotoxin that is 25,000 times stronger than a solution that prevents conduction in frog nerve, they continue to transmit impulses. It is clear that some characteristic of their nerves renders newts (and also puffer fish) resistant to their own toxin.

The investigation of the exact way in which tetrodotoxin blocks conduction in nerves has required experiments of much precision and delicacy. Such experiments have been undertaken by John Wilson Moore and Toshio Narahashi of the Duke University School of Medicine and by others. Before taking up this work, it will be helpful to review the way in which a nerve impulse is relayed along

the axon from the spinal cord to a neuromuscular junction. The essential conditions for the process are differences in the concentration of sodium and potassium ions inside and outside the axon, and an axon membrane capable of undergoing rapid transient changes in permeability to these ions. In a resting nerve the concentration of potassium ions is higher inside the axon than it is in the external fluid. The concentration of sodium ions is the reverse: it is higher in the external fluid than it is inside the axon. At this time the axon membrane is almost impermeable to sodium ions, but its permeability to potassium ions is sufficient to allow them to diffuse outward. As a result there is an electric potential difference of about 60 millivolts

(thousandths of a volt) across the axon membrane and the inside of the axon is negative with respect to the external fluid. When the axon is stimulated, the resting potential at the point where the impulse arises is reduced and the two phases of the action potential begin. In the first phase the axon becomes permeable to sodium ions, allowing them to pass inward through the membrane. The axon remains in this state for one or two milliseconds, during which time the sign of the potential difference across the membrane is reversed: the interior of the axon is about 50 millivolts positive with respect to the external fluid. In the second phase of the action potential the axon becomes comparatively impermeable to sodium ions and highly perme-

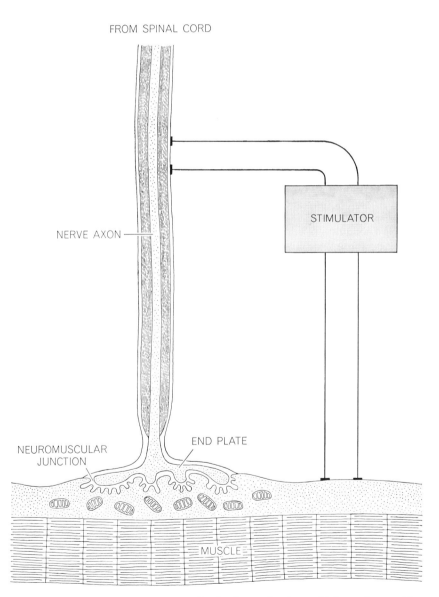

FROM SPINAL CORD

STIMULATOR

NERVE AXON

NEUROMUSCULAR JUNCTION

END PLATE

MUSCLE

NERVE AXON AND MUSCLE are stimulated to locate the site at which a toxin acts. After a toxin is injected into an animal a nerve axon to a muscle is stimulated. If the muscle fails to contract, the toxin may be acting on axon, muscle or the junction of the two. If direct muscle stimulation does produce contraction, the toxin's site must be the axon or junction.

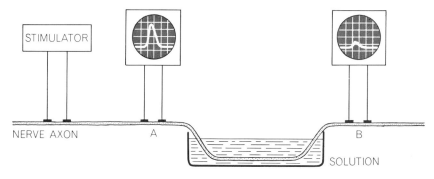

IMMERSED NERVE AXON demonstrates tetrodotoxin's effect on nerve conduction. The axon is stimulated (*left*), producing an impulse marked by a change in the membrane's electric potential that is recorded (*A*). At point *B* the same change is not recorded when tetrodotoxin is present in the solution bathing the nerve between recording points.

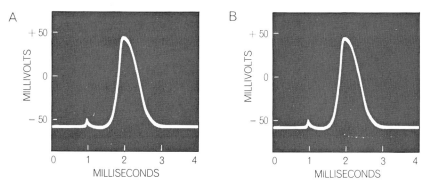

NERVE-IMPULSE RECORDS from two points along the axon are exactly the same when the axon is bathed in a solution similar to the blood surrounding it in normal conditions.

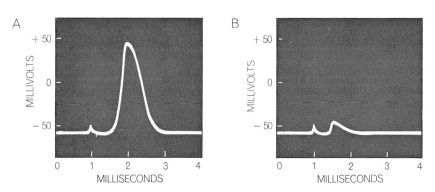

WITH TETRODOTOXIN present in the solution bathing the axon, stimulation produces a normal change in electric potential at the first point; a change is not recorded at second.

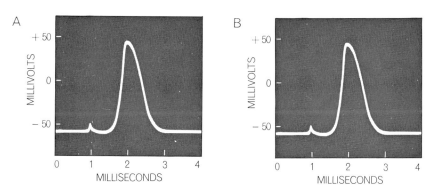

AFTER TETRODOTOXIN immersion the axon is washed and placed in a bloodlike solution. Following stimulation a normal change in potential is recorded at both axon points.

able to potassium ions. Potassium ions pass outward from the inside of the axon and the resting potential difference is restored.

This process, which moves down the axon like a wave, can be blocked by drugs and toxins. Most commonly such agents make the axon membrane equally permeable to both sodium and potassium ions. The potential difference therefore cannot be maintained, which is to say that the membrane is depolarized. Tetrodotoxin, however, has a much more specific effect on the movement of ions. It is for this reason that it provides a new tool for the study of nerve function.

The specific manner in which tetrodotoxin blocks conduction in nerves was learned from experiments with large axons dissected from a squid or a lobster. In these experiments the axon was studied by means of an arrangement called a voltage clamp, with which it is possible to regulate the potential difference across the axon membrane at a desired level. This is done by means of an external voltage source controlled by a feedback amplifier and connected to electrodes on the inside and outside of the membrane. The current that flows through an area of the membrane under the influence of the fixed voltage is measured with a separate amplifier. Such measurements have been made of the electric current flowing through axons that have been immersed in a fluid resembling the one that normally surrounds the axon except that it contains a low concentration of tetrodotoxin. Under these conditions the current carried by sodium ions that move into the axon is first reduced and then completely abolished. The current outward from the axon carried by potassium ions is not affected.

By what kind of mechanism does tetrodotoxin accomplish its selective effect on the flow of ions? It has been demonstrated that when the toxin is present inside an axon rather than outside it, the movement of sodium ions is not suppressed. The toxin must therefore be outside the membrane in order to produce its effect. In other experiments it has been shown that this effect is not restricted to sodium ions. Certain other ions have been substituted for sodium in the fluid surrounding an isolated axon, and it has been found that these ions function as sodium ions do in carrying a current inward across the membrane. When tetrodotoxin is present in such a fluid, however, the current is suppressed just as it is when it is carried by ions of

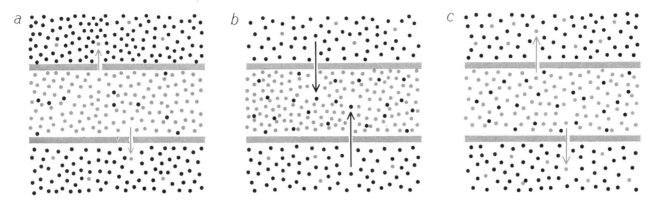

ION DIFFUSION through the axon membrane establishes an electric potential. Because of concentration gradients, sodium ions (*black dots*) tend to diffuse from outside the axon inward; potassium ions (*colored dots*) tend to diffuse from inside outward. A resting axon is more permeable to potassium than to sodium; this creates a resting potential (*a*). An action potential begins when the nerve is stimulated. First the membrane becomes permeable to sodium ions; they move inward (*b*). After sodium "gates" close, the membrane becomes permeable to potassium ions, which move out (*c*). Action potential ends with reduced potassium permeability.

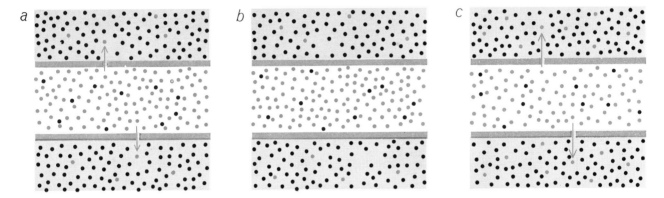

BLOCKED ION DIFFUSION occurs when a nerve axon is placed in a bloodlike solution containing tetrodotoxin. The diffusion of potassium ions outward from the axon is not affected; a normal resting potential is established (*a*). Following stimulation that normally initiates an action potential, the membrane fails to become highly permeable to sodium ions; their passage is blocked (*b*). The second phase of an action potential is not blocked by tetrodotoxin; potassium ions diffuse normally through the axon membrane (*c*).

sodium. Apparently the toxin's selective effect is not on the ions but on the membrane through which the ions move to carry current inward.

With certain other agents it is possible to block the movement of potassium ions that carries a current outward across the membrane. These agents do not abolish the current flowing inward. To be sure, all these experiments measuring the flow of ions across an axon membrane rely on the voltage clamp and hence study nerve processes under highly artificial circumstances. The British physiologists A. L. Hodgkin and A. F. Huxley have shown, however, that measurements made under these conditions are in accord with knowledge of the normal nerve impulse. It follows from the voltage clamp experiments, then, that there must be two kinds of "gate" in the nerve membrane, one controlling the inward movement of sodium ions during the first phase of the action potential and one the outward movement

of potassium ions during the second phase. Exactly how tetrodotoxin produces its effect on the membrane is not known. It may attach itself to the membrane in such a way as to keep the gates from opening or it may block gates that are open.

In addition to the gates through which sodium and potassium ions pass there is in the axon another mechanism for moving sodium and potassium ions: the ion pump. A nerve is capable of transmitting impulses for hours at a time, but ultimately lost potassium ions must be replaced and sodium ions that have penetrated the axon must be expelled. This is accomplished by the ion pump. The routes taken by sodium and potassium ions associated with it are altogether separate from the gates that operate during an action potential.

When conduction in a nerve axon is blocked, signals to and from the central nervous system are not transmitted; if the block occurs in a sensory nerve, local

anesthesia results. Most of the common local anesthetics—for example cocaine—block conduction by mechanisms that are less specific than the mechanism of tetrodotoxin. Applied to an isolated frog nerve, tetrodotoxin is about 150,000 times more effective in blocking conduction than cocaine is. Nevertheless, tetrodotoxin is not a good local anesthetic. One reason is that when tetrodotoxin is injected in the vicinity of a nerve, it does not remain in contact with the nerve tissue but diffuses away. Moreover, it is carried by the blood to other tissues; there it can be toxic and may even cause death. Workers in several laboratories have undertaken to learn what group of atoms in the tetrodotoxin molecule is responsible for its specific effect on nerve conduction. If this active group can be identified, it should be possible to synthesize new compounds that have the favorable characteristics of other local anesthetics and that contain the active group of tetrodotoxin.

The Author

FREDERICK A. FUHRMAN is professor of experimental medicine and director of the Max C. Fleischmann Laboratories of the Medical Sciences at Stanford University. The laboratories are the multidiscipline teaching laboratories of the university's School of Medicine. Fuhrman is a physiologist, having received a Ph.D. in physiology from Stanford in 1943 after study at Oregon State University, the University of Freiburg and the University of Washington. He began his work on the Stanford faculty in 1945 as an instructor in physiology. His research activities have dealt with the effects of drugs and toxins on the metabolism of tissues; the effect of temperature on metabolism; the pathophysiology of frostbite, and the mechanism of active transport of sodium and the effect of drugs and toxins on the process. Fuhrman's present research, in which he has the collaboration of his wife, Geraldine Jackson Fuhrman, who is also a physiologist, is on the effects of various toxins from poisonous animals.

Bibliography

BASIS OF TETRODOTOXIN'S SELECTIVITY IN BLOCKAGE OF SQUID AXONS. J. W. Moore, Mordecai P. Blaustein, Nels C. Anderson and Toshio Narahashi in *Journal of General Physiology*, Vol. 50, No. 5, pages 1401–1411; May, 1967.

THE STRUCTURE OF TETRODOTOXIN. R. B. Woodward in *Pure and Applied Chemistry*, Vol. 9, No. 1, pages 49–74; December, 1964.

TARICHATOXIN-TETRODOTOXIN: A POTENT NEUROTOXIN. H. S. Mosher, F. A. Fuhrman, H. D. Buchwald and H. G. Fischer in *Science*, Vol. 144, No. 3622, pages 1100–1110; May 29, 1964.

SCIENTIFIC
AMERICAN January 1958, Vol. 198, No. 1, pp. 60-64

OFFPRINT 1081

BARBITURATES

by Elijah Adams

They are among the most useful of all drugs. In small doses
they act as sedatives; in larger doses they induce sleep; in
still larger doses they are able to produce deep anesthesia.

The barbiturates are the most versatile of all depressant drugs. They can produce the whole range of effects from mild sedation to deep anesthesia—and death. They are among the oldest of the modern drugs. Long before reserpine and chlorpromazine, the barbiturates were being used as tranquilizers. Indeed, phenobarbital has been called "the poor man's reserpine." Had phenobarbital been introduced five years ago instead of half a century ago, it might have evoked the same burst of popular enthusiasm as greeted Miltown and its fashionable contemporaries. Not that the vogue of the barbiturates is any less spectacular, in terms of total production and consumption. The U. S. people take an estimated three to four billion doses of these drugs per year, on prescription by their physicians. The barbiturates rank near the top of the whole pharmacopoeia in value to medicine. They are also a national problem.

It was nearly a century ago that a young assistant of the great chemist August Kekulé in Ghent made the first of these compounds. In 1864 this young man, Adolf von Baeyer, combined urea (an animal waste product) with malonic acid (derived from the acid of apples) and obtained a new synthetic which was named "barbituric acid." There are several stories about how it got this name. The least apocryphal version relates that von Baeyer and his fellow chemists went to celebrate the discovery in a tavern where the town's artillery garrison was also celebrating the day of Saint Barbara, the saint of artillerists. An artillery officer is said to have christened the new substance by amalgamating "Barbara" with "urea."

Chemists proceeded to produce a great variety of derivatives of barbituric acid. The medical value of the substances was not realized, however, until 1903, when two other luminaries of German organic chemistry, Emil Fischer and Joseph von Mering, discovered that one of the compounds, diethylbarbituric acid, was very effective in putting dogs to sleep. Von Mering, it is said, promptly proposed that the substance be called veronal, because the most peaceful place he knew on earth was the city of Verona.

Within a few months of their report, "A New Class of Sleep-Inducers," physicians in Europe and the U. S. took up the new drugs enthusiastically. More and more uses for them were discovered. Veronal (barbital) was soon followed by phenobarbital, sold under the trade name Luminal. In all, more than 2,500 barbiturates were synthesized in the next half-century, and of these some two dozen won an important place in medicine. By 1955 the production of barbiturates in the U. S. alone amounted to 864,000 pounds—more than enough to provide 10 million adults with a sleeping pill every night of the year.

As is true of most drugs, we still do not know how the barbiturates work or exactly how their properties are related to their chemistry. The basic structure is a ring composed of four carbon atoms and two nitrogens [see diagrams on page 2513]. Certain side chains added to the ring increase drug potency; in some instances the addition of a single carbon atom transforms an inactive form of the compound into an active one. Empirical analysis of the thousands of barbiturates has given us some practical information about relations between structure and activity. But by and large the mode of action of the drugs is an unsolved problem.

We know a great deal, however, about the action itself. We can follow it best by examining the successive stages of the drugs' depressant effect on the central nervous system [see "Anesthesia," by Henry K. Beecher; SCIENTIFIC AMERICAN, January, 1957]. From this standpoint the barbiturates can be considered together as a group, for the differences among them are not fundamental and concern such matters as the speed and duration of the effect.

In small doses these drugs are sedatives, acting to reduce anxiety and to relieve psychogenic disorders—for example, certain types of hypertension and gastrointestinal pain. In this respect the barbiturates are yesterday's tranquilizers. They have now taken second place in popularity as sedatives to the newer tranquilizers.

At three to five times the sedative dose, the same barbiturate produces sleep. Barbiturates are still by far the most widely used drugs for this purpose, as millions of us know. Hardly any hospital patient has escaped his yellow or red capsule at evening, for it is an article of clinical faith that the patient needs chemical assistance to achieve sleep in his new environment. Another large block of users are the chronic travelers by plane and train. Finally there are the thousands of sufferers from insomnia who take the drugs habitually.

Many persons find barbiturate-induced sleep as refreshing as natural sleep. Others awake with a hangover, feeling drowsy, dizzy and suffering a headache. Tests show that, with or without symptoms, the barbiturates reduce efficiency: six to eight hours after a sleeping dose of sodium pentobarbital (Nembutal) the subjects perform below par on mental and memory tests. The various drugs act differently as sleep-

	FORMULA	NAMES	CHIEF USES	DURATION
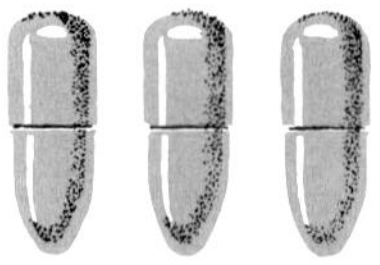	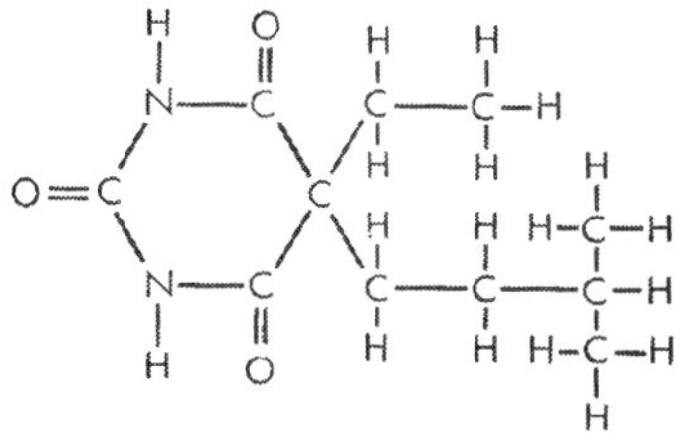	AMOBARBITAL (AMYTAL)	SEDATIVE HYPNOTIC	INTER- MEDIATE
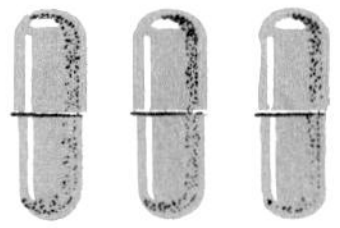	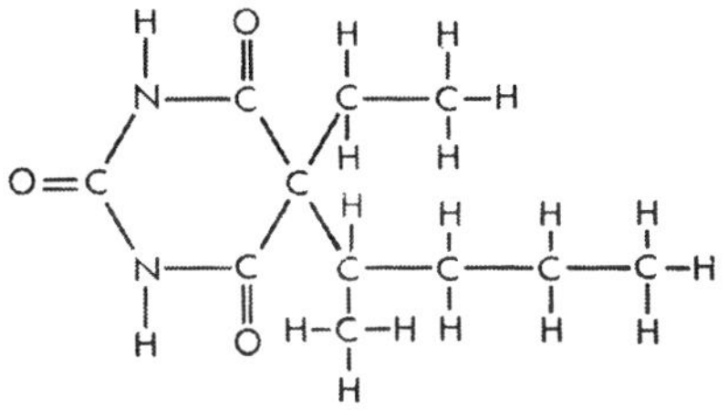	PENTOBARBITAL (NEMBUTAL)	HYPNOTIC	SHORT- ACTING
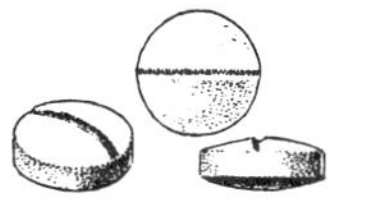	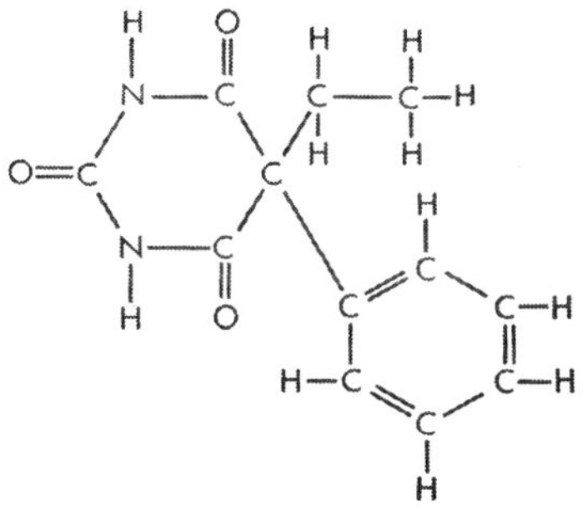	PHENOBARBITAL (LUMINAL)	HYPNOTIC SEDATIVE ANTICONVULSANT	LONG- ACTING
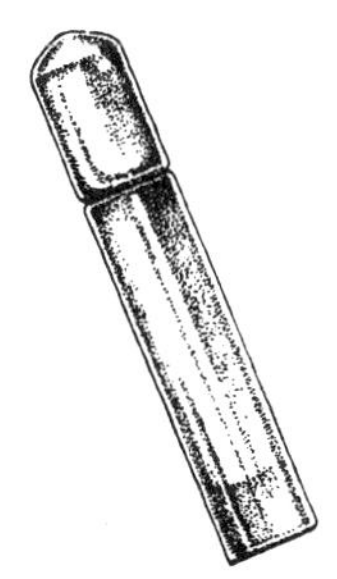	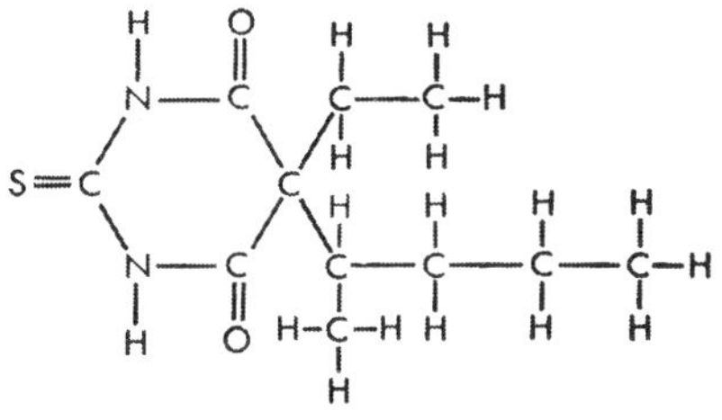	THIOPENTAL (PENTOTHAL)	ANESTHETIC	ULTRA SHORT- ACTING
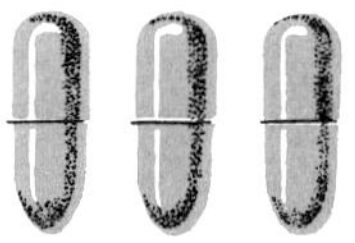	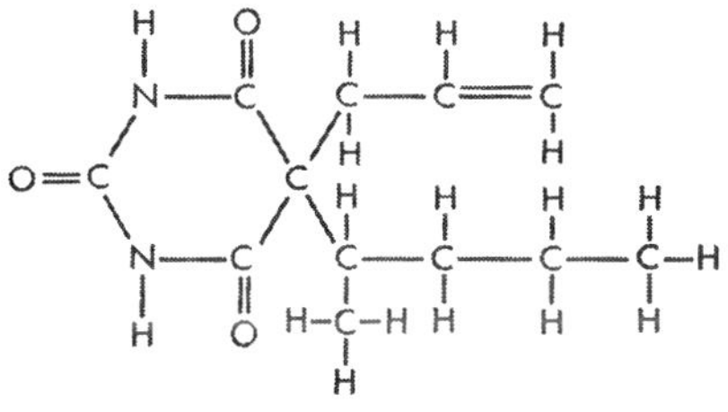	SECOBARBITAL (SECONAL)	HYPNOTIC	SHORT- ACTING

FIVE BARBITURATES are depicted at the left as they are commonly manufactured. As shown in the column of chemical structures, four of these drugs differ only in the chains of atoms attached to the carbon atom at the right side of their basic ring structure. The permutability of the basic barbiturate structure makes possible variations in speed and duration of its effect on the body.

BARBITURIC ACID (*right*), is made by combining urea (*left*) and malonic acid (*right*) with the elimination of water (*colored rectangles*). The barbiturate families arise from substitution of other substances for hydrogens at position 5 in the basic barbiturate structure.

producers. Some last for only three hours or less, others for six hours or more. The shorter-acting barbiturates (sodium pentobarbital, secobarbital) are appropriate for insomniacs who have trouble falling asleep; the longer-acting ones (barbital, phenobarbital) for people who go to sleep easily but awake after four to six hours. The latter drugs, however, are more likely to produce a hangover.

In large doses a barbiturate acts as an anesthetic. Not only does the patient become unconscious, but his spinal cord reflexes are depressed so that the muscles are relaxed and manageable for surgery. Like the gaseous anesthetics, the barbiturates depress the cerebral cortex first, then lower brain centers, next the spinal cord centers and finally the medullary centers controlling blood pressure and respiration.

The fast-acting barbiturates produce anesthesia more rapidly than ether: the patient passes from the waking state to anesthetic coma in a few moments. Sodium thiopental is the most widely used. It has important advantages over a gaseous anesthetic such as ether. Injected intravenously, it works rapidly, avoids the sense of suffocation, requires no special equipment, is free from the explosion hazard and from respiratory complications. A barbiturate anesthetic has, however, an outstanding disadvantage: the dose necessary for good muscular relaxation may seriously reduce oxygen supply to the tissues by depressing the brain center that drives respiration. Consequently for a long operation the barbiturate is often combined with a gaseous anesthetic; for a short one the dose is reduced and combined with a specific muscle-relaxing drug that has no brain-depressant action.

There are two other interesting uses of the barbiturates. One of them is in the field of psychology. As Henry Beecher observed in his article on anes-

thesia in SCIENTIFIC AMERICAN, an anesthetic can provide "planned access to levels of consciousness not ordinarily attainable except perhaps in dreams, in trances or in the reveries of true mystics." During World War II the barbiturates, particularly thiopental, were used for analysis and therapy for many thousands of GIs, who by this means relived and verbalized traumatic battle experiences which had been buried beyond voluntary recollection. The inhibition-relieving action of these drugs has also been employed by the police—in which application the press has given them the name of "truth serum," although they are neither a serum nor a guaranteed truth-producer.

The other important use of the drugs is for the control of epileptic convulsions. Certain of the barbiturates—phenobarbital, mephobarbital and methabarbital—can prevent or stop these seizures by depressing brain activity. Barbiturates can control not only the generalized convulsions of genuine (idiopathic) epilepsy—a disease afflicting almost a million persons in the U. S.—but also seizures induced by stimulating drugs or by bacterial toxins such as tetanus. The barbiturates and the convulsant drugs act in opposite fashion, and, curiously, each is used as an antidote for the other. Acute barbiturate poisoning is often treated with a stimulating drug such as pentylenetetrazol (Metrazol) or picrotoxin to bring the patient out of his coma; if the dose of the stimulant turns out to be too strong, producing convulsions, this in turn is treated with a dose of a fast-acting barbiturate!

The toxic effects of barbiturates are subtle and sometimes unpredictable. For some patients even a comparatively small dose may be dangerous. The body gets rid of barbital and phenobarbital chiefly by excretion in the urine; for a person with damaged kidneys, therefore, these drugs become toxic. Pentobarbital and secobarbital, two of the most widely

used barbiturates, are broken down in the liver. Given to a patient with a poorly functioning liver, they may produce a far longer sleep than desired.

Next to carbon monoxide, the barbiturates are the most popular suicide poison in the U. S. They account for one fifth of all the cases of acute drug poisoning, and most of these are suicide attempts. The barbiturates are not, as a matter of fact, a very efficient suicide agent: only about 8 per cent of those poisonees who arrive at hospitals die. But they are widely known and readily available, and they produce from 1,000 to 1,500 deaths in the U. S. each year.

Some of these deaths, though self-inflicted, are accidental. A British physician first called attention some years ago to a specific and probably common hazard. The person takes a small dose to go to sleep, and later, half asleep and confused, he swallows another, lethal dose. Some physicians now warn barbiturate-users not to keep their bottle of tablets on a night table, where they may stretch out a hand to take more while in the comatose state.

There is a comfortable margin of safety between the ordinary sleeping dose (a tenth of a gram for the average adult) and a definitely toxic dose (more than half a gram). The lethal dose is usually a gram and a half or more. Acute barbiturate poisoning has to be treated promptly. Unfortunately it is often not recognized in time, because the victim is thought to be merely in a deep sleep. The first step in treatment is to strengthen the victim's breathing, in a respirator if necessary. And a stimulant may have to be administered to restore the activity of the brain centers.

Are the barbiturates habit-forming? This much-debated question has been answered rather conclusively by recent studies. They can indeed produce addiction and chronic intoxication. The two chief criteria of addiction to a drug are a heightened tolerance to it and physical dependence on it, so that removal of the drug produces withdrawal symptoms. A morphine addict, for example, may be able to take many times the dose that would be lethal for a normal person, and he becomes acutely sick if the drug is stopped. Several years ago Havelock Fraser, Harris Isbell and their associates at the U. S. Public Health Service hospital for drug addicts in Lexington, Ky., made a thorough study of whether the barbiturates had these properties. Their investigation included experiments with human subjects who were given large doses of barbiturates over a period of months and then abruptly taken off the drug.

They found that the barbiturates acted as addicting drugs in every respect—physical and psychic. The men behaved like chronic alcoholics: they neglected their appearance and hygiene, became confused and quarrelsome, showed unpredictable mood swings and lost physical coordination and the mental discipline necessary for simple games. After abrupt withdrawal of the drug, the subjects began within a few hours to show signs of increasing apprehension and developed weakness, tremors, nausea and vomiting. In the next five days most of the subjects had convulsions like those of epilepsy and an acute psychosis such as alcoholics suffer, with delirium and violent hallucinations.

The Lexington investigators also made similar tests on dogs. They too exhibited withdrawal symptoms. In their "canine delirium" the dogs would stare at a blank wall and move their heads, eyes and ears as if responding to imaginary animals, people or objects; even while alone in a cage a dog would growl as if being attacked.

Stories in the press have greatly exaggerated the extent of barbiturate addiction in the U. S. "Thrill pills," "goof balls," "wild geronimos," "red devils" (secobarbital), "yellow jackets" (sodium pentobarbital), "blue heaven" (amobarbital)—all these terms certainly have a currency in a limited circle of addicts, but the number of addicts is not nearly so large as some of the stories have alleged. There are probably not more than 50,000 barbiturate addicts, compared with a million chronic alcoholics. Nevertheless, in view of the easy access to the barbiturates, the public does need to be alerted to their addictive property.

The saving fact is that it takes extraordinarily heavy use of these drugs to produce addiction. Subjects who have taken a fifth of a gram (twice the usual sleeping dose) every night for a year have shown no withdrawal symptoms after stopping the drug. In contrast, morphine, taken in the usual hospital doses for as short a time as 30 days, produces definite physical dependence. Moderate use of the barbiturates, in the doses prescribed by physicians, will not lead to addiction. Those who become addicts are probably, in the main, drug-users who turn to the barbiturates because they cannot get narcotics, alcoholics who seek relief from alcohol withdrawal and, in general, abnormal personalities who are addiction risks for any intoxication that will give psychic relief. Whether stricter Federal laws are needed to control misuse of the barbiturates has been a matter of considerable controversy.

Biologists look forward to the day when progress in medicine will make all present drugs, including the barbiturates, obsolete. Better understanding and treatment of the personal and social causes of anxiety should reduce our present reliance on chemical aids to tranquility and sleep. Meanwhile the barbiturates can teach us much about the functions of the brain and so help lead toward that more tranquil day.

SODIUM BARBITAL (*lower right*) can be made from barbital (*shown at top of diagram in its two forms*) by addition of sodium hydroxide (NaOH) and the elimination of water.

The Author

ELIJAH ADAMS was recently appointed professor and director of the pharmacology department of the Saint Louis University School of Medicine. A graduate of the Johns Hopkins University, he took his M.D. at the University of Rochester in 1942. He served as an Army physician during World War II, but on his return to civilian life he left medicine for research in enzyme biochemistry. Adams, who has been teaching pharmacology at the New York University College of Medicine, still finds biochemistry his main interest, but believes that the fields are merging "as biochemistry tends to become more physiologic and pharmacology more biochemical."

Bibliography

THE DISTRIBUTION IN THE BODY AND METABOLIC FATE OF BARBITURATES. J. Raventós in *The Journal of Pharmacy and Pharmacology*, Vol. 6, No. 4, pages 217–235; April, 1954.

REPORT ON BARBITURATES. The New York Academy of Medicine Committee on Public Health, Subcommittee on Barbiturates, in *Bulletin of the New York Academy of Medicine*, Vol. 32, No. 6, pages 456–481; June, 1956.

SYMPOSIUM ON SEDATIVE & HYPNOTIC DRUGS. The Williams & Wilkins Company, 1954.

SCIENTIFIC
AMERICAN   July 1959, Vol. 201, No. 1, pp. 113-121   OFFPRINT 1082

ALKALOIDS

by Trevor Robinson

This ill-defined group of plant compounds includes many that
are both useful and toxic. Though most of them strongly affect
human physiology, their functions in plants are still obscure.

The alkaloids are a class of compounds that are synthesized by plants and are distinguished by the fact that many of them have powerful effects on the physiology of animals. Since earliest times they have served man as medicines, poisons and the stuff that dreams are made of.

The alkaloid morphine, the principal extract of the opium poppy, remains even today "the one indispensable drug." It has also had an illicit and largely clandestine history in arts and letters, politics and crime. Quinine, from cinchona bark, cures or alleviates malaria; colchicine, from the seeds and roots of the meadow saffron, banishes the pangs of gout; reserpine, from snake root, tranquilizes the anxieties of the neurotic and psychotic. The coca-leaf alkaloid cocaine, like morphine, plays Jekyll and Hyde as a useful drug and sinister narcotic. In tubocurarine, the South American arrow poison, physicians have found a powerful muscle relaxant; atropine, said to have been a favorite among medieval poisoners, is now used to dilate the pupils of the eyes and (in minute doses!) to relieve intestinal spasms; physostigmine, employed by West African tribes in trials by ordeal, has come into use as a specific for the muscular disease myasthenia gravis. Aconitine is catalogued as too toxic to use except in ineffective doses. On the other hand, caffeine and nicotine, the most familiar of all alkaloids, are imbibed and inhaled daily by a substantial fraction of the human species.

Our self-centered view of the world leads us to expect that the alkaloids must play some comparably significant role in the plants that make them. It comes as something of a surprise, therefore, to discover that many of them have no identifiable function whatever. By and large they seem to be incidental or accidental products of the metabolism of plant tissues. But this conclusion somehow fails to satisfy our anthropocentric concern. The pharmacological potency of alkaloids keeps us asking: What are they doing in plants, anyway? Investigators have found that a few alkaloids actually function in the life processes of certain plants. But this research has served principally to illuminate the subtlety of such processes.

The pharmacology of alkaloids has inspired parallel inquiry in organic chemistry. Some of the greatest figures in the field first exercised their talents on these substances. But nothing in the composition of alkaloids has been found to give them unity or identity as a group. The family name, conferred in an earlier time, literally means "alkali-like." Many alkaloids are indeed mildly alkaline and form salts with acids. Yet some perfectly respectable alkaloids, such as ricinine (found in the castor bean), have no alkaline properties at all. Alkaloids are often described as having complex structures. The unraveling of the intricate molecules of strychnine and morphine has taught us much about chemical architecture in general. Yet conine, the alkaloid poison in the draught of hemlock that killed Socrates, has a quite simple structure. Nor is there much distinction in the characterization of alkaloids as "nitrogen-containing compounds found in plants." Proteins and the amino acids from which they are made also fit this definition.

From the chemical point of view it begins to seem that alkaloids are in a class of compounds only because we do not know enough about them to file them under any other heading. Consider the vitamin nicotinamide, the plant hormone indoleacetonitrile and the animal hormone serotonin. All these compounds occur in plants, and all contain nitrogen. We would call them alkaloids except that we have learned to classify them in more descriptive ways. As we come to know the alkaloids better, we may select other substances from this formless group and assign them to more significantly defined categories.

Though all alkaloids come from plants, not all plants produce alkaloids. Some plant families are entirely innocent of them. Every species of the poppy family, on the other hand, produces alkaloids; the opium poppy alone yields some 20 of them. The Solanaceae present a mixed picture: tobacco and deadly nightshade contain quantities of alkaloids; eggplant, almost none; the potato accumulates alkaloids in its foliage and fruits but not in its tubers. Some structurally interrelated alkaloids, such as the morphine group, occur only in plants of a single family. Nicotine, by contrast, is found not only in tobacco but in many quite unrelated plants, including the primitive horsetails. Alkaloids are often said to be uncommon in fungi, yet the ergot fungus produces alkaloids, and we might classify penicillin as an alkaloid had we not decided to call it an antibiotic. However, alkaloids do seem to be somewhat commoner among higher plants than among primitive ones.

Some 50 years ago the Swiss chemist Amé Pictet suggested that alkaloids in plants, like urea and uric acid in animals, are simply wastes—end products of the metabolism of nitrogenous compounds. But the nitrogen economy of most plants is such that they husband the element, reprocessing nitrogenous compounds of all sorts, including substances such as ammonia which are poisonous to animals. Indeed, many plants have evolved elaborate symbiotic arrangements with bacteria to secure additional nitrogen from the air. From the evolutionary standpoint the tying-up of valuable nitrogen in alkaloids seems an inefficient arrangement.

More recently investigators have come to regard alkaloids not as end products but as by-products thrown off at various points along metabolic pathways, much as substandard parts are rejected on an assembly line. That is to say, alkaloids arise when certain substances in the plant cell cross signals and make an alkaloid instead of their normal product. This idea is certainly plausible when it is applied to the alkaloids formed by the action of the commonest enzymes on the commonest metabolites. The alkaloid trigonelline, for example, is found not only in many plant seeds but also in some species of sea urchins and jellyfish. It is merely nicotinic acid with a methyl group (CH_3) added to it. Now nicotinic acid is one of the commonest components of plant cells. Compounds that can donate methyl groups are also common, as are the enzymes that catalyze such donations. A "confused" enzyme, transferring a methyl group to nicotinic acid instead of to some other substance, could thus form trigonelline by mistake [see illustration at top of page 2519]. Nicotine, which has equally wide distribution, may likewise be produced by everyday biochemical processes.

The more frequent occurrence of alkaloids in higher plants suggests another idea. More highly evolved organisms have obviously made more experiments in metabolism. Some alkaloids may represent experiments that never quite worked. Others may have originated as intermediates in once-useful processes that are no longer carried to completion. Since most alkaloids seem neither to help nor to hurt the plant, natural selection has not operated for or against them. Thus the modern plant that produces alkaloids may do so for no other reason than the persistent pattern of its genes.

Such explanations for the presence of alkaloids in plant tissues find support in what we know about the synthesis of these substances. In 1917 the noted British chemist Sir Robert Robinson showed that the structures of scores of alkaloid molecules could be built up from amino acids by postulating reactions of a few simple types: dehydration, oxidation and so on. For example, he showed that the amino acid tyrosine could easily be transformed into the alkaloid hordenine. Even the complex molecule of reserpine could be built up,

according to his scheme, from tyrosine and the amino acid tryptophan, plus a methylene group [see illustration on page 2520]. More recently Robert B. Woodward of Harvard University has proposed that the same three substances, through another series of reactions, may yield the extremely complex molecule of strychnine.

During the past 40 years considerable experimental evidence has accumulated to show that these reactions are not just paper-and-pencil chemistry, but actually occur in nature. Robinson himself correctly predicted the structures of several highly complex alkaloids before these structures were worked out. Later experimenters have shown that enzyme-containing plant extracts can promote amino acid-alkaloid transformations such as the tyrosine-hordenine synthesis. Other investigators, by simply mixing together the postulated precursors of certain alkaloids, have obtained compounds of approximately the correct structure even in the absence of enzymes. Tracer experiments have furnished additional support for Robinson's scheme. If labeled amino acids are injected into alkaloid-producing plants, the plants produce labeled alkaloids. Moreover, the alkaloids contain labeled atoms at just the points that theory predicts.

We know, however, that many steps intervene between the introduction of a labeled precursor and the production of a labeled alkaloid. Moreover, a compound that yields an alkaloid when it is injected in high concentration may not be the normal precursor. Some intermediates go to form all sorts of things, and may only get to alkaloids by quite devious routes. The problem of alkaloid biosynthesis resolves itself into the task of establishing the point at which the alkaloid-producing process diverges from the other metabolic processes of the plant. In principle we might feed various labeled compounds to a plant and ascertain whether the labeled material shows up only in alkaloids or in other substances as well. But we must decide which intermediate compounds we are going to feed. It is fruitless to test a versatile intermediate like glucose, which enters into many processes. In a sense, therefore, we must know our intermediate before conducting the experiment that will identify it. One way of breaking out of this impasse may be to work backward by feeding the alkaloid itself to the plant. By building up high concentrations of alkaloid in the plant's tissues we can perhaps block the alkaloid "production line" and thus cause the im-

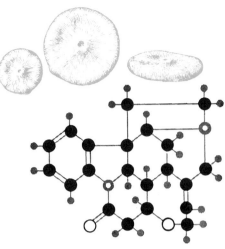

STRYCHNOS NUX-VOMICA
STRYCHNINE

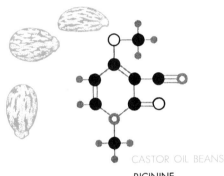

CASTOR OIL BEANS
RICININE

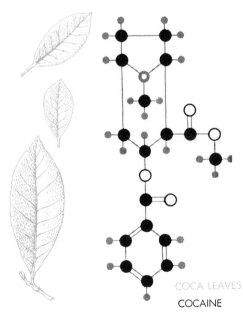

COCA LEAVES
COCAINE

● CARBON
○ OXYGEN
◎ NITROGEN
• HYDROGEN

ALKALOIDS show great structural variety. Depicted in this chart are molecules of nine typical alkaloids together with

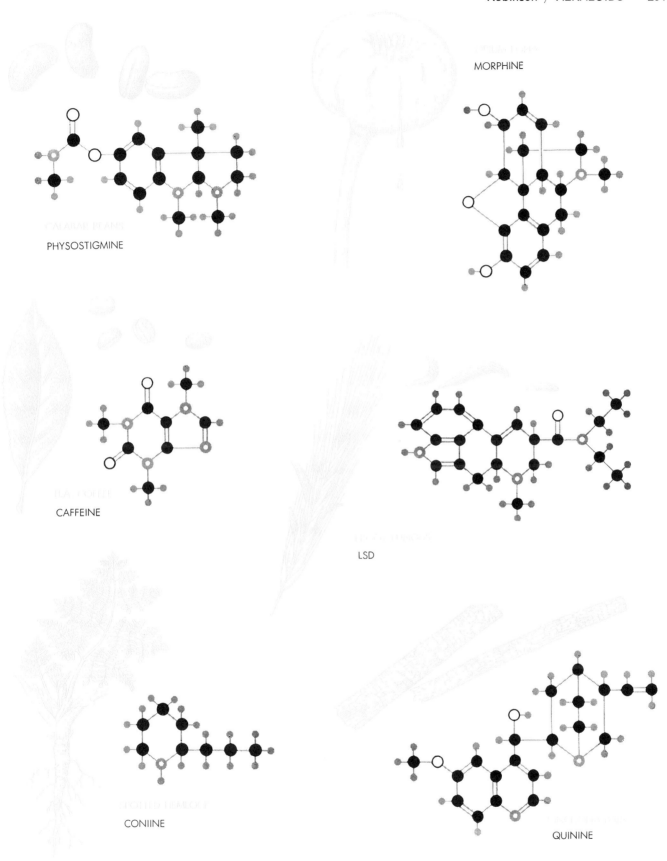

MORPHINE

PHYSOSTIGMINE

CAFFEINE

LSD

CONIINE

QUININE

the plants or plant substances from which they derive. At left is the key to these diagrams and those elsewhere in this article. Strychnine, a violent poison, is one of the most complex alkaloids; coniine, the poison which killed Socrates, one of the simplest. Physostigmine, a West African "ordeal poison," is now used to treat the muscular disease myasthenia gravis; LSD (lysergic acid diethylamide) produces delusions resembling those of schizophrenia. Ricinine is one of the few alkaloids that exert little effect on human beings.

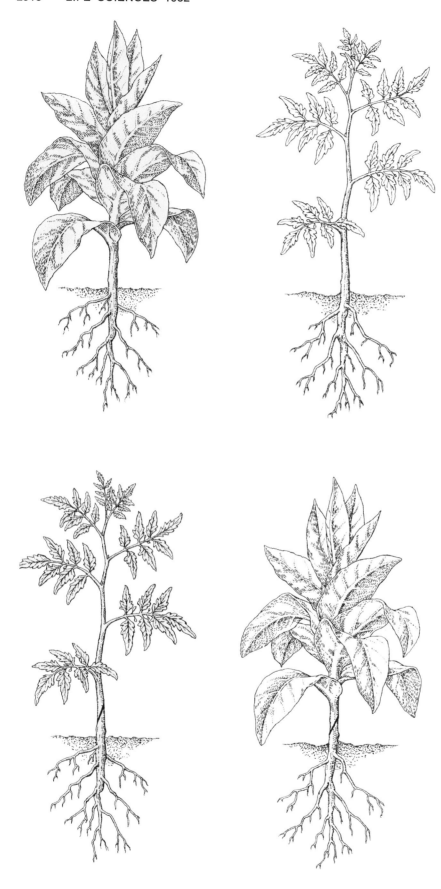

GRAFTING EXPERIMENT indicates that nicotine has no effect on plants. Tobacco plant (*top left*) produces nicotine (*color*) in its roots; the alkaloid then migrates to the leaves. Tomato plant (*top right*) produces no nicotine. A tomato top grafted to a tobacco root (*bottom left*) becomes impregnated with nicotine with no apparent ill effects; tobacco top grafted to tomato root (*bottom right*) is unaffected by the absence of alkaloid. Similar grafting experiments with other alkaloid-producing plants have with few exceptions yielded similar results.

mediate precursors of the alkaloid to accumulate to the point where they can be identified.

With this information in hand, we can go on to inquire which enzymes transform these precursors into alkaloids and whether these enzymes function only in alkaloid formation or in other metabolic processes as well. If we find, for example, that a certain enzyme catalyzes the transfer of a methyl group to an alkaloid precursor but does not function in other methylations, we will have to regard this alkaloid synthesis as a definitely programmed process, and not as a mere aberration. Our present sparse knowledge strongly suggests that at least some alkaloids are programmed. Thus ricinine contains a nitrile group (CN), which rarely occurs in living organisms. If the formation of this group is catalyzed by an enzyme that normally does some other job, we have no indication of what the other job might be. From intimate understanding of this kind we may yet help the plant physiologist to discover what there is about different plants that causes one to make reserpine, while another makes strychnine from the same starting materials.

Of course the study of any metabolic process involves not only the synthesis but also the breakdown of the substances involved. If a given alkaloid is just a waste product or by-product, it has no future and there is no breakdown to be considered. In this case it may simply accumulate in the plant's tissues. For example, quinine piles up in the bark of the cinchona tree, and nicotine in the leaves of the tobacco plant. Some ingenious grafting experiments have furnished additional evidence that many alkaloids, once synthesized, become inert and play no further role in the plant's metabolism. The tobacco plant, for example, manufactures nicotine in its roots, whence the alkaloid migrates to the leaves. However, if we graft the top of a tobacco plant to the roots of a tomato plant, which produces no nicotine, the tobacco flourishes despite the absence of the alkaloid. Conversely, a tomato top grafted to a tobacco root becomes impregnated with nicotine with no apparent ill effects.

But the alkaloids in plants are not always inactive. Hordenine, for example, is found in high concentrations in young barley plants, and gradually disappears as the plant matures. By the use of tracers Arlen W. Frank and Leo Marion of the National Research Council of Canada have found that the disappearing hordenine is converted into lignin, the "plastic" that binds the cellulose fibers in the structure of plants. To be sure, not all plants employ hordenine as an intermediate in making lignin [see "Lignin," by F. F. Nord and Walter J. Schubert; SCIENTIFIC AMERICAN, October, 1958]. But it is gratifying to find at least one case in which an alkaloid performs an identifiable function. Similarly, Edward Leete of the University of Minnesota has shown that in some plants nicotine serves as a "carrier" for methyl groups which it ultimately donates to other molecules.

Such modest findings are a far cry from the first grand-scale function assigned to alkaloids a century ago by the great German chemist Justus von Liebig. Since most alkaloids are alkaline, he proposed that plants use them to neutralize deleterious organic acids by forming salts with them. Many alkaloids do, in fact, occur in plants as salts of organic acids. But no one could explain why alkaloid-producing plants should elaborate poisonous acids when closely related plants manage to get along without either the acids or their metabolic antagonists. The question "Why?" still persists. Today it stimulates more modest but sometimes quite intriguing proposals.

Some experiments by the French physiologist Clément Jacquiot suggest a variant of Liebig's neutralization theory. Jacquiot has shown that the tannin produced in cultures of oak cells inhibits cell growth. The alkaloid caffeine counteracts the effects of the tannin and allows growth to proceed. Unfortunately this suggestion merely replaces one question with another, since the function of tannins in plants is itself unknown. The suggestion has another flaw in that oak cells produce no caffeine of their own. But here, at any rate, is one case in which caffeine is good for something other than providing a pleasant stimulant for coffee-drinkers.

Botanists and ecologists have speculated that the bitter taste of some alkaloids may discourage animals from eating a plant that contains them, and that poisonous alkaloids may kill off pathogenic organisms that attack the plant. One species of wild tomato does produce an alkaloid that protects it against Fusarium wilt, a common fungus disease of the cultivated tomato. However, the "protection" idea must be handled with caution because of its anthropocentric bias. What is unpalatable or poisonous to a man may be tasty and nourishing to a rabbit or a cutworm.

More recently the concept of chelation, the process by which certain organic molecules "sequester" the atoms

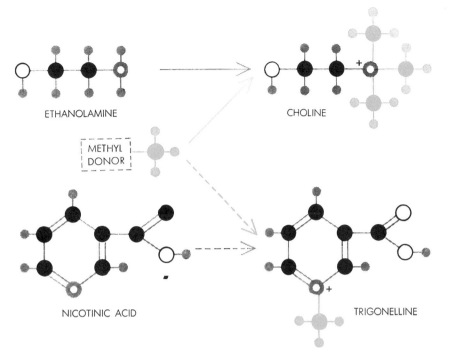

METABOLIC ACCIDENTS may account for the synthesis of some alkaloids. Changing one compound to another by adding methyl groups (*color*), as in the ethanolamine-choline transformation, is a common biochemical process. If a methyl group were added "by mistake" to nicotinic acid, a common plant substance, the alkaloid trigonelline would result.

of metals, has suggested another possible alkaloid function. The structures of some alkaloids should permit them to act as chelating agents. The structure of nicotine is temptingly similar to that of dipyridyl, a common chelating agent for iron [*see illustration at bottom of this page*]. Such alkaloids may help a plant select one metal from the soil and reject others. Alternatively, they might facilitate the transport of the metal from the roots, where it is absorbed, to the leaves, where it is utilized. Quite a few alkaloids, including nicotine, migrate from roots to leaves, but no one has yet determined whether any of them carry metals along with them.

The structures of many alkaloids resemble those of hormones, vitamins and other metabolically active substances. This resemblance suggests that such alkaloids may function as growth regulators. In a way this hypothesis fits in with the chelation theory, since certain important growth regulators seem to owe their activity to their chelating capacity. Some alkaloids do affect growth processes: Alkaloids from the seeds of certain lupines can inhibit the germination of seeds of related species that produce no alkaloids. Presumably they help the former species to compete successfully with the latter. Similar alkaloids may function as "chemical rain gauges" which prevent germination until sufficient rain has fallen to leach them away

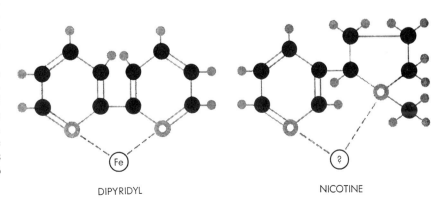

STRUCTURE OF NICOTINE resembles that of dipyridyl, a compound that can "chelate" or bind atoms of iron. This similarity suggests that nicotine, and perhaps other alkaloids, may function as chelating agents in some plants. Whether they actually do so is not yet known.

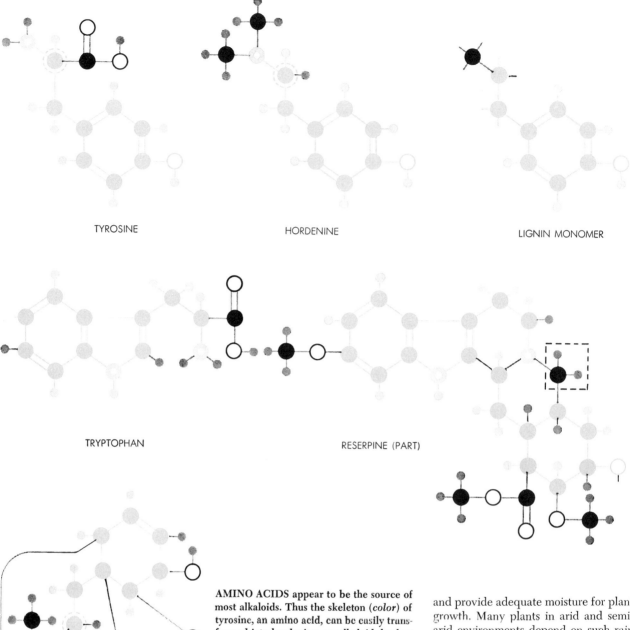

TYROSINE

HORDENINE

LIGNIN MONOMER

TRYPTOPHAN

RESERPINE (PART)

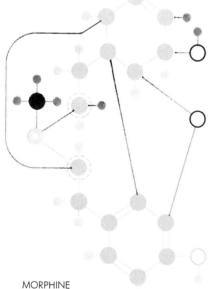

MORPHINE

AMINO ACIDS appear to be the source of most alkaloids. Thus the skeleton (*color*) of tyrosine, an amino acid, can be easily transformed into hordenine, an alkaloid; broken circles indicate the "labeled" atoms which confirm the synthesis. Two tyrosine skeletons similarly form morphine, as shown at left; the molecule shown here is distorted (*see diagram on page 2517*) to emphasize its derivation. Tyrosine and tryptophan, another amino acid, could join with a methylene group (*broken square*) to form part of the molecule of reserpine, an alkaloid tranquilizer. Hordenine is one of the few alkaloids known to undergo further metabolism in plants; it is converted into one of the units that form the long-chain molecules of lignin, an essential structural material in many plants. The complete structure of the lignin monomer is not known.

and provide adequate moisture for plant growth. Many plants in arid and semiarid environments depend on such rain gauges for survival [see "Germination," by Dov Koller; SCIENTIFIC AMERICAN Offprint 117].

No one theory can account for the functions of so heterogeneous a group of compounds. The steroid compounds, a much smaller and structurally a far more homogeneous group, play a wide variety of physiological roles. Future research will probably reveal an even greater functional diversity among the alkaloids. Certainly we must learn a good deal more about the functions of a few alkaloids before we can safely propose generalizations about all of them.

The Author

TREVOR ROBINSON teaches in the department of bacteriology and botany at Syracuse University. "I received my A.B. at Harvard in 1950," he says, "majoring in a hodgepodge field called 'biochemical sciences'—a smattering of courses on a number of scientific subjects. As I intended to be a high-school teacher, I stayed on at Harvard for an A.M. in science education. Despite the supposed need for science teachers nobody wanted to hire me, so I took some more advanced courses at the University of Massachusetts. There I became so interested in biochemical research that I gave up the idea of high-school teaching and took an M.S. in chemistry. Next I went to the department of biochemistry at Cornell University, taking my Ph.D. in 1956. While at Cornell I became entranced by the fantastic array of peculiar compounds found in plants. In addition to alkaloids I have studied plant tannins, another peculiar class of materials. In quite a different vein, I have been carrying on research on the inactivation of dilute enzyme solutions by ionizing radiation."

Bibliography

THE ALKALOIDS. K. W. Bentley. Interscience Publishers, Inc., 1957.

THE PLANT ALKALOIDS. Thomas Anderson Henry. P. Blakiston's Son & Co. Inc., 1939.

THE STRUCTURAL RELATIONS OF NATURAL PRODUCTS. Robert Robinson. Oxford University Press, 1955.

SCIENTIFIC
AMERICAN   October 1967, Vol. 217, No. 4, pp. 81-90   OFFPRINT 1083

THE STRUCTURE OF ANTIBODIES

by R. R. Porter

The basic pattern of the principal class of molecules that neutralize
antigens (foreign substances in the body) is four cross-linked chains.
This pattern is modified so that antibodies can fit different antigens.

It has been known for millenniums that a person who survives a disease such as plague or smallpox is usually able to resist a second infection. Indeed, such immune people were often the only ones available to nurse the sick during severe epidemics. A general understanding of immunity had to await the discovery that microorganisms are the causative agents of infectious disease. Then progress was rapid. A key step was taken in 1890 by Emil Von Behring and Shibasaburo Kitasato, working in the Institute of Robert Koch in Berlin. They showed that an animal could be made immune to tetanus by an injection of the blood serum obtained from an animal that had survived the disease and had

developed immunity to it. Serum is the clear fluid that is left behind when a blood clot forms; it contains most of the blood proteins. Thus immunity to tetanus is a function of a substance or substances in the blood. These substances were named antibodies.

Antibodies are produced by all vertebrates as a defense against invasion by certain foreign substances, known collectively as antigens. The most effective antigens are large molecules such as proteins or polysaccharides (and of course the microorganisms that contain these molecules). The demonstration of the appearance of antibodies in the blood is most dramatic if the antigen is a lethal toxin or a pathogenic microorganism:

the immune animals live and the non-immune die when injected with the antigen. Innocuous substances such as egg-white protein or the polysaccharide coat of bacteria, however, are equally effective as antigens. The antibodies formed against them can be detected by their ability to combine with antigen. This can be shown in many ways. Perhaps the simplest demonstration is provided by the precipitate that appears in a test tube when a soluble antigen combines with antibody contained in a sample of serum. The most remarkable aspect of this phenomenon is the specificity of the antibody for the antigen injected. That is, the antibody formed will combine only with the antigen injected

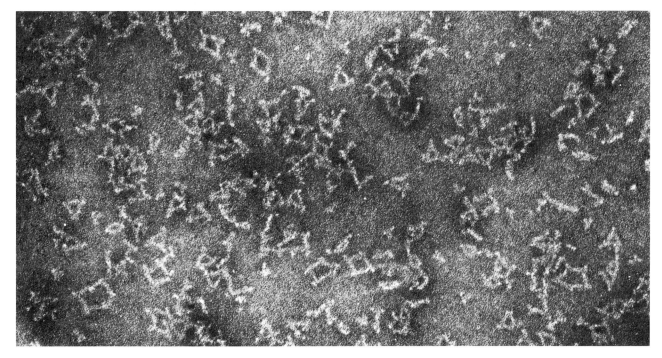

ANTIBODIES BOUND TO ANTIGENS are depicted in this electron micrograph made by Michael Green and Robin Valentine of the National Institute for Medical Research in London. The antigen itself is too small to be visible, but it evidently acts as the coupling agent that binds antibody molecules together to form the various multisided structures. The magnification is about 275,000 diameters.

FRAGMENT ANTIGEN BINDING (Fab)

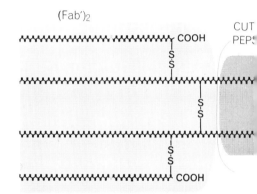

(Fab')₂

CUT BY PAPAIN

CUT PEPS

IMMUNOGLOBULIN GAMMA, the chief class of antibody, is a protein molecule consisting of four polypeptide chains held together by disulfide (S—S) bonds. The two light chains are identical, as are the two heavy chains. Depending on the source, the light chains contain from about 210 to 230 amino acid units; the heavy chains vary from about 420 to 440 units. Thus the lengths, the spacing between disulfide bonds and enzyme cleavage points shown here are approximate. The enzyme papain splits the molecule into three fragments (*above*): a fragment that forms crystals (Fc) and two fragments (Fab) that do not crystallize but contain the antigen binding sites. Approximately half of each Fab fragment (*color*) is variable in amino acid composition. Site 191 is genetically variable. When immunoglobulin gamma is split by the enzyme pepsin (*right*), the Fab fragments remain bonded together (Fab')₂ because the cleavage occurs on the other side of the central disulfide bond.

or with other substances whose structure is closely related.

Numerous different antibodies can be formed. Although an individual animal may respond poorly, or perhaps not at all, to a particular antigen, there is no known limit to the number of specific antibodies that one species, for example the rabbit, can synthesize. Conceptually there is a great difference between the capability of one species to synthesize a very large but limited number of antibodies and the capacity to synthesize an infinite number, but an experimental decision as to which is correct is not possible at present.

All antibodies are found in a group of related serum proteins known as immunoglobulins. The challenge to the protein chemist lies in the fact that antibody molecules are surprisingly similar even though they possess an enormous range of specific combining power. Although it is clear that there must be significant differences among antibodies, no chemical or physical property has yet been found that can distinguish between two antibody molecules: one able to combine specifically, say, with an aromatic compound such as a benzene derivative and the other with a sugar, although the benzene compound and the sugar have no common structural features. Antibodies of quite unrelated specificity appear to be identical,

within the limits of present experimental techniques, except, of course, in their specific combination with antigen.

An antibody can be isolated from the serum of an immunized animal only by using the special property of allowing it to combine with the antigen, freeing the complex from the other serum proteins and then dissociating and separating the antibody and antigen. This can be done by allowing a precipitate to form, washing the precipitate well with salt solution and then suspending the precipitate in weak acid. Under these conditions the antibody-antigen precipitate will dissolve and dissociate, and the antibody and antigen can be separated from each other to yield the purified antibody. As we shall see, however, even this purified material usually contains a variety of antibody molecules that differ slightly in their molecular structure.

If an animal has not been immunized, it will still have a good concentration of immunoglobulin in its blood, usually about 1 percent by weight. This material is believed to be made up of many thousands of different antibodies

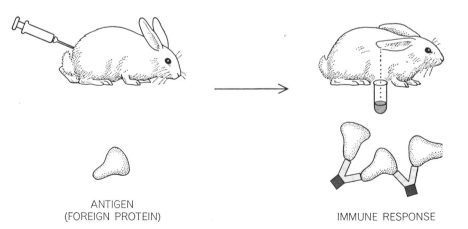

ANTIGEN
(FOREIGN PROTEIN)

IMMUNE RESPONSE

MIXTURE OF SIMILAR ANTIBODIES can be produced by injecting a rabbit or other animal with a purified antigen, typically a large protein of foreign origin. In response the animal produces antibodies, primarily immunoglobulin gamma, that are able to bind specifically to the antigen. Evidently a given antigen provides many different binding sites, thus giving rise to many different antibody molecules. If blood is removed from the animal

FRAGMENT CRYSTALLINE (Fc)

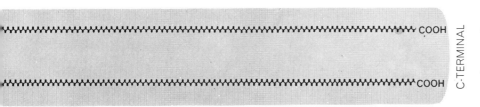

Fc

against microorganisms the animal has encountered during its lifetime or against other antigenic substances that accidentally entered its body. Evidence that this view is correct comes from experiments in which small animals have been born and raised in an entirely germ-free environment. Under these conditions the immunoglobulin content of the blood is much lower, perhaps only 10 percent of the immunoglobulin in the blood of a normal animal, suggesting that mild infections are the main source of antigens.

The immunoglobulins can be isolated from serum by the usual methods of protein separation. Hence the protein chemist has available for study two general kinds of immunoglobulin fraction: a complex mixture of many antibodies and purified antibodies that have been isolated by virtue of their specific affinity for the antigen. It would seem to be a relatively straightforward task, after the great progress made in the techniques of protein chemistry in recent years, to carry out detailed studies of such material and pinpoint the differences. Clearly structural differences responsible for the specific combining power of antibodies must exist among them and should become apparent.

Major difficulties have arisen, however, because the immunoglobulins have been found to be a very complex mixture of molecules and the complexity is not necessarily due to the presence of the many different kinds of antibody. One difficulty is that there are three main classes of immunoglobulins distinguished chemically from one another by size,

carbohydrate content and amino acid analysis. Antibodies of any specificity can be found in any of the classes; hence there is no correlation between class and specificity. The class present in the largest amounts in the blood and the most easily isolated is called immunoglobulin gamma. Since most of the work has been done with this material I shall limit my discussion to it.

Immunoglobulin gamma has a molecular weight of about 150,000, corresponding to some 23,000 atoms, of which a carbohydrate fraction forms no more than 2 or 3 percent. Chemical studies have shown that the immunoglobulin gamma molecule is built up of four polypeptide chains, which, as in all proteins, are formed from strings of amino acids joined to one another through peptide bonds. The four chains are paired so that the molecule consists of two identical halves, each consisting of one long, or heavy, chain and one short, or light, chain. The four chains are held to one another by the disulfide bonds of the amino acid cystine [see illustration at top of these two pages]. If the disulfide bonds are split, the heavy and light chains are still bound to each other. If, however, they are put in an acid solution or one containing a substance such as urea, they dissociate and can be separated by their difference in size.

Immunoglobulin gamma molecules can also be split by proteolytic enzymes such as papain, which breaks the molecule into three pieces of about equal size. Two, known as Fab (for "fragment antigen binding"), appear to be identical, and the third, known as Fc (fragment crystalline), is quite different. Fab is so named because it will still combine with the antigen although it will not pre-

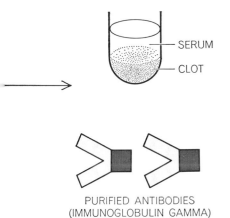

PURIFIED ANTIBODIES
(IMMUNOGLOBULIN GAMMA)

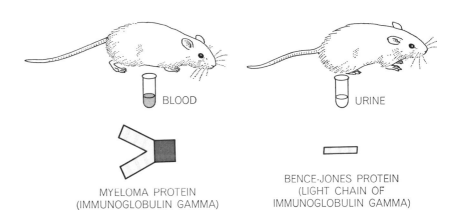

MYELOMA PROTEIN
(IMMUNOGLOBULIN GAMMA)

BENCE-JONES PROTEIN
(LIGHT CHAIN OF
IMMUNOGLOBULIN GAMMA)

and allowed to coagulate, antibodies can be isolated from the serum fraction. Even when purified by recombination with the original antigen, immunoglobulin gamma molecules produced in this way vary slightly.

IDENTICAL ANTIBODY-LIKE MOLECULES are produced in large numbers by mice and humans who suffer from myelomatosis, a cancer of the cells that synthesize immunoglobulin. These abnormal immunoglobulins, all alike, can be isolated from the animal's blood (left). Often an abnormal protein also appears in the urine (right). Called a Bence-Jones protein, it seems to be the light chain of the abnormal immunoglobulin.

cipitate with it. Each F*ab* fragment carries one combining site; thus the two fragments together account for the two combining sites that each antibody molecule had been deduced to possess. The F*c* fragment prepared from rabbit immunoglobulin gamma crystallizes readily, but neither the F*ab* fragments nor the whole molecule has ever been crystallized.

Since crystals form easily only from identical molecules, it was guessed that the halves of the heavy chain that comprise the F*c* fragment are probably the same in all molecules and that the complexity is mainly in the F*ab* fragments where the combining sites are found. The enzyme papain, which causes the split into three pieces, can hydrolyze a

great variety of peptide bonds, and yet only a few in the middle of the heavy chain are in fact split; it looks as if in the F*ab* and F*c* fragments the peptide chains are tightly coiled in such a way that the enzyme cannot gain access. This suggests a picture in which three compact parts of the molecule are joined by a short flexible section near the middle of the heavy chain.

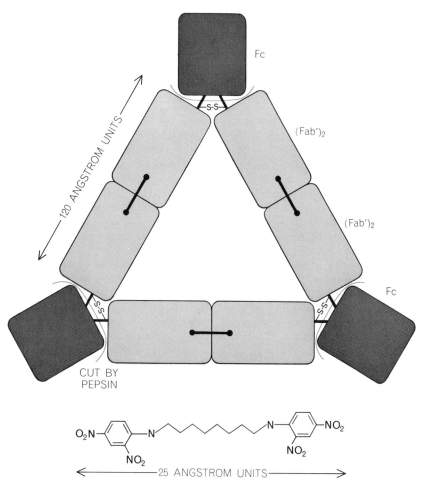

ANTIBODY-ANTIGEN COMPLEX seen in electron micrographs (*below and page 2523*) is thought to have this triangular structure. Below it, drawn to a large scale, is the synthetic antigen: an eight-carbon chain with a dinitrophenyl group at each end. Three antigen molecules appear able to bind together three immunoglobulin gamma molecules.

The full structure of a protein molecule showing the arrangement in space of the peptide chains and the positioning of the amino acids along them can at present only be achieved by X-ray crystallography. Such work has been started at Johns Hopkins University with the F*c* fragment. Electron microscopy, however, can provide much information about the shape of protein molecules, and successful electron microscope studies have been made recently with rabbit antibodies. When the antibodies are free, no clear pictures are obtained, which suggests that the molecules have a loose structure that is without definite shape. If they are combined with antigen, however, good pictures can be made. Michael Green and Robin Valentine of the National Institute for Medical Research in London prepared antibodies in rabbits that would combine with a benzene derivative known as a dinitrophenyl group. This can be done, as Karl Landsteiner showed many years ago, by injecting into the rabbit a protein on which dinitrophenyl groups have been substituted. Antibodies are formed, some of which combine specifically with the substituent dinitrophenyl coupled onto other proteins or into smaller molecules.

Green and Valentine investigated the smallest compound carrying two dinitrophenyl groups that would crosslink two or more antibody molecules. This proved to be an eight-carbon chain with a dinitrophenyl group at each end. This material does not form a precipitate with antibody, but with the electron microscope one can see ringlike structures that appear to contain three to five antibody molecules [*see illustrations on page 2523 and at left*]. The antigen molecule is not visible. The three-component structure is believed to consist of three antibody molecules linked by three molecules of invisible antigen. The lumps protruding from the corners are thought to be F*c* fragments. This interpretation is supported by using the proteolytic enzyme pepsin to digest off the F*c* fragment, leaving two F*ab* molecules held together by a disulfide bond and referred to as (F*ab'*)$_2$. When these (F*ab'*)$_2$ molecules are combined

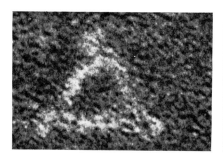 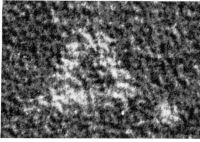

EFFECT OF PEPSIN COMPLEX is demonstrated in electron micrographs taken by Green and Valentine. In the normal complex formed by immunoglobulin gamma and the dinitrophenyl compound (*left*) a typical triangular structure contains a small lobe, or lump, at each corner, which is thought to be the F*c* part of the immunoglobulin molecule. If the antibody is first treated with pepsin, which splits off the F*c* fragment, the remaining (F*ab'*)$_2$ molecule still reacts with the antigen but the corner lobes are missing (*right*).

with antigen, rings are formed as before, but the lumps at the corners are now gone, confirming the idea that they were indeed the Fc part of the molecule.

Since most interest centers on the antibody combining site, the next problem to solve is whether the site is to be found in the light chain, which is entirely in the Fab fragment, or in the half of the heavy chain that is also present, or whether the site is formed by both chains together. It has not been possible to get a clear answer to this problem because the chains cannot be separated except in acid or urea solutions; this causes a partial loss of the affinity for antigen, which is not recovered even after the acid or urea is removed. Present evidence suggests that the heavy chain is the most important but that the light chain plays a role. This may be because it actually forms a part of the site or because it helps to stabilize the shape that the heavy chain assumes and hence plays a secondary role that may be only partially specific.

In any case, the field is clear for a direct attempt at comparative studies of the chemical structure of the light chain as well as of the half of the heavy chain that lies in the Fab part of the molecule. The shape and hence the specificity of the combining site must depend on the configuration of the peptide chains of the Fab fragment; this is believed to be determined only by the sequence of the different amino acids in the chain. Therefore it is reasonable to expect that if the amino acid sequence is worked out for the Fab half of the heavy chain and perhaps also for the light chain, then in some sections sequences will be found that determine the configuration of the combining site and that will be characteristic for each antibody specificity. Attempts to carry out such sequence studies, however, seemed unattractive because of convincing evidence that all preparations of immunoglobulin gamma—even samples of purified antibodies obtained by precipitation with a specific antigen—were actually mixtures of many slightly different molecules with presumably different amino acid sequences.

Although the complexity of immunoglobulin gamma (and of the other classes of immunoglobulins) has presented investigators with a most difficult puzzle, considerable progress has now been made in solving much of it [*see illustration on this page*]. First, there are two kinds of light chain, named kappa and lambda, but in any one molecule both light chains are of the same type.

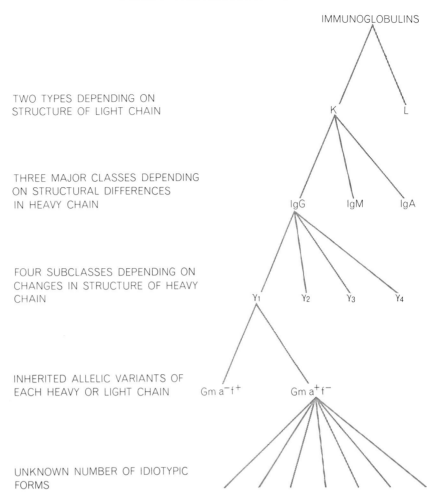

IMMUNOGLOBULINS

TWO TYPES DEPENDING ON STRUCTURE OF LIGHT CHAIN — K   L

THREE MAJOR CLASSES DEPENDING ON STRUCTURAL DIFFERENCES IN HEAVY CHAIN — IgG   IgM   IgA

FOUR SUBCLASSES DEPENDING ON CHANGES IN STRUCTURE OF HEAVY CHAIN — γ_1 γ_2 γ_3 γ_4

INHERITED ALLELIC VARIANTS OF EACH HEAVY OR LIGHT CHAIN — Gm $a^- f^+$ Gm $a^+ f^-$

UNKNOWN NUMBER OF IDIOTYPIC FORMS

SUBDIVISIONS OF HUMAN IMMUNOGLOBULIN presented investigators with a difficult problem to unravel. For simplicity, subdivisions are shown for only one branch at each level. The abbreviation "IgG" stands for immunoglobulin gamma, the antibody found in largest amounts and the one most easily isolated. Idiotypic forms are apparently unique to individual animals and may involve alterations in both the light and the heavy chains.

The molecules containing kappa chains are known as *K* type and those with lambda chains as *L* type. Then in some species (probably in all) the immunoglobulin gamma class contains several subclasses; four have been identified in human gamma globulin. The subclasses differ in their heavy chains, which carry not only the characteristic features of the class but also small differences that distinguish the subclasses. In any one individual, molecules will be found of both *K* and *L* type, and they belong to all the subclasses. In addition each of the kinds of chain shows differences, known as allelic forms, that are inherited according to Mendelian principles. In an individual homozygous for this property only one allelic form of, say, the kappa chain will be present, but in a heterozygous individual there will be two forms of the kappa chain. It scarcely need be stressed that all these phenomena lead to a very complex mixture of molecules of immunoglobulin gamma in the serum of any

individual. Yet there is still another kind of complexity termed idiotypic. In certain circumstances it is possible for an animal to synthesize antibody molecules that are unique to itself, distinct from other antibody molecules of the same specificity in other individuals of the same species—and distinct from all other immunoglobulins in its own blood.

Perhaps the most remarkable aspect of all of this is that the complexity seems to bear no relation to the structure of the antibody combining site. As far as we know at present, any antibody specificity may be found on any of these many different kinds of molecule.

All such variations are likely to be based on differences in amino acid sequence, and already some differences relating to subclass and allelic changes have been identified. The structural differences are so small, however, that it is not possible to separate out single kinds of molecule by the methods available for the fractionation of proteins.

Thus it was a great step forward when it was recognized that in certain forms of cancer, immunoglobulin molecules of apparently a single variety appear in the blood. Such immunoglobulins have only one type of light chain and one subclass of heavy chain, and each chain belongs to one or the other allelic form. As far as we know each chain has only one amino acid sequence and therefore belongs to only one idiotypic form.

The disease responsible for this unique production of antibody is known as myelomatosis. Observed in both mice and men, it is a cancer of the cells that synthesize immunoglobulin, often those in the bone marrow. Apparently a single cell, one of the great number that synthesize immunoglobulins, starts to divide rapidly and leads to an excessive production of a single kind of immunoglobulin. This provides evidence, incidentally, that the complexity of immunoglobulin molecules arises from their synthesis by many different kinds of cells. These abnormal immunoglobulins are known as myeloma proteins. Because they are often present in the blood in a concentration several times higher than all the other immunoglobulins together, they can be isolated rather easily.

Moreover, in about half of all myeloma patients an abnormal protein appears in the urine in large amounts. This substance was first observed by Henry Bence-Jones at Guy's Hospital in London in 1847 and has been known ever since as Bence-Jones protein. Its nature, however, was not recognized until five years ago, when Gerald M. Edelman and J. A. Gally of Rockefeller University and independently Frank W.

Putnam of the University of Florida showed that Bence-Jones protein is probably identical with the light chains of the myeloma protein in the serum of the same patient. Because Bence-Jones proteins can be obtained easily, without any inconvenience to the patient, they were the first materials used for amino-acid-sequence studies.

Although complete sequences have been worked out for only two Bence-Jones proteins in the mouse and only three human Bence-Jones proteins, perhaps 20 more have been partially analyzed. A remarkable fact has emerged. It seems that all Bence-Jones proteins of the same type have exactly the same sequence of amino acids in the half of the molecule that ends in the chemical group COOH (hence known as the C-terminal half) but show marked variation in the half that ends in the group

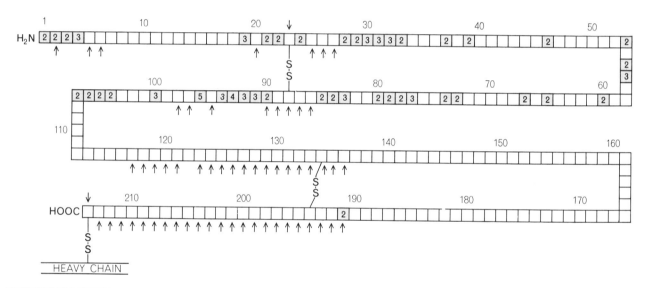

IMMUNOGLOBULIN LIGHT CHAIN, represented by analyses of human Bence-Jones proteins of the K type, has 214 amino acid units. Colored squares show where amino acids have been found to vary from one protein to another; blank squares show where no variation has yet been found. Numbers in the squares indicate how many different amino acids have been identified so far at a given site. Arrows mark positions where a particular amino acid has been found in at least five different proteins. Complete amino acid sequences are now known for three human Bence-Jones proteins and partial sequences for about 20 others. All variations occur in the first half of the chain with one exception, the variation at position 191. This is related to the allelic, or inherited, character of light K chains, hence differs from the alterations in the variable half of the chain. The diagram is based on one recently published by S. Cohen of Guy's Hospital Medical School in London and C. Milstein of the Laboratory of Molecular Biology in Cambridge.

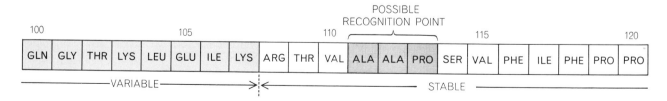

MIDDLE PART OF LIGHT CHAIN, as determined for one human Bence-Jones protein (K type), includes the amino acids at positions 111, 112 and 113 that are common to both K- and L-type Bence-Jones proteins of humans and to K-type Bence-Jones proteins of mice. It has been suggested that the section of the gene coding for this sequence may provide a special "recognition point" for the joining of two different genes responsible for the variable and stable sections of the light chain or, possibly, for providing a mechanism to change the amino acid sequence in the variable section (see illustration on page 2530).

ALA	ALANINE	LYS	LYSINE
ARG	ARGININE	PHE	PHENYL-
GLN	GLUTAMINE		ALANINE
GLU	GLUTAMIC ACID	PRO	PROLINE
GLY	GLYCINE	SER	SERINE
ILE	ISOLEUCINE	THR	THREONINE
LEU	LEUCINE	VAL	VALINE

NH₂ (the N-terminal half). Of 107 amino acid positions in this half, at least 40 have been found to vary. No two Bence-Jones proteins have yet been found to be identical in the N-terminal half, so that the possibility of molecular variation is clearly great. Given the possibility of variation at 40 sites and supposing that only two different amino acids can occupy these sites, it would be possible to construct 2^{40}, or more than 10 billion, different sequences. Actually as many as five different amino acids have been found to occupy one of the variable sites [*see upper illustration on opposite page*].

The amino acid sequence studies of the heavy chain are less advanced than those with the Bence-Jones proteins because the material is more difficult to obtain and is more than twice the length. Results with the heavy chain of two human myeloma proteins, however, have shown them to have many differences in sequence for more than 100 amino acids from the N-terminal end, whereas the remainder of the chain appears to be identical in both cases. Accordingly it seems certain that the heavy chains will show the same phenomena as the light chains; it is possible that the length of the variable section in both chains will be similar.

Inasmuch as both variable sections are in the F*ab* fragment of the molecule it seems obvious that these sections must participate in creating the many different antibody combining sites. All the work discussed here has been done with myeloma proteins, and since each has a single amino acid sequence in both heavy and light chains, it would follow that each will be a specific antibody against one of an untold number of different antigenic sites. The chances, therefore, of finding a myeloma protein in which antibody specificity is directed to a known, well-defined antigenic site seemed small. Nevertheless, several myeloma proteins have recently been found to possess antibody-like activity against known antigens. A comparison of the sequences of their heavy and light chains may give a lead as to where the combining site is located.

It has been believed with good reason that myeloma proteins are typical of normal molecules of immunoglobulin gamma, each being a homogeneous example of the many different forms present. It thus seemed likely that any attempt to determine the amino acid sequence of immunoglobulin gamma from a normal animal would be impossible, especially in the variable region that is

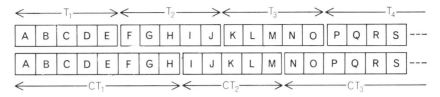

DETERMINATION OF AMINO ACID SEQUENCE in the polypeptide chains of proteins depends on the use of enzymes that cleave the chains into short fragments next to particular amino acids. The sequence in the resulting fragments can then be established. Thus trypsin might split a chain into fragments T_1, T_2, T_3 and T_4. Another enzyme, chymotrypsin, might split the same chain into fragments CT_1, CT_2 and CT_3. Since these fragments must overlap one can establish their order unequivocally and thereby the sequence of the entire chain.

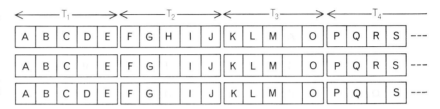

IMMUNOGLOBULIN SEQUENCE should be amenable to analysis even though a particular antibody sample might contain a variety of slightly different molecules. Slight variations at certain positions (*color*) should not prevent the ordering of similar fragments.

of particular interest. One would expect normal immunoglobulin gamma to be a mixture of many thousands of different molecules, each with a different sequence in the variable region.

Amino acid sequences of polypeptide chains are found by using enzymes to break the chains into pieces from 10 to 20 amino acids long. It is then possible to work out the sequence of each piece. By using different enzymes the original chain can be broken at different places, with the result that some pieces overlap. This provides enough clues for the whole sequence to be put together, rather like a one-dimensional jigsaw puzzle [*see upper illustration above*]. When the protein is pure, there is only one order of amino acids possible, and all the sequences of the individual fragments will fit into it.

One can see that if this method were attempted with a protein that was in fact a mixture of many slightly different proteins, each with a different sequence, a hopelessly confusing picture would probably result. The work with the myeloma protein suggested, however, that there would be a constant part as well as a variable part, and it seemed worthwhile to see what progress could be made in determining at least the constant part. Work at Duke University showed that the whole of the F*c* section of the heavy chain of normal rabbit immunoglobulin gamma gave a coherent sequence and was therefore part of the stable section, as had been expected. Recent work in our laboratory has now

shown that the coherence continues well into the other half of the heavy chain. Although the work is far from complete, it seems possible that a full sequence will be established right through the entire heavy chain. Variations have been picked up in a number of positions and no doubt many more will be found, but the results are not completely confusing, as might have been expected if normal immunoglobulin gamma were a mixture of many thousands of myeloma proteins, each with substantially different sequences in the variable parts of the chain. The conflict between the results with the myeloma proteins and the recent results with normal rabbit immunoglobulin gamma may be more apparent than real.

What does all this mean in terms of the structure of antibodies and their power to combine specifically with antigens? The phenomenon of a variable section and a stable section in both heavy and light chains is extraordinary and is unique to immunoglobulins; the variable section is in the part of the molecule known to contain the combining site. It therefore seems certain that this must be the basis of the specific configuration of the combining site.

It should be emphasized that all this work is very incomplete. In another year or so it will undoubtedly be much easier to see just how different one myeloma protein is from another in both the heavy and the light chains. It may be that the differences between any two

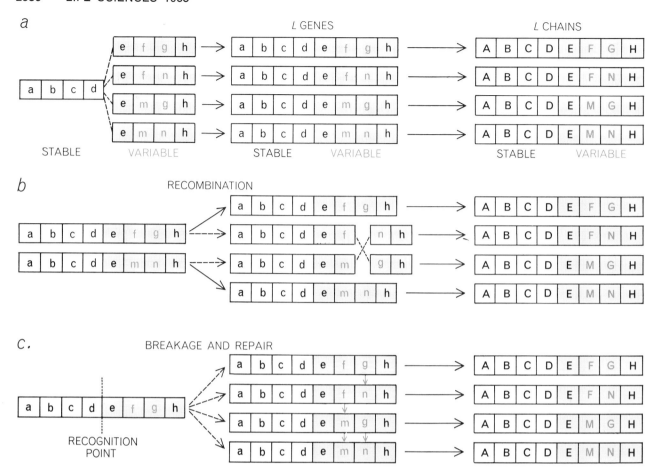

VARIABILITY OF IMMUNOGLOBULIN MOLECULES has been explained by three principal hypotheses. The simplest (*a*) suggests that one gene codes for the stable section of each chain and that a great number, perhaps hundreds of thousands, code for the variable section. A second idea (*b*) is that several genes are divided into stable and variable sections and that the latter inter-change parts during cell division. A third proposal (*c*) suggests that there may be a recognition point in the gene (*see lower illustration on page 2528*) and that an enzyme partially splits, or breaks, the gene on the variable side of that point. When repaired by other enzymes (*arrows*), mistakes are made, thus giving rise to many different amino acid sequences in the antibody molecule.

will on the average be small, so that for a mixture of many molecules the amino acid in any one position will be common to 80 or 90 percent of the molecules [*see lower illustration on page 2529*]. Presumably this shows how it is possible to find a comprehensible amino acid sequence in normal immunoglobulin gamma.

It may also be, however, that myeloma proteins are not quite typical of normal antibodies. Because they are the result of a disease they may exaggerate a normal phenomenon. Although they are invaluable in drawing attention to a fundamental mechanism, they may mislead us by exhibiting greater variability than is present in normal immunoglobulin gamma.

Whatever the answer, the existence of a stable section and a variable section, which has been shown so clearly in the Bence-Jones proteins and which also occurs in the heavy chains, is a remarkable phenomenon. The mechanism of its biological origin has aroused intense interest. Many hypotheses have been put forward, but there are perhaps three principal ones [*see illustration above*].

A straightforward mechanism would be to have a single gene coding for each stable section in the antibody molecule and as many genes as necessary (tens of thousands or hundreds of thousands) coding for the variable sections. The cell would also be provided with a means for fusing the product of the two kinds of gene to construct the complete immunoglobulin molecule. (In this as in the other suggestions, the presence of an antigen would somehow trigger production of the appropriate antibody.)

A second proposal invokes the concept of genetic recombination, which involves the exchange of parts of genes. One can imagine several genes that are divided into a stable portion and a variable portion. During cell division, when genes are pairing and duplicating, the variable portion would interchange sections, thereby giving rise to many different genes capable of coding the variable parts of the antibody molecule.

The third suggestion visualizes that the gene for, say, the light chain may contain a "recognition point" midway in its structure [*see lower illustration on page 2528*]. This might provide a specific attachment site for an enzyme that can split the nucleic acid of the gene only on the side coding for the variable section. When the broken portion is repaired by other enzymes, mistakes are made, thereby giving rise to many different sequences of nucleotides—the nucleic acid building blocks that embody the genetic message. These differences are then translated into different amino acid sequences in the variable portion of the antibody molecule.

There is no clear answer as to which methods, if any or all, are the operative mechanisms, but a continuation of the structural studies may provide a clearer understanding. When this understanding is attained, it should lead to ideas about how to change, stimulate or suppress immune reactions as medical practice requires and therefore should be of great practical value as well as solving one of the most intriguing problems in biology.

The Author

R. R. PORTER was recently appointed Whitley Professor of Biochemistry at the University of Oxford and director of a Medical Research Council Unit for research into immunochemistry that is being set up at the university. Previously he had been for seven years Pfeizer Professor of Immunology at St. Mary's Hospital Medical School of the University of London. Porter, who was elected a Fellow of the Royal Society in 1964, studied at the University of Liverpool and the University of Cambridge. From 1949 to 1960 he was a member of the scientific staff at the National Institute for Medical Research in London. He lists his recreations as walking and fishing.

Bibliography

IMMUNOGLOBULINS. Julian B. Fleischman in *Annual Review of Biochemistry*, Part II, Vol. 35, pages 835–872; 1966.

IMMUNOGLOBULINS. E. S. Lennox and M. Cohn in *Annual Review of Biochemistry*, Part I, Vol. 36, pages 365–406; 1967.

THE STRUCTURE OF IMMUNOGLOBULINS. R. R. Porter in *Essays in Biochemistry: Vol. 3*, edited by P. N. Campbell and G. D. Greville. Academic Press; Fall, 1967 (in press).

SCIENTIFIC
AMERICAN October 1967, Vol. 217, No. 4, pp. 94-102   OFFPRINT 1084

VISUAL ISOLATION IN GULLS

by Neal Griffith Smith

Some species of gulls live together and look alike, yet they do not
interbreed. How do the species remain isolated? Experiments in the
Arctic indicate that they do so by recognizing subtle visual signals.

Gulls look remarkably alike. That was the problem. Differences in appearance among the large gulls of the genus *Larus* can be subtle: a slight variation in size or a change in the color of the wing tips or of the eye and the small fleshy ring around the eye. Observing differences of this kind, an ornithologist discriminates among species of the genus. The problem arose from the fact that the gulls are equally discriminating. In some places *Larus* species that seem virtually indistinguishable nest side by side, yet they do not interbreed. How do gulls of one species avoid interbreeding with gulls of another?

The question of how species acquire and maintain their identity has received much attention in the century since the publication of Charles Darwin's *Origin of Species*. It is now well established that geographic isolation between populations is of prime importance in initiating the process by which species arise. Indeed, the gulls of the Northern Hemisphere have been cited as a classic example supporting this concept.

The common ancestor of the *Larus* gulls probably emerged in the Siberian region. As these gulls spread to the east and west, simple geographic distance began to inhibit the flow of genes between the most distant populations. By the time these populations had spread around the hemisphere and overlapped in western Europe, their respective ge-

netic backgrounds were different enough so that hybrids between them were at some disadvantage; thus they did not interbreed. The advance and retreat of the ice during the Pleistocene epoch caused a further fracturing and recombination of these circumpolar gull populations. In some cases the differences evolved were not critical enough to confer a disadvantage on hybrids; thus rejoined populations interbred.

It seems clear that the mechanisms by which species discriminate among one another evolved gradually during the process of species formation. In the Canadian Arctic, and probably elsewhere in the north, the ice intruded between various gull populations at different times and for different lengths of time. Accordingly the isolating mechanisms were likely to be at different stages of development in different populations. By studying populations in such an area one can uncover what these mechanisms are.

It is one thing to identify differences in the appearance of two closely related species living side by side and infer that the differences function as a barrier to interbreeding. It is quite another to demonstrate that these features are actually utilized in the isolation of species. To explain how such features work is still another step. This article is primarily concerned with the last two problems. It also considers the evolutionary history of the *Larus* gulls, because the elucidation

of a feature that is utilized in species isolation can suggest what the species were like in the past and how the isolation mechanisms evolved.

The four species of *Larus* gulls I have been studying comprise the Canadian portion of the complex of gull populations around the North Pole. The fact that the four species do not interbreed has been clearly established by other workers and myself. All four gulls have a white body and a gray back and wings. The largest in body size is the glaucous gull (*Larus hyperboreus*). The tips of its wings are white, the iris of its eye is yellow and the fleshy ring around the eye is an even brighter yellow. Colonies of glaucous gulls are found throughout the polar area; in the eastern part of the Canadian Arctic they usually nest on cliffs. A more familiar species is the herring gull (*L. argentatus*), the only one of the four that breeds in the continental U.S. It is a medium-sized bird with wing tips that are partly black and partly white. Like the glaucous gull, the herring gull has a yellow iris; its eye-ring, however, is orange. In the Arctic this species usually nests on the ground in marshy areas. About the same size and coloration is Thayer's gull (*L. thayeri*), except that in this species the iris of the eye is dark brown and the eye-ring is a reddish purple. Thayer's gull nests almost exclusively on towering cliffs. The smallest of the four species (although not by very much) is Kumlien's gull (*L. glaucoides*), which also nests on cliffs. It is most like Thayer's gull: its eye-ring is reddish purple but the iris varies from clear yellow to dark brown. Its wing tips also vary in their amount of gray.

The common breeding grounds of these gulls are difficult to visit, and not

HEADS OF *LARUS* GULLS are almost identical except for the color of the eye and its encircling fleshy ring. At top on the opposite page are two Kumlien's gulls (*Larus glaucoides*), a species in which the iris varies from clear yellow to mottled brown. The eye-ring is reddish purple. Below appear two Thayer's gulls (*L. thayeri*). The eye-ring is the same as it is in Kumlien's gulls but the iris tends to be darker in this species. Next is a herring gull (*L. argentatus*), the only one that nests in the continental U.S. The glaucous gull (*L. hyperboreus*), shown last, has an eye-ring of yellow, which distinguishes it from the herring gull. Smallest gull is at top; largest at bottom. The painting was made by Guy Tudor.

much has been known about them. When I began my work, the evidence was that no one area was shared by all four species. In the course of trying to find such an area I spent three seasons (April to September) in the Canadian Arctic, during which I covered just under 2,000 miles by dogsled and canoe. During this time I studied three of the gulls (glaucous, Kumlien's and herring) I found nesting together on the south side of Baffin Island and a different trio (glaucous, Thayer's and herring) on nearby Southampton Island. Finally I discovered all four species nesting together on the east side of Baffin Island. It was never easy to find the ground-nesting herring gulls in association with the cliff-nesting species. Nesting on cliffs evolved as an adaptation against predators such as foxes; apparently competition with the other gulls for nesting sites has re-sulted in the herring gulls' occupying poorer sites. Nevertheless, where the surface allowed it and where the birds were safe from predators in a place such as a rocky islet, all the gulls would nest together.

There were a number of factors, for instance the habitat differences I have mentioned, that tend to reduce the possibility of mixed matings in the areas shared by different populations of gulls; here, however, I shall discuss only differences in external appearance among the species that function as major isolating mechanisms. In 1950 Finn Salomonsen, a Danish ornithologist, suggested that the color of the eye-ring might serve as a signal for differentiation between Kumlien's gull (reddish-purple eye-ring) and the glaucous gull (yellow eye-ring). Although I tested the possible significance of all the differences in the gull's external appearance (with the exception of size), I concentrated on the color of the eye-ring.

In order to study the gulls closely it was necessary to catch them. At first I did so by stretching over a ledge a large fishnet under which food was placed. When the gulls were under the net, an Eskimo assistant and I rushed forward and dropped it, pinning the gulls to the ground. This was obviously an inefficient method, and later I used the drug tribromoethanol. Capsules of the drug were inserted into pieces of meat; after eating the meat the gulls quickly became anesthetized. In this way more than 1,800 gulls were trapped. After the gulls had been drugged they were immobilized with a surgical rubber band that pinned their legs and wings to their bodies, and colored leg bands were put on them to

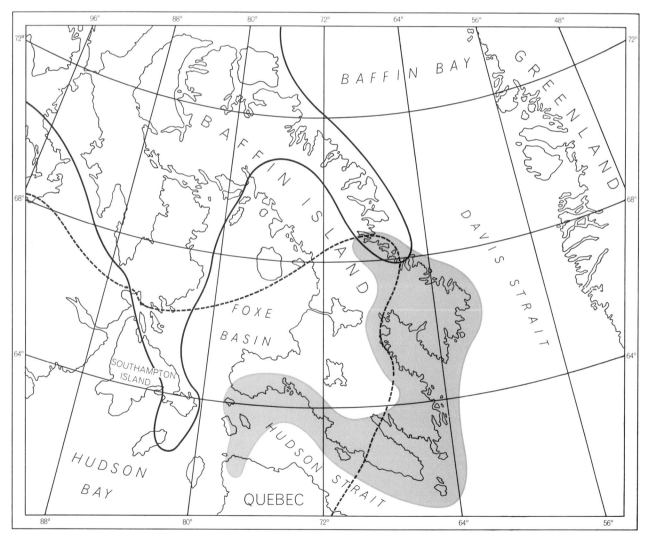

BREEDING RANGES of large *Larus* gulls lie in the eastern Canadian Arctic. Thayer's gulls (*black line*) and Kumlien's gulls (*colored area*) usually were found nesting in colonies on sea cliffs. The glaucous gull nests throughout this region; it was observed both on cliffs and on level ground. Herring gulls (*broken line*), a ground-nesting species, were found with the others only on rocky islands. Before discovering all four gulls on the east coast of Baffin Island, the author studied some species on Southampton Island.

make it possible to recognize individuals. Sex was determined by measuring bill, feet and wings; the males are usually larger. The determinations were confirmed by the subsequent behavior of the gulls.

One of my first thoughts had been that if markings and coloration play a role in the gulls' mating behavior, it should be possible to demonstrate it by changing these features artificially. This I now undertook to do. To change the color of the eye-ring I applied oil paint with a thin brush. The wing-tip pattern was changed with white or black ink after first wiping the feathers with alcohol so that the ink would penetrate. Judging from the behavior of the painted gulls neither of these procedures caused any physical irritation. On the other hand, when I attempted to change the color of a gull's back by spraying it with paint, the feathers stuck together and the gull tried repeatedly to remove the paint.

In my first season, after observing the behavior of individual pairs of glaucous, Kumlien's and herring gulls in a colony in southern Baffin Island, I captured a small group of the gulls. The eye-ring of each one was changed to the color of a different species. Over the yellow ring of the glaucous gull, for example, I painted a ring of reddish purple. All the female birds had copulated with males before the experiment but none had laid eggs. When the females returned to their nests, they were accepted by their mates. In the days that followed, however, the males would no longer mount, in spite of intense solicitation by the females. In all cases where the female's eye-ring color had been changed the pair did not remain together. Five of the males whose mates had been painted formed pairs with nonaltered females in adjacent territories. Copulation ensued, and after two weeks all the new pairs had eggs. The females I had painted left the colony.

In contrast to these findings, changing the eye-ring of a mated male gull appeared not to affect a pair's behavior. The females accepted their altered mates and the males responded to the soliciting behavior of the females. In the one case where both individuals of a mated pair were changed the results were exactly the same as they were when only the female was changed.

The results looked promising. Although the number of individuals involved was small (33 females and 30 males) and some important controls were lacking, I now had a working

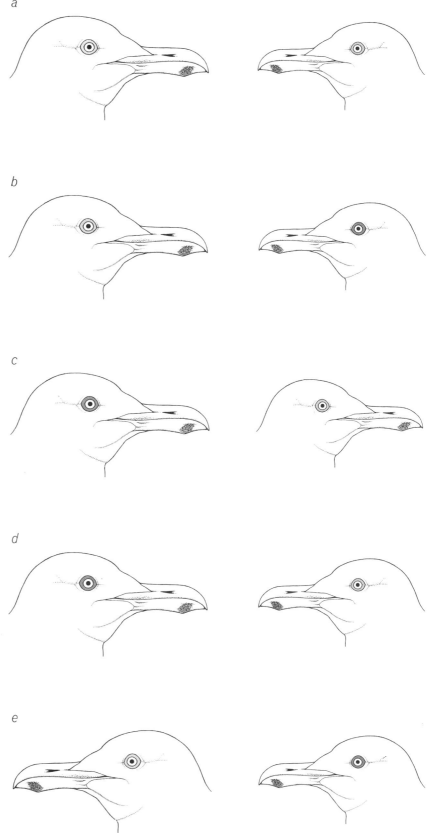

MATING BEHAVIOR OF GULLS changes when their appearance is artificially altered. Even when individuals of like species nest nearby, males and females of the same species normally mate (*a*). Such pairs still form even if an unmated female's eye-rings are painted to look like those of another species (*b*). Painted the same way, an unmated male fails to obtain a mate of his species (*c*). If the male is mated when his eye-rings are altered, his mate remains with him (*d*). When same change is made in the female, they usually separate (*e*).

hypothesis, namely that in some way the eye-ring color of the females functioned as a stimulus for mounting by their mates and that this reaction was keyed to differences among the species.

The program for the next two seasons was to repeat the eye-ring experiments with the necessary controls and to explore the hypothesis in greater detail. Was it the fleshy eye-ring alone or the entire eye that functioned as a stimulus? Was the important factor color or was it contrast? In answering these questions the critical species would be Thayer's gull, with its dark eye-ring and dark iris. It was also of prime importance to test the function of eye-ring color and other physical features with unmated gulls. There was reason to believe that the females choose the males, and it seemed unlikely that a mixed pair would form only to separate later because copulatory behavior was disrupted. There should also be an earlier isolating barrier.

In the experiments of these two seasons there were three major control groups: gulls that were drugged but not painted, gulls that were drugged and painted with their own color or pattern and gulls that were not captured but whose behavior was observed. My earlier findings with mated female gulls were confirmed in experiments that also shed light on the question of color v. contrast. A group of Kumlien's gulls was captured and their reddish-purple eye-ring was painted over either with a light color (yellow, orange or white) or a dark one (red or black). When the female gull had been painted with a light color, copulation usually stopped and the pair separated; in this regard an eye-ring painted white was the most effective. Dark colors had no significant effect. Exactly the reverse was true when the herring gull (orange eye-ring) and the glaucous (yellow) were painted in the same way: dark colors inhibited copulation and light colors had no significant effect. Among the broken pairs were a number of female glaucous gulls whose yellow eye-ring had been changed to orange. This fitted the other results rather nicely, the orange eye-ring of the herring gull being darker than the yellow one of the glaucous gull.

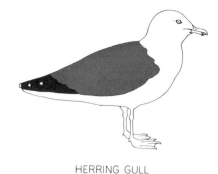

HERRING GULL

EXTERNAL FEATURES vary among species. Each gull shown above differs from the

When the same procedures were tried with Thayer's gull, however, there was no significant change in behavior. In this species it is not the eye-ring that stands out against the bird's white head, as it does in the other three species, but the entire orbital region—both the iris and the eye-ring are dark. This suggested that the orbital region as a whole functioned as a stimulus.

One could not paint the eye to change its color, but painting the reddish-purple eye-ring white reduces the contrast of the orbital region against the white head. After thus "erasing" the eye-ring I painted a larger one on the feathers around the eye. In making this "super-eye-ring" I used on various gulls the same assortment of colors as I had in the other experiments. One might think of the painted circle as the "eye-ring," the white feathers between it and the eye as the "iris" and the actual iris as the new "pupil." This may seem a bit far-fetched, and I do not mean to imply that this is what the gull sees, but the fact remains that in a significant number of cases where the female had been given a light-colored super-eye-ring copulation was inhibited and pairs separated. Apparently the stimulus to copulation was the contrast pattern of the ringed eye against the white head: dark color against white in Kumlien's and Thayer's gull and light color against white in the herring and glaucous gull.

In the course of these experiments I observed that a male occasionally mounted his altered mate but did not attempt copulation even when the female prodded his breast or rubbed her tail against his anal region. Earlier I had observed that copulation was invariably preceded by such tactile stimulation on the part of the females. I concluded that successful copulation probably involves both visual and tactile stimuli.

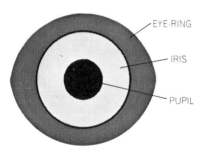

HERRING GULL

EYE-RING

IRIS

PUPIL

GLAUCOUS GULL

THAYER'S GULL

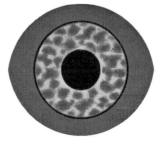

KUMLIEN'S GULL

RINGED EYES of various colors, photographed in black and white, exhibit different color values. The orange eye-ring of the herring gull shows darker than the yellow ring of the glaucous gull. Darker still is the reddish-purple ring of Thayer's and Kumlien's gulls. In Thayer's gulls brown irises enforce the contrast of the orbital region against the white head; the iris of Kumlien's is lighter. Eye-head contrast acts as an interspecies barrier among gulls.

GLAUCOUS GULL   THAYER'S GULL   KUMLIEN'S GULL

others in size, in the coloration of the orbital region, back and wings and in the pattern of the wing tips. The author's experiments suggest that the wing-tip pattern serves to supplement the signal of eye-head contrast in preventing the formation of mixed pairs.

(Auditory stimuli may also be involved, but this was not tested.) It appeared that the tactile stimuli were supplemental to other stimuli rather than independent of them; the eye-head contrast of the females played the major role.

Did the eye-head contrast play a role in the formation of pairs as well as in copulatory behavior? The eye-rings of a group of unmated female gulls were changed to determine if this made a significant difference in the species of the males with which they paired. The results showed no difference between this group and control groups. After a week or two, however, pairs in which the females had been given the "wrong" eye-head contrast separated; the males would not copulate. This of course further supported the role of eye-head contrast.

When the same experiment was performed with unmated male gulls, the results were quite different. In one instance 91 percent of the male Thayer's gulls that had been changed by the super-eye-ring technique to the light-eyed condition failed to obtain mates of their own species. An experiment with glaucous and herring gulls also showed that if the eye-head contrast of unmated males was "wrong," they were significantly less successful in obtaining mates of the same species than the controls were.

The results suggested that in pair formation it is indeed the females that choose the males, and that they select males with an eye-head contrast like their own. In other words, the same feature works in two ways to isolate the species: in the males it serves the purpose of pair formation and in females the purpose of copulatory behavior. What role is played by other external differences? The color of the mantle (back and wings) of the *Larus* gulls varies from one species to another.

Among the four species I studied the differences in mantle coloration were not pronounced; still it seemed worthwhile to attempt an evaluation of mantle color as a possible signal. As I have indicated, however, spraying paint on a gull's back has too great an effect on the gull's behavior to make for a sound experiment, and the role played by this feature remains obscure.

Tests of the wing-tip pattern displayed at rest suggest that this feature does function as a signal in species discrimination during pair formation. There was no significant change in behavior after alteration of the wing tips of female gulls, whether mated or unmated. On the other hand, alteration of the wing tips of unmated males indicated that the wing-tip pattern functions as a stimulus to pair formation in combination with the eye-head contrast. This was shown by the fact that female gulls chose males with both "right" wing-tip pattern and eye-head contrast over males with only the "right" eye-head contrast. The wing-tip pattern alone is apparently not utilized in species discrimination during pair formation.

In several of the experiments male Thayer's gulls painted to appear light-eyed had been chosen by glaucous females. Since the females had the "wrong" eye-head contrast for the males, no copulation resulted and these mixed pairs did not remain together. After 59 Thayer's-glaucous pairs had formed I captured all but three of the glaucous females and altered them to the "right" contrast. Ten days later all 56 male Thayer's gulls had been observed to mount their altered mates, and about two weeks later 55 of the pairs had eggs. (One pair did not remain together.) Heavy ice on the rocks unfortunately forced me to abandon these colonies; I was never able to return to them. Before leaving I did collect several eggs, and

they contained well-developed embryos. It may be that the mixed pairs produced hybrid offspring.

From the start of my experiments it had been clear that there was a strong correlation between the behavior that resulted from changing the eye-ring and the gonadal cycle of the gull. Two identical experiments, one performed 16 days before the first eggs were laid and the other 12 days later, yielded strikingly different results. I considered initially as a working hypothesis that the main component of the pair bond was the attachment of the individuals to each other, and that during and after the egg-laying period the main component of the bond became the attachment of the individuals to the nest and the eggs. This hypothesis could account for certain pairs of gulls that had remained together even though the males had failed to respond to the solicitations of their altered mates. It could not, however, explain instances in which males continued to mount their mates after they had been painted and before egg-laying had begun. Moreover, the hypothesis offered no answer to the crucial question of what the physiological basis for the male's behavior is.

The solution to the problem was found in the relation between the internal physiological state of the male (indicated by the weight of the testes) and the number of times a pair had copulated. All but 12 of the 168 pairs of gulls that had remained together after the female had been given a different eye-ring had copulated six or more times before she had been painted. This number of copulations could be correlated with a certain weight of the testes attained in the male's gonadal cycle. I concluded that a gull whose testes had developed to the critical weight or

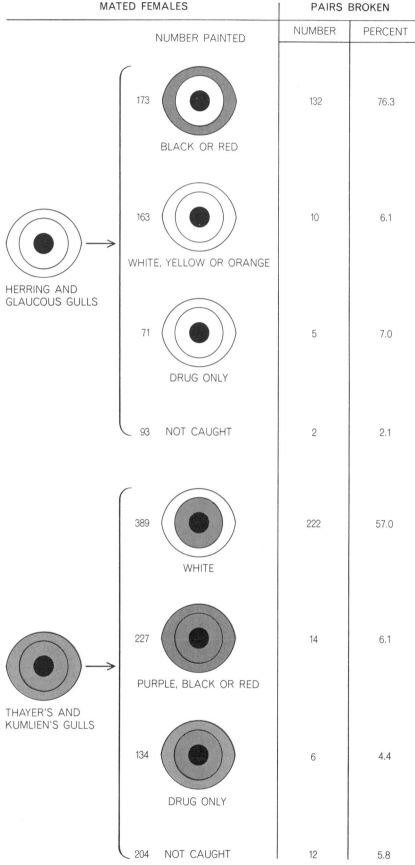

MATED FEMALES	PAIRS BROKEN	
NUMBER PAINTED	NUMBER	PERCENT
HERRING AND GLAUCOUS GULLS		
173 BLACK OR RED	132	76.3
163 WHITE, YELLOW OR ORANGE	10	6.1
71 DRUG ONLY	5	7.0
93 NOT CAUGHT	2	2.1
THAYER'S AND KUMLIEN'S GULLS		
389 WHITE	222	57.0
227 PURPLE, BLACK OR RED	14	6.1
134 DRUG ONLY	6	4.4
204 NOT CAUGHT	12	5.8

SUMMARIZED FINDINGS document experiments with mated female gulls. Pairs separated in most cases where the female's eye contrast was changed; this feature appears to be a major stimulus to the male in copulatory behavior. Because a drug was used in capturing gulls, one control group was drugged and not painted. Another group, not captured, was observed. Some gulls were painted with eye-rings of their own color as a further control.

beyond it would respond to a mate whether or not her eye-ring had been changed. The most telling evidence was that if the female's eye-ring was changed at a time before the critical weight was reached, the testes of her mate did not increase in weight—in fact, they diminished!

It is fairly well substantiated that gonadal development in many species is stimulated by changes in the daily cycle of daylight and darkness. On arriving in the Arctic in summer gulls are subjected to periods of daylight lasting almost 24 hours. This factor alone, however, could not cause the gulls' testes to develop beyond the level attained at the end of pair formation. Certain other stimuli must interact with light, and one of them—probably the most important one —is the presence of a mate with the proper eye-head contrast.

Although I have no evidence for it, it seems likely on logical grounds that a similar mechanism functions in females during pair formation. Once a pair bond is formed and a series of hormonal events is activated, inhibition of the female's gonadal development does not occur, even when the original stimulus—the eye-head contrast of the male—is removed. In the female, as in the male, the interaction of stimuli and the hormonal background at different times in the season provides a species-isolating mechanism that appears to be wholly effective.

Thus far we have been considering the two questions raised at the beginning: What are the visual factors involved in species discrimination among the gulls, and how do these factors affect reproductive behavior? At this point I should like to take up the matter of how the mechanisms that isolate species have evolved. In this regard it is instructive to examine the natural variation in iris and wing-tip color that occurs in one species: Kumlien's gull.

The eye-ring color differs little among Kumlien's gulls and Thayer's gulls but the amount of dark pigment in their irises varies considerably. Kumlien's gull is by far the more variable, ranging from individuals with completely dark irises to those with completely clear eyes of yellow. I divided this variation into six classes, Class 1 being the darkest and Class 6 the clearest. On the south coast of Baffin Island, Kumlien's gulls live together with herring gulls and glaucous gulls. In this locale Kumlien's gulls with clear irises were almost entirely absent; they occupied classes from Class 1 to Class 4 or Class 5.

On the east coast of Baffin Island, where Kumlien's gulls nested with Thayer's gulls. and glaucous gulls, the situation was reversed. There almost all the Kumlien's gulls fell into the last three classes, being clear-eyed or nearly so.

This pattern can be explained in terms of the natural selection of the variations that will reduce the possibility of mixed mating. According to my experiments, the contrast of the eye-ring and iris against the white head is the chief factor in species discrimination among gulls. To avoid mixed pairings, then, selection favored dark-eyed individuals where Kumlien's gulls nested with the light-eyed herring gulls and light-eyed individuals where Kumlien's gulls nested with the dark-eyed Thayer's gulls. Apparently the dark eye-ring of Kumlien's gull has been adequate for species recognition between Kumlien's gull and the yellow-eye-ring glaucous gull. The orange eye-ring of the herring gull affords a darker contrast, however, and where herring gulls and Kumlien's gulls nest together the dark iris of the latter reinforces the eye-head contrast. It is interesting to note that in Greenland, Kumlien's gulls have light eyes; there herring gulls are not found and glaucous gulls are.

The amount of dark pigment in the iris of Kumlien's gull is highly correlated with the amount of pigment in the wing tips. Individual Kumlien's gulls with light irises, as found on the east coast of Baffin Island, have white wing tips; those with dark eyes, as found on the south coast of the island, have dark blotches on their wing tips. It has been suggested that this variation in wing-tip pattern is the result of hybridization between Thayer's gulls and Kumlien's gulls, but that is not the case. The two species are most unlike each other where they nest together; they are very much like each other where they do not live together but where each is associated with glaucous gulls and herring gulls. The explanation for the variation of the wing tip is simply that it reflects the correlation between the pigment in the iris and the wing tip and the results of selection for differences in iris color in different populations.

In the course of my earlier work I had come to the conclusion that female gulls chose males that in eye-head contrast and wing-tip pattern were most like themselves. This created a problem, because it implied that the female knows what it looks like. A series of observations and one experiment on the east

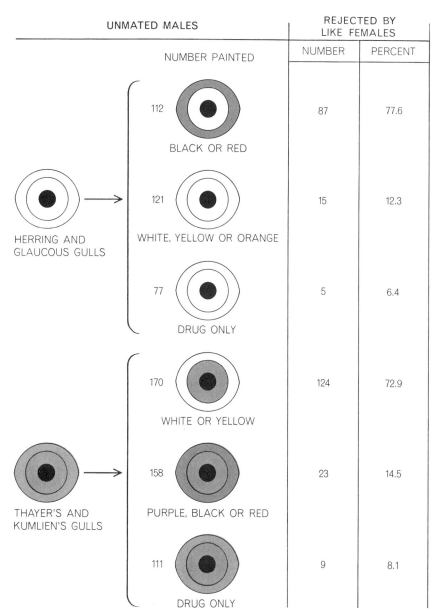

UNMATED MALES			REJECTED BY LIKE FEMALES	
		NUMBER PAINTED	NUMBER	PERCENT
HERRING AND GLAUCOUS GULLS	112	BLACK OR RED	87	77.6
	121	WHITE, YELLOW OR ORANGE	15	12.3
	77	DRUG ONLY	5	6.4
THAYER'S AND KUMLIEN'S GULLS	170	WHITE OR YELLOW	124	72.9
	158	PURPLE, BLACK OR RED	23	14.5
	111	DRUG ONLY	9	8.1

MATED MALES			PAIRS BROKEN	
		NUMBER PAINTED	NUMBER	PERCENT
HERRING AND GLAUCOUS GULLS	164	DARK	4	.02
THAYER'S AND KUMLIEN'S GULLS	51	LIGHT	3	.06

RESULTS of experiments in which the eye-ring of male gulls was altered indicate (*top*) that in pair formation the female chooses the male and that eye-head contrast is a factor in the choice. Changing the eye contrast of the male after a pair was formed (*bottom*) did not produce a change in mating behavior: nearly all the pairs of gulls remained together.

2540 LIFE SCIENCES 1084

coast of Baffin Island provided an escape from this dilemma and also showed how responsive to very slight evolutionary pressures the visual isolating mechanisms are. The experiment was one in which I had hoped to induce mixed matings between Kumlien's gulls and Thayer's gulls by painting the eye-rings of unmated male Kumlien's gulls black to increase the contrast. The males were chosen not by Thayer's gull females, however, but by females of their own species. I concluded that other features, perhaps the wing-tip pattern, were the critical ones in discrimination between the two species.

Then further investigation of Kumlien's gulls in this area where they overlapped with Thayer's gulls revealed a curious phenomenon. Although the majority of Kumlien's gulls in the area had clear yellow irises, there were many individuals (37 percent) with various amounts of iris pigmentation. If the gulls are viewed as two groups, one with iris pigmentation and one without it, a pattern emerges: in each group there is a striking preponderance of matings between individuals that look alike. Outside the overlap area mating is essentially random with respect to the presence or absence of iris pigmentation.

With this information in mind the results of the black-eye-ring experiment can be interpreted very differently. In this area dark-eyed males normally are chosen by dark-eyed females and clear-eyed males by clear-eyed females. Yet in the experiment even though 92 clear-eyed females were available for mating, they chose only four from the group of 41 clear-eyed males that had been painted to have a darker orbital region; 29 of these males were chosen by dark-eyed females. None of the 26 males with dark irises ringed in black were picked by light-eyed females. It can be seen that by increasing the eye-head contrast of unmated male Kumlien's gulls in this area, one can predict the iris coloration of the eventual mate.

Thus there was both observational and experimental evidence that the female Kumlien's gulls of this area were discerning very slight differences in the eye-head contrast of prospective mates. This mating system probably evolved as a result of two primary pressures. The presence of small numbers of herring gulls in the overlap zone probably provided pressure to maintain the delicate balance between clear-eyed and dark-eyed Kumlien's gulls. Secondly, in order to avoid mixed pairings with Thayer's gulls, selection favored individual Kumlien's gulls that perceived slight eye-head contrast differences. The mating system among Kumlien's gulls, in which like mates with like, is a by-product of such selection. This view is supported by the fact that increasing the eye-head contrast of unmated males just outside the overlap zone had no detectable effect, whereas the same alterations within the zone produced major changes in the mating system.

At first, of course, it is difficult to imagine how female gulls manage to choose mates that look like themselves. Presumably they do not actually see themselves. (Mirrors are rare in the Arctic.) The answer may nonetheless be quite simple. It is known that many birds "imprint" on their parents soon after birth, and that they choose mates that look like their parents. Possibly gulls do the same. If eye color in gulls is inherited (as seems likely, although genetic information is lacking), then female gulls choose mates that look like themselves simply because they are looking for mates that look like their parents, and in most cases they themselves look like their parents. This hypothesis suggests that the Kumlien's gull mating system may simply represent an intensification of the normal process, that is, a female chooses a male most like her parents in eye-head contrast and wing tips.

To understand the evolution of these visual signals that function as reproductive isolating mechanisms one can examine the distribution of the large *Larus* gulls throughout the Northern Hemisphere. With the exception of the glaucous gull, all the other *Larus* gulls that overlap with the herring gull and are reproductively isolated from it have the contrast pattern of a dark eye against a white head. The populations that apparently hybridize with herring gulls have dark eye-rings but light irises. As I have indicated, dark eye-rings without dark irises are insufficient as isolating mechanisms against the orange-eye-ring herring gulls. The important point here is that the darkening of the eye region (principally the eye-ring) begins to develop in an isolated population. If the population becomes genetically so different from the one from which it was separated that on coming together again the two populations remain distinct, selection will favor a further increase in the darkening of the eye region, specifically a darkening of the iris. The end result is a sealing off of gene exchange.

The Author

NEAL GRIFFITH SMITH is a zoologist with the Smithsonian Tropical Research Institute, operated in the Panama Canal Zone by the Smithsonian Institution. He went to the Smithsonian in 1964, after receiving a Ph.D. from Cornell University. His undergraduate work was done at St. John's University in New York. Smith writes: "My research interests center on experimental investigations in the field of evolution of specific adaptations in animals, particularly birds. Put less ponderously, I like to manipulate birds to discover how certain 'tricks' they have evolved work and why they have evolved. I am currently investigating some tricks evolved by avian brood parasites such as tropical cowbirds and cuckoos and the evolution of the counteradaptations by their hosts. The strategies involved are complex and fascinating; among them are egg mimicry and behavior polymorphism."

Bibliography

ANIMAL SPECIES AND EVOLUTION. Ernst Mayr. Harvard University Press, 1963.

THE EVOLUTION OF BEHAVIOR IN GULLS. N. Tinbergen in *Scientific American*, Vol. 203, No. 6, pages 118–130; December, 1960.

THE STUDY OF INSTINCT. Nikolaas Tinbergen. Oxford University Press, 1950.

SCIENTIFIC
AMERICAN   November 1967, Vol. 217, No. 5, pp. 62-72   OFFPRINT 1085

LYSOSOMES AND DISEASE

by Anthony Allison

Lysosomes are organelles of the living cell that contain digestive enzymes. They play an important part in normal life processes, and there is evidence that they are also involved in pathological ones.

In the 19th century, when advances in microscopy made possible the study of pathology at the cellular level, it became clear that changes in diseased tissue depend on changes in the growth and function of cells. In recent years the electron microscope and the centrifuge have made for increasingly effective investigations at the subcellular level. The aim of the cell biologist is to identify the enzymes that catalyze cellular reactions and find out in what organelles of the cell each reaction takes place. A specific task for the pathologist is to identify and localize the primary error in function that may lead to a secondary disorder in other organelles and systems; in other words, to learn how disease processes may begin with a specific malfunction in an organelle. In certain disorders, for example, the primary damage is to the energy-producing reactions in the mitochondria; in others it is to the protein-synthesizing apparatus of the ribosomes.

When lysosomes were first recognized as distinct organelles, it was clear that they would be of special interest in subcellular pathology. It was in 1949 that Christian de Duve and his colleagues at the Catholic University of Louvain realized that certain enzymes were segregated within particles in the cytoplasm; it was not until 1955 that they identified the particles visually. Lysosomes are small baglike organelles, usually spherical and about a quarter of a micron (a four-thousandth of a millimeter) in diameter, containing a variety of enzymes that among them break down all the major constituents of living things: proteins, carbohydrates, fats and nucleic acids. (Most of the enzymes function more efficiently under slightly acid conditions, and so they are known collectively as acid hydrolases.) As one might expect, such enzymes are involved in a

wide range of normal and disease processes [see "The Lysosome," by Christian de Duve; SCIENTIFIC AMERICAN Offprint 156]. They digest things that enter the cell; they can also break down part or all of the cell itself or tissues outside the cell. It now appears that in doing so they are implicated in the development and the death of tissues, in diseases such as silicosis and gout, in cell division, in the immune process and thus perhaps in autoimmune disease, and in damage to chromosomes. Their effect on chromosomes suggests that lysosomes may possibly play a central role in the induction of cancer.

Lysosomal enzymes, like other proteins, are synthesized in ribosomes within the folded membranes of the endoplasmic reticulum [see illustration on page 2547]. In the outer part of a series of vesicles known as Golgi apparatus the lysosomal enzymes are packaged into organelles surrounded by single lipoprotein membranes. These "nascent granules" develop into "primary lysosomes" in which the enzymes are stored in an inactive form, ready for use.

Once the identity of lysosomes was established it soon became apparent that a number of rather different looking cytoplasmic bodies familiar to cell biologists are lysosomes. The granules that are characteristic of the cytoplasm of white blood cells are perhaps the most typical lysosomes, but all animal cells studied so far, with the exception of red blood cells, have organelles containing some of the characteristic hydrolytic enzymes and falling within the general definition of lysosomes. Similar organelles have been observed in plants, including fungi and yeasts. Bacteria do not contain lysosomes in the forms recognized in higher organisms, but hydrolytic enzymes with properties like those

of lysosomal enzymes can be released from bacteria by certain procedures. In other words, the presence of lytic, or digestive, enzymes that are normally enclosed in membranes but can be released by appropriate stimuli seems to be a widespread characteristic of living organisms.

The Functions of Lysosomes

The simplest and the most obvious role of a packet of digestive enzymes within a cell is in necrosis and autolysis, the death and self-dissolution of tissues. The membrane of the primary lysosome simply dissolves, liberating the enzymes to consume the cell. This can follow wounding or other damage to tissue or it can take place naturally in the course of development: when the corpus luteum of the ovary degenerates, for instance, or when a tadpole loses its tail. Any time a tissue or an organ is isolated from its supply of oxygen or nutrients under sterile conditions it breaks down rapidly; the large molecules such as proteins, lipids, nucleic acids and carbohydrates are digested, and there is evidence that enzymes from lysosomes play a part in the process. Rudolf Weber of the University of Berne found that the concentration of lysosomal enzymes in the tail of an amphibian increases before metamorphosis, and the concentration of these enzymes—which as catalysts are

LYSOSOMES glow orange-red in the fluorescence photomicrograph on the opposite page, made by M. R. Young and the author. The cultured monkey-kidney cells were vitally stained (stained without being killed) with acridine orange, which concentrates in lysosomes. Lower concentrations of the dye in the nuclei produce the green fluorescence.

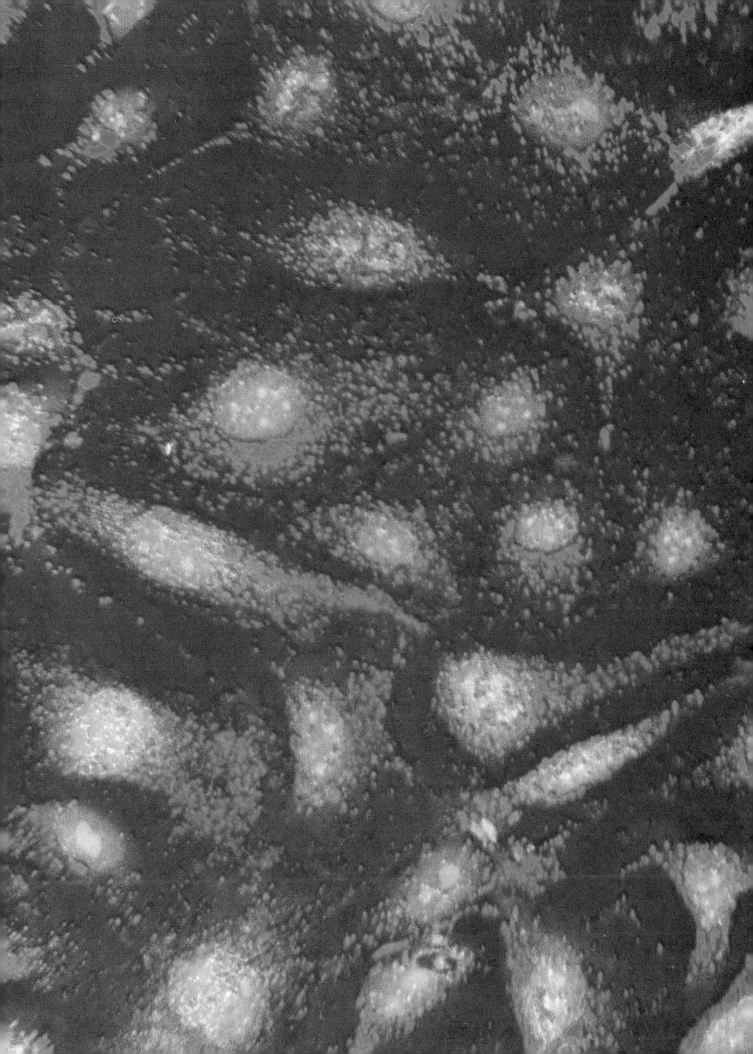

not consumed in the process of tissue digestion—increases as the tail is resorbed.

Soon after lysosomes were first described it was suggested that the release of their enzymes into the cytoplasm or outside the cell might be the primary event that accounted for many other and different types of tissue damage. More detailed scrutiny showed that biochemical changes in other organelles sometimes preceded those that could be demonstrated in lysosomes; the conviction grew that lysosomal release was a secondary effect. It now seems that both situations can prevail. In some cases it is virtually certain that the primary event is an increase in the permeability of the lysosomal membrane, with the consequent release of hydrolytic enzymes. In other cases the lysosomal changes may indeed be secondary to reactions in other systems.

A specialized form of autolysis occurs in the cells of starved protozoa or mammals: organelles such as ribosomes and mitochondria somehow become incorporated in cytoplasmic vacuoles, or membrane-enclosed cavities, that fuse with lysosomes. The contents of the resulting "autophagic vacuoles" are thereupon digested, but the cell may often survive. Autophagy appears to be a mechanism by which, under unfavorable conditions, part of the cell's substance —presumably currently unneeded elements—can be broken down and its constituents utilized to provide energy or material needed to maintain the life of the cell.

The second obvious function of lysosomes is in intracellular digestion. This process, as Élie Metchnikoff recognized more than half a century ago, follows essentially the same course in single-celled animals and in what he termed the "phagocytic" cells of mammals. Metchnikoff postulated then that bacteria ingested by phagocytic cells were digested by "ferments" contained in the granules visible in those cells. In 1960 James G. Hirsch and Zanvil A. Cohn of the Rockefeller Institute showed that the granules of the white blood cells called leukocytes do contain the enzymes characteristic of De Duve's lysosomes—that they are lysosomes. Foreign particles such as bacteria are ingested by the process now known as endocytosis [see illustration on page 2547]. The membrane folds inward to form a pocket, the edges of which fuse to enclose the particles in "phagosomes." The primary lysosomes become attached to the phagosomes and discharge enzymes into them, forming "secondary lysosomes." Here the particles are more

or less completely digested; indigestible material remains segregated from the cytoplasm within "residual bodies," which may remain in the cell for a long time or may fuse with the cell wall and so discharge their contents.

The digestion of bacteria by leukocytes is of course an important form of defense against disease. Not surprisingly, bacteria have developed adaptations that enable them to survive lysosomal attack. Several bacteria elaborate poisons that kill leukocytes before the bacteria are themselves destroyed; others, including those causing tuberculosis, have thick, waxy coats that resist attack by the enzymes. There are rare inherited diseases in which the lysosomal system is ineffective. In one condition, the Chédiak-Higashi syndrome, the lysosomal granules are abnormally large and ingested bacteria may not be killed. Comparable conditions exist in animals: in mink with the Aleutian, or blue, gene and in certain albino cattle [see "The Kinship of Animal and Human Diseases," by Robert W. Leader; SCIENTIFIC AMERICAN, January]. In another disease the enzyme leukocyte oxidase is inactive; bacteria or viruses are phagocytized but are not digested in the usual manner. In much the same way the inactivity of individual lysosomal enzymes may be responsible for certain "storage" diseases in which a metabolic product that should be broken down in the cell is not and therefore accumulates in pathological amounts.

Particles and Lysosomes

Most of the lysosomal disease processes investigated so far involve situations in which the untimely rupture of the lysosomal membrane leads to some form of cell damage. At the National Institute for Medical Research in London my colleagues and I have spent much time studying the effects of small particles of silica, asbestos and other substances on cells. It is relatively easy to follow the fate of these particles by microscopic examination, and so one can be sure that the initial effects are lysosomal; moreover, the particles have simple chemical compositions, so that the number of reactions they can initiate is limited. The results bear not only on diseases such as silicosis and asbestosis but also on gout and perhaps inflammatory processes in general. The fact that exposure to asbestos particles is associated with the development of some kinds of cancer also provides one of the links between lysosomes and carcinogenesis.

Most foreign particles taken into the human body by inhalation—such as the carbon particles that remain for years within phagocytic cells in the lungs of people who breathe smoke-polluted air— are innocuous. However, certain inhaled particles, including several crystalline forms of silica (silicon dioxide), stimulate a severe reaction in the lungs, ending with the deposition of fibrous tissue that can lead to marked impairment of lung function. The reaction proceeds in two stages. First, the inhaled silica particles are ingested by phagocytic cells in the lungs; these cells die, releasing the particles, which are taken up by other cells with the same result. Second, the repeated death of phagocytic cells stimulates a reaction of fibroblasts, cells that synthesize and lay down nodules of collagen fibers.

We analyzed these mechanisms separately by studying each stage independently in tissue culture. When phagocytic cells in culture take up silica particles, they are rapidly killed. When the cells ingest nontoxic particles of the same size and shape, such as diamond dust or titanium dioxide dust, they survive. We can show that lysosomal enzymes are discharged into vacuoles containing either the toxic or the nontoxic particles, but that the nontoxic particles and enzymes remain within the lysosomes for long periods, whereas silica particles and enzymes escape rapidly into the cytoplasm [see bottom illustration on page 2548]. The key factor seems to be the readiness with which silica particles react with lysosomal and other membrane systems. We illustrate this by suspending red blood cells with particles of different materials that have the same size and surface area. The red cells are lysed by particles of many types of crystalline silica—but not by one type that also fails to damage phagocytic cells or to stimulate fibrosis in experimental animals, or by other nontoxic particles.

The reactivity of silica particles appears to be due to the fact that silicic acid is formed on their surface. Silicic acid is unusual in having hydroxyl groups that can form powerful hydrogen bonds with suitable acceptor molecules—including certain groups in the phospholipids and proteins that are characteristic of cellular membranes. Such hydrogen bonding is sufficient to account for the disruption of lysosomal (or red-cell) membranes. Evidence in support of this interpretation comes from the observation, made some years ago by H. W. Schlipköter and his colleagues at the University of Düsseldorf, that the polymer polyvinylpyridine-N-oxide protects cells against the toxic effects of silica. This polymer is taken up into lysosomes along with the silica. The oxygen atoms of polyvinylpyridine-N-oxide readily form hydrogen bonds with the hydroxyl groups of silicic acid, and thereby prevent the latter from bonding with, and thus attacking, lysosomal membranes.

That explains why silica is so toxic to phagocytic cells: the particles are taken up into lysosomes and damage lysosomal membranes through hydrogen-bonding interactions. The second problem remains: How does the death of phagocytic cells stimulate the deposition of collagen fibers? A. G. Heppleston and J. A. Styles of the University of Newcastle upon Tyne have found that when phagocytic cells in culture are incubated with silica particles, they release a factor that, when added to a second culture of fibroblasts, stimulates the synthesis of connective tissue fibers in it; the release of such a factor could perhaps account for the fibrous nodules of silicosis. The chemical structure of this factor is still not known, but certain polymers of the sugar galactose are known to stimulate fibrogenesis, and it is of interest that these polymers are all taken up into lysosomes.

Prolonged exposure to asbestos dust can lead to asbestosis, a disease that, like silicosis, is associated with the deposition of fibrous tissue in the lungs. A second hazard was recognized in 1960 by C. J. Wagner and his colleagues in South Africa: the development of malignant tumors in asbestos workers. Many of these tumors are mesotheliomas, an unusual type arising from the mesothelium, the layer of cells that lines the body cavities. Similar tumors are found in animals that have received injections of asbestos particles in the chest cavity, even when the particles have been purified to remove possibly carcinogenic hydrocarbon contaminants. Electron-microscope and other studies show that asbestos particles are taken up into lysosomes, from which some of them later escape. Wagner has also found that injections of silica particles into the thoracic cavities of rats lead to malignant tumors of the thymus gland.

Another example of pathology due to the effects of particles on cells is provided by gout, in which a metabolic defect causes small crystals of sodium urate to accumulate in the joints and produce pain, swelling and other symptoms. Several investigators have shown that the crystals of sodium urate are ingested by

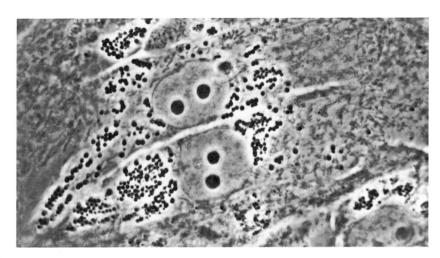

PHASE-CONTRAST microscopy shows lysosomes as small black granules because of their different light-refracting properties. These living kidney cells in culture are enlarged 1,000 diameters. Lysosomes clump near nuclei; mitochondria (*dark gray*) lie in the background.

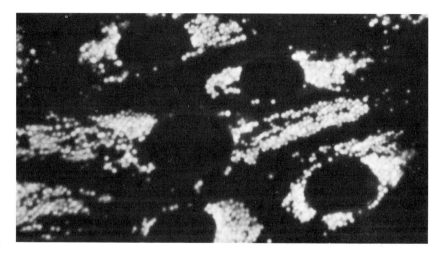

FLUORESCENCE microscopy shows lysosomes as brightly glowing particles. These kidney cells were photographed after being treated with methylcholanthrene, a cancer-inducing substance that is concentrated in lysosomes. Nuclei and mitochondria do not fluoresce.

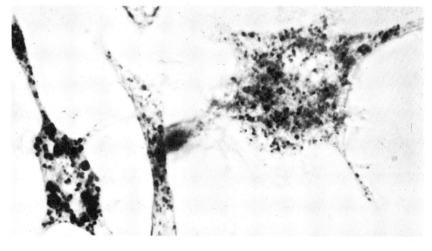

GOMORI METHOD, which depends on the presence in lysosomes of the enzyme acid phosphatase, yields a reaction product, lead sulfide, that stains lysosomes black. The photomicrograph shows mouse macrophages, large phagocytic cells, enlarged 1,600 diameters.

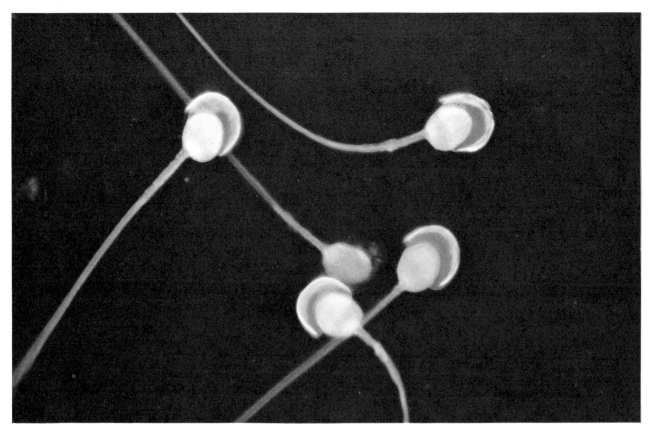

SPERMATOZOA of a guinea pig were vitally stained with acridine orange and then photographed by fluorescence microscopy. The nuclei are green and the acrosomes, small sperm-cell organelles that contain some of the same enzymes as lysosomes, are orange.

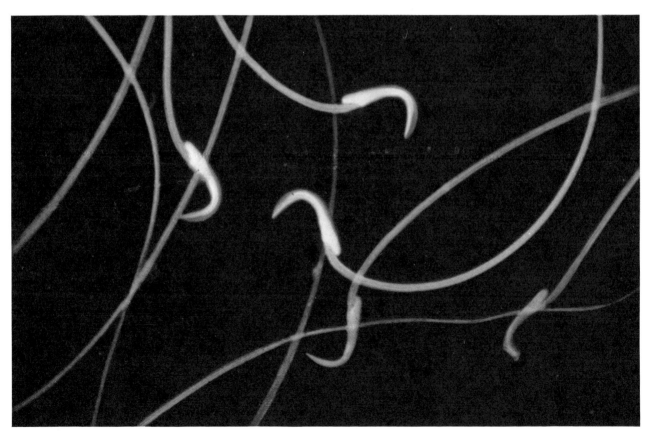

ACROSOMES of long, thin rat spermatozoa are similarly stained. The enzymes in acrosomes break down the protective film around an egg and then apparently disrupt "cortical granules" in the egg, thus initiating a process that leads to cleavage and development.

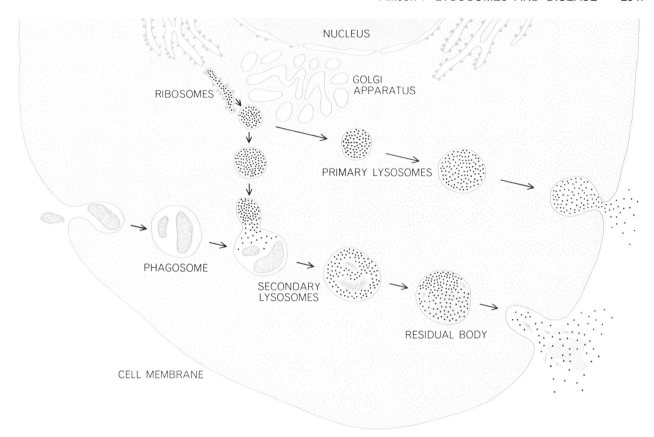

LYSOSOMAL ENZYMES (*black dots*) are synthesized, like other proteins, in ribosomes of the endoplasmic reticulum and are then packaged within the Golgi apparatus into small baglike particles called nascent granules. These develop into primary lysosomes, which sometimes secrete their contents outside the cell (*middle right*) but in most circumstances store the enzymes until they are required for intracellular digestion. Particles taken into the cell by the amoeboid process called endocytosis (*left*) are enclosed in membrane-bounded vesicles called phagosomes. Primary lysosomes fuse with the phagosomes and discharge digestive enzymes into them, forming secondary lysosomes. The particles are thereupon digested. Undigested material remains enclosed in residual bodies, which may persist in the cell for some time or may fuse with the cell membrane and so excrete their contents (*bottom right*).

phagocytic cells and that this leads to the formation of factors producing pain and other manifestations of inflammation. There is evidence that lysosomal enzymes may be released in the process.

The role of lysosomes in inflammation in general is a difficult problem. Inflammation is marked, as ancient observers realized, by four signs: swelling, heat, redness and pain. It is a common reaction in tissues damaged as a result of infection, burns or the presence of a noxious substance, and in many disease processes. The swelling, redness and warmth are caused by enlargement of the blood vessels supplying the inflamed area and the passage of fluid containing protein (which is normally confined within blood vessels) through the walls of blood vessels into the spaces between them. One of the proteins, fibrin (which provides the structural basis of blood clots), is deposited in the tissue spaces. White blood cells also leave the vessels and move into the tissues. Several chemical mediators are liberated from damaged cells or formed in inflamed tissues.

Among these are factors that increase the permeability of blood vessels, including histamine, serotonin and bradykinin, a peptide produced from serum proteins by enzyme action. There are also pain-producing factors and in some cases pyrogens, which bring about a rise in body temperature.

In this complex situation it is difficult to know which reactions are primary and which take place later. It seems clear, however, that lysosomes are implicated in some of the reactions. Basic protein fractions from leukocyte lysosomes, isolated by Hassan Zeya and John K. Spitznagel of the University of North Carolina, were shown by Aaron Janoff of New York University to effect the release of histamine from certain cells. John Herion and Hussein Saba of North Carolina showed that such fractions inhibit a step in blood coagulation. S. Y. Ali and C. H. Lack of the Royal National Orthopaedic Hospital in London found that a factor from lysosomes activates the enzyme plasminogen, which in turn breaks down fibrin. Lysosomal proteases

(protein-digesting enzymes) probably contribute to tissue damage in inflammation and participate in the formation of bradykinin. Spitznagel and his colleagues have shown that lysosomes contain a pyrogen, although it may not be the only factor that is responsible for fever. The role of lysosomes in inflammation is now under intensive study and more precise definition is likely to emerge in the next few years.

Effects of Dyes and Drugs

Among the substances that are selectively concentrated in lysosomes are a number of soluble dyes and drugs. The localization of the dyes makes lysosomes clearly distinguishable in photomicrographs [*see illustration on page 2543*] and localization of the drugs offers some suggestive evidence as to their mode of action. Among the drugs that are concentrated in lysosomes are the antimalarial compounds chloroquine, quinine and quinacrine. When malaria parasites take hemoglobin from red blood cells,

1   2   3

SILICOSIS, a lung disease, begins with the endocytosis of silica (silicon dioxide) particles by phagocytic cells in the lung (1). Once inside secondary lysosomes, the silica particles damage the lysosomal membrane through a hydrogen-bonding reaction between silicic acid and the membrane (2). The lysosomal enzymes kill the cell, releasing silica particles to be taken up by other cells (3). The dying cells elaborate a factor that somehow stimulates specialized cells to synthesize fibrous nodules that can impair lung function.

they ingest it into their lysosomes much as leukocytes ingest bacteria, and so it may be impairment of the lysosomal mechanism that explains the effects of some antimalarial drugs.

The precise effects of some drugs on the lysosomal membrane are known. Cortisone and related corticosteroids, for example, stabilize the membrane, making it less liable to rupture; this may explain the broad anti-inflammatory effects of these drugs. A number of steroids and other compounds weaken the membrane, making it more permeable. One such compound is carbon tetrachloride, which is used against certain worms that parasitize the intestines of mammals. The digestive tract of these worms is a tube of amoeba-like cells, each of

which ingests food and breaks it down in lysosomes. Small doses of carbon tetrachloride damage the worm-cell lysosomes, and the massive release of hydrolytic enzymes kills the worms.

Dame Honor Fell and John Dingle of the Strangeways Research Laboratory in Cambridge have found that high doses of vitamin A added to cultures of chick embryo rudiments bring about the release of large amounts of lysosomal enzymes; the lysosomal protease that is thus released digests the intercellular matrix.

Lysosomes apparently also take part in the action of certain hormones. Soon after thyrotrophic hormone is administered, for instance, there is an increased uptake of colloidal substances into the

lysosomes of thyroid cells; the colloid is digested and then the synthesis and secretion of thyroid hormones proceeds. Again, some effects of insulin on blood sugar are counteracted by glucagon; liver lysosomes show marked changes after glucagon is administered.

Some compounds that become concentrated in lysosomes produce fetal abnormalities. If the dye trypan blue or the detergent Triton WR-1339 is administered to pregnant animals, it is taken up by the lysosomes of cells in the placenta, the membrane that nourishes the fetus, and a high incidence of abortions and congenital defects results. High doses of vitamin A, which have been shown to decrease the stability of the lysosomal membrane, produce similar results.

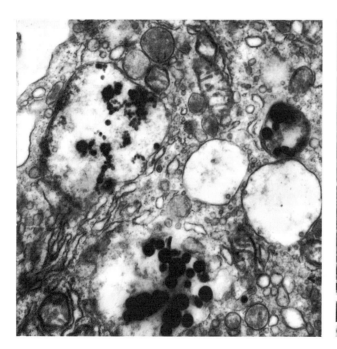 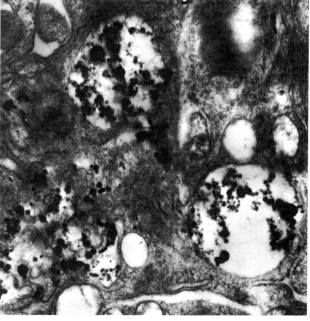

PHAGOCYTIC CELL that has recently taken up silica particles (dark spheres) into phagosomes is enlarged 25,000 diameters in an electron micrograph made by M. J. Birbeck and the author (left). Primary lysosomes are free in the cytoplasm just above the phago-some at upper left or are about to discharge enzymes into the lower phagosome. In a micrograph made 18 hours after uptake of silica (right), the lysosomal membranes have been damaged. Silica is escaping into the cytoplasm, which is now visibly disorganized.

Whether the abnormalities are due simply to a lack of nutrition for the embryo or to some other effect is still not established; moreover, it is likely that some other drugs produce fetal abnormalities by nonlysosomal mechanisms.

There is a category of compounds that tend to make the skin sensitive to the action of light. Some years ago my colleagues Ian Magnus and M. R. Young and I noticed that many of these photosensitizing compounds accumulate in the lysosomes of living cells, and we proceeded to study the phenomenon in experimental animals and in cultured cells. The experiments with cells led to the development of a convenient way to damage various components of the living cell selectively.

We incubate the cells in the dark with substances such as neutral red, anthracene or porphyrin, all of which are concentrated in lysosomes, and then illuminate the cells with light of a wavelength absorbed only by the photosensitizing substance, not the living material. In this way photooxidative damage to lysosomal membranes is achieved, without any immediate damage to the cell membrane or nucleus or to the rest of the cytoplasm. (If enough enzyme is released from lysosomes, the cells die after a short delay.) Other photosensitizing dyes, such as Janus green or acriflavine, selectively sensitize mitochondria or nuclei respectively. Still others, including eosin and rose bengal, do not enter most cells at all. They collect on the surface and in the presence of light cause damage to the cell membrane, killing the cells.

The induced photosensitization procedure has several advantages over other methods of selectively destroying cellular organelles. Photons of radiation produce photochemical reactions only where they are absorbed, and the relatively long wavelengths we use are not absorbed by most cell constituents. Ultraviolet or X rays, on the other hand, attack many biological constituents, and even if one aims a very narrow beam at the nucleus or cytoplasm, it can damage the cell membrane.

Lysosomes and Cell Division

One of the features of cells stained with acridine orange is that mitotic, or dividing, cells have comparatively few lysosomes, and these lysosomes lie on the periphery of the cell instead of near the nucleus, as is usually the case. This suggests that the breakdown of lysosomes may act as a trigger for mitosis in cells that are prepared for it. To establish that the relation is causal, however,

it is necessary to show that agents increasing the permeability of lysosomes induce mitosis in cells that would not otherwise divide, whereas agents stabilizing the lysosomal membrane can prevent this mitogenic effect.

These questions have been studied in experiments with lymphocytes, a species of white cell involved in immune responses. Lymphocytes from the peripheral blood do not ordinarily undergo mitosis in a cell culture, but if they are treated with phytohemagglutinin, an extract of red kidney beans, they are "transformed": they enlarge, synthesize nucleic acids and protein, and then divide. A similar mitogenic effect is produced by a factor extracted from the pokeweed, by a vegetable-oil fraction called phorbol A, by certain bacterial toxins and by antibodies against the lymphocytes themselves. The mode of action of these agents has been studied primarily by Kurt Hirschhorn, Gerald Weissmann and Rochelle Hirschhorn at the New York University School of Medicine and independently by Livio Mallucci and me in London. It has been shown that at least some of the agents—the bacterial toxins, the antilymphocyte antibodies and phorbol A—increase the permeability of lysosomal membranes. The mitogenic effects, moreover, can be prevented by cortisone and chloroquine, which are known to stabilize the membranes. It seems, therefore, that mitosis is ordinarily inhibited by some kind of "repressor" substance and that the lysosomal mechanism is involved in "derepressing" it.

The phenomenon of lymphocyte transformation assumes biological significance with the demonstration that if lymphocytes have been sensitized by prior exposure to a particular foreign substance, or antigen, transformation follows a second exposure to the antigen. The sensitized cells multiply and can participate in a particular type of immune response that is carried out by the lymphocytes themselves instead of by free immunoglobulin molecules. Cell-mediated immune responses of this type are important in the reaction against foreign cells or tumor cells. It is of interest that the transformed cells contain more lysosomes than normal lymphocytes; the lysosomes may participate in the reactions by which the foreign cells are destroyed. These reactions are important because they prevent the transplantation of kidneys and other organs from one human to another, so that it would be useful to understand the underlying mechanisms and control them.

The animal egg is another cell that

does not normally divide unless it is stimulated to do so, as in the case of fertilization. The spermatozoon has a special organelle, the acrosome, that becomes detached as soon as the egg is penetrated. E. F. Hartree and I found the acrosome shows the same uptake of dyes as lysosomes and contains several enzymes characteristic of lysosomes [see illustrations on page 2546]. These include hyaluronidase and proteases, which break down the gelatinous material around the egg. In the outer part of the egg itself there are "cortical granules" with the same enzymes and staining reactions as lysosomes. Apart from their role in facilitating the penetration of the sperm head into the egg, enzymes from the acrosome seem to initiate disruption of the cortical granules, beginning a chain reaction that spreads rapidly around the egg. (The cortical granules in frog eggs can be disrupted physically, as by puncturing the membrane of the egg with a needle, to stimulate parthenogenetic cleavage in the absence of a sperm. In sea urchin eggs parthenogenetic cleavage can be induced by other agents, including antibodies that react with cortical granules.) After disruption of the cortical granules the outer layers of the egg are broken down, a new membrane resistant to enzymatic attack is formed underneath and various synthetic reactions proceed, culminating in cleavage.

There are some hints as to how hydrolytic enzymes might, paradoxically, initiate synthetic reactions. Alberto Monroy and his colleagues at the University of Palermo have found that ribosomes from unfertilized sea urchin eggs synthesize very little protein, whereas ribosomes from fertilized eggs synthesize freely. Brief exposure of the former to proteases stimulates protein synthesis, apparently by removing an inhibitor that is a protein or a peptide. Whether the derepression that follows the release of enzymes is entirely due to the destruction of inhibitors or whether other factors are involved is still not known.

Chromosome Damage

The photosensitization technique has the advantage that by regulating the amount of radiation one can control accurately the amount of lysosomal enzyme release and consequent cell damage. Gillian Paton and I found that if a moderate amount of enzyme is released in human body cells by this technique, chromosome breaks and rearrangements appear. We concluded that strands of deoxyribonucleic acid, the genetic ma-

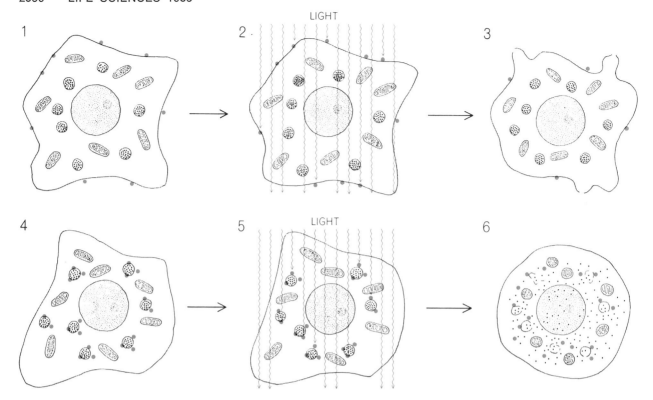

PHOTOSENSITIZATION offers a means of selectively killing parts of the cell. Some dyes, such as eosin, remain associated with the cell membrane (1). Exposure to visible light of a wavelength absorbed by the dye (2) damages the membrane, causing the death of the cell and the extrusion of cytoplasm (3). Other dyes are taken into the cell; neutral red becomes concentrated in the lysosomes (4). Exposure to light of a wavelength absorbed by such a dye (5) damages the lysosomal membrane, releasing enzymes that only later cause the cell to swell up and die (6). Photons of light that miss lysosomes pass through the cell without causing damage.

terial, had been attacked by the enzyme deoxyribonuclease (DNAase) from lysosomes.

Lysosomal DNAase has been studied intensively by Giorgio Bernardi of the Center for Research on Macromolecules in Strasbourg and others. It has some interesting properties. The enzyme molecule has two active sites, and so it can attack both strands of a double helix of DNA at the same time [*see illustration on page 2552*]. While breaks in one strand of a DNA double helix are known to be very efficiently repaired, breaks of the type induced by lysosomal DNAase are not; they are more likely to persist. (Although the lysosomal enzyme works best under acidic conditions, there is considerable activity at physiological hydrogen-ion concentrations, provided that the concentration of salts is not too high.)

We have found that when isolated chromosomes are incubated in the presence of lysosomal DNAase, breaks are produced; enzymes attacking protein or ribonucleic acid remove some material from chromosomes, but they do not produce breaks. Apparently the linear integrity of chromosomes is maintained by the DNA helix. We do not have direct evidence that lysosomal DNAase can enter nuclei and break chromosomes in living cells, although our recent observations strongly suggest that this is the case.

Lysosomes and Cancer

Lysosomal induction of chromosomal changes may help to explain the origin of certain types of cancer. The agents known to produce cancer fall into three major classes: physical, chemical and viral. The important physical agents are ultraviolet and other ionizing radiations. The chemical carcinogens are numerous and structurally diverse, including polybenzenoid hydrocarbons, various organic compounds containing nitrogen, female sex hormones given in excess under certain conditions, metals and asbestos and silica particles. Many cancer-producing viruses are known, and they differ in their properties. Some contain DNA, others RNA; some multiply in the nucleus, others in the cytoplasm; some contain lipid and are formed in close association with cell membranes, others do not contain lipid and are assembled independently of cell membranes.

In this situation a central problem in cancer research is to find some common reaction in cells that can be triggered by such a bewildering variety of cancer-inducing agents. To be sure, there may be several different cancer-producing mechanisms. Perhaps, however, there is only one, and in our present state of knowledge we should follow Occam's principle and try to reduce the number of hypotheses about cancer induction to a minimum.

Chromosomal changes might be the common factor. Most authorities agree that transformation from normal to malignant cells must be regarded as a mutation, in the sense that progeny cells inherit the abnormal growth potential and other properties of parental malignant cells. It can therefore be concluded that some change in genetic material occurs. Mutations are usually produced in one of two ways: point mutations involve substitutions of one or a few base pairs of DNA, and often lead to formation of a single abnormal protein; chromosomal mutations involve the breakage of chromosomes and their reunion with pieces (much larger than those affected by point mutations) deleted, inverted or duplicated. There are reasons for believing that changes of the second type are more relevant to carcinogenesis.

It has been known for many years that the chromosome constitution of malignant cells is often abnormal. Most authorities feel that the marked abnormalities often seen in fully developed or transplanted tumors, involving unusual numbers of chromosomes and other changes, are secondary modifications in cells that are already malignant, that although these abnormalities may contribute to such properties of the tumor cells as lack of differentiation and invasiveness, they cannot be invoked to explain the original transformation. Other evidence suggests, however, that chromosomal anomalies can precede and lead to malignancy.

An example is the chromosomal abnormality associated with a human cancer of blood cells, chronic myeloid leukemia. This is a partial deletion of a small chromosome, No. 21; the deletion can precede detectable malignancy and may in some cases have been produced by ionizing radiation. Other examples come from two rare inherited diseases of children, known after their discoverers as Fanconi's anemia and Bloom's syndrome. Children with either of these conditions have a greatly increased tendency to develop cancer of the blood or epithelial tissues. Several groups of investigators have found that when cells from these patients are cultured, they show numerous chromosome breaks and other aberrations. It has been suggested that this chromosome breakage is due to an abnormal release of lysosomal enzymes, although it is not yet certain whether this is true. It is at least clear that the tendency in human subjects for chromosome breakage to develop spontaneously is correlated with an increased risk of cancer. Agents increasing the incidence of malignancy also increase the rate of production of chromosome rearrangements.

The central question, then, is whether the chromosomal abnormalities are the cause of malignancy or simply a concurrent phenomenon. If the latter is the case, then one would have to postulate that many different processes can independently produce both malignant transformation and chromosomal aberrations. It is simpler to accept as a working hypothesis that chromosomal aberrations are induced by carcinogenic factors and are themselves responsible for the malignant process. The main argument against such a view is that if cells from early cancers are studied, they often seem to be normal. Still, deletions that are quite large in terms of the number of genes involved might be overlooked by current cytological procedures.

This line of reasoning explains why the involvement of lysosomes in chromosome breakage may be important in

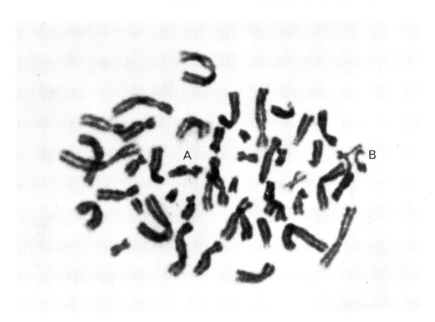

HUMAN CHROMOSOMES, enlarged 2,000 diameters in this photomicrograph from the author's laboratory, were broken by enzyme from lysosomes damaged by photosensitization effects. There is a simple break (A) and another that was followed by a relocation (B).

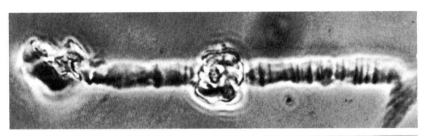
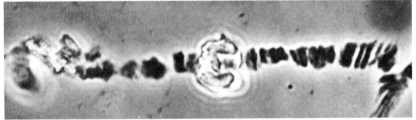
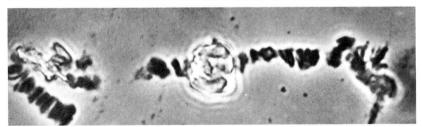
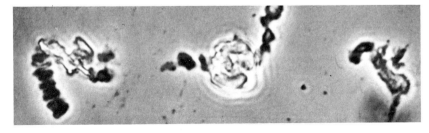

GIANT CHROMOSOME of the midge Chironomus is attacked by highly purified lysosomal deoxyribonuclease (DNAase) in this series of phase-contrast photomicrographs made in collaboration with M. Lezzi and Giorgio Bernardi. The enzyme attacked the DNA at vulnerable sites and within 15 minutes had broken the chromosome in two places.

certain types of tumor induction, and why we have examined the effects of various cancer-producing agents on lysosomes. As I have indicated, asbestos and silica particles certainly affect lysosomes. The polybenzenoid hydrocarbon carcinogens such as benzopyrene and methylcholanthrene are concentrated in the lysosomes of living cells. Small amounts of these hydrocarbons produce cancers in mice if they are followed by treatment with so-called cocarcinogens. The most effective cocarcinogen is phorbol A, which Weissmann has shown to increase the permeability of isolated lysosomes. The only steroids that are effective in cancer production are the female sex hormones and the synthetic analogue diethylstilbestrol, all of which weaken lysosomal membranes. Release of enzymes from lysosomes also follows moderate doses of ultraviolet or ionizing radiation in several experimental systems. Cancer-producing viruses all produce chromosomal aberrations, although these may be incidental to the transformation from normal to malignant cells.

My colleagues and I have found that if selective damage to lysosomes is produced in embryonic cells by the photosensitization technique, a small proportion of these cells develop malignant potential: they produce tumors when they are injected into young animals of the same genetic constitution. Their efficiency in tumor production is low compared with that of certain viruses, but the results suggest that lysosomal damage represents one way by which tumors can be induced. Many years ago the Italian geneticist Adriano A. Buzzatti-Traverso found that if the eggs of the fruit fly *Drosophila* were treated with neutral red and light, some cells of the hatched flies would be normal while others had abnormal chromosomes. The mechanism was not understood at the time, but it may well have been lysosomal enzyme release, and the results are remarkably analogous to the genetic changes expected in tumor induction. I have mentioned the inherited abnormality of lysosomal structure and function in the Chédiak-Higashi syndrome. Many affected children who survive early bouts of recurrent infection develop malignant disease of the spleen and lymph glands. This again suggests a connection between lysosomes and susceptibility to cancer. The possibility is not yet excluded, however, that these children harbor a cancer-producing virus or other organism.

This raises the general problem of whether all tumors are due to viruses, with other cancer-inducing factors such as chemicals or radiation merely facilitating the action of latent viruses. Unfortunately there is no way of testing this hypothesis; it has not been possible to obtain animals free of viruses, even by rearing them in germ-free tanks. Moreover, although the *presence* of a virus can be demonstrated, it is never possible formally to prove the *absence* of viruses. Since only hypotheses that can be tested are productive, the concept that all tumors are virus-induced is simply not useful at present, and one is justified in looking for alternative hypotheses. The suggestion that lysosomes may be involved in certain types of cancer induction provides a working hypothesis that can be tested at various points. It has helped to let in some new light on an old and obscure problem, but it would be premature to conclude that it offers a definitive solution.

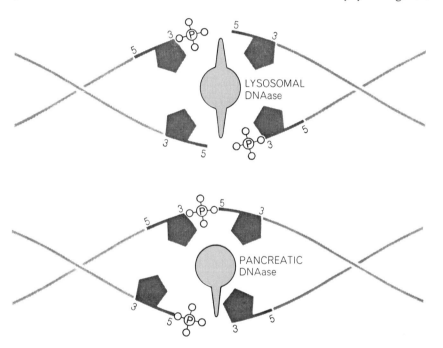

DOUBLE HELIX of DNA is completely cleaved by lysosomal DNAase molecule (*color*), which has two active sites and can cut both strands (*top*). The DNA backbone consists of sugar molecules (*pentagons*) connected by phosphate groups; the lysosomal enzyme leaves a phosphate group attached to the so-called 3-carbon of a sugar molecule. DNAase from the pancreas cuts only one strand (*bottom*), releasing 5-phosphate instead of 3-phosphate.

The Author

ANTHONY ALLISON has spent most of his professional career with the Medical Research Council of Great Britain. He is now head of the division of cell pathology at the council's Clinical Research Centre in London. Allison received a Ph.D. and a medical qualification from the University of Oxford. For several years he worked on human population genetics, traveling extensively in various parts of the world to get material. Recently he has been attempting to apply similar principles to populations of cells in single organisms, particularly in relation to the origin of cancer. Allison describes himself as "interested in far too many things." He adds that he is "a firm believer in the necessity for communicating scientific approaches and results to nonscientists, especially to children in high school."

Bibliography

FUNCTIONS OF LYSOSOMES. Christian de Duve and Robert Wattiaux in *Annual Review of Physiology*, Vol. 28, pages 435–492; 1966.

LYSOSOMES: CIBA FOUNDATION SYMPOSIUM. Edited by A. V. S. de Reuck and Margaret P. Cameron. Little, Brown and Company, 1963.

THE ROLE OF LYSOSOMES IN THE ACTION OF HORMONES AND DRUGS. A. C. Allison in *Advances in Chemotherapy: Vol. III*; Academic Press (in press).

SCIENTIFIC
AMERICAN   November 1967, Vol. 217, No. 5, pp. 112-120   OFFPRINT **1086**

THE FUNGUS GARDENS OF INSECTS

by Suzanne W. T. Batra and Lekh R. Batra

Several kinds of insects live only in association with one kind
of fungus, and vice versa. In some instances the insect actively
cultivates the fungus, browsing on it and controlling its growth.

Anyone with at least a passing interest in biology is aware that fungi, being plants that lack chlorophyll and cannot conduct photosynthesis, live on other organisms or on decaying organic matter. It is less well known that many insects are similarly dependent on fungi. Indeed, there are insects that tend elaborate gardens of a fungus, controlling the growth of the plant according to their own specialized needs.

Some insect species are always found in association with a certain fungus, and some fungi only with a certain insect. Such complete interdependence is called mutualism. The mutualistic partners of insects are not limited to fungi; they include other microbial forms such as bacteria and protozoa. In some cases the insect feeds on its partner or on the partner's own partly digested food. In others the partner lives in the insect's alimentary tract and digests food the insect cannot digest for itself; frequently the partner supplies an essential constituent that is deficient in the insect's diet, such as nitrogen or a vitamin. Some mutualistic partners serve more than one of these functions.

The insects that are mutualistic with microbial organisms are divided into two groups. In one group the fungus, bacterium or protozoon lives inside the insect, either in the alimentary tract or in specialized cells. In the other group a fungus lives in the insect's nest. Here we shall discuss the relations between fungi and insects in the latter group, leading up to those insects that actively cultivate fungus gardens. Such relations have been studied for more than a century, but they offer many new possibilities for investigation. One wants to know more about the physiology of the relations, about how the partners interact at the molecular level. A deeper knowledge of the physiological mecha-

nisms would undoubtedly clarify how the mutualistic partnership came to be established in the course of evolution. It might also have important by-products. For example, much work has been done on the possibility of using fungi that are harmful to insects as a means of selectively controlling insect pests; such work might be advanced by knowing more about the relation between insects and beneficial fungi. As a second example, those insects that control the growth of fungi in gardens may do so by means of antibiotic substances that might well be useful to man.

The first kind of insect-fungus relation we shall take up centers on the tumor-like galls that sometimes appear on the bud, leaf or stem of a plant. These galls develop when certain insects deposit their eggs in the plant and somehow cause it to form an abnormal tissue which then nourishes the larva that emerges from the egg [see "Insects and Plant Galls," by William Hovanitz; SCIENTIFIC AMERICAN, November, 1959]. The galls caused by the mosquito-like midges of the family Itonididae also contain fungi. Growing parasitically on the gall tissue, these fungi usually form a thick layer on the inside of the gall. They appear at an early stage of the gall's development, and a single species of fungus is consistently found in association with larvae of each midge species. Many of these fungi, however, also grow independently of the insect. How fungus and insect come to be together in the gall is not known, but some workers believe the female midge deposits spores of fungus at the time she lays her eggs. Many of the fungus galls are caused by insects that feed by sucking plant sap, and except for a few cases it is unlikely that the fungus acts directly as a source of

food. It may be that the fungus assists the insect indirectly by partly breaking down the gall tissue so that the insect can digest it.

Many plants bear insect-fungus galls but only a few of the fungi have been identified. Some galls we have studied in our laboratory at the University of Kansas are leaf-blister galls on several kinds of goldenrod and aster caused by at least nine species of the midge *Asteromyia* (all of them associated with the fungus *Sclerotium asteris* in the U.S.), and flower-bud galls of the broom (*Cytisus*) caused by the midge *Asphondylia cytisii* (associated with the fungus *Diplodia* in the U.S. and Europe).

In contrast to the casual association between insects and fungi in plant galls, several species of the fungus *Septobasidium* and various scale insects (family Coccidae) that inhabit the fungal tissue coexist in a manner that is clearly mutualistic. *Septobasidium* resembles a thick lichen: it clings tightly to the leaves or branches of trees. The scale insects that inhabit this fungus in some way modify its growth in their vicinity, giving it a different texture or color; as a result some colonies of *Septobasidium* have a mottled surface.

The relation between this fungus and the insects that colonize it has been described by John N. Couch of the University of North Carolina. The insect, which feeds on sap, is attached to the tree by its sucking tube. The mycelium of the fungus—its thick mat of fine threads—shelters the insect from the weather and shields it from birds and parasitic wasps. In turn a few of the insects are penetrated by specialized threads, called haustoria, that extract nourishment from the insects' blood. Scale insects characteristically ingest more sap than they need; the fungus may take advantage of this fact by uti-

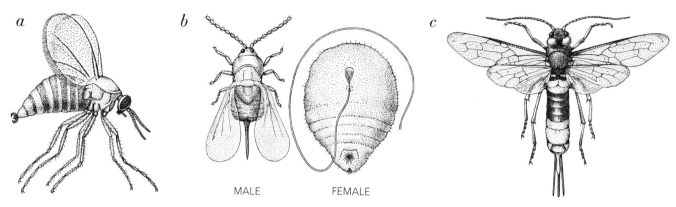

a   b   c

MALE   FEMALE

INSECT GARDENERS comprise unrelated species. Depicted here are representatives of six groups of insects that nest with a fungus. At left is a gall midge of the genus *Lasioptera*; its larvae probably feed on plant material that a fungus has partly digested. The scale insect *Aspidiotus osborni* (*b*) lives on trees under a protective canopy of fungus. The wood wasp *Sirex gigas* (*c*) deposits eggs cov-

lizing such nutrients while they are still circulating in the insect's body.

Septobasidium is distributed by scale insects as well as nourished by them; it does not live independently in nature. When the insects are young, some of them crawl on the surface of the fungus and become covered with spores. A few of the contaminated insects migrate to new areas on the tree, where they insert their sucking tube. Shielded by new mycelium, they survive. They are

also invaded, however, by haustoria from the developing spores, and as a result they do not attain maturity and reproduce. The new mutualistic colony is nonetheless able to continue because uncontaminated insects now find shelter in the mycelium. In some species of *Septobasidium* the entire process has apparently been made more efficient by the development of hollow "insect houses" that attract and hold the migrating scale insects. Thus the fungus

furnishes a shelter for the insects and the insects provide both a food supply and a means of dispersal for the fungus. Some insects are sacrificed for the benefit of the colóny as a whole; both fungus and insect benefit at the expense of the tree.

A different kind of mutualism has been observed involving on the one hand the wood wasps of the genera *Sirex*, *Tremex* and *Urocerus* and on the other the fungi *Stereum* and *Daedalea*.

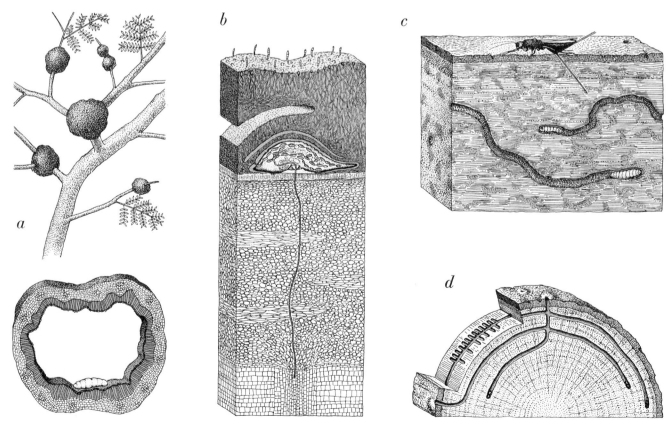

a   b   c

d

INSECT NESTS contain a fungus (*color*). The nests were made by the insects at the top of these two pages, or by an insect of the same group. The plant gall that encloses larvae of the midge is lined with a fungus parasitic on the plant (*a*). *Septobasidium* fungus sends threads into some of the insects it shelters and extracts nourishment from their blood (*b*). The burrows of wood

d

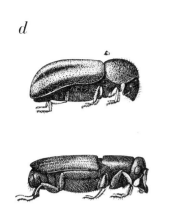

e

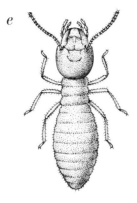

f

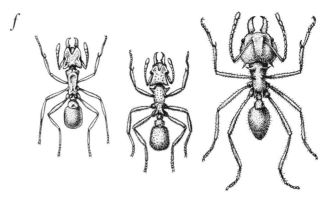

ered with fungus in moist wood. Ambrosia beetles of the genera *Trypodendron* (*d, top*) and *Crossotarsus* (*d, bottom*) carry spores from which grow their fungus. The termite *Odontotermes gurda-*

spurensis (*e*) fertilizes its fungus garden. Ants of the genera *Cyphomyrmex, Trachymyrmex* and *Atta* (*f, left to right*) actively cultivate fungus gardens. The insects are not drawn to the same scale.

These common fungi sometimes build a semicircular "bracket" out from a tree. The adult female wood wasp deposits her eggs in moist wood by means of a long, slender ovipositor, and at the base of this organ are tiny pouches that contain fungus cells called oidia. When the egg is deposited, oidia cling to it. Then the mycelium of the fungus grows into the wood, and when the wasp larva emerges from the egg it follows the path of the mycelium. The fungus partly di-

gests the wood before it is eaten by the larva. In the laboratory wood wasp larvae have been reared on a diet consisting only of *Stereum,* but whether the fungus is essential to the insect's nutrition is not known. Both in nature and in the laboratory the fungus grows well without the help of any insect.

The wood wasp nonetheless acts as an agent for the dissemination of *Stereum.* A larva that later develops into a female has organs that ensure the pres-

ervation of the fungus. These organs are tiny pits hidden in folds between the wasp's first and second abdominal segments; in them bits of fungus are trapped in a waxy material. Here an inoculum of fungus remains viable, whereas fungus in wood is inactivated when the wood eventually dries out. When the larva metamorphoses into a pupa, the organs that hold the fungus are discarded. Then, when the adult female wasp emerges from the pupal skin, tiny flakes

e

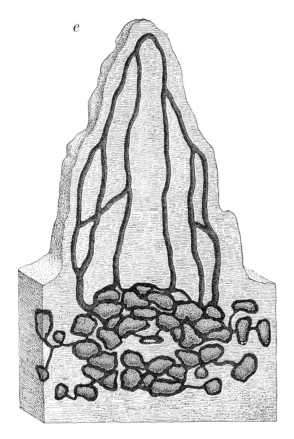

f

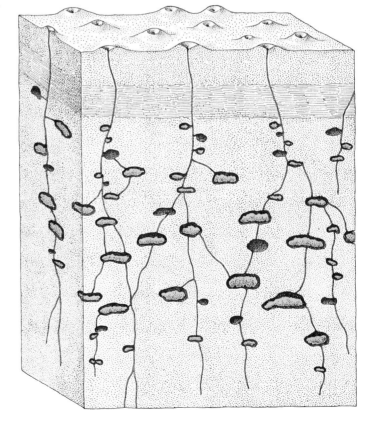

wasp larvae are made in wood infested with *Stereum* fungus, propagated by the insect (*c*). Fungus carried by an ambrosia beetle grows inside the insect's tunnels in timber (*d*). Mounds of earth

in which the termite nests contain fecal material permeated with fungus (*e*). *Atta texana* cultivates a garden of fungus in a huge underground nest (*f*). Here also the nests are not drawn to scale.

FUNGUS-GARDENING ANT *Mycetosoritis hartmani* is shown in its underground nest feeding on a "kohlrabi body." These bodies are made up of bromatia, particles that consist of the swollen tips of the filaments of the fungus and that form only in the presence of the insect. Surrounding the ant are the filaments themselves. Photograph was provided by John C. Moser of the U.S. Forest Service.

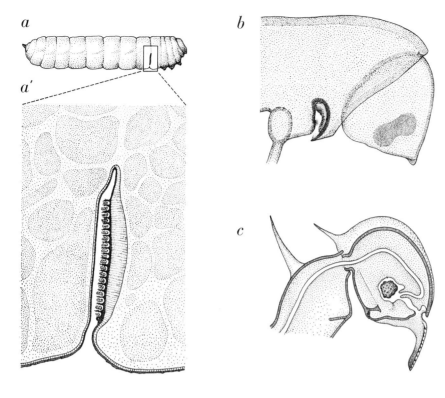

FUNGUS CONTAINERS of three insects are depicted with the fungus indicated in color. Larvae of wood wasps that develop into females carry an inoculum of fungus behind the thorax (*a*). A section through the first two abdominal segments of the larva shows one of the fungus-filled organs (*a'*). The ambrosia beetle, shown here in longitudinal section, conveys spores of fungus in pockets located at the base of its front legs (*b*). A longitudinal section of the head of a fungus-growing ant reveals the pouch in which it transports fungus (*c*).

borings. The males of some species assist the females with tunnel excavation. Inside the ambrosia beetle's tunnel system one finds, depending on the species, either separate niches, each enclosing a single glistening larva or a pearl-like egg, or several larvae sharing an enlargement of the tunnel. In many species the adult insect, on emerging from the pupal skin and proceeding to feed on the mass of ambrosia fungus lining the tunnel, rocks back and forth in a curious manner. What this does is force fungus spores into the mycangia before the insect flies away to found its own nest.

Each species of ambrosia beetle is normally associated with only one species of ambrosia fungus. Ambrosia fungi are pleomorphic: they can readily change, when their growth medium is changed, from a fluffy moldlike form to a dense yeastlike form. In the mycangia and the tunnels of the ambrosia beetles the yeastlike form prevails. Recently we have discovered that ambrosia beetles can also change the form of other fungi from the moldlike form to the yeastlike one. This is a significant phenomenon, and we shall be returning to it.

The most conspicuous and most destructive of the fungus-growing insects are termites. The termites that cultivate fungi are native to the Tropics of Africa and Asia. In West Africa it is estimated that the yearly cost of repairing the damage done by these insects to buildings is equal to 10 percent of the buildings' value. In addition to wood the insects eat growing and harvested crops and objects made of rubber, leather and paper; they destroy documents, works of art, clothing and even underground cables. The enormous mounds that some termite species build for nests interfere with farming and hinder road construction. If they are incompletely destroyed, the insects rebuild them.

Many species in the genera *Macrotermes* and *Odontotermes* make their nests in spectacular steeple-like mounds of hardened earth, which in Africa reach a height of as much as 30 feet. Other species, in the genus *Microtermes*, are completely subterranean, and if it were not for their mating flights and the damage they do, they would be quite inconspicuous. Each nest contains a white, sausage-like queen and a king, usually enclosed together in a protective cell of earth; the much smaller workers, soldiers and young termites (nymphs) of various ages and both sexes are found through-

of fungus-impregnated wax that have fallen from the organs become lodged in the moist pouch at the base of the ovipositor. The flakes now give rise to a fungal mycelium, which develops the oidia that will coat the eggs. The transfer of the fungus from generation to generation of wood wasps is thereby assured.

Ambrosia beetles also carry fungi within their bodies. The numerous species of these wood-boring insects (families Scolytidae, Platypodidae and Lymexylonidae) cannot survive without ambrosia fungi (several genera of Ascomycetes and "imperfect" fungi). In their external skeleton are small pockets called mycangia (literally "fungus containers"); these pockets always contain a supply of viable fungus spores. When an ambrosia beetle tunnels into wood, spores are dislodged from the mycangia, and soon a mass of velvety fungus lines the interior of the tunnel. On this "ambrosia," which concentrates in its cells nutrients that have been extracted from the wood, the beetles feed.

Ambrosia beetles have been the subject of considerable research because they destroy much timber all over the

world. Their tunnels, sometimes called "shot holes," extend deep into the sapwood of trees and are surrounded by streaks of a stain manufactured by the enzymatic action of the fungus. The beetles most often attack hardwoods, preferring trees that have been weakened by drought, disease or fire, or fallen timber that is moist and filled with sap. Attracted by the odor of fermenting sap, the beetles fly upwind, usually at dusk; it is easy to collect them in the evening around a newly felled log. They are similarly attracted by the yeasty smell of beer and beer drinkers, and it is also convenient to collect them at a beer picnic! Some kinds bore into beer and wine kegs, which is why in Europe they are called "beer beetles."

The tunnels of ambrosia beetles can be distinguished from those of other wood-boring insects by a black or brown discoloration of the wood around the neat circular tunnel opening. There is, moreover, no wood dust or fecal matter inside the tunnel. When the beetles are excavating, fine wood particles, sometimes mixed with the insect's brown feces, accumulate outside the tunnel entrance. The beetle does not as a rule eat wood, and it rids its nest of wood

out the nest. At certain seasons winged male and female reproductives (future kings and queens) are also present. In each nest are one or several fungus gardens, the number and shape depending on the species of termite. Material collected by the workers is chewed and swallowed, and the partly digested fecal material is deposited on a fungus garden when the workers return to the nest. In the nests of some species there is a single large mass of fungus garden; in one of the nests of *Odontotermes obesus* that we studied in India the mass was two feet in diameter and weighed 60 pounds. In other nests many fungus gardens one or two inches long are scattered along burrows throughout the nest. Each fungus garden is enclosed in a close-fitting cavity lined with a mixture of saliva and dirt. The chambers of some species are ventilated by an elaborate system of vertical conduits ex-

tending to the surface of the nest [see "Air-conditioned Termite Nests," by Martin Lüscher; SCIENTIFIC AMERICAN, July, 1961].

The fungus gardens look like a gray or brownish sponge or may be convoluted like a walnut meat. They are moist but usually firm and brittle, and are permeated by threadlike mycelium. Scattered over the surface and inside the pores of the gardens are numerous minute, glistening, pearly white spherules composed of masses of rounded fungus cells.

The role of the fungi in the nutrition of the fungus-growing termites is not clear. It is well known that many common Temperate Zone termites cannot digest the cellulose in the wood they eat but rely on certain cellulose-digesting protozoa that live in their intestines to do it for them. No protozoa live in the gut of the fungus-growing

termites; therefore it seems likely that the fungi growing in the gardens break down cellulose and may provide vitamins also. In fact, these termites soon die on a diet restricted to wood. The termites continually eat away the fungus gardens as they add fresh fecal material. There is thus a communal interchange of food, the fecal material being partly broken down by fungi in the garden, then eaten again and redeposited in the garden for further digestion by the fungi.

The white spherules are frequently picked up by the workers and moved to other parts of the fungus garden or are sometimes eaten. The king, queen, young nymphs and soldiers are apparently fed saliva by the workers; no trace of fungus or plant material can be found in their digestive tract. Some winged reproductive termites contain material from the fungus gardens, with which

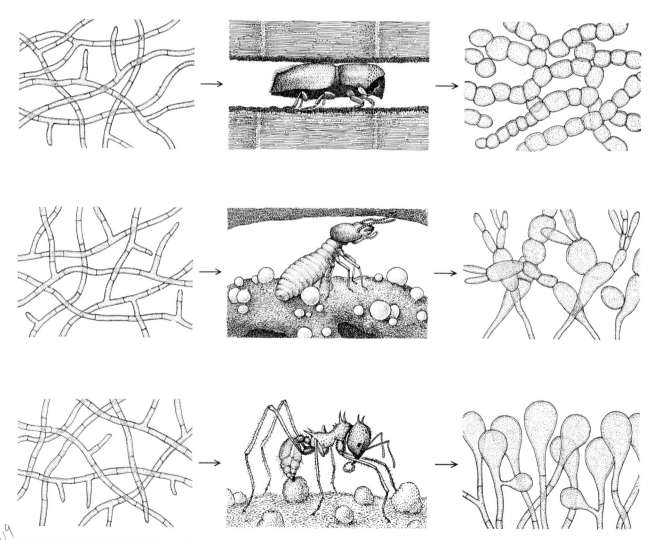

FUNGUS IS TRANSFORMED in being cultivated by insects. Under ordinary conditions in the laboratory fungus associated with the ambrosia beetle is threadlike (*top left*). In the beetle's tunnel, where it is continually grazed, the fungus is denser and more like a yeast (*top right*). The fungus-growing termites also modify their fungi (*middle*), which they fertilize, lick and enclose in mud. Under these conditions white spherules appear. The fungus of a fungus-growing ant (*bottom*) looks much like the others when grown in a culture; in association with the ant, which licks, manipulates and defecates on its garden, the fungus has thickened tips.

they may begin a new garden when they start a new nest.

Several genera of unrelated fungi grow in the fungus gardens of termites; the most abundant are species of the mushroom *Termitomyces*. The various species of *Termitomyces* are found only in nests of fungus-growing termites. Somehow the fruiting ·structure—the mushroom—of the fungi is not allowed to grow from fungus gardens while they are being actively tended by termites. If the termites are removed or die, however, some of the spherules grow into the mushrooms of *Termitomyces*.

In Africa and southern India termites in many nests simultaneously remove the outer layers of their fungus gardens and spread crumbs of them in a thin layer on the ground during the rainy season. Soon *Termitomyces* mushrooms appear, and after their spores have been disseminated by the wind the termites come to the surface to collect fungus that may have grown from a mixture of spores from many nests. It is believed that in this way the termites provide for the cross-fertilization of their fungi, as man does for corn and other crops.

Closely resembling the fungus-growing termites in behavior but only distantly related to them are the fungus-growing ants (tribe Attini). Here we have an example of convergent evolution, in which two nearly unrelated animals and their fungi occupy a very similar ecological niche. These ants are found only in the Western Hemisphere, and most of them are tropical. Some species are found in the deep South and the Southwest of the U.S.; a small species (*Trachymyrmex septentrionalis*) is found in sandy areas near the Atlantic coast as far north as Long Island.

Atta texana, which lives in eastern Texas and southern Louisiana, does considerable damage to citrus groves, gardens and plantations of young pine trees by cutting leaves from them for its fungus gardens. This ant is known locally as the "town ant," and it inhabits an ant metropolis that is sometimes 50 feet across and 20 feet deep. (One student of these ants, John C. Moser of the U.S. Forest Service, opens their nests with a bulldozer!) Other species of fungus-growing ants build smaller nests; some are so small that they are extremely difficult to find.

The fungus-growing ants probably represent the most advanced stage in the evolution of fungus gardening because they feed only on the fungus, and they actively cultivate it. The workers, depending on the species, collect cater-pillar excrement, fallen flower anthers and other soft plant debris as well as leaves cut from trees. Rather than eating the material the ant cuts it into pieces and adds them to a fungus garden in the nest. The ants' gardens look somewhat like those of the termites: they are gray, flocculent masses of finely divided moist plant material loosely held together by threads of mycelium. In the underground chambers of the nest the fungus garden often is suspended from the roots of plants. Scattered over the surface of the older parts of the garden are white specks just visible to the unaided eye. These specks, called kohlrabi bodies, are clusters of bromatia, the swollen tips of the filaments forming the mycelium. The kohlrabi bodies look much like the white spherules found in the gardens of termites. When the ants are not feeding on the bromatia, they lick them.

The flying, nest-founding ant queen carries a small pellet of fungus in a pouch below her mouthparts, much as the ambrosia beetle carries fungus in its mycangia. In starting a new nest the young queen grows a small fungus garden on her excreta; with this she feeds the first worker larvae. When the workers are mature, they leave the nest to gather the material with which the garden is enlarged. They feed bits of bromatia to the larvae that nestle in the mycelium of the gardens.

As long as the ants actively tend the garden the fungus does not develop fruiting structures, but mushrooms of four genera have been found growing from abandoned nests of some species or have been cultured from fungus gardens in the laboratory. We do not know how the ants control the growth of their fungi so that they produce bromatia and nothing else; perhaps it is by constantly "pruning" away excess growth of the mycelium. It is also possible that the excreta and the saliva of the ants, which are deposited on the fungus garden, contain some substance that influences the growth of the mutualistic fungus and inhibits the development of the many spores that accidentally enter the nest.

With the fungus-growing ants our brief survey of the mutualistic associations between insects and fungi ends. It can be seen that there are two distinct kinds of relation. In the gardens of wood wasps, ambrosia beetles, termites, ants and probably those of midge ·galls the fungus extracts nourishment from a substrate and the insect feeds either on the fungus, the substrate predigested by the fungus or both. The fun-

BEETLE TUNNELS are marked by the dark fungus that lines the interior surface. The photograph above shows a cross section of excavations made in wood by ambrosia beetles. Dark circles are niches at right angles to the tunnel that hold larvae. The beetles have penetrated through the bark into the sapwood; in this way they destroy felled timber.

gus is prevented from producing sexual fruiting structures but is supplied by an insect partner with a suitable habitat and a means of dispersal. In the colonies of the fungus *Septobasidium* and scale insects the situation is reversed: the insect feeds on the substrate and nourishes the fungus, and the fungus provides shelter for its castrated insect partner.

Insects are unique among animals in having developed mutualistic relations with fungi. This may have come about because so many insects and fungi share the same tiny habitats. Moreover, most insects are equipped to carry living spores of fungi, either in their gut, in folds between their joints that contain waxy secretions or among their bristles.

The fungus-gardening insects convey into their nest the spores of many fungi other than the one on which they depend. If the insects are removed, the alien fungi will grow and soon overrun the nest; they do not grow in the nest, however, when the insects live there. Apparently the fungus-gardening insects either secrete or excrete antibiotic substances that prevent the growth of alien fungi. The substances may also act to transform the mutualistic fungi, causing either ambrosia, spherules or bromatia to appear rather than sexual fruiting structures.

In the case of the termite, which licks the spherules of its fungus and encloses the fungus garden in saliva-moistened mud, the saliva may contain the substances in question. We have tested the effect of adding saliva taken from termites to a culture of their fungus, which under ordinary growing conditions in the laboratory does not produce spherules. After saliva was added to the culture spherules grew; the saliva also

FUNGUS-GARDENING TERMITE of the genus *Odontotermes* is photographed crawling on the surface of its garden. The round white objects are spherules of the fungus, which arise only in gardens that are tended and fertilized by termites. The insect in the picture is a soldier defending the nest; it produces a copious supply of pungent saliva for this purpose.

inhibited the growth of alien fungi. We have performed other tests on the excreta that ants deposit on their nest gardens. Although we have found that the excreta inhibit the growth of certain bacteria, much experimental work remains to be done. The saliva of the ant, which also licks its fungus, may help to form bromatia.

Spores of the ambrosia beetle's fungus remain in the yeastlike form while they are carried by the insect, and it is possible that the waxy secretion might also affect the form of fungi in the tunnel, where the beetle and the fungus are in close contact. On the other hand, it has been shown that the mutualistic fungi of some species of the beetle can be modified to the yeastlike form by certain physical conditions and in the absence of the beetle. These physical conditions duplicate conditions in the beetle's tunnel, where the feeding insect steadily mows the tips of the fungus as they grow.

In nature the presence of a living insect partner is necessary to maintain ambrosia, spherules or bromatia, but these forms can be produced in the laboratory on special mediums in the absence of the insects. When the mutualistic fungi are grown on ordinary carbohydrate-rich laboratory mediums, fluffy mycelium and sometimes sexual fruiting bodies appear. If the same fungi are grown on acid mediums that are rich in amino acids, and are exposed to more than .5 percent carbon dioxide, then bromatia, spherules or ambrosia are formed. These cultural conditions apparently resemble conditions in the nests of the insects. The fungi of some ambrosia beetles also become ambrosial if they are repeatedly scraped or are grown at low temperatures. Clearly the problem of how insects control the growth of a fungus partner remains an intriguing one.

The Authors

SUZANNE W. T. BATRA and LEKH R. BATRA are biologists at the University of Kansas; Mrs. Batra is a research associate in the department of entomology and her husband is associate professor of botany. Mrs. Batra was graduated from Swarthmore College in 1960 and obtained a Ph.D. from the University of Kansas in 1964. At Kansas she has been primarily studying the behavior and ecology of various social and solitary wild bees. Her husband, a native of India, received a Ph.D. from Cornell University in 1958 and then taught for two years at Swarthmore. At Kansas he teaches mycology and plant pathology. He is currently working at the National Fungus Collection of the U.S. Department of Agriculture in Beltsville, Md.

Bibliography

AMBROSIA FUNGI: EXTENT OF SPECIFICITY TO AMBROSIA BEETLES. Lekh R. Batra in *Science*, Vol. 153, No. 3732, pages 193–195; July 8, 1966.
FUNGUS-GROWING ANTS. Neal A. Weber in *Science*, Vol. 153, No. 3736, pages 587–604; August 5, 1966.
SYMBIOSIS AND SIRICID WOODWASPS. E. A. Parkin in *Annals of Applied Biology*, Vol. 29, No. 4, pages 268–274; August, 1942.
TERMITES: THEIR RECOGNITION AND CONTROL. W. Victor Harris. Longmans, Green and Co. Ltd., 1961.
TRAILS OF THE LEAFCUTTERS. John C. Moser in *Natural History*, Vol. 76, No. 1, pages 32–35; January, 1967.

SCIENTIFIC
AMERICAN December 1967, Vol. 217, No. 6, pp. 19-27

OFFPRINT 1087

INFECTIOUS DRUG RESISTANCE

by Tsutomu Watanabe

Bacteria can suddenly become resistant to several antibacterial drugs. The resistance is transferred from one strain to another by an "episome" that carries the genes for multiple resistance.

The advent of sulfonamide drugs and antibiotics brought with it the promise that bacterial disease might be brought under control, but that promise has not been fulfilled. Although many infections respond dramatically to chemotherapy, tuberculosis, dysentery and typhoid fever continue to be endemic in many parts of the world; cholera and plague erupt periodically; staphylococcal infections persist in the most advanced medical centers. One major reason is that the disease organisms have developed resistance to the drugs.

Until recently it was assumed that the appearance of drug-resistant bacteria was the result of a predictable process: the spontaneous mutation of a bacterium to drug resistance and the selective multiplication of the resistant strain in the presence of the drug. In actuality a more

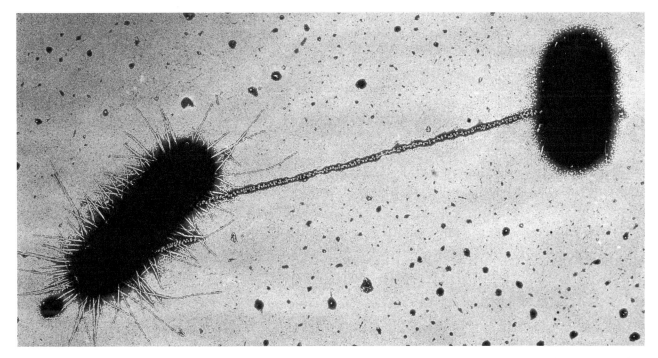

R FACTOR, the particle that imparts infectious drug resistance, is transferred from one bacterial cell to another by conjugation. The various forms of conjugation are thought to be effected by way of thin tubules called pili. In this electron micrograph made by Charles C. Brinton, Jr., and Judith Carnahan of the University of Pittsburgh a male *Escherichia coli* cell (*left*) is connected to a female bacterium of the same species by an *F* pilus, which shows as a thin white line in the negatively stained preparation. Numerous spherical bacterial viruses, or phages, adhere to the *F* pilus. The cells have been magnified about 20,000 diameters.

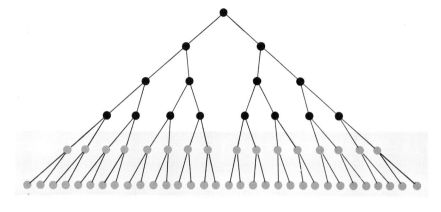

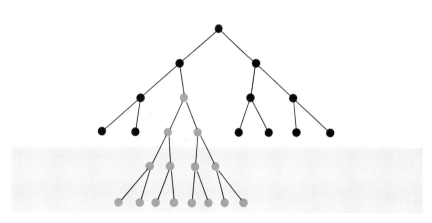

DRUG RESISTANCE involves a change in the genetic material of a bacterial cell. The change from drug-sensitive cell (*black*) to drug-resistant cell (*color*) is not induced by the presence of the drug (*light color*), as was once thought (*top*). It is the result of a spontaneous mutation that gives rise to cells that survive in the drug environment (*bottom*).

In 1955 a Japanese woman recently returned from Hong Kong came down with a stubborn case of dysentery. When the causative agent was isolated, it turned out to be a typical dysentery bacillus of the genus *Shigella*. This shigella was unusual, however. It was resistant to four drugs: sulfanilamide and the antibiotics streptomycin, chloramphenicol and tetracycline. In the next few years the incidence of multiply drug-resistant shigellae in Japan increased, and there were a number of epidemics of intractable dysentery.

The familiar process of mutation and selection could not explain either this rapid increase in multiple resistance or a number of other findings concerning the dysentery epidemics. For one thing, during a single outbreak of the disease resistant shigellae were isolated from some patients and sensitive shigellae of exactly the same type from other patients. Even the same patient might yield both sensitive and resistant bacteria of the same type. Moreover, the administration of a single drug, say chloramphenicol, to patients harboring a sensitive organism could cause them to excrete bacteria that were resistant to all four drugs. Then it was found that many of the patients who harbored drug-resistant shigellae also harbored strains of the relatively harmless colon bacillus *Escherichia coli* that were resistant to the four drugs. It was impossible, on the other hand, to obtain multiple resistance in the laboratory by exposing sensitive shigellae or *E. coli* to any single drug; multiply resistant mutants could be obtained only after serial' selections with each drug in turn, and these mutants, unlike the ones taken from sick patients, multiplied very slowly.

Taken together, these characteristics of the resistant shigellae suggested to Tomoichiro Akiba of Tokyo University in 1959 that resistance to the four drugs might be transferred from multiply resistant *E. coli* to sensitive shigellae within a patient's digestive tract. Akiba's group and a group headed by Kunitaro Ochiai of the Nagoya City Higashi Hospital thereupon confirmed the possibility by transferring resistance from resistant *E. coli* to sensitive shigellae—and from resistant shigellae to sensitive *E. coli*—in liquid cultures. Other investigators demonstrated the same kind of transfer in laboratory animals and eventually in human volunteers. Clearly a new kind of transferable drug resistance had been discovered. What, then, was the mechanism of transfer? There were three known mechanisms of genetic transmis-

ominous phenomenon is at work. It is called infectious drug resistance, and it is a process whereby the genetic determinants of resistance to a number of drugs are transferred together and at one stroke from a resistant bacterial strain to a bacterial strain, of the same species or a different species, that was previously drug-sensitive, or susceptible to the drug's effect. Infectious drug resistance constitutes a serious threat to public health. Since its discovery in Japan in 1959 it has been detected in many countries. It affects a number of bacteria, including organisms responsible for dysentery, urinary infections, typhoid fever, cholera and plague, and each year it is found to confer resistance to more antibacterial agents. (What may be a related form of transmissible drug resistance has been discovered in staphylococci and may be responsible for "hospital staph" infections.) Quite aside from its importance to medicine, the study of infectious drug resistance is making significant contributions to microbial genetics by illuminating the complex and little understood relations among viruses, genes and

the particles called episomes that lie somewhere between them.

If an antibacterial drug is added to a liquid culture of bacteria that are sensitive to the drug, after a while all the cells in the culture are found to be resistant to the drug. Once it was thought that the drug must somehow have induced the resistance. What has actually happened, of course, is that a few cells in the original culture were already resistant; these cells survive and their daughter cells multiply when the sensitive majority of bacteria succumb to the drug [*see illustration above*]. The resistance was not induced by the drug but was the result of a spontaneous mutation. Bacteria, like higher organisms, have chromosomes incorporating the genetic material, 'and from time to time a gene—perhaps one controlling drug resistance—undergoes a mutation. The mutation of a drug-sensitivity gene occurs only once in 10 million to a billion cell divisions, and when it occurs it alters a cell's sensitivity to one particular drug or perhaps two related drugs.

sion in bacteria that had to be considered as possibilities.

One was transformation, which involves "naked" deoxyribonucleic acid (DNA), the stuff of genes. DNA can be extracted from a donor strain of bacteria and added to a culture of a recipient strain; some of the extracted genes may "recombine," or replace homologous

genes on chromosomes of the recipient bacteria, thus transferring a mutation from the donor to the recipient [see top illustration below]. In this way, for example, streptomycin-sensitive bacteria can become streptomycin-resistant.

Transformation occurs in a number of different bacteria, and it can occur spontaneously as well as experimentally. Because only small fragments of DNA are

taken up by bacteria in transformation, however, it is seldom that more than two different drug-resistance genes are transferred together. It requires optimal laboratory conditions, moreover, for transformation to occur at a significant frequency, and such conditions are not likely to prevail in nature.

Another mechanism of gene transmission is transduction, in which genes are

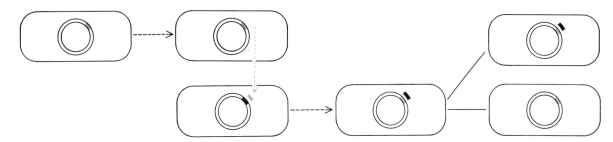

TRANSFORMATION is a form of genetic transmission in which deoxyribonucleic acid (DNA) extracted or excreted from a donor cell (top) enters a recipient cell (bottom) and is incorporated into its chromosome. In this way a mutated gene (color) controlling resistance to a drug may be transferred to a drug-sensitive cell, replacing a homologous gene, which is unable to replicate and dies out.

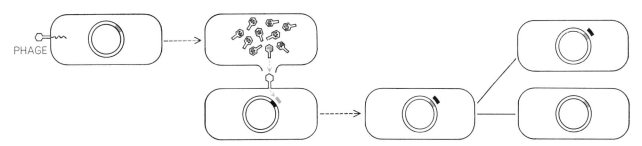

TRANSDUCTION is effected by phage, or bacterial virus. Phage DNA enters a cell (left) and directs the synthesis of new phage, killing the cell (second from left). A bit of bacterial DNA (color), perhaps a mutated gene that causes drug resistance, may be incorporated inside a newly formed phage, be carried to a sensitive cell (bottom) and "recombine," or replace a gene on the chromosome.

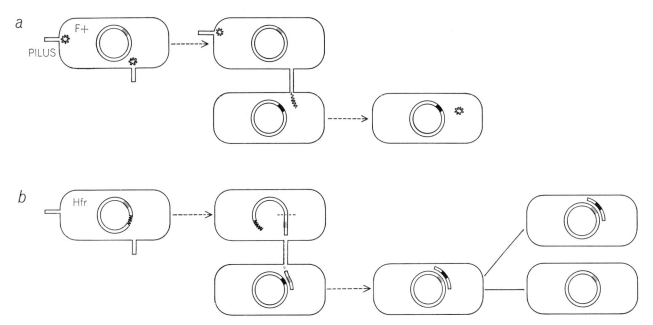

SEXUAL MATING is a form of conjugation. If a fertility factor (F) is in the cytoplasm of a male (F+) cell (a), it is transferred alone through a pilus to a female (F−) cell. In an Hfr cell (b) the F is incorporated in the chromosome. Cell-to-cell contact causes part or all of the chromosome, perhaps including a mutation for drug resistance (color), to pass to a female cell and recombine.

carried from one bacterial cell to another by infecting phages, or bacterial viruses. Transduction occurs when a phage, reproducing inside a cell by taking over the cell's synthesizing machinery, incorporates a bit of the bacterial chromosome within its protein coat "by mistake." When the phage subsequently infects a second cell, the bacterial genes it carries may recombine with homologous genes on the second cell's chromosome. The phage in effect acts as a syringe to bring about what in transformation is accomplished by the movement of naked DNA [*see middle illustration on preceding page*]. Transduction takes place in a variety of bacteria, but at a very low frequency. Genes for resistance can be transduced like other genes, but it is unlikely that more than two resistance genes could be transferred together because the small transducing phage can carry only a short segment of bacterial chromosome.

The third type of genetic transmission in bacteria is conjugation: a direct contact between two cells during which genetic material passes from one cell to the other. Transfer by conjugation occurs primarily from male to female cells of certain groups of bacteria. The male bacteria carry a fertility factor, the F factor, that is ordinarily located in the cytoplasm of the cell but may become integrated into the chromosome. When the F is cytoplasmic, the male cells are called F^+. In such cells the F is readily transferred to female (F^-) cells by conjugation, but it is transferred alone. When the F factor is integrated into the bacterial chromosome, it serves to "mobilize" the chromosome. That is, the chromosome, which in bacteria forms a closed loop, opens and portions of it can pass by conjugation to a female cell, recombine with the female chromosome and thereby endow the female bacterium with traits from the male. Because this transfer occurs with a high frequency in male cells with an integrated F, such cells are called *Hfr*, for "high frequency of recombination" [*see bottom illustration on preceding page*].

The F factor is what is generally called an episome: a genetic element that may or may not be present in a cell, that when present may exist autonomously in the cytoplasm or may be incorporated into the chromosome, and that is neither essential to the cell nor damaging to it. An episome is something like a virus without a coat; indeed, some bacterial viruses can become "temperate" and exist as harmless episomes inside certain bacterial cells [see "Viruses

and Genes," by François Jacob and Elie L. Wollman; SCIENTIFIC AMERICAN Offprint 89].

Until recently the actual route of transfer was not known. In 1964 Charles C. Brinton, Jr., of the University of Pittsburgh and his colleagues proposed that the F factor or the F-mobilized chromosome passes from one cell to the other through a thin tubular appendage, the F pilus, that is formed on both F^+ and *Hfr* cells by the presence of the F factor. Another kind of pilus, the Type 1 pilus, is seen on female cells as well as male cells, but the two can be distinguished: the F pilus is the site of infection by certain phages, and so the phages cluster along the F pili, marking them clearly in electron micrographs [*see top and middle illustrations on page 2567*].

If a male chromosome transferred to a female cell by conjugation carries drug-resistance genes, these genes may be incorporated into the female chromosome. Experiments with sexual mating showed that drug-resistance genes are in fact sometimes scattered along bacterial chromosomes. Rather long segments—sometimes the entire length—of the chromosome can be transferred in sexual mating, and so it is possible for several resistance genes to be transferred in a single mating event.

In 1960 we took up the study of the resistant shigellae in my laboratory at the Keio University School of Medicine. It soon became clear that the mechanism of transfer of multiple resistance was not transformation, because sensitive strains were not made resistant by DNA extracted from the resistant bacteria. It was not transduction, because it could not ordinarily be effected by cell-free filtrates of the resistant cultures.

There was strong evidence that some form of conjugation must be responsible. Microscopic examination of a mixed culture of sensitive and resistant bacteria revealed pairing between the different kinds of cells. When a mixed liquid culture was agitated in a blender to break off any cell-to-cell contact, and the culture was then diluted to prevent further pairing, the transfer of resistance ceased. If the mechanism of resistance transfer was conjugation, however, it was not the familiar process of sexual mating. For one thing, it occurred between F^- cells. Moreover, two observations showed that unlike the transmission of traits by sexual mating the transfer did not involve the chromosome itself. First, we noted that known chromosomal traits of certain strains, such as the inability to syn-

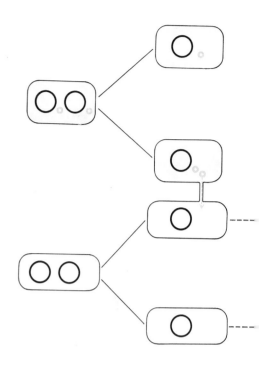

INFECTIOUS DRUG RESISTANCE, another form of conjugation, involves transfer of the R (resistance) factor. A cell of a re-

thesize particular substances, were not usually transferred along with the drug-resistance traits. Second, we noted that the recipient cells became resistant immediately after the transfer occurred, whereas chromosomal drug resistance is ordinarily expressed only after the original drug-sensitivity genes have been lost in the course of cell division through the process known as segregation.

We concluded that the factor responsible for infectious drug resistance was an extrachromosomal element, which we called the R factor (for "resistance"). A number of experiments have confirmed the cytoplasmic nature of these factors. They are obtained by bacteria only by infection from other R-factor-carrying cells, never by spontaneous mutation. They can be eliminated from cells by treatment with acridine dyes; F factors can be eliminated in the same way when they are in the cytoplasm of F^+ cells but not when they are incorporated into the

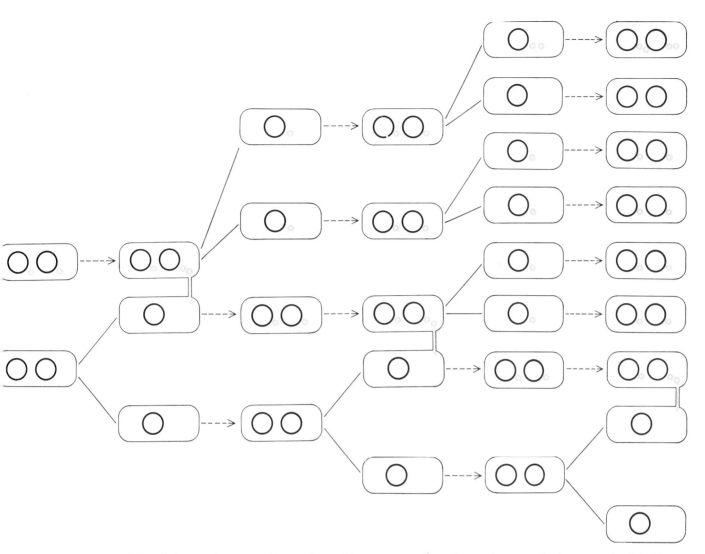

sistant strain (*light color*) comes in contact with one of a sensitive strain (*white*); one of its R factors (*color*) replicates and a copy passes through a pilus to the sensitive recipient. The procedure is repeated as cells come in contact. In the course of cell division an R factor is sometimes lost. The diagram is highly schematic; the actual sequence of replication and transfer is not established.

chromosome of *Hfr* cells. Finally, consider what happens when one adds a small number of bacteria with R factor to a culture of drug-sensitive cells. There is a rapid increase in the relative number of drug-resistant cells; in 24 hours or so the culture is almost completely resistant. This must be owing to the rapid infectious spread of R factors to the once sensitive bacteria, because it occurs at a much faster rate than the overall growth of the culture [*see top illustration on next page*]. Since chromosome replication is synchronized with cell division, the R factor must be replicating faster than the chromosomes and must therefore replicate outside the chromosome, in the cytoplasm.

Although the R factor is usually located in the cytoplasm, in rare instances it is integrated into the chromosome, and when that happens it is transferred together with some chromosomal genes. Such behavior suggests that the R factor,

like the F factor, is episomal in nature. Both of them may be of selective advantage to the cells in which they exist, the F factor by making for genetic variability and the R factor of course by providing drug resistance. When they are not providing an advantage, they at least do the host cells no harm; they are symbionts rather than harmful parasites. Their behavior is similar to that of a temperate, or nonvirulent, phage, and it may be that both are descended from bacterial viruses. Unlike viruses, they cannot exist at all outside the cell; they are obligatory intracellular symbionts with even less biological function than viruses, which are usually considered to be on the borderline between living and nonliving matter.

There is a further major point of similarity between the F and the R factor, and that is their method of transfer. In London, Naomi Datta of the Royal

Postgraduate Medical School, A. M. Lawn of the Lister Institute of Preventive Medicine and Elinor Meynell of the Medical Research Council observed in 1965 that most R factors induce the formation of pili that are shaped like F pili and attract the same phages as F pili: apparently they *are* F pili [*see bottom illustration on page 2567*]. When bacteria that have such pili and are able to transmit multiple resistance are severely agitated in a blender, the pili are sheared off. Such "shaved" cells are unable to transfer the R factor; later, when the F pili have been regenerated, the cells are once again infectious. It now appears that both R factors and any chromosomal genes mobilized by R factors are transferred by the F pili or another closely related kind of pili.

The big difference between the transfer of F factors and the transfer of R factors is in the frequency with which they occur. In a mixed culture of male

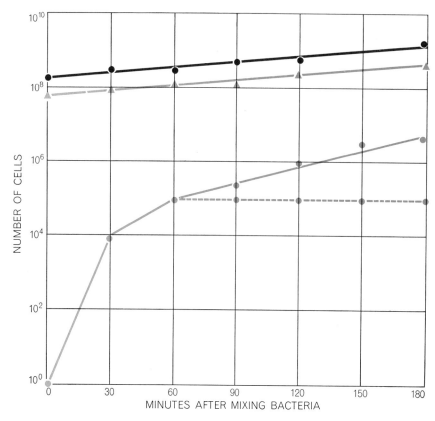

TIME COURSE of *R*-factor transfer is shown. Equal volumes of cells of a donor *E. coli* strain infected with *R* factors (*triangles*) and of an initially sensitive recipient strain (*black dots*) were mixed. Sampling at intervals traced the increase in the number of resistant *E. coli* (*colored dots*). After one hour some of the culture was removed, agitated to break off cell-to-cell contact and diluted to prevent pairing. In the diluted culture there was no increase in the number of resistant cells (*broken line*), indicating that conjugation was the mechanism of transfer. The data are from David H. Smith of the Harvard Medical School.

percent. (If this were not the case, *R* factors could hardly multiply so rapidly in a newly infected culture.) They lose this high competence after several cell-division cycles. The explanation seems to be that most *R* factors form a "repressor" substance that somehow inhibits the formation of *F* pili. Cells that are newly infected with such *R* factors contain no repressor, and so *F* pili are initially induced at a high frequency. Later, as the repressor accumulates, the formation of the pili is inhibited.

It is now possible to describe what happens when bacteria with the *R* factor come into contact with a population of drug-sensitive bacteria [*see illustration on preceding two pages*]. A few *R* factors are transmitted by conjugation from donor cells bearing pili into the cytoplasm of recipient cells, which immediately become resistant. The transfer process is repeated from cell to cell, and the normal process of cell division also contributes to the rapid proliferation of multiple resistance in the recipient population. From time to time an *R* factor is lost. Both the rate of transfer and the rate of loss vary in different strains of bacteria and *R* factors, thus accounting in part for the fact that naturally occurring multiple resistance is much more common in some bacteria that are susceptible to infectious drug resistance than in others.

For several years we have been seeking to map the various elements of an *R* factor as one maps the genes of a chromosome. To do this we capitalize on the fact that although *R* factors are not normally transferred by transduction, it is possible to transduce them under carefully controlled conditions. If we grow large phages in a culture of bacteria with *R* factors, a few of the phages pick up entire *R* factors and are capable of transferring them to recipient cells. If we use small phages, there is room for only part of the *R* factor to be incorporated inside their protein coats and transduced. Some of the transduced particles impart drug resistance but lack the ability to replicate or to be transferred by conjugation; others lack determinants of one or more of the multiple drug resistances. By calculating the frequency with which various segments of the *R* factor are transduced together, we can determine their relative distance from one another and so visualize the structure of the *R* factor we are studying.

and female bacteria the transfer of nearly 100 percent of the *F* factors or *F*-mobilized chromosome, as the case may be, occurs within an hour. In a culture of drug-resistant (donor) and drug-sensitive (recipient) bacteria, on the other hand, only 1 percent or less of the donor cells transfer their *R* factors in an hour. The low frequency of transfer is due to the relative scarcity of cells with *F* pili in a culture of bacteria carrying *R* factors. Bacteria that have newly acquired the *R* factor, on the other hand, can transfer it at a very high frequency—almost 100

The map is not yet conclusive, but we think the factor is circular and that it has a segment—the resistance-transfer factor, or RTF—that controls replication

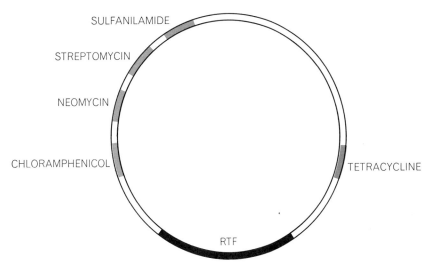

MAP OF *R* FACTOR, still tentative, shows a closed loop. There are five determinants (*color*) of resistance to five different drugs. There is also a determinant, the resistance-transfer factor (*black*), that controls the ability of the *R* factor to replicate and be transferred.

and transferability, as well as segments determining resistance to each of five types of drug [*see bottom illustration on opposite page*]. We have suggested that the R factors originate when a resistance-transfer factor picks up resistance genes from some bacterial chromosome and that the two then form a single episomal unit. E. S. Anderson and M. J. Lewis of the Central Public Health Laboratory in London have advanced a different view. They consider that the resistance-transfer factor and the set of resistance determinants exist as two separate units, which on occasion become associated to form R factors.

Since R factors are self-replicating units, carry genetic information and can recombine with bacterial chromosomes, it is safe to assume that they are composed of DNA. This is confirmed by the fact that R factors, like nucleic acids in general, are inactivated by ultraviolet radiation and by the decay of incorporated radioactive phosphorus. At the Walter Reed Army Institute of Research and at Keio, Stanley Falkow, R. V. Citarella, J. A. Wohlhieter and I were able to isolate the DNA of R factors by density-gradient centrifugation. A first attempt to separate R-factor DNA from that of E. coli was unsuccessful, suggesting that the densities of the two DNA's are very similar. We then selected as the host cell the bacterium *Proteus mirabilis,* which was known to have a DNA of unusually low density and to be subject to infectious drug resistance.

When DNA extracted from *Proteus* carrying the R factor is centrifuged in a solution of cesium chloride, two satellite bands of DNA appear in addition to the band characteristic of the bacterial DNA [*see illustration on next page*]. These bands disappear if the *Proteus* loses its R factors spontaneously or if they are eliminated by the acridine dye treatment, and so we conclude that the bands do represent the R-factor DNA. Analysis of this fraction by column chromatography shows that it is typical double-strand DNA. It is possible that R factors contain components other than DNA, but this is not likely in view of the fact that entire factors are transduced and transducing phages incorporate only DNA.

The original finding that infectious drug resistance affected four unrelated drugs implied that some factor was altering the cell membrane, reducing its permeability and thereby barring all the drugs from their normal sites of action inside the cell. The finding that there are separate resistance determinants for the various drugs, however, indicated that

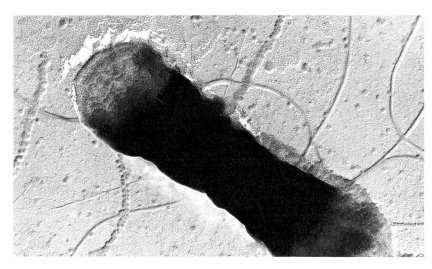

MALE BACTERIUM, an *E. coli* infected with phage, has *F* pili. They are thin fibers, here hidden below the spherical phage particles. The thin fibers without phages are Type 1 pili and the thick fibers are flagella. The preparation has been enlarged about 30,000 diameters.

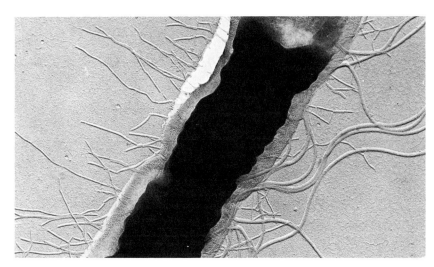

FEMALE *E. COLI*, which lacks the *F* factor, also lacks *F* pili. It does have both the Type 1 pili and flagella, which are organelles of locomotion for the cell. The electron micrograph, like the others on this page, was made by Toshihiko Arai in the author's laboratory.

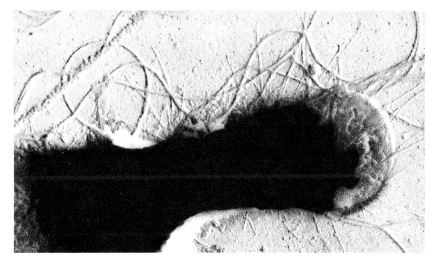

E. COLI WITH *R* FACTOR, although a female cell, does carry an *F* pilus, the phage-covered fiber at top left. It also has Type 1 pili and flagella. The most common type of *R* factor initially induces the formation of *F* pili, but it also tends to repress them later.

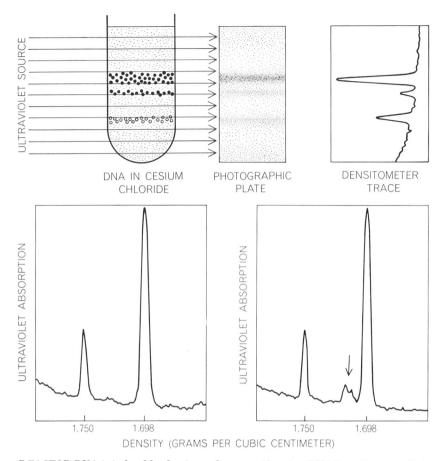

R-FACTOR DNA is isolated by density-gradient centrifugation. DNA from *Proteus* cells is suspended in a cesium chloride solution and spun in a high-speed centrifuge. The cesium chloride establishes a density gradient and the DNA forms bands in the solution according to its density. The DNA pattern is photographed in ultraviolet, which is absorbed by DNA, and the photograph is scanned by a densitometer (*top*). The densitometer trace derived from *Proteus* without R factor (*bottom left*) shows a band at 1.698 grams per cubic centimeter that is characteristic of *Proteus* DNA and a reference band at 1.750. The trace from *Proteus* with R factor (*right*) has extra bands (*arrow*) at 1.710 and 1.716, representing R-factor DNA.

each determinant had its own mode of action. Permeability may be involved in the case of some drugs, but it is now clear that other processes are at work. S. Okamoto and Y. Suzuki of the National Institute of Health in Japan and Mrs. Datta and P. Kontomichalou in Britain have shown that bacteria bearing various R factors synthesize particular enzymes that inactivate specific drugs, thereby rendering them harmless to the bacteria.

The public health threat posed by infectious drug resistance is measured by the range of bacterial hosts it affects, the number of drugs to which it imparts resistance and the prevalence of certain practices in medicine, agriculture and food processing that tend to favor its spread. R factors can be transferred not only to shigellae but also to *Salmonella*, one species of which causes typhoid fever; to *Vibrio cholerae*, the agent of cholera; to the plague bacillus

Pasteurella pestis and to *Pseudomonas aeruginosa*, which causes chronic purulent infections. In addition, more than 90 percent of the agents of urinary tract infections, including *E. coli*, *Klebsiella*, *Citrobacter* and *Proteus*, now carry R factors.

(These organisms are all gram-negative bacteria; R factors seem not to be transferable to the gram-positive bacteria, which include streptococci and staphylococci. A somewhat similar form of transmissible resistance has been discovered in staphylococci, however. There are cytoplasmic genes, or plasmids, in some staphylococci that determine the production of penicillinase, an enzyme that inactivates penicillin. Richard P. Novick of the Public Health Research Institute in New York and Stephen I. Morse of Rockefeller University recently showed that these plasmids can be transduced to drug-sensitive staphylococci both in the test tube and

in laboratory animals. The actual clinical importance of this process remains to be determined.)

The R factors seem to be acquiring resistance genes for an increasing number of antibiotics. The original factors, it will be remembered, imparted resistance to sulfanilamide, streptomycin, chloramphenicol and tetracycline. In 1963 G. Lebek of West Germany discovered a factor that causes resistance to these four drugs and also to the neomycin-kanamycin group of antibiotics. In 1965 Mrs. Datta and Kontomichalou reported a new determinant of resistance to aminobenzyl penicillin (ampicillin). In 1966 H. W. Smith and Sheila Halls of the Animal Health Trust in Britain found factors imparting resistance to the synthetic antibacterial drug furazolidone. This year David H. Smith of the Harvard Medical School reported R-factor-controlled resistance to gentamycin and spectinomycin. We must assume that additional drug-resistance determinants will appear and proliferate as new antibiotics come into use.

This is implicit in the mechanism of infectious resistance. R factors are common in *E. coli*, which are often present in the intestinal tracts of human beings and animals. When a person or an animal becomes infected with a susceptible disease organism, the R factor is readily transferred to the new population. Although the frequency of transfer of R factors is not high even in the laboratory, and is reduced by the presence of bile salts and fatty acids in the intestine, recipient bacteria bearing the R factor are given a selective advantage as soon as drug therapy begins, and they soon predominate.

In addition to being ineffective and helping to spread resistance, "shotgun" treatment of an infection with drugs to which it is resistant causes undesirable side effects. It is therefore important to culture the causative agent, determine its drug-resistance pattern and institute treatment with a drug to which it is not resistant; that is the only way to combat the multiple-resistance strains. As more is learned about the R factor, new forms of therapy may be developed—possibly utilizing the acridine dyes, which attack drug-resistant as well as sensitive cells and can also eliminate R factors from cells.

In many parts of the world antibiotics are routinely incorporated in livestock feeds to promote fattening and are also used to control animal diseases. Anderson and Mrs. Datta have shown clearly that the presence of antibiotics in livestock exerts a strong selective pressure in favor of organisms—particularly salmonellae—with R factors and plays an important role in the spread of infectious resistance. Meat and other foodstuffs are also treated with antibiotics and synthetic drugs as preservatives in many countries, and this too may help to spread R factors and carry them to man. Unless we put a halt to the prodigal use of antibiotics and synthetic drugs we may soon be forced back into the preantibiotic era of medicine.

One final note. Typhoid, cholera and plague bacilli are obviously much more difficult to combat if they are resistant to drug therapy. There are grounds for believing that the military in some countries are investigating the potentialities of R factors as weapons of bacteriological warfare.

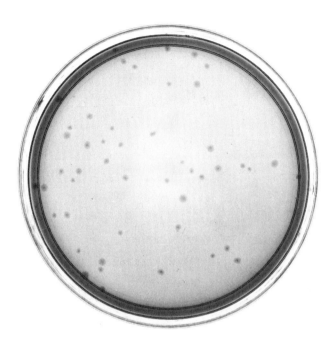
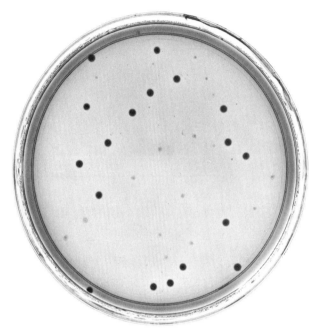

SENSITIVITY TEST conducted in Smith's laboratory at Harvard demonstrates infectious drug resistance. A culture of *Salmonella typhimurium* with an R factor controlling resistance to four drugs is mixed with drug-sensitive *E. coli*. A portion of the mixed culture is immediately plated on a medium containing the drugs (*left*). Only *Salmonella* colonies (*gray*) appear. After the mixed culture has incubated, the plating procedure is repeated, and now *E. coli* colonies (*black*) grow as well (*right*): the R factor was transferred.

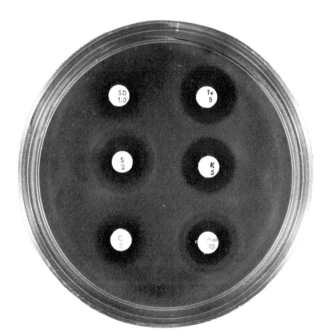
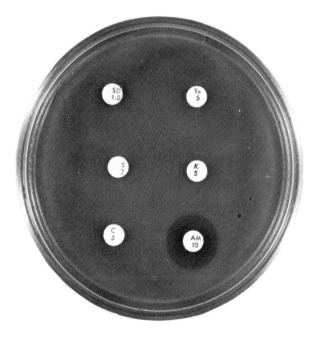

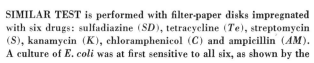

SIMILAR TEST is performed with filter-paper disks impregnated with six drugs: sulfadiazine (*SD*), tetracycline (*Te*), streptomycin (*S*), kanamycin (*K*), chloramphenicol (*C*) and ampicillin (*AM*). A culture of *E. coli* was at first sensitive to all six, as shown by the dark zones around each disk where the bacteria have been killed (*left*). After the culture was incubated with a strain of *Klebsiella*, taken from a patient, that was resistant to all the drugs but ampicillin, the *E. coli* too were resistant to all but ampicillin (*right*).

The Author

TSUTOMU WATANABE is associate professor of microbiology at the Keio University School of Medicine in Tokyo, where he received his M.D. degree in 1948. He spent a year as an exchange student in radiobiology and bacteriology at the University of Utah in 1951 and 1952 and also worked as a research associate in the department of zoology at Columbia University in 1957 and 1958. He writes that his special field of study is "microbial genetics, particularly the genetics of bacterial drug resistance. Through the studies on infectious drug resistance I have become interested in various episomes and plasmids (parasitic or symbiotic agents) of bacteria and also in the evolution of microorganisms."

Bibliography

EVOLUTIONARY RELATIONSHIPS OF R FACTORS WITH OTHER EPISOMES AND PLASMIDS. T. Watanabe in *Federation Proceedings*, Vol. 26, No. 1, pages 23–28; January–February, 1967.

INFECTIVE HEREDITY OF MULTIPLE DRUG RESISTANCE IN BACTERIA. Tsutomu Watanabe in *Bacteriological Reviews*, Vol. 27, No. 1, pages 87–115; March, 1963.

SCIENTIFIC
AMERICAN   December 1967, Vol. 217, No. 6, pp. 118-125

OFFPRINT 1088

THE WATER BUFFALO

by W. Ross Cockrill

This gentle beast is a source of power and food for a substantial
fraction of humanity. It has much to recommend it for these uses,
yet it has been studied far less than many other domestic animals.

The water buffalo has long been a major servant of man in the East. The plow it pulls across the padi, or rice, fields today differs in no detail from the one it pulled three centuries ago. The animal plods through its task slowly, but it is sure of foot and easy to manage. Moreover, it is versatile: it not only provides draft power for tilling the padi fields and for many other purposes but also supplies a rich milk and a meat that compares favorably with beef.

A large proportion of the world's water buffaloes live in the countries of South Asia, including some where there is a chronic shortage of food. The water buffalo's contribution to the food supply is a considerable one, and we in the Food and Agriculture Organization of the United Nations (with the support of the Commonwealth of Australia) are studying the animal in order to learn how that contribution might be enhanced. Little is known about the water buffalo's physiology and full potential as a domestic animal; even the elephant, the yak, the camel and the llama are better understood. Still, the water buffalo is an exceptionally productive animal; it is capable of performing work and producing milk on a diet consisting only of stubble. In this regard it surpasses all other animals.

The water buffalo of Asia (*Bubalus bubalis*) is at times confused with its distant cousin the wild buffalo of Africa (*Syncerus caffer*) and even with the American bison (*Bison bison*), but it is not closely related to these animals. It differs from the cattle of the East in a number of particulars; for example, it has no dewlap or hump. It is also distinguished by its proclivity for wallowing in mud or water. Left to itself, it is of nocturnal inclination.

The various breeds of the water buf-

falo are customarily separated into two groups: river buffaloes, which are found mostly in India, and variants of the swamp buffalo, which are most plentiful in the great rice-growing lands east and south of Burma. Among the river buffalo breeds are the Murrah, Surati and Jaffarabadi. The desi, which means "local," is the most common type. It is a hardy, nondescript mongrel, the result of many generations of haphazard crossbreeding. River buffaloes prefer clean running water to mud, and they are better suited to milk production than the swamp buffalo.

The swamp buffalo is the animal par excellence for the complex labors of the padi fields. As a domestic animal it is a single breed that closely resembles the Arni, or wild buffalo, of India and South China (from which all breeds of the water buffalo are supposed to have originated). Variants of the swamp buffalo range from the massive Thai buffalo, which weighs upward of 2,000 pounds, to the small carabao of the Philippines, which averages about 900 pounds. Even in limited geographical areas it varies widely in size. The swamp buffalo is normally slate black, having a dark skin and a sparse coat of black or gray hair. Below its jaw is a white chevron of unpigmented skin and hair; this marking is repeated lower down on the chest. Such stripes are peculiar to the swamp buffalo, the Surati breed of river buffalo and certain wild breeds, among them the Arni of India. Occasionally the hair of an animal will be tinged with red. White buffaloes, which have a pink skin and white or yellowish hair, are quite common. They are not true albinos; some pigment is present in their eyes, horns, hooves and mouth tissues. The whiteness appears to be a recessive hereditary characteristic. Probably about 5 percent of all the swamp buffaloes are white; in

some areas—for example northern Thailand—the white buffaloes account for as much as 30 percent of the buffalo population.

In hot climates swamp buffaloes must have almost unlimited access to water. Buffaloes are not noticeably tolerant of heat, and they can suffer extreme discomfort if they are exposed for any length of time to the direct rays of the sun. They need to wallow uninterruptedly during the heat of the day. Few animals convey an impression of such blissful contentment as a swamp buffalo immersed to the nostrils in a mud wallow or standing ecstatically in a downpour of tropical rain.

In China and India the water buffalo has a particularly long history. The river buffalo is known to have been domesticated in India by 2500 B.C. and the swamp buffalo in China about 1,000 years later. From these two centers buffaloes have spread around the world. Their migration has not, however, been rapid; most of it appears to have taken place during comparatively recent times. In Egypt, for example, there are now as many buffaloes as there are cattle (well over 1.5 million), but buffaloes do not appear in the paintings and frescoes of the pharaohs' tombs. Introduction of the animal seems to have been delayed until about A.D. 800. Even in Cambodia, which is in the heart of buffalo territory, the animal may not have arrived in any numbers until the 15th century. At that time the Cambodian Khmer civilization fell to the Thai conquerors, who were very likely utilizing large numbers of swamp buffaloes for draft purposes.

It is estimated that there are in the world today about 100 million buffaloes. This may well be a considerable underassessment, since accurate totals are not available for many of the 20-odd countries where buffaloes are plentiful. The buffalo thrives in many places where it is truly exotic. There are at least 40,000 of the animals in Italy, where they supply the milk for mozzarella cheese. Brazil has almost 70,000. Ranchers on the island of Marajó in the Amazon delta consider the animal more productive than any other in that environment, where widespread flooding is common and marshy conditions prevail. Hong Kong has an indigenous population of about 2,000 small, sturdy swamp buffaloes, and it imports many thousands a year from China and other countries for slaughter.

Not all attempts to introduce the buffalo have been successful. The animal was brought to England by the Earl of Cornwall, the brother of Henry III, but it did not flourish. Efforts have been made for five centuries to establish buffaloes in African countries south of the Sahara, but usually the animal has died out. A more spectacular failure occurred in Australia, where buffaloes were introduced in the first half of the 19th century. Perhaps due to a lack of proper managerial skills—the animal is nervous

WATER BUFFALO WALLOWS in a pool of muddy water. Wallowing is a necessity for water buffaloes in hot climates; they can-not tolerate exposure to the direct rays of the sun. The animal's horns are a formidable defense but are seldom used in attack.

with strangers and does not adapt easily to European handlers—large numbers of the imported buffaloes were allowed to become feral. Today in the Northern Territory of Australia there are perhaps 200,000 "wild" swamp buffaloes.

In their native Eastern environment buffaloes are tractable; indeed, it is this characteristic that basically accounts for their domestication. They are difficult to herd or drive but easy to lead; they seldom kick or use their massive horns for attack. In many countries buffaloes are customarily handled by small children. It is a common sight to see children leading or riding these enormous beasts to wallowing places, lying at full length on their backs, getting into the water with them, washing them down, cleaning their ears and eyes and nostrils with care and complete fearlessness. This same imperturbable docility makes the buffalo an ideal animal for the slow, plodding work of the padi fields. Its sureness of foot, due to an unusual flexibility of the fetlock and pastern joints above its large cloven hoof, enables it to step carefully over the bunds: the low mud walls that bound each padi field. So easy is the animal to guide that usually the only means of controlling it is a single rein fastened to a nose thong or even looped casually around the great horns.

All breeds of the buffalo have horns. They vary in size and conformation from the heavy, backswept horns of the swamp buffalo to the tightly curled ones of the Murrah breed. The fact that the animals are rarely polled purposely is in itself a sufficient comment on the danger of the horns. In some countries buffaloes may have to defend themselves against tigers and other predators, and their horns can provide a formidable defense. The animals do not as a rule fight among themselves. When they do, it is likely to be a sort of pushing match, more a trial of strength than a sustained battle. On occasion, however, they will fight to the death and can only be dissuaded by tossing bundles of burning straw between them.

Buffaloes of all breeds are competent work animals where speed is not an essential requirement. In the rice-growing lands they are used to plow the soil and, when the fields have been flooded, to harrow and puddle the lumpy earth to the proper consistency for planting the rice shoots. They work with ease hock-deep in liquid mud. Often the buffaloes are rested between planting and harvest by being turned out to open forest land to fend for themselves. The

owner of a buffalo has no difficulty recognizing it at harvest time, when it is reclaimed and put to work pulling sheaves to the threshing floor in carts or light bamboo sledges.

The buffalo is widely employed in most Eastern countries in a primitive but surprisingly effective system of threshing rice. A rattan mat is laid over a baked mud floor and sheaves are loaded onto it with pitchforks. Then two, three or more buffaloes plod in a tight circle, trampling the grain from the stalks as the herdsman forks the rice stalks under their hooves.

Buffaloes still operate simple mills that express oil from seeds and juice from sugarcane. In India, along with oxen and camels, they raise water from wells and, by involved irrigation systems, send it coursing down shallow channels. Sometimes the large bucket in which the water is raised is made out of buffalo hide.

Crude two-wheeled carts drawn by one or two buffaloes can move heavy loads, again where speed is not important. In the crowded bazaar areas of cities such as Peshawar in Pakistan and Calcutta, Ahmadabad and Madras in India, buffaloes drag their heavy loads through teeming crowds with ponderous imperturbability. The bells that hang from the animals' necks give warning of their approach, and it is well to get out of their way. With a load of a ton or

more and a cart that has no brakes a sudden pull-up is impossible.

In Ceylon and a few other countries where buildings of mud brick or wattle-and-daub are still common, water buffaloes are used to puddle clay to the correct consistency. The buffalo can be used as a riding animal and as a pack animal. In Taiwan the traffic halts on a main highway to allow the passage of a buffalo train—four loaded trucks drawn by a plodding buffalo on a narrow-gauge line.

No source of power is quite as cheap as buffalo power. Rice is the staple food of half the world's population, and in the immemorial pattern of padi production the buffalo is an integral part of the slow, meticulous, backbreaking monotony of the struggle for existence. It is with justification that the animal is called the living tractor of the East. Where time is of little consequence, the buffalo can be turned out to pasture between planting and harvest and still be a profitable animal. Tractors, like most other machines, are economic only as long as they are in continual use. Alternating periods of activity and rest are feasible for the living tractors but not for the rust-prone variety.

It is in India and as a source of milk that the buffalo is seen to greatest advantage. It is responsible for more than half of the milk produced in that country. Buffalo milk contains about twice as

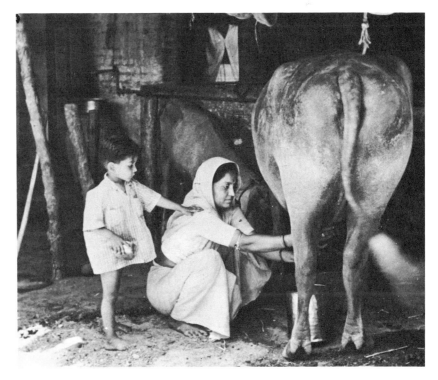

BUFFALO IS MILKED by a woman of India, where more than half of the milk supply is furnished by water buffaloes. Water buffalo milk is twice as rich in butterfat as cattle milk.

much butterfat as cattle milk. On occasion the fat content may be as high as 15 percent; the overall average is probably 7 percent or a little more. The nonfat-solid content compares well with that of cattle milk, varying between about 9 and 10.5 percent. Depending on the breed and on management practices the daily milk yield of a buffalo ranges from the two to four quarts produced by an actively working draft female to the occasional 16 quarts of an exceptional dairy animal. Although buffaloes usually yield less milk than cattle maintained under similar conditions, the product for general purposes is as good as cattle milk and for certain purposes is much better. The buffalo is the main source of ghee, a dehydrated and clarified form of butter that is as widely used for cooking in India as olive oil is in the Mediterranean countries. Because the milk is so rich it lends itself to judicious dilution. When powdered skim milk and water are added, the resulting product is attractive and palatable. A rich and nutritious yogurt is made from buffalo milk, as are candy and excellent ice cream. In the Philippines, where the working swamp buffalo is also used as a dairy animal, the milk is made into a soft cheese.

A number of cooperative dairy farms have been formed in India. The Haringhata farm, which is near Calcutta, has in addition to cattle 3,500 buffaloes; it professes to be the largest dairy farm in the world. Another enterprise, located at Anand in the state of Gujarat, is one of the most remarkable of all cooperative movements. It is the Kaira District Cooperative Milk Producers' Union Ltd., managed by V. Kurien together with a team of specialists in nutrition, breeding and health. If India had a hundred cooperatives like this one, the world would

hear less of food crises and famines. More than 80,000 farmers in some 500 villages make up the cooperative. The usual holding is no bigger than three acres. It supports a family of five owning two or three buffaloes (of the Surati breed) and possibly also several cattle. The Surati are favored for their docility and for their ability to convert rough fodder into rich milk. The men tend the cattle, but the care of the buffaloes is entrusted exclusively to the women and children. Each day the cooperative sends 200,000 pounds of buffalo milk to Bombay in rail tankers donated to the cooperative by New Zealand. In addition, its milk-processing plant can produce sizable quantities of butter, condensed milk, ghee, cheese, casein, dried milk powder and baby food. Besides health and breeding services for the animals, the cooperative supports schools, libraries and hospitals. It demonstrates

what can be done when basic principles of nutrition, breeding and health control are applied to the management of buffaloes.

In its reactions to disease the water buffalo is an enigma. It is highly susceptible to certain infections and notably resistant to others. To rinderpest, hemorrhagic septicemia, anthrax and foot-and-mouth disease buffaloes are even more susceptible than cattle, or so it often appears. Yet the degree of susceptibility varies, and it should be noted that one of the principal reasons for the survival of the animal in Egypt is its resistance to rinderpest, which in the past has decimated the cattle population. Complicating the problem is the fact that some vaccines producing a satisfactory degree of immunity in cattle initiate a poor antibody response in buffaloes.

The great famines that have occurred

BUFFALOES PER 1,000 PEOPLE

LESS THAN 10

10 TO 49

50 TO 99

OVER 100

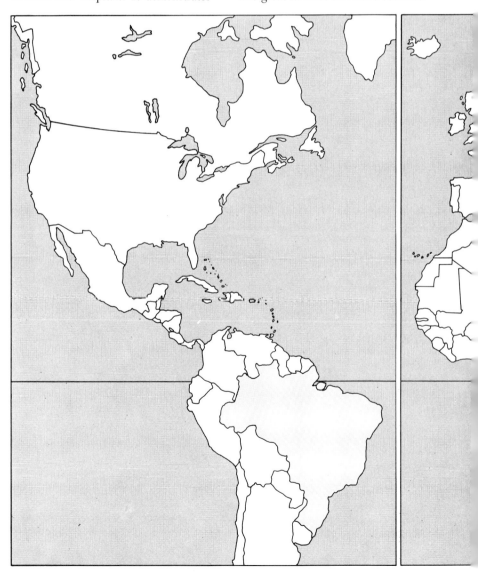

WORLD DISTRIBUTION of water buffaloes (which total about 100 million) indicates that they are most plentiful in South Asia and exist in appreciable numbers as far afield as Europe and Brazil. The animal flourishes on ranches in the Amazon valley. In Italy

in the East, some of them within living memory, have been supported, perhaps sometimes even started, by outbreaks of disease among livestock. Since the buffaloes are a mainstay of rice production, when a killing disease such as rinderpest or a crippling one such as foot-and-mouth disease strikes at the time of planting or harvest, there is no source of draft power save man himself.

Mastitis and tick-borne diseases are much less common in buffaloes than in cattle, but the incidence of tuberculosis varies. In some parts of India buffaloes have been found to be more heavily infected than cattle. Heavy parasitic infestations—for example liver fluke—are common in buffaloes.

The gaps in our knowledge of this animal become most obvious when we consider it as a meat producer. In many of the lands of the buffalo millions of people are vegetarian from choice, religious conviction or sheer economic necessity. Cereals are all-important in these areas, and the production of animal protein for human consumption is low on the list of priorities. Many countries prohibit the slaughter of buffaloes younger than 12 years of age, except when they are infertile, intractable or incapable of work. An enormous waste results from the practice of depriving some calves of the commercially valuable milk and allowing them to starve to death. Tens of thousands of calves representing a potential source of high-quality veal are thus lost.

The widespread ban in India on the slaughter of sacred cows does not apply to buffaloes; however, in India and elsewhere the buffalo is accorded an extraordinary degree of care and consideration. It is treated almost as a member of the family. Indeed, the animal is as important as a member of the family in the contribution it makes to the family's welfare. So narrow can be the margin between subsistence and starvation that the loss of a single buffalo can be disastrous.

The water buffalo is remarkable for its longevity. Instances are quite often cited of animals living to the age of 40, and it is common to find buffaloes still working at 20. When the pace of the buffalo slows below the level at which the animal is economic to maintain, its owner—particularly in Buddhist countries—will often refuse to slaughter it and salvage something in meat value. The buffalo is allowed to live in peaceful retirement until death overtakes it.

In the East buffalo meat is generally a by-product. The animal is slaughtered only after years of labor or milk production or both. Abattoir facilities in developing countries tend to be inadequate: killing methods are clumsy, the meat is

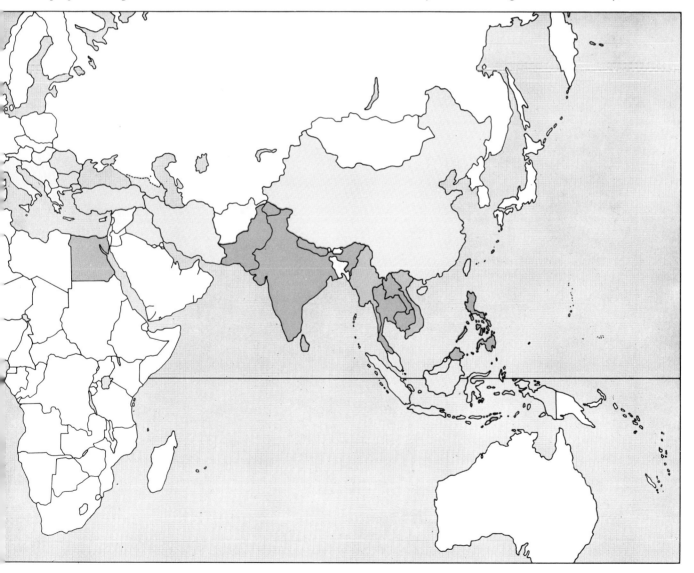

it provides milk for mozzarella cheese; there and in Bulgaria and Yugoslavia it is also reared for meat. The water buffalo has been introduced into England, France, Germany, Spain, South Africa and Australia, but for various reasons it did not become established.

BUFFALOES PULL A HARROW across a flooded padi, or rice, field in Indonesia. Between the padi fields are bunds, or low mud walls. One advantage of the water buffalo is that it is able to walk securely in mud and step over the bunds without damaging them.

poorly butchered and handled and it is usually sold without being properly aged. If the buffalo were treated as a meat-producing animal and reared for slaughter at less than two years, meat of excellent quality could be obtained. The weight of the dressed carcass is usually 45 to 47 percent of the live weight, but in Italy, Bulgaria and Yugoslavia, which are among the few countries where the buffalo is reared for meat, a dressed carcass weight of 51 percent is not uncommon. The quality of the meat is good, and its flavor is generally indistinguishable from that of beef, although it is somewhat coarser in appearance. The fat, due to the absence of the pigment carotene, is pure white.

In parts of Nepal and Thailand the hide of the buffalo is eaten. It is cut into small strips and after prolonged boiling is dried in the sun and stored. When it is cooked in deep fat, it makes delicious "buffalo chips" that might well find a market in the West.

The hide makes excellent leather for shoe soles, belts or any other purpose where a thick, tough product is required. Reins and lassos are made from strips of buffalo hide that have been twisted, woven and softened with fat. They are prized possessions and are handed down from father to son. With modern machinery the hides can be split to yield a thin, strong product that when it is processed and dyed compares favorably with any other type of leather.

If modern methods of selective breeding were widely applied to the water buffalo, in place of the haphazard breeding and crossbreeding that obtain almost everywhere it is found, the potential of the buffalo could be realized. The buffaloes of Thailand are bigger and stronger than those of almost any other country. This is probably due to the fact that a system of selective breeding has been practiced there for many years. The best of the male animals are bought by the government, which sells or lends them to the villages for breeding purposes.

The buffalo is a sluggish and seasonal breeder. With good management a calf every year is possible; two in three years is normally the case. Sexual activity is seldom seen; under natural conditions mating commonly takes place in the dusk or dark. Under modern management, however, the animal's breeding behavior is similar to that of cattle, although artificial insemination procedures call for certain adjustments. Estrus in buffaloes can be regulated by controlling the temperature of the building in which they are housed and also the degree of their exposure to light. If such procedures were instituted, it would be possible to reduce seasonal fluctuations in breeding that result in a majority of calves being born in the last quarter of the year, which of course gives rise to alternating periods of glut and scarcity in milk.

Even under the prevailing conditions of haphazard breeding, inept management and casual nutrition the buffalo is an outstandingly productive animal. It apparently has a unique ability to digest and assimilate the poorest of roughage. It is quite common to see buffaloes and cattle grazing together during the dry season in pastures where one would say there was barely enough sustenance to maintain life. The buffaloes will be in good condition, the cattle thin to the point of emaciation. Padi straw is the buffalo's staple diet. Dairy buffaloes receive extra feed in the form of concentrates and cut green herbage, when they are available, but the animal is able to work and to produce milk on the poorest of diets.

It is unfortunate that cattle protagonists are often violently antibuffalo; there is a place for both animals in this hungry world, and it is probable that in tropical and subtropical countries the buffalo has a greater protein potential

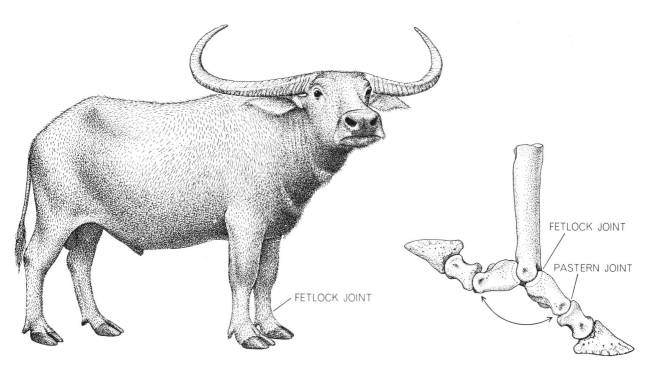

A MAJOR BREED of water buffalo is the swamp buffalo (left). All other domestic breeds are classified as river buffaloes. Both the swamp buffalo and river buffaloes are of the species *Bubalus bubalis*. The swamp buffalo varies widely in size but can weigh more than 2,000 pounds. Its fetlock and pastern joints (right) are exceptionally flexible, which is what enables the animal to move easily in mud. Swamp buffaloes are distinguished by heavy backswept horns and two white chevrons on the chest. They are sometimes confused with the wild buffalo of Africa (*Syncerus caffer*) and the American bison (*Bison bison*), but are not closely related to either of them.

than cattle. Countries in the Temperate Zone that have no buffaloes and that could provide the requisite management skills and quarantine security might usefully consider introducing some of the animals from selected areas on an experimental basis.

The buffalo has been profoundly important to the agricultural economy of bygone civilizations. Given the necessary scientific support it can make an outstanding contribution to our own. It is not an obsolete or vanishing animal. It can take a big part in meeting the needs of the expanding human population for increased supplies of meat and milk, and for the working power to produce more food.

BUFFALOES DRAW A HEAVY LOAD to a cotton mill in Pakistan. Each crude two-wheeled cart carries nearly a ton of cotton. Bells hanging from the neck of the buffaloes warn of their approach; the carts have no brakes and a sudden halt is impossible. The tightly curled horns of the animals identify them as river buffaloes of the Murrah breed. Water buffaloes are tractable and calm.

The Author

W. ROSS COCKRILL is employed by the Food and Agriculture Organization of the United Nations and is stationed at the FAO headquarters in Rome. A 1935 graduate of the Royal College of Veterinary Surgeons, Cockrill worked for a time in general practice in his native country of Scotland, did research and as a government veterinary officer engaged in the control of bovine tuberculosis, foot-and-mouth disease and other diseases of economic importance to the livestock industry. He became a Fellow of the Royal College of Veterinary Surgeons in 1950 and is also a doctor of veterinary medicine at the University of Zurich. During World War II he served with the Royal Air Force as a flying officer. Since joining the FAO he has traveled widely and during the past 15 years has visited and worked in some 60 countries.

Bibliography

A COMPILATION OF AVAILABLE DATA ON THE WATER BUFFALO. Bradford Knapp. International Cooperation Administration, 1967.

SEASONAL VARIATIONS IN REACTION TIME AND SEMEN QUALITIES OF BUFFALO BULLS. N. S. Kushwara, D. P. M. Mukherjee and P. Buhattacharya in *Indian Journal of Veterinary Science and Animal Husbandry,* Vol. 25, Part 4, page 327; 1955.

THE WATER BUFFALO IN INDIA AND PAKISTAN. David C. Rife. International Cooperation Administration, 1959.

COMPLEMENTARY TEST PATTERNS are used to demonstrate the color changes that take place when a negative afterimage of a brightly colored stimulus is formed on the retina. A negative after-image of the pattern above resembles the pattern below and vice versa. A positive afterimage of either test pattern retains the colors of the original. Cover one pattern when looking at the other.

SCIENTIFIC
AMERICAN   October 1963, Vol. 209, No. 4, pp. 84-93   OFFPRINT 1089

AFTERIMAGES

by G. S. Brindley

Recent experiments have elucidated the mechanism of these
curious visual phenomena and have provided several new
insights into the photochemistry of normal human vision.

Look steadily at the small cross in the center of the four colored disks at the top of the opposite page; after about 10 seconds look at a blank sheet of white paper. You should see against the colorless background a set of colored disks similar to those you have just been looking at. The colors of the new pattern will, however, appear to be quite different; in fact, the array will probably look more like a pale, washed-out copy of the test pattern at the bottom of the page. Now if you look steadily at the cross in the center of this illustration for 10 seconds and then look at the white paper, you will see a pale version of the test pattern at the top of the page. In both cases what you have "seen" against the uniformly white surface is a negative afterimage. As in a photographic negative, the dark parts of the original look bright and the bright parts look dark. The colors of the afterimage, however, are complementary to those in the original.

A complementary, or negative, afterimage achieved by this method normally fades into invisibility in about 15 seconds. It can usually be brought back, however, if as soon as it has disappeared the eyes are closed for a few seconds and then reopened. A negative afterimage can be prolonged in this manner for a minute or more. In the laboratory negative afterimages produced under special conditions have survived for as long as 20 minutes.

Positive afterimages are much more transient, seldom lasting longer than five or 10 seconds; for this reason they are also more difficult to see than negative afterimages. With luck a positive afterimage can be produced by the following method: In direct sunlight look again at the cross in the center of the top illustration on the opposite page. This time, after only two or three seconds, close

your eyes and cover them with your hands without pressing on them. For about the first five seconds you should be able to see a faint pattern similar in shape and color to the test pattern. If this method fails, you may succeed in seeing a positive afterimage in yet another way: Sit indoors in a position from which you can see a window with bright sky behind it. Close and cover your eyes for about half a minute. With your eyes still covered, face the window; then uncover them and look steadily at the intersection of two crosspieces in the window. After about three seconds of steady looking close and cover your eyes once more. For the first five or 10 seconds you will almost certainly see a bright, positive afterimage of the same shape as the window. As in all positive afterimages, bright will appear as bright, dark as dark and colors faded but true.

The classification of afterimages into negative and positive is widely applicable but does not exhaust all the possibilities. Another type of afterimage can be produced by looking steadily for a few seconds at an unshaded light bulb. For a minute or so afterward you will be able to see, either in darkness or against a uniform white surface, an afterimage that passes through several stages of varying bright colors, although the light bulb is perfectly white. In this case no simple relation exists between the color of the afterimage and that of the inducing stimulus. It nonetheless remains true that at any given stage the color of the afterimage seen in darkness is complementary to the color seen at the same stage against a white background.

It is easy to give a rough explanation of afterimages. The negative ones are presumably due to an insensitivity or "fatigue" of some part of the visual system, caused by previous strong stimula-

tion; the positive ones, to the persistence of some of the stimulatory effects of bright light after the light has ceased to shine. To explain why the color of an afterimage seen in darkness is similar to the color of the inducing stimulus, whereas that seen against a white background is complementary, we must suppose the three mechanisms involved in the reception of the three primary colors (red, green and blue) of normal human vision can be fatigued, and can show persistent excitation, independently of one another.

Explanations of this kind were suggested as early as the middle of the 18th century, and one was developed in detail by Hermann von Helmholtz in 1858. The theory proposed by Helmholtz was hampered, however, by the absence at that time of any means of detecting which parts of the visual pathway leading from the retina to the brain were involved in either the fatigue or the persisting excitation. It was not until 1940 that Kenneth Craik of the University of Cambridge provided the first such means. Craik succeeded in making one of his eyes temporarily blind for about a minute by pressing firmly on it to cut off its blood supply; he discovered that a bright light that fell on the eye while it was blind could cause a negative afterimage to appear when he looked at a uniform white surface after recovering from the blindness. This experiment proved that at least part of the fatigue responsible for negative afterimages occurs in the eye and not in the brain, since during the period of blindness the brain could not have received any messages from the eye.

Efforts to locate more precisely the site at which afterimages originate met with little success following Craik's discovery until a few years ago, when my

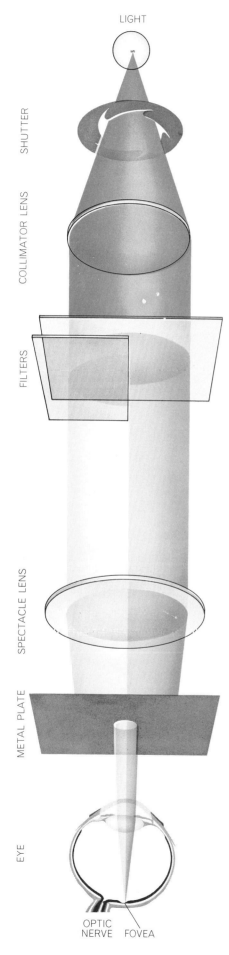

LIGHT

SHUTTER

COLLIMATOR LENS

FILTERS

SPECTACLE LENS

METAL PLATE

EYE

OPTIC NERVE   FOVEA

colleagues and I at the University of Cambridge took up the search. Our experiments have since yielded a twofold profit: in addition to explaining the origin of several types of afterimage, they have made it possible to deduce some of the elusive properties of the receptive cone pigments on which our color vision depends.

One of the principal obstacles we encountered in our search was the difficulty of trying to resolve any afterimage into its photochemical and nervous constituents. Experiments with animals suggested that both these mechanisms were capable of producing afterimages, either in combination or independently. It is a well-known fact, for example, that the receptive pigments of the rods and cones are chemically altered, or "bleached," by light and that after strong illumination it is several minutes before they are restored to their original state. During this period either the lack of pigment where it ordinarily would be or the presence of a temporary product of the photochemical reaction could cause a residual insensitivity to new stimuli or a persistent excitation in the dark. We also knew from experiments with animals that the response of a network of nerve cells—either in the retina or in the brain—may last for many seconds after the stimulus has ceased. Moreover, it had been shown by several experimenters that the nervous response made to one stimulus was capable of influencing or even suppressing the response made to other stimuli coming many seconds later [see "Inhibition in Visual Systems," by Donald Kennedy; SCIENTIFIC AMERICAN Offprint 162].

The extremely difficult problem of resolving an afterimage into its photochemical and nervous components could of course be circumvented if it were possible to find an example in which only nervous effects or only photochemical effects are involved. So far attempts to identify an afterimage of purely nervous

EXPERIMENTAL SETUP was used by the author to test a person's ability to discriminate between two slightly differing flashes of light by means of their immediate sensations and then by means of their late afterimages. Gradations in brightness of the two half-fields were obtained by substituting various combinations of neutral filters (*center*). Subjects were instructed to trigger the shutter (*top*) that delivered the flashes only when they were looking at the dividing line between the two half-fields, thus ensuring that the flashes fell directly on color-sensitive foveal region of the retina (*bottom*).

origin have not been successful, but conditions have been found under which some appear to arise purely from photochemical reactions.

In our search for an afterimage that depends only on the photochemical effects of the inducing stimulus we made use of a basic law of photochemistry: the Bunsen-Roscoe law, which states that the photochemical effects on any two light stimuli are identical if the products of their strength and the length of time they operate are equal. On this basis it might seem that if two bright flashes of light, one lasting a second and the other lasting a hundredth of a second but being 100 times brighter than the first, were compared by eye, they would be indistinguishable. Obviously this is not the case; the eye easily establishes that the first flash is longer and the second brighter. We are forced to conclude that the two flashes must have different effects on at least some of the nerve cells along the visual pathway. From experiments in which the electrical activity of nerve cells in the retina and brain of various animals have been recorded an even stronger conclusion can be drawn: Two flashes of this sort almost certainly differ in their effects on every nerve cell that responds to them at all.

If we now examine the negative afterimages of these two flashes, we find that during the first 15 seconds the afterimages differ in color and strength. After 15 seconds, however, they are absolutely indistinguishable. This is not simply because we are poor at distinguishing one kind of late afterimage from another (a late afterimage is defined as one that is more than 15 seconds old); if the second flash is made not 100 times brighter but 80 or 120 times brighter, its late afterimage is easily distinguished from that of the first flash. Thus the Bunsen-Roscoe law is valid for this type of late afterimage, since two flashes consisting of the same total amount of light produce identical late afterimages. This suggests that the late negative afterimage of a brief, bright stimulus must depend only on its photochemical effects and not at all on its immediate effects on nerve cells.

A simple experiment conducted in our laboratory corroborated this hypothesis. The test situation is shown in the illustration at the left. Two flashes of light were presented to a test subject, who was asked to judge which of the two flashes was the brighter, first by means of the immediate sensations they produced, and then by means of their late afterimages. The flashes appeared simultaneously on the left and right halves of a circular field with an apparent diameter of two degrees, and the subjects were instructed to press the trigger that delivered the flashes only when they were looking at the dividing line between the two half-fields. In this way we made certain that the flashes fell directly on the fovea, the small central depression on the retina that contains thousands of color-sensitive cones but few, if any, rods. Since it was the total amount of light in each flash that we were interested in, we had to take special care to incorporate the time factor into all our calculations. Accordingly we amended the conventional unit of luminance—candelas per square meter—to read candelas per square meter times seconds in order to obtain our unit of the amount of light in a flash.

The graph at the right presents the results of this experiment. When the intensity of the dimmer half-field was between 1/300 and 100 of our units, the ratio of brightness between the two flashes required for discrimination by means of their immediate sensations was small and approximately constant. When the intensity of the dimmer half-field rose above 100 units, however, this ratio became much greater, so that two flashes, one of which was 10 times brighter than the other, gave identical immediate sensations.

When the two flashing lights were compared on the basis of their late afterimages, another mechanism for discrimination was revealed. Although a flash of less than 30 units did not produce an afterimage lasting long enough for any judgment to be made, flashes above 100 units were discriminated much more readily by their afterimages than by their immediate sensations. In fact, it was possible to discriminate between two slightly differing flashes by means of their late afterimages for intensities as high as 100,000 units! Above this threshold the ratio of brightnesses required for discrimination rises sharply, so that an afterimage of a 1.5-million-unit flash cannot be distinguished from the afterimage of a flash 10 times brighter.

This greater capacity for discriminating slight differences in light intensity by means of late negative afterimages fits in nicely with our hypothesis that such afterimages are purely photochemical in origin. Presumably two differing flashes that present the same immediate sensations must have an identical effect on nerve cells throughout all the later stages of the chain of vision. Since the same two flashes do not necessarily produce the same afterimages, however, we can conclude that the total information re-

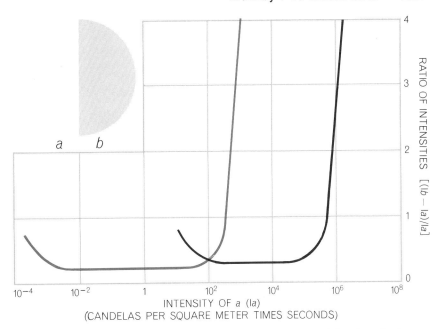

DISCRIMINATION between two slightly differing flashes by means of their late afterimages (*black curve*) proved to be more precise at high intensities than discrimination by immediate sensations (*colored curve*). The dimmer half-field (*a*) in the inset at top left corresponds to the double filter in the illustration on the opposite page; the brighter half-field (*b*), to the single filter. Flashes of less than 30 units do not produce late afterimages.

garding light intensity capable of being received by the pigments of the retina is greater than the nerve circuitry of the visual pathway can transmit instantaneously; but under suitable conditions it can transmit additional information later, in the form of an afterimage.

The only two cases in which the late afterimages of widely dissimilar flashes were indistinguishable occurred when both flashes contained the same amount of light (measured in candelas per square meter times seconds) or when both were so bright they were able to bleach nearly all the receptive pigments of the foveal cones. We know from the independent measurements of W. A. H. Rushton of the University of Cambridge that under the conditions of our experiment a flash of 1.5 million units is able to bleach about 98 per cent of the green-sensitive and red-sensitive pigments of the foveal cones [see "Visual Pigments in Man," by W. A. H. Rushton; SCIENTIFIC AMERICAN Offprint 139]. A stronger flash could do no more than increase this amount to 100 per cent, so that it was not surprising that any pair of flashes above 1.5 million units produced indistinguishable afterimages.

These observations led us to examine another peculiar property of human foveal afterimages. In 1955 W. A. Hagins discovered while working at the University of Cambridge that a single very brief flash of light, however bright,

could never bleach more than half of a sample of rhodopsin, the receptive pigment of the rods. A second flash delivered within a millisecond had no bleaching effect, but if the flash was delivered several tens of milliseconds later, it would bleach half of the remaining rhodopsin, leaving only a quarter of the original sample. The behavior of the foveal cone pigments appears to be quite similar. A flash of light lasting only two-tenths of a millisecond, however bright, never produces an afterimage as strong as the one produced by a very bright flash lasting many milliseconds. Even if we add to the first flash another similar flash coming about a quarter of a millisecond later, the afterimage produced by the two flashes together does not differ from that produced by the first alone. If, however, the second flash comes four milliseconds later, it makes the afterimage appreciably stronger. From this evidence I argued in 1959 that human cone pigments must share the property discovered by Hagins for rhodopsin. It has since been verified experimentally by Rushton that they do indeed possess this property.

The analogy between the behavior of the rod pigment rhodopsin and that of the various cone pigments involved in the production of colored afterimages extends even further. Molecules of rhodopsin in a rod are not packed at random; they all lie with their chromophore, or color-absorbing, groups perpendicular to the long axis of the rod [*see bottom illustration on page 2586*]. Evidence for this is that rods are dichroic: they absorb light differently according to its direction of incidence or of polarization. Plane-polarized light that strikes a rod from the side is absorbed if its electric vector is perpendicular to the long axis of the rod but not if its electric vector is along the long axis of the rod. The question now arises whether the chromophore groups are free to rotate in the plane perpendicular to the axis of the rod or are absolutely fixed. For rhodopsin this question was answered by Hagins: They are free to rotate, as is shown by the failure of the rods to become dichroic for light striking them from the end after partial bleaching with plane-polarized light also delivered from the end. Our experiments with afterimages have provided us with an answer to the corresponding question for the receptive pigments of the cones. If a flash of unpolarized light is of a strength sufficient to bleach a large fraction of the receptive pigment, and if the chromophore groups are fixed, then a flash of plane-polarized light of the same energy would bleach less because those molecules whose chromophores are parallel to the electric vector of the light would not be affected by it. Following the same line of reasoning, we would expect that if the chromophore groups were fixed, strong plane-polarized and unpolarized flashes of the same energy would give clearly differing afterimages. They do not, however, even if the flashes are as short as 200 microseconds. Therefore the chromophore groups of the cone pigments must be free to rotate, and the time required for rotation must be small compared with 200 microseconds. Flashes bright enough to bleach a large fraction of the receptive pigment in less than 200 microseconds are difficult to produce in the laboratory, but when the technical difficulty is overcome (perhaps by the use of lasers or exploding wires), it may be possible to ascertain how much less than 200 microseconds is the time required for this rotation.

We are now in a position to explain several related phenomena pertaining to the progressive blurring of human foveal afterimages. A flash that is strong enough to bleach nearly all the receptive pigment of the foveal cones produces a negative afterimage that remains visible

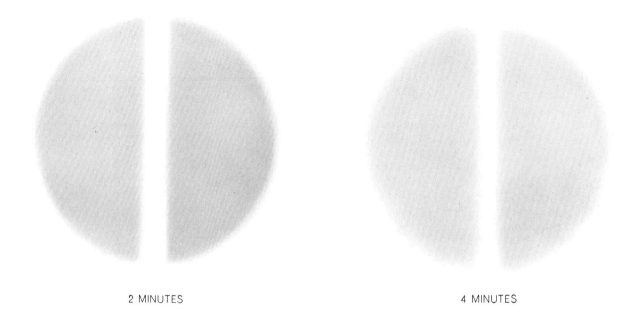

2 MINUTES   4 MINUTES

8 MINUTES   16 MINUTES

PROGRESSIVE BLURRING of a late negative afterimage is represented here. The stimulus was an extremely bright green flash presented to the fovea as a disk with an apparent diameter of 1.5 degrees crossed by a vertical dark bar 10 minutes of arc in width.

for about 20 minutes. If the source of the flash has sharp outlines and if these are properly focused on the retina, they will appear clearly in the afterimage, at least for the first two minutes. After the end of the second minute, however, such fine details will become progressively blurred, and after 15 minutes only the coarsest aspects of the shape of the stimulus will be resolvable in the afterimage. This progressive blurring is represented in the illustration on the opposite page. In this case the stimulus used was a disk of green light with an apparent diameter of 1.5 degrees crossed by a dark bar 10 minutes of arc wide. The resulting negative afterimage was pink, the complementary color to green. After two minutes this afterimage still showed the sharp edges of the stimulus, although a faint pink cloud had begun to invade the surrounding white. After four minutes the sharp edges had dissolved, the central gap had narrowed and the cloud continued to spread outward. After eight minutes the central gap could be seen only with difficulty, and after 16 minutes it disappeared entirely.

Another demonstration of the progressive blurring of a foveal afterimage was obtained by placing a grating between the stimulus and the eye in order to measure the length of time the individual bars of the grating could be resolved in the afterimage [*see top illustration on this page*]. We tested this technique using three different types of light stimulus: a deep red foveal stimulus, which produces green negative afterimages (probably depending on the cone pigment erythrolabe); a green foveal stimulus, which produces pink negative afterimages (probably depending on the cone pigment chlorolabe), and a green extrafoveal stimulus, which produces gray negative afterimages (probably depending on the rod pigment rhodopsin). In the extrafoveal afterimages the finest gratings could never be resolved, but there was little progressive blurring, so that almost as fine a grating could be resolved at 15 minutes as at one minute. Both kinds of foveal afterimage showed progressive blurring, which took place faster in those produced by the green stimulus. For both kinds the speed of the progressive blurring was unaffected by light conditions between the time of the stimulus and the examination of the afterimage many seconds later; the speed was the same whether the eye was kept in darkness, in steady light or in alternating darkness and light. This constant rate could not have been merely a result of the faintness of the late afterimages; the two-minute afterimage of the weak stim-

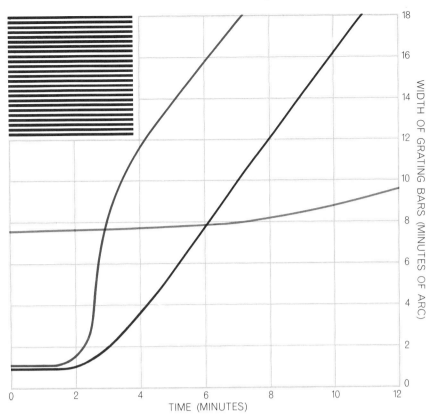

GRATINGS WERE SUBSTITUTED (*inset at top left*) for the neutral filter nearest the viewer in the experimental setup shown on page 2582 in order to measure the rate of blurring of several types of late negative afterimage. The colored curve indicates this rate for the afterimage of a deep red foveal stimulus, the black curve for the afterimage of a green foveal stimulus and the gray curve or the afterimage of a green extrafoveal stimulus.

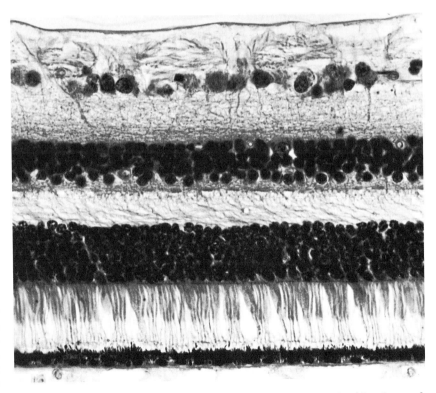

SECTION OF HUMAN RETINA is magnified about 450 diameters in this micrograph made, as are those on the opposite page, by Ben S. Fine of the Armed Forces Institute of Pathology. Light enters the retina from the top and passes through several layers of nerve cells before striking the rods and cones, the spindly vertical structures near the bottom.

ROD in this electron micrograph is magnified about 24,000 diameters. The layered structures are the photoreceptor segments.

CONE in this electron micrograph is located in an extrafoveal region of the retina. Magnification is approximately 13,000 diameters.

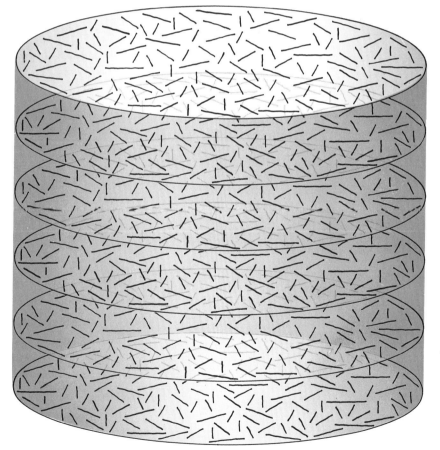

VISUAL PIGMENTS are probably arranged in roughly circular layers in both rods and cones. In this highly schematic drawing the chromophore, or color-absorbing, groups of the pigment molecules are the short black lines lying in planes perpendicular to the long axis of the receptor. The protein parts of the molecules are not shown. Experiments with afterimages suggest that the chromophore groups must be free to rotate within their planes.

ulus can be fainter than the 15-minute afterimage of a strong stimulus, but it is always very much sharper.

If, as I have argued, the persisting change responsible for these late negative afterimages is photochemical and originates in the cones, then it may be either a lack of receptive pigment in the bleached cones or the presence of some substance produced by the action of light on the receptive pigment. Lack of receptive pigment can hardly be supposed to afflict other than bleached cones and therefore cannot on any simple hypothesis explain the progressive blurring. A substance produced by the action of light might, however, diffuse from the place where it originated and influence the sensitivity of other structures. If so, these structures must also be cones, because if they were nerve cells engaged in transmitting messages from the cones to the brain, an afterimage would spread in the blurring process much more slowly between two points on opposite sides of the foveal center than between two otherwise similar points on the same side of it. No such discontinuity of spread at the foveal center is observed. To account for the colors of the late afterimages we must assume that the product derived from erythrolabe affects predominantly the sensitivity of red-sensitive cones and the one derived from chlorolabe affects predominantly the sensitivity of green-sensitive cones. To account for the fact that the green negative afterimages of deep red stimuli become blurred faster than the pink negative afterimages of green stimuli we must assume that the product derived from erythrolabe diffuses faster through the retina. For both substances the diffusion rates required to account for the blurring are plausible; that is, they are substantially smaller than those of small organic molecules in water.

This explanation of the progressive blurring of foveal afterimages also provides us with an interpretation of a hitherto unaccountable afterimage phenomenon known as the green halo. When the negative afterimage of an orange-red flash is seen against a uniform white background about eight minutes after the stimulus is administered, the central part of the afterimage is the same pink color as the late afterimages of green or yellow stimuli. Surrounding the pink center, however, is a ring of green whose color is the same as that of a late afterimage of a deep red stimulus [see illustration on page 2587]. We know from Rushton's experiments the orange-red stimulus bleaches both erythrolabe and chlorolabe; if we assume that because of the difference in their diffusion rates the chlorolabe product stays mainly in the part of the retina illuminated by the stimulus, and that the erythrolabe product mainly diffuses away into the surrounding region, the green halo is explained.

We still know very little about the chemical nature of these diffusible products or about their possible function in normal vision. We do know that when light acts on the rod pigment rhodopsin, it splits it into a protein called opsin, which remains fixed in the rods, and a substance called retinene 1, which can diffuse out of the rods in which it was formed. The only receptive cone pigment whose chemistry has been investigated is iodopsin, which George Wald of Harvard University extracted from chicken retinas and showed to be made up of retinene 1 combined with a protein different from that contained in rhodopsin. It seems reasonable to guess that each of the human cone pigments chlorolabe and erythrolabe is, like rhodopsin and iodopsin, made up of retinene 1 and a specific protein and is split by light, yielding retinene 1 as its diffusible product; indeed, Paul K. Brown of Harvard has recently obtained direct evidence in favor of this idea. If, however, we accept the evidence from afterimages that different diffusible products are liberated by light acting on chlorolabe and erythrolabe, then we must conclude either that chlorolabe and erythrolabe are not both retinene-1 derivatives or that the diffusible substances responsible for late afterimages are secondary products (not directly derived from the receptive pigments but nonetheless liberated in quantities determined by the quantities of receptive pigments bleached). Regardless of whether these substances are directly derived from chlorolabe and erythrolabe or are secondary, the specific action that each has on the sensitivity of its own kind of cone suggests that they may form a part of the least understood link in the chain of excitation leading from absorbed light quanta to the brain, the link by which a photochemical change occurring in a very small number of molecules of receptive pigment—at most seven and perhaps only one—provokes a cone to send a signal to the nerve cells of the retina.

GREEN HALO appears around the late negative afterimage of a very bright orange-red flash. A stimulus of this type bleaches both erythrolabe and chlorolabe. According to the author's hypothesis the erythrolabe product, which is responsible for the green in the afterimage, diffuses faster than the chlorolabe product, which is responsible for the pink.

The Author

G. S. BRINDLEY is lecturer in physiology at the University of Cambridge. He received a B.A. from Cambridge in 1947 and a medical degree from the University of London in 1950. After a year's residency in ophthalmology at the London Hospital he did research on visual problems of aviation for the Royal Air Force Institute of Aviation Medicine from 1952 to 1954. In 1960 he spent a year doing research at the University of California at Berkeley and in 1962 he worked at the Johns Hopkins Medical School.

Bibliography

THE DISCRIMINATION OF AFTER-IMAGES. G. S. Brindley in *Journal of Physiology*, Vol. 147, No. 1, pages 194–203; June, 1959.

PHYSIOLOGY OF THE RETINA AND THE VISUAL PATHWAY. G. S. Brindley. Edward Arnold (Publishers) Ltd, 1960.

TWO NEW PROPERTIES OF FOVEAL AFTER-IMAGES AND A PHOTOCHEMICAL HYPOTHESIS TO EXPLAIN THEM. G. S. Brindley in *Journal of Physiology*, Vol. 164, No. 1, pages 168–179; October, 1962.

SCIENTIFIC
AMERICAN   May 1956, Vol. 194, No. 5, pp. 48-52

OFFPRINT 1090

THE EYE AND THE BRAIN

by R. W. Sperry

If the optic nerve of a newt is cut and its eye is turned through 180 degrees, the nerve regenerates and the animal sees upside down. Such results deeply affect our picture of how the nervous system develops.

Probably no question about the behavior of living things holds greater general interest than the age-old issue: Heredity versus Learning. And none perhaps is more difficult to investigate in any clear-cut way. Most behavior has elements of both inheritance and training; yet each must make a distinct contribution. The problem is to separate the contributions. We can take vision as a case in point. An animal, it is often said, must learn to see. It is born with eyes, but it matures in the use of them. The question is: Just where does its inborn seeing ability end and learning begin? To put the matter another way: Exactly what equipment and instinctive skills are we born with?

This article is an account of experiments which have given some new insight into the heredity-learning question. The behavior studied is vision, and the story begins 31 years ago.

In 1925 Robert Matthey, a zoologist of the University of Geneva, delivered to the Society of Biology in Paris an astonishing report. He had severed the optic nerve in adult newts, or salamanders, and they had later recovered their vision! New nerve fibers had sprouted from the cut stump and had managed to grow back to the visual centers of the brain. That an adult animal could regenerate the optic nerve (and even, as Matthey reported later, the retina of the eye) was surprising enough, but that it could also re-establish the complex network of nerve-fiber connections between the eye and a multitude of precisely located points in the brain seemed to border on the incredible. And yet this was the only possible explanation, for without question the newts had regained normal vision. They would stalk a moving worm separated from them by a glass wall in their aquarium; they were able to see a small object distinctly and follow its movements accurately.

A long series of confirmations of Matthey's discovery followed. He transplanted an eyeball from one newt to another, with good recovery of vision. Leon S. Stone and his co-workers at Yale University transplanted eyes successful-

ly from one species of salamander to another, and grafted the same eye in four successive individuals in turn, each of which was able to use the eye to regain its vision. Eventually experimenters found that fishes, frogs and toads (but not mammals) also could regenerate the optic nerve and recover vision if the nerve was cut carefully without damage to the main artery to the retina.

The optic nerve of a fish has tens of thousands of fibers, most or all of which must connect with a specific part of the visual area of the brain if the image on the retina is to be projected accurately to the brain. The newt, whose retina is less fine-grained than a fish's, has fewer optic fibers, but still a great many. The system is analogous to a distributor's map with thousands of strings leading from a focal point to thousands of specific spots on the map. How can an animal whose optic fibers have all been cut near the focal point re-establish this intricate and precisely patterned system of connections? Matthey found that the regenerating fibers wound back into the

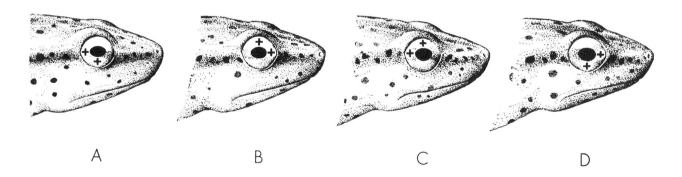

EYE OF THE NEWT was turned in various ways by the experiments described in this article. In A the normal position of the eye is marked with crosses. In B the eye has been turned so that its front-back and up-down axes are inverted. In C the eye on the op-posite side of the head has been transplanted to the side shown with its up-down axis inverted. In D the eye on the opposite side of the head has been transplanted to the side shown with its front-back axis inverted. In each case the operation is done on both eyes.

brain in what looked like a hopelessly mixed up snarl. Yet somehow, from this chaos, the original orderly system of communications was restored.

Two possible explanations have been considered. The one that was long regarded as the more plausible is that the connections are formed again by some kind of learning process. According to this theory, as the cut nerve regenerates a host of new fibers, branching and crawling all over the brain, the animal learns through experience to make use of the fiber linkages that happen to be established correctly, and any worthless connections atrophy from disuse.

The second theory is that each fiber is actually specific and somehow manages to arrive at its proper destination in the brain and reform the connection. This implies some kind of affinity, presumably chemical, between each individual optic fiber and matching nerve cells in the brain's visual lobe. The idea that each of the many thousands of nerve fibers involved has a different character seemed so fantastic that it was not very widely accepted.

These were the questions we undertook to test: Does the newt relearn to see, or does its heredity, forming and organizing its regenerated fibers according to a genetic pattern, automatically restore orderly vision?

Our first experiment was to turn the eye of the newt upside down—to find out whether this rotation of the eyeball would produce upside-down vision, and if so, whether the inverted vision could be corrected by experience and training. We cut the eyeball free of the eyelids and muscles, leaving the optic nerve and main blood vessels intact, then turned the eyeball by 180 degrees. The tissues rapidly healed and the eyeball stayed fixed in the new position.

The vision of animals operated on this way was then tested. Their responses showed very clearly that their vision was reversed. When a piece of bait was held above the newt's head, it would begin digging into the pebbles and sand on the bottom of the aquarium. When the lure was presented in front of its head, it would turn around and start searching in the rear; when the bait was behind it, the animal would lunge forward. (Since its eyes are on the side of the head, a newt can see objects behind it.) As color-adapting animals, the newts with upside-down eyes even adjusted their color to the brightness above them instead of to the dark background of the aquarium bottom. Besides seeing everything up-

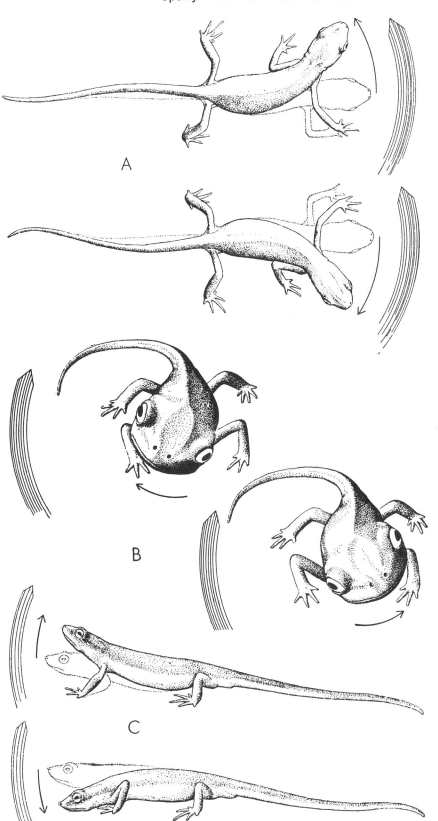

RESPONSE OF THE NEWT to moving objects varies with the operations depicted on the opposite page. The first newt in each of the three pairs of animals on this page is normal. When an object (*thick arrows*) is moved past the newt, the animal turns its head in the same direction (*thin arrows*). The second newt in each pair represents the behavior of the animal after one or more of the operations. The response of the second newt in A corresponds to operations B and D on the opposite page; in B, to operations B and C; in C, to C and D.

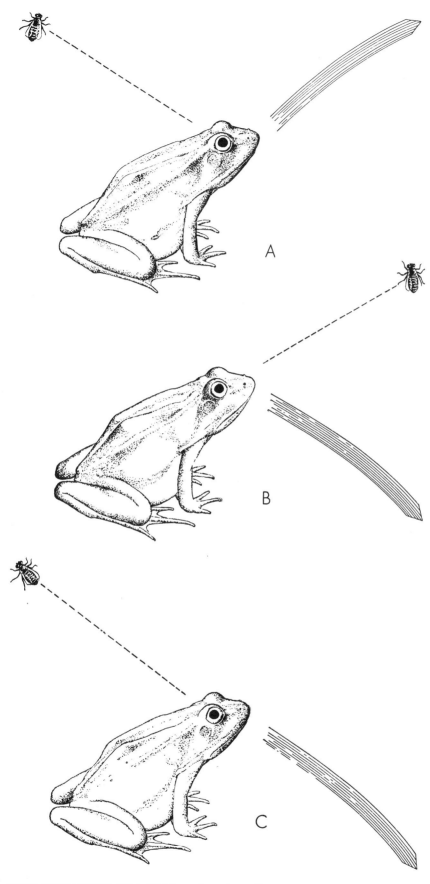

side down and backward, the animals kept turning in circles, as if the whole visual field appeared to be whirling about them. Human subjects who have worn experimental lenses that invert the visual field have reported that any movement of the head or eyes tends to make everything seem to whirl around them.

The operated newts never relearned to see normally during the experiment. Some were kept with their eyes inverted for as long as two years, but showed no significant improvement. However, when rotated eyes were turned back to the normal position by surgery, the animals at once resumed normal behavior. There was no evidence that their long experience with inverted vision had brought about any change in the functioning of the central nervous system.

A second experiment bore out further the now growing suspicion that learning probably was not responsible for the recovery of vision by newts whose optic nerves had been cut. This time we rotated the eyeball and severed the optic nerve as well. The object was to find out whether the regenerating nerve fibers would give the newt normal vision, inverted vision or just a confused blur.

During the period of nerve regeneration the animals were blind. The first visual responses began to reappear about 25 to 30 days after the nerve had been cut. From the beginning these responses were systematically reversed in the same way as those produced by eye rotation alone. In other words, the animals again responded as if everything was seen upside down and backward. In these animals also the reversed vision remained permanently uncorrected by experience.

In another series of experiments we cut the optic nerves of the two eyes and switched their connections to the brain. Normally each optic nerve crosses to the side of the brain opposite the eye. We connected the cut nerve to the brain lobe on the same side. The result was to make the animals behave after regeneration as if the right and left halves of the visual field were reversed. That is, the animals responded to anything seen through one eye as if it were being viewed through the other eye. This switch too was permanent, uncorrected by experience. Frogs and toads responded to the experiment in the same way as newts.

By rotating the eyeball less than 180 degrees (*e.g.*, a 90-degree turn), and by combining eye transplantation from one side to the other with various degrees of rotation, we produced many

SAME OPERATIONS ON A FROG produce these effects when the animal strikes a fly. In A the fly is above and behind a frog whose eyes have been turned by operation D on page 2588; the animal strikes in the direction shown by the thick arrow. In B the eyes have been turned by operation C. In C the eyes of the frog have been turned by operation B.

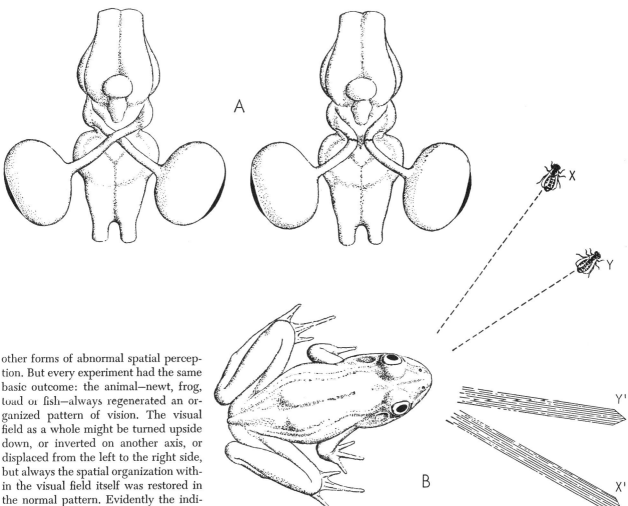

OPERATION ON THE OPTIC NERVES of a frog produced the effect shown at the lower right. At upper left the eyes of the frog are joined to the brain by the optic nerves. In the operation, which is depicted at top center, the nerves were cut and rejoined so that they did not cross. When a fly was at X, the frog struck at X'; when it was at Y, the frog struck at Y'.

other forms of abnormal spatial perception. But every experiment had the same basic outcome: the animal—newt, frog, toad or fish—always regenerated an organized pattern of vision. The visual field as a whole might be turned upside down, or inverted on another axis, or displaced from the left to the right side, but always the spatial organization within the visual field itself was restored in the normal pattern. Evidently the individual nerve fibers from the retina, after regeneration, all regained their original relative spatial functions in projecting the picture to the brain.

This orderly restoration of the spatial relations could hardly be based on any kind of learning or adaptation, under the conditions of our experiments. Animals don't *learn* to see things upside down and backward or reversed from left to right: reversed vision is more disadvantageous than no vision at all. The results clearly demonstrated that the orderly recovery of correct functional relations on the part of the ingrowing fibers was not achieved through function and experience, but rather was predetermined in the growth process itself.

Apparently the tangle of regenerating fibers was sorted out in the brain so as to restore the orderly maplike projection of the retina upon the optic lobe. If we destroyed a small part of the optic lobe after such regeneration, the animal had a blind spot in the corresponding part of its visual field, just as would be the case in normal animals. It was as if each regenerated fiber did indeed make a connection with a spot in the brain

matching a corresponding spot in the retina.

It follows that optic fibers arising from different points in the retina must differ from one another in some way. If the ingrowing optic fibers were indistinguishable from one another, there would be no way in which they could re-establish their different functional connections in an orderly pattern. Each optic fiber must be endowed with some quality, presumably chemical, that marks it as having originated from a particular spot of the retinal field. And the matching spot at its terminus in the brain must have an exactly complementary quality. Presumably an ingrowing fiber will attach itself only to the particular brain cells that match its chemical flavor, so to speak. This chemical specificity seems to lie, as certain further experiments indicate, in a biaxial type of differentiation which produces unique arrays of chemical properties at the junction places.

Such chemical matching would account for recognition on contact, but how does a fiber find its way to its destination? There is good reason to believe that the regenerating fibers employ a shotgun approach. Each fiber puts forth many branches as it grows into the brain, and the brain cells likewise have widespreading branches. Thus the chances are exceedingly good that a given fiber will eventually make contact with its partner cells. We can picture the advancing tip of a fiber making a host of contacts as it invades the dense tangle of brain cells and their treelike expansions. The great majority of these contacts come to nothing, but eventually the growing tip encounters a type of cell surface for which it has a specific chemical affinity and to which it adheres. A chemical reaction then causes the fiber tip to stop advancing and to form a lasting functional union with the group of cells, presumably roughly circular in

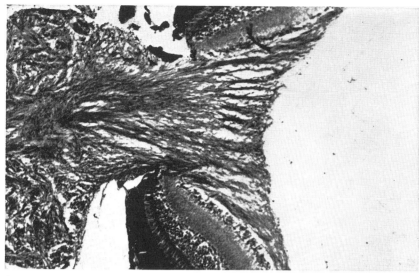

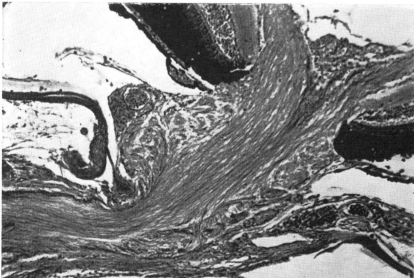

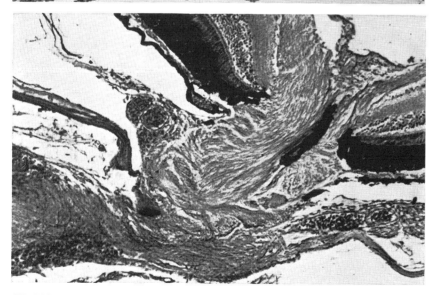

OPTIC NERVE of *Bathygobius soporator*, a fish of the goby family, was cut and allowed to regenerate. The regenerated nerve is shown in these three photomicrographic sections. In each photograph the eye is toward the right and the nerve runs from right to left. The top photograph shows a section of one nerve; the bottom two photographs show different sections of the same nerve. In all three sections the nerve fibers are tangled. Despite this apparent disorganization the fishes from which the sections were taken could see normally.

formation, which constitutes the spot in the brain matching the fiber's source spot in the retina.

The experiments on vision have been found to apply equally to other parts of the central nervous system. Normal function can be recovered through regeneration by general sensory nerves in the spinal cord, by the vestibular nerve in the ear mediating the sense of equilibrium and by other sensory and motor nerve circuits.

All the experiments point to one conclusion: the theory of inherent chemical affinities among the nerve fibers and cells is able to account for the kinds of behavior tested better than any hypothetical mechanism based on experience and learning. There is no direct proof of the theory, for no one has yet seen evidence of the chemical affinity type of reaction among nerves under the microscope. But an ever-growing accumulation of experimental findings continues to add support to the chemical theory.

We return to our original question: How big a role does heredity play in behavior? The experiments cited here show that in the lower vertebrates, at least, many features of visual perception—the sense of direction and location in space, the organization of patterns, the sense of position of the visual field as a whole, the perception of motion, and the like—are built into the organism and do not have to be learned. More general experiments suggest that the organization of pathways and associations in the central nervous system must be ascribed for the most part to inherent developmental patterning, not to experience. Of the thousands of circuit connections in the brain that have been described, not one can demonstrably be attributed to learning. Whatever the neural changes induced in the brain by experience, they are extremely inconspicuous. In the higher animals they are probably located mainly in the more remote byways of the cerebral cortex. In any case they are superimposed upon an already elaborate innate organization.

The whole idea of instincts and the inheritance of behavior traits is becoming much more palatable than it was 15 years ago, when we lacked a satisfactory basis for explaining the organization of inborn behavior. Today we can give more weight to heredity than we did then. Every animal comes into the world with inherited behavior patterns of its species. Much of its behavior is a product of evolution, just as its biological structure is.

The Author

R. W. SPERRY is a psychobiologist in the department of biology at the California Institute of Technology. He graduated from Oberlin College in 1935, and in 1941 took a Ph.D. in zoology at the University of Chicago. He then did several years of research at Harvard University and at the Yerkes Laboratories of Primate Biology. From 1946 to 1952 he taught anatomy at the University of Chicago, and from 1952 to 1954 he was chief of developmental neurology at the National Institutes of Health. Since 1954 he has been Hixon Professor of Psychobiology at Cal Tech. His work has covered various aspects of the functioning of the central nervous system, including the mechanisms involved in perception, learning and memory.

Bibliography

MECHANISM OF NEURAL MATURATION. R. W. Sperry in *Handbook of Experimental Psychology.* John Wiley & Sons, Inc., 1951.

PATTERNING OF CENTRAL SYNAPSES IN REGENERATION OF THE OPTIC NERVE IN TELEOSTS. R. W. Sperry in *Physiological Zoology,* Vol. 21, No. 4, pages 351-361; October, 1948.

SCIENTIFIC
AMERICAN   January 1968, Vol. 218, No. 1, pp. 21-27

OFFPRINT 1091

EARLIER MATURATION IN MAN

by J. M. Tanner

Over the past 100 years people have been not only getting bigger but also bigger earlier. By the same token there has been a dramatic decline in the age of puberty.

It is often remarked that children seem to be bigger than they used to be. It is true; measurements of boys and girls made in various parts of the world over the past 100 years show that the average size of children at all ages has increased markedly. It is perhaps not so widely recognized that children also reach physical maturity at an earlier age. Here again the facts are clear: during the past 50 to 100 years children in North America, Europe, Japan and at least parts of China have come to puberty progressively earlier. Today children in many communities are reaching puberty three years sooner—in some communities five years sooner—than they once did. The average age at which growth stops

is also lower; it is exceptional nowadays for a man to grow more than a small fraction of an inch after he is 19. The worldwide trend toward earlier maturity may be leveling off in the U.S.; elsewhere it shows little sign of doing so.

One naturally wants to know what factors of heredity and environment have given rise to this trend. There are strong indications of the causes. As with all trends in human biology, however, the causes are complex. Ordering them into those that are more important and those that are less so is still largely a matter of personal judgment.

The evidence for the earlier maturation of children comes from two sources: records of the age at which girls have

first menstruated and records of children's height and weight. A demonstration that the children of today are larger than children in the past could simply mean that today adults are proportionately bigger. Indeed, there is evidence that the fully grown adults of the present are larger than the fully grown adults of 100 years ago. The increase in adult size, however, is much less than the increase we find in children. The increase in the size of children must therefore also be related to earlier maturation.

The records documenting the change in children's size between 1880 and 1960 come from several European countries, the U.S., Canada, Japan, Hong Kong, New Zealand and New South

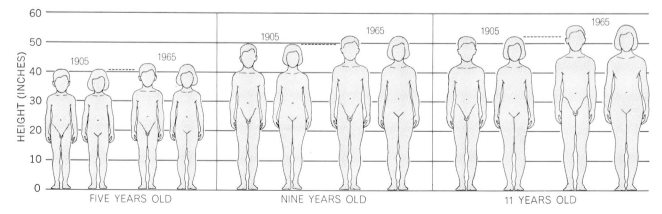

INCREASED SIZE OF CHILDREN stems mainly from earlier maturation. A boy and girl aged five in 1965, and of average economic circumstances, were taller by about two inches than their counter-parts of a half-century ago; nine-year-olds of 1965 averaged some three inches taller and 11-year-olds nearly four inches taller. The figures are based on measurements made in the U.S. and Europe.

Wales in Australia. There are also figures for London and Glasgow and much earlier but also scrappier data on the height of English boys. Extensive records have been kept in Sweden; for example, we can compare measurements made there in 1883 with those made in

1938. For 1938 we have measurements of 8,500 Swedish schoolchildren between the ages of seven and 19. As is usually the case, the children receiving only elementary education (as judged by their course of study) were slightly smaller than children of the same ages who

continued in school. These differences appear very small, however, when the height and weight of the 1938 group are compared with the height and weight of Swedish children of the same ages in 1883. The 15-year-old boys of 1938 were on the average some five inches

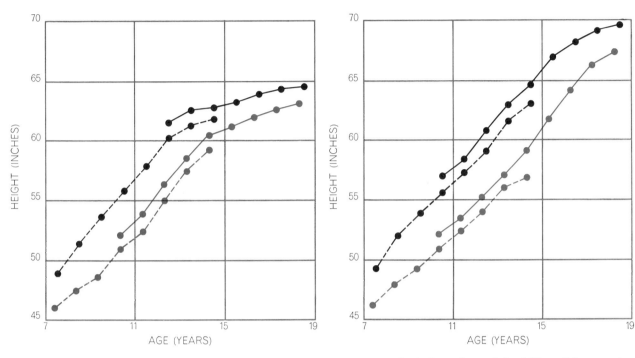

HEIGHT GAIN is exemplified by measurements of Swedish schoolchildren in 1938 (*black lines*) compared with a like group in 1883 (*colored lines*). By 18 the girls (*left*) had stopped growing; the boys (*right*) had not. Some of the children did not continue in school after the age of 14. As is usually the case, these children did not grow quite as tall as those whose education continued.

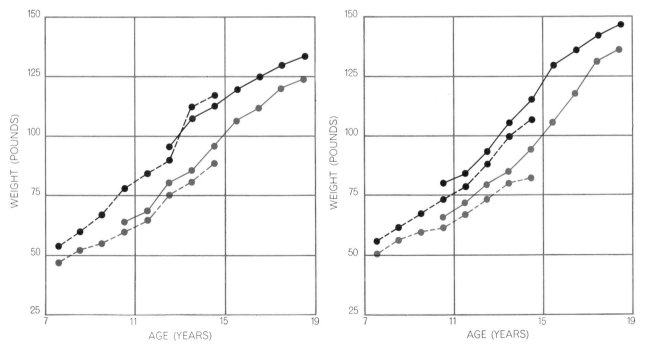

WEIGHT GAIN of the same girls (*left*) and boys (*right*) was approximately proportional to the gain in height between 1883 (*color*) and 1938 (*black*). Body proportions have not altered; the change has been in the size of children rather than their shape. The four graphs shown here are based on the data of Birger Broman, Gunnar Dahlberg and A. Lichtenstein of the University of Uppsala.

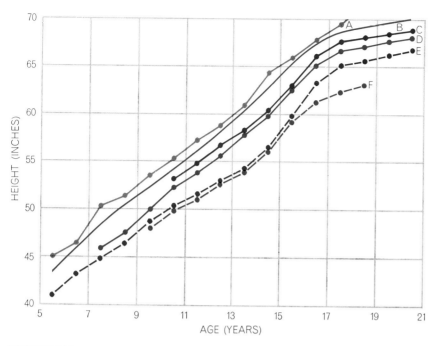

HISTORICAL TREND shows itself in the height of English boys who were measured at different times during the 19th and 20th centuries. The largest boys are those of the upper class (from Birmingham) in 1955 (*A*). They were slightly taller than the average for all boys in Britain in 1958 (*B*). Students of 1878 from the upper classes were somewhat shorter (*C*). They surpass boys of 1874 in the nonlaboring class (*D*), who were taller than those of 1874 in the laboring class (*E*). The shortest of all were factory boys of 1833 (*F*). The figure is based on the data of E. M. B. Clements for 1955; the author for 1958; Charles Roberts, W. Fergus, G. F. Rodwell and Francis Galton for 1874 and 1878, and S. Stanway for 1833.

taller than their 1883 counterparts. By the age of seven the 1938 children have an edge corresponding to about a year and a half's growth. The difference between the two groups is less when growth is completed than it is in the growing period; it nevertheless exists. The girls of 18 in 1938 were nearly two inches taller than the girls of 18 in 1883. (It should be said that the last comparison, between groups still in school at 18, involves the best-off part of the population and is therefore a rather special case.)

The trend in size that is apparent in the Swedish data also appears in the records from other countries. The data from Europe and North America are all in good agreement; they indicate that beginning in 1900 or a little earlier children in average economic circumstances aged five to seven have increased in height between half an inch and three-quarters of an inch per decade. For those 10 to 14 the increase in height has been between an inch and an inch and a half per decade. In Glasgow, for example, present-day five-year-olds are about two inches taller than the five-year-olds of 1906; nine-year-olds are some three inches taller than their 1906 counterparts and 11-year-olds nearly four

inches taller. In Iowa nine-year-old girls are also about three inches taller now than they were in 1900. The gain in weight and in other bodily dimensions is approximately proportional to the gain in height. Thus there has been little or no change in overall shape.

Only scanty measurements exist of children of preschool age. The last large series of measurements from London (between 1954 and 1959) shows most of the change concentrated between ages eight and 14, with little difference at five, six or seven. Otherwise the data, such as they are, indicate that in Europe and America the increase in size shows up directly after birth and may even be greater from ages two to five than it is later on.

The upward trend in size has been slowed from time to time by the famines of economic crisis and, to a greater extent, the famines of war. In Moscow during the 1940's the height of 13-year-old boys declined as much as an inch as an effect of the war. Apart from such interruptions the trend has been steady. It apparently began, at least in England, quite some time before 1880. Charles Roberts, writing in 1876, said that "a factory child of the present day at the

age of nine years weighs as much as one of 10 years did in 1833.... Each age has gained one year in forty years." Nearly all data agree in indicating that the trend in Europe not only is continuing but also has in most areas been more marked in the past 20 years than in the preceding 40. In Japan it seems that, whereas the school-age gain is nearly up to the European values, the gain before age six has been less than in Europe.

The fact that the trend is leveling off in the U.S. may indicate that the best-off children are maturing at something approaching the maximum rate. The children studied at the Fels Research Institute for the Study of Human Development in Yellow Springs, Ohio, between 1946 and 1966 showed only a small gain in height compared with the historical trend; among eight-year-olds the increase was only .1 inch per decade for the girls and .2 inch per decade for the boys. In other data as well as these there is a difference between the sexes. Probably this is because boys react more than girls to various stresses, ranging from malnutrition in Central Europe to the effects of radiation at Hiroshima. When circumstances improve, boys therefore tend to respond more than girls.

If the stature of full-grown people had remained constant over the years, Roberts' implied interpretation of the size increase of children would be correct: the entire gain in the children's height would be due to earlier maturation. The Glasgow five-year-old of 1950 would be (not look like, but be) the Glasgow six-year-old of 1900 and the 11-year-old of 1950 would be the 12½-year-old of 1900. Since the size of adults has increased, however, factors other than earlier maturation must enter in.

There has been some dispute about the adult height increase. At the turn of the century men reached their full height at around 26, whereas in Europe and America today they do so at 18 or 19. Quite recent data show that university students in France attain full growth at an even earlier age. The students who were born in 1925 were still growing between 17 and 18, and those born in 1933 had nearly stopped growing at that age. (At 17 the difference in height between the two groups was an inch and a half; at 20 the difference was only a quarter of an inch.) Since growth now stops at a different age, one cannot compare the height of 20-year-olds in 1960 with those in 1900. One must either use the 26-year-olds of 1900 for comparison or at least make allowance for the gain in height between ages 20 and 26 at that time. Fortunately there are sets of measurements that make this possible.

The classic series of measurements is from Norway, and it was reported in 1939 by Vilhelm Kiil of Oslo. This superb archive makes it clear that in Norway final adult height increased little (less than .4 inch) during the 70 years from 1760 to 1830. During the next 45 years, from 1830 to 1875, the gain was somewhat greater, around .6 inch, or about an eighth of an inch per decade. From 1875 to 1935 the gain per decade doubled, to a quarter of an inch. The tradition of the Norwegian archive has been continued by Ludwig Udjus of the Norwegian army medical corps, whose figures show a still greater gain, about a third of an inch per decade, from 1922 to 1962. In Norway, then, the trend in adult size seems to be continuing. Figures for Sweden show about the same increase for the 20th century.

In Holland records have been kept of the height of men called up at 19 or 20 for the civil militia and remeasured at the age of 25. This series of measurements extends from 1819 to 1902. The adult (25-year-old) height actually dropped slightly from 1820 to 1860, but in the 100 years since 1860 it is estimated to have risen to about the same extent as it has in Norway and Sweden.

The British data are not as useful as these. Nonetheless, Joan Ward of the Loughborough University of Technology has shown that miners of the 1950's were some 1.18 inches taller than a simi-

lar group in 1943. The increase from about 1930 to 1960 in this occupational group approaches one inch per decade. This is almost certainly greater than the increase in the better-off segments of the population. In most western European countries, however, the data are in excellent agreement; they point to an increase in adult height of between a quarter and a third of an inch per decade from about 1870 to the present day. In general, adults are from 2½ to 3½ inches taller now than they were in 1870.

In the U.S. between 1940 and 1960 the increase in size was greater for adult Negroes than it was for adult whites: a quarter of an inch per decade compared with an eighth of an inch. During the 25 years prior to 1940 the gain for whites was probably nearer a quarter of an inch per decade. It is of much interest that the data from Norway, the Netherlands and Denmark show little gain in height until 1860, and an accelerated gain from 1880 up to the present. We shall discuss the possible reasons for the trend below; suffice it here to say that it is believed that in France the real wages of laborers began to rise around 1850, and that in England the diet of laborers began to improve around 1815 but progressed very slowly until around 1850, after which it rapidly got better. Mortality in England began declining around 1840; the decline became marked after 1860. European laborers of the 16th and 17th centuries may well have been shorter than village people of medieval times. When the population expansion started (around 1750 in England), it seems likely that the social conditions of most man-

ual workers grew worse for at least 100 years.

The size of full-grown people is still increasing, at least in Europe. For example, French university students in 1951 were a quarter of an inch taller than the 1941 group. In the U.S. this trend, like the trend toward earlier maturation, may be flattening out, at least in the better-off segments of the community. It has been shown that students entering Harvard University from private schools in 1958 were only a quarter of an inch taller than the group from private schools entering Harvard in the 1930's. On the other hand, 1958 entrants from public schools, a group not as well off on the average, were taller by nearly 1½ inches than their 1930's counterparts. When the stature of public school and private school entrants of the same year are compared, however, one finds a smaller difference between the groups in 1958 than those in the 1930's: .4 inch compared with 1.6 inches. No difference at all is found between girls from public schools and girls from private schools entering Wellesley College in 1958. They were only .2 inch taller than girls entering in the 1930's.

Although the uptrend in adult size accounts for a part of the increase we find in children, it is clear that a far greater part of the increase is related to the earlier age at which children mature. The trend toward earlier maturation is perhaps best shown by statistics on the age of menarche, or first menstrual period. (The age of the first appearance of pubic hair in boys is a less reliable index

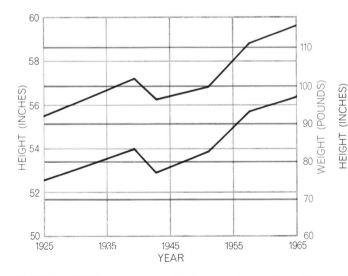

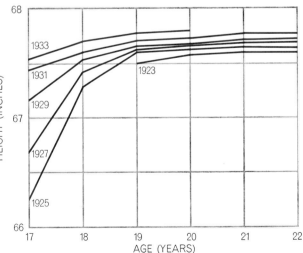

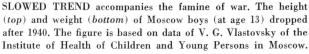

SLOWED TREND accompanies the famine of war. The height (top) and weight (bottom) of Moscow boys (at age 13) dropped after 1940. The figure is based on data of V. G. Vlastovsky of the Institute of Health of Children and Young Persons in Moscow.

CHANGED TREND emerges from measurements of university students in France, divided according to birth date. The more recently born are closer to maximal height at 17. Data are taken from Maurice Aubenque of the National Institute of Statistics in Paris.

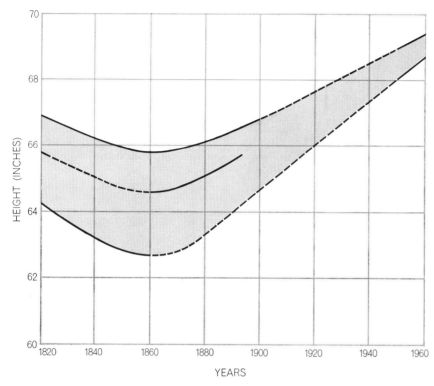

CHANGE IN MAXIMUM SIZE is illustrated by the height of conscripts in the Dutch civil militia at the age of 18 (*bottom line*), 19 (*center line*) and 25 (*top line*), when they had reached full growth. Between 1820 and 1860 the height of conscripts dropped; during the century since then it is estimated (*dotted line*) to have risen. The figure is based on data analyzed by V. M. Oppers of the Department of Health Statistics of the city of Amsterdam.

of puberty.) It is known from studies in which individuals were followed through childhood until menarche that, in a given population, age at menarche is distributed in the bell-shaped curve called the normal distribution. This makes it possible to estimate the mean age of menarche on a "cross-sectional" basis. What one does is select a proper sample—for instance a sample of all the schools in a certain area—and then ask every girl in the sample group whether or not she has experienced her first menstrual period. Ideally all girls between the ages of nine and 17 should be interrogated. Most large-scale modern studies are carried out in this way. An equally valid procedure is the "longitudinal" study, in which every child is checked repeatedly until menarche.

A procedure that may not give a valid estimate of the mean in cross-sectional data, but that has been much used in the past, is to inquire of all girls in a school at what age they first menstruated. Apart from the difficulty of exact recollection by those who had attained menarche several years earlier, a more important bias is introduced if there remain any girls who have not yet menstruated. Such girls will exhibit high values for menarcheal age, and if these

values are omitted the mean age obtained is spuriously low.

The early studies of the age of menarche suffered from disadvantages of both sampling and technique. Most records prior to 1920 concern hospital patients, who may be a biased sample of the population at large. Worse, these data give the age at which menarche was thought to have occurred by women interrogated five, 10 or even 20 years after the event. It is true that some women are able to recall fairly accurately the age of menarche even when they are over 30, but many state frankly that they may be in error by a year or more.

In spite of such defects, the data we have on menarcheal age are impressively consistent. Records are available for Britain, many other European countries and parts of America. The main conclusion is perfectly clear: girls have experienced menarche progressively earlier during the past 100 years by between three and four months per decade. On this basis puberty is attained 2½ to 3½ years earlier today than it was a century ago. The trend in height and weight at the age of puberty is in good agreement with this figure, the 11-year-old children of today having the size of 12-year-olds 30 or 40 years ago.

The decline in menarcheal age over the years is exemplified by figures for various groups in Britain beginning about 1820. At that time the mean age of menarche for working-class women in Manchester was 15.7 years. In 1830 obstetric patients in London's University College Hospital gave a mean age of 15.4 years. In 1880 girls of the London "middle class" yielded a mean age of 15.0. It is particularly interesting that in 1820 the mean age of menarche for the girls of Manchester who had completed their education, and thus represent the middle and upper class, was 14.6; for this group menstruation began more than a year earlier than for working-class women of the same place and time. Today middle- and upper-class women have an average age at menarche of 12.9 years. For this economic group, then, the rate of change is about half the rate for the general population.

In Poland a similar difference appears between women in Warsaw, who were and are relatively well off, and women in rural areas, whose diet, hygiene and general circumstances are poor by comparison. The Warsaw women show a per-decade decline in the age of menarche of only 2.8 months. The country women, on the other hand, had a very late menarche in 1890, and since then it has become earlier by 5.4 months per decade. Even so, there remains a difference of nearly two years between the two economic groups.

The earliest menarche that has been recorded is for girls in Cuba. In 1963 the mean age for girls mainly of Negro descent was 12.3 years; the figure for girls mainly of white descent was 12.4. The authors of the study from which these figures are taken remark that the Negro girls live under poorer economic circumstances than the whites; if circumstances were matched, the menarche of Negro Cubans might be even more advanced. It has recently been shown that Chinese girls in Hong Kong also begin menstruation early; even the very poor attain menarche as early as most Europeans, whose circumstances are far better. Among Europeans, Italian girls appear to have the earliest menarche, but the data for this group are not extensive. Apparently eastern Europeans mature somewhat earlier than western Europeans, distinctly so when economic circumstances are matched. Comfortably-off Americans are slightly ahead of western Europeans. Well-off Africans are not much later than Europeans. Badly-off Africans (the South African Bantu) are certainly later. The only known group that nowadays experiences menarche as late as many Europeans of a century ago did are the Bundi of New Guinea, with an average menarcheal age of 18.8 and no girl menstruating before the age of 17.

There is at present little sign that girls are maturing at something like the earliest possible age. It appears that the 140-year trend toward earlier menarche may continue for at least another decade or two. Extrapolation backward, however, is clearly an impossibility. There is scant real information available for dates before 1800. Nonetheless, Hippolitus Quarinonius, writing in 1610 of Austria at that time, says: "The peasant girls in the landschaft in general menstruate much later than the daughters of the townsfolk or the aristocracy, and seldom before their seventeenth, eighteenth or even twentieth year. For this reason they also live much longer than the townsfolk and aristocratic children and do not become old so early. The townsfolk have usually borne several children before the peasant girls have yet menstruated. The cause seems to be that the inhabitants of the town consume more fat food and drink and so their bodies become soft, weak and fat and come early to menstruation in the same way as a tree. which one waters too early produces earlier but less well-formed fruit than another."

What is it that has caused the trend toward earlier puberty and larger size? In seeking causal factors we must first distinguish between the trend toward greater height and weight in children and the lesser trend seen in adults. It is probable that environmental changes of various kinds are chiefly responsible for the change in menarcheal age and for that portion of the larger size in children that reflects earlier maturation. The trend in adult height, on the other hand, may be due at least as much to genetic factors as to environmental ones.

Of the environmental factors better nutrition is the most obvious. In periods of acute starvation growth is certainly delayed and puberty temporarily postponed. When such starvation ends, the child's rate of growth accelerates. He may or may not reach his normal growth curve, depending on the severity and duration of the malnutrition. Where there is chronic malnutrition it is fairly certain that there is not only a great delay in maturation but also a stunting of adult height.

If better nutrition is the major cause,

then one would expect that the trend toward greater size in childhood and earlier puberty would be weaker for better-off children than for poor ones, on the grounds that in most industrialized countries the circumstances of the poor have altered more than those of the rich during the past 100 years. As we have seen, this expectation is confirmed by the course of menarcheal age in England and Poland. The most recent data from England and Scotland indicate no significant differences in the age of menarche of girls whose fathers are in different occupational groups (which reflect, at least approximately, differences in income). Yet the recent data on menarcheal age from Hong Kong show a difference of nine months between rich and poor. Presumably this discrepancy is due to the fact that the poor in Hong Kong are much worse off than the poor in England and Scotland.

Why is menarcheal age so early in Hong Kong and Cuba? It has been shown (by studies of sisters and identical and nonidentical twins) that when the environment is good, the age at menarche is controlled by genetic factors. We must therefore suppose that unless some unidentified climatic difference is responsible, the genetic threshold for Chi-

nese and perhaps Cubans is below that for the English and by the same token that the threshold for eastern Europeans is perhaps below that for western Europeans.

One of the most convincing arguments for nutritional causes is the example of the Lapps, who had virtually the same average age of menarche (16.5 years) from 1870 to 1930 while maintaining intact their nomadic way of life. During the same period the age of menarche among the neighboring settled farmers declined by nearly two years.

In the past a hot climate was considered a potent cause of early menstruation. The evidence for this seems chiefly anecdotal, and nobody nowadays supposes that climate exerts more than a very minor influence, if any at all. Little more can be said until equally well-nourished groups living under greatly different climatic conditions are available for study. At present Eskimos and Nigerians have the same menarcheal age. Their diets are perhaps too dissimilar, however, for a difference related to climate to be distinguished. The studies of Phyllis Eveleth in Brazil indicate that American girls growing up in the hot environment of Rio de Janeiro but retaining their American nutritional habits

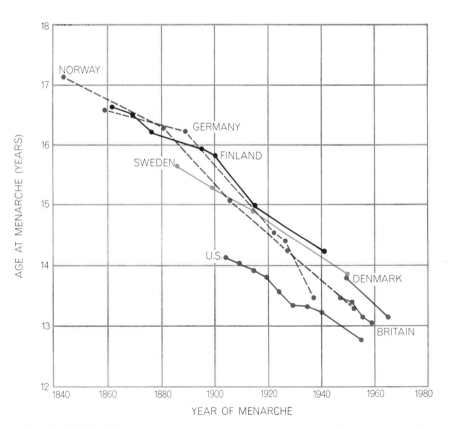

AGE AT MENARCHE, or first menstrual period, has declined in the U.S., Britain and Europe. Girls are estimated to begin menstruation between 2.5 and 3.3 years earlier on the average than a century ago. The age of menarche is an index of the rate of physical maturation.

have a mean menarcheal age strictly comparable to that of the same socioeconomic group in the U.S. To be sure, the mean world temperature has been rising since the 1850's (until, it would seem, about 1940, when the trend began to reverse). It nevertheless seems unlikely on present evidence that this warming-up process has contributed significantly to the trend toward earlier menarche.

It has also been suggested that an increasing emphasis on sex is responsible for the earlier menarche, presumably as a result of some kind of psychosocial stimulation. Again there is no evidence to support this dubious view. In Sweden and in Hungary girls educated in coeducational schools have been compared with girls educated at single-sex schools, with absolutely no difference detected. Whether this constitutes a fair test of school-days sex activity, however, remains to be established. Should psychosexual stimulation actually be the cause, it would have to start in nursery school.

The one thing that all authors find significantly related to age at menarche is the number of children in the family. The larger the number, the later the menarche and the less the height and weight at all ages, both of the earlier- and the later-born children. In Czechoslovakia the age of menarche for girls with one sibling or none is 14.3 years; for those with six or more siblings it is 14.6. In England the same relation holds.

The most obvious explanation for this sibling-number effect is nutritional; the more mouths there are to feed and children there are to care for, the less well the feeding (and perhaps the general care) may be done. An alternative or supplementary explanation might be that children with more siblings get more diseases. The effect of childhood disease on growth, however, is uncertain. Most childhood diseases are without effect on the growth of well-nourished children. Chronic disease in malnourished children may be another matter; possibly the decline of chronic disease may have helped to create the trend in children's growth. Even here the evidence is largely negative; the suppression of malaria in heavily parasitized populations has not noticeably increased the rate or amount of growth. Until we have much more evidence we can hardly lay the trend, either in children or in adults, at the door of the decline in disease.

In circumstances where severe malnutrition is chronic, the final height children attain is undoubtedly affected. Nevertheless, the trend in adult height may have in whole or in part a genetic explanation, as was first suggested by Gunnar Dahlberg of the University of Uppsala. Suppose there exists some degree of dominance in the genes governing human stature, so that on the average the offspring of a tall parent and a short one have a height that is not halfway between the two parental heights but a little closer to the height of the tall parent. If there is such a genetic dominance, people with a variety of the genes influencing height (heterozygotes) would on the average be slightly taller than people without such variety (homozygotes). A trend that gave rise to an increase of heterozygotes in the population would then give rise to an increase in height. Such a trend may exist in the gradual decline of intermarriage of members of the same village community. In western European communities, at least, outbreeding has increased at a fairly steady rate ever since the introduction of the bicycle.

There is some direct evidence, although it is perhaps not completely conclusive, that outbreeding does lead to increased stature in man. Frederick S. Hulse of the University of Arizona found in 1957 that the full-grown sons of parents who had come from different Swiss villages were taller than the sons of parents from the same village. The difference was nearly an inch. Similarly, Arthur P. Mange of the University of Massachusetts has shown that among the Hutterites, an inbred religious sect in North America, persons whose parents represented the degree of inbreeding of first cousins were on the average about 1.4 inches shorter than persons whose parents were unrelated. In different regions of France the height of adults also was found to be significantly and inversely correlated with the degree of inbreeding in the region. Genetic causes can explain at least part of the trend in adult height, but whether they can explain all of it is doubtful. On the basis of the figures given above, the change toward outbreeding seems unlikely to be responsible for an increase of more than .8 inch per generation. Natural selection, on the assumption that tall people have more marriageable offspring than short people, operates too slowly to account for this trend, and in any case such evidence as we have indicates that tall people have fewer children rather than more. Thus the trend toward larger size and earlier maturation in man has been firmly established, and although some of its causes are known, some remain to be investigated.

The Author

J. M. TANNER is professor of child health and growth at the Institute of Child Health of the University of London. He attended Marlborough College, University College of the South-West and St. Mary's Hospital Medical School before coming to the U.S. in 1941 to study medicine as a wartime Rockefeller Student, obtaining his M.D. from the University of Pennsylvania Medical School in 1944. He has since acquired three more degrees from the University of London: a D.P.M. in 1946, a Ph.D. in 1953 and a D.Sc. in 1957. From 1946 to 1948 he taught human anatomy at the University of Oxford, and from 1948 to 1955 he was a lecturer in physiology at the Sherrington School of Physiology of the University of London. He joined the Institute of Child Health in 1956. For the past 15 years, he writes, "I have been conducting two longitudinal studies of growth of healthy children, with particular emphasis on change in body composition during growth and on the development of endocrine function. I am currently engaged in an extensive study of growth disorders in children, in particular with a clinical trial of growth hormone administration in children with dwarfism."

Bibliography

THE SECULAR TREND IN GROWTH AND MATURATION AS REVEALED IN POLISH DATA. H. Milicer in *Tijdschrift voor Sociale Geneeskunde*, Vol. 44, pages 562–568; 1966.

THE SECULAR TREND IN GROWTH AND MATURATION IN THE NETHERLANDS. V. M. Oppers in *Tijdschrift voor Sociale Geneeskunde*, Vol. 44, pages 539–548; 1966.

THE SECULAR TREND IN THE GROWTH AND DEVELOPMENT OF CHILDREN AND YOUNG PERSONS IN THE SOVIET UNION. V. G. Vlastovsky in *Human Biology*, Vol. 38, No. 3, pages 219–230; September, 1966.

SCIENTIFIC
AMERICAN January 1968, Vol. 218, No. 1, pp. 36-42

OFFPRINT **1092**

HOW PROTEINS START

by Brian F. C. Clark and Kjeld A. Marcker

The chain of amino acid units that constitutes a protein molecule
begins to grow when a variant of one of the standard amino acids
is delivered to the site of synthesis by a specific transfer agent.

Over the past 15 years a tremendous amount of information has been amassed on how the living cell makes protein molecules. Step by step investigators in laboratories all over the world are clarifying the architecture of specific proteins, the nature of the genetic material that incorporates the instructions for building them, the code in which the instructions are written and the processes that translate the instructions into the work of construction. With the information now available experimenters have already synthesized a number of protein-like molecules from cell-free materials, and the day seems not far off when we shall be able to describe, and perhaps control, every step in the making of a protein.

How is the building of a protein initiated? Until recently this question seemed to create no special problems. Given a supply of the amino acids from which a protein is made, the cell assembles them into a polypeptide chain that grows into a protein molecule, and it did not appear that the cell used any special machinery to start the construction of the chain. We have now learned, however, that the cell does indeed possess a starting mechanism. With the discovery of this mechanism it has become possible to study in detail the first step in the production of a protein molecule.

In order to discuss this new development we must first review the general features of protein synthesis by the cell. Proteins are made up of some 20 varieties of amino acid. A protein molecule consists of a long chain of amino acid units, typically from 100 to 500 or more of them, linked together in a specific sequence. The instructions for the particular order in each protein (the cell manufactures hundreds of different proteins) reside in the chainlike molecule of deoxyribonucleic acid (DNA). The DNA

molecule consists of units called nucleotides; each nucleotide contains a side group of atoms called a base, and the sequence of bases along the DNA chain specifies the sequence for amino acids in the protein. There are four different bases in DNA: adenine (A), guanine (G), thymine (T) and cytosine (C). A "triplet" (a sequence of three bases) constitutes the "codon" that specifies a particular amino acid. The four bases taken three at a time in various sequences provide 64 possible codons; thus the four-letter language of DNA provides a vocabulary that is more than sufficient to designate the 20 amino acids. (In fact, some amino acids can be indicated by more than one codon.)

DNA does not guide the construction of the protein directly. Its message is first transcribed into the daughter molecule called messenger ribonucleic acid (mRNA). Messenger RNA also has four bases; three of them (A, G and C) are the same as in DNA, but the fourth, taking the place of thymine, is uracil (U). The RNA molecule is generated from DNA by a coupling process based on the fact that U couples to A and G couples to C. Thus during the transcription of DNA into RNA the four bases A, G, T and C in DNA give rise respectively to U, C, A and G in RNA [see illustration on following page].

The coded message is then read off the messenger RNA and translated into the construction of a protein molecule. This process takes place on the cell particles known as ribosomes, and it requires the assistance of smaller RNA molecules called transfer RNA (tRNA) that bring amino acids to the indicated sites. Each transfer RNA is specific for a particular amino acid, to which it attaches itself with the aid of an enzyme. It possesses an "anticodon" corresponding to a particular codon on the messenger RNA

molecule. The ribosome moves along the messenger RNA molecule, reading off each codon in succession, and in this way it mediates the placement of the appropriate amino acids as they are delivered. As the amino acids join the chain they are linked together through peptide bonds formed by means of enzymes.

The decipherment of the genetic code for protein synthesis began in 1961 when Marshall W. Nirenberg and J. Heinrich Matthaei of the National Institutes of Health synthesized a simplified form of messenger RNA, composed of just one type of nucleotide, and found that it could generate the formation of a protein-like chain molecule made up of one variety of amino acid. Their artificial messenger RNA was the polynucleotide called "poly-U," containing uracil as the base. When it was added to a mixture of amino acids, extracts from cells of the bacterium *Escherichia coli* and energy-supplying compounds, it caused the synthesis of a polypeptide chain composed of the amino acid phenylalanine. Thus the poly-U codon (UUU) was found to specify phenylalanine.

This breakthrough quickly led to the identification of the codons for a number of other amino acids by means of the same device: using synthetic forms of messenger RNA. The experiments suggested that the initiation of synthesis of a protein was a perfectly straightforward matter. It appeared that the first codon in the messenger RNA chain simply called forth the delivery and placement of the specified amino acid and that no special starting signal was required. In 1964, however, Frederick Sanger and one of the authors of this article (Marcker) discovered a peculiar form of an amino acid, in combination with its transfer RNA, that threw entirely new light on the situation.

Using extracts from the *E. coli* bacterium, we were studying the chemical characteristics of the combination of the amino acid methionine with its specific tRNA. In the course of this study we decided to investigate the breakdown of the compound by pancreatic ribonu-clease, an enzyme known to split RNA chains at certain specific bonds [*see top illustration on page 2604*]. To facilitate identification of the products we labeled the methionine in advance with radioactive sulfur, and after treatment of the methionine-tRNA compound with

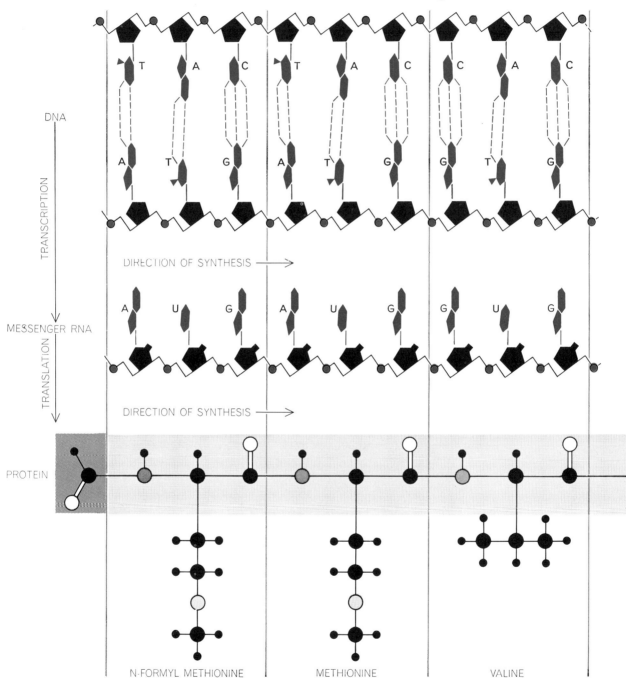

TRANSMISSION OF GENETIC INFORMATION takes place in two main steps. First the linear code specifying a particular protein is transcribed from deoxyribonucleic acid (DNA) into messenger ribonucleic acid (RNA). The code letters in DNA are the four bases adenine (A), thymine (T), guanine (G) and cytosine (C). Hydrogen bonds (*broken lines*) between the complementary bases A–T and G–C hold the two strands of the DNA molecule together. The strands, which run antiparallel, consist of alternating units of deoxyribose sugar (*pentagons*) and phosphate (PO$_3$H). The code letters in messenger RNA duplicate those attached to one strand of the DNA except that uracil (U) replaces thymine. In RNA the sugar is ribose. In the second step of the process messenger RNA is translated into protein. The code letters in RNA are read in triplets, or codons, each of which specifies one (or sometimes more) of the 20 amino acids that form protein molecules. It has now been found that the codon AUG can specify a modification of methionine known as formyl methionine, which signals the start of a protein chain. Inside the chain AUG specifies ordinary methionine. The codon GUG, which codes for valine inside the chain, can also specify formyl methionine and initiate chain synthesis.

the enzyme we separated the products by means of electrophoresis, the technique that segregates electrically charged molecules according to their charge, size and shape. As was to be expected, one of the products was the compound known as methionyl-adenosine, a combination of methionine with the terminal adenosine portion of the tRNA molecule. But we also found, to our surprise, that the products included a considerable amount of a formylated variety of this compound, that is, a variation in which a formyl group (CHO) replaced a hydrogen atom in the amino group

(NH₂) of the molecule. It turned out that this was by no means an artifact of the treatment to which the original compound had been subjected; growing cells proved to contain a high proportion of formylated methionine tRNA.

It was immediately evident that formylated methionine must occupy a special position in the protein molecule. The attachment of the formyl group to the amino group would prevent the amino group from forming a peptide bond [see illustration below]. Consequently the formylated amino acid must be an end unit in the protein molecule. Since an

amino group forms the "front" end of protein molecules when they are being assembled, formylated methionine must constitute the initial unit of the molecule.

We were able to separate the methionine tRNA of E. coli into two distinct species, and found that only one can be formylated. The formylatable species constitutes about 70 percent of the bacterium's methionine tRNA [see bottom illustration on following page]. Recent work in our laboratory at the Medical Research Council in Cambridge has established that the compound is formylated (at methionine's amino group) only after the amino acid has become attached to the tRNA molecule. The donor of the formyl group is 10-formyl tetrahydrofolic acid, and the reaction is catalyzed by a specific enzyme that acts exclusively on the combination of methionine with the formylatable species of tRNA.

Our laboratory and others have proceeded to analyze the initiation of protein formation by several experimental techniques. We began by testing a number of different synthetic messenger RNA's for their ability to bring about synthesis of a polypeptide incorporating methionine. Only two of the synthetic polynucleotides we tried proved to be capable of doing this. One contained the bases uracil, adenine and guanine (poly-UAG); the other had only uracil and guanine (poly-UG). We found that in a mixture of amino acids and other cell-free materials where only the formylatable species of methionine tRNA was present, either poly-UAG or poly-UG would cause the synthesis of a polypeptide with methionine in the starting position—and only in that position. Surprisingly, this was true even when no formyl group was attached to the methionine-tRNA compound. We had to conclude that the formylatable version of the tRNA for methionine possessed a special adaptation that helped it to function as a polypeptide-chain initiator.

A thorough search was made for formylated varieties of other tRNA's: that is, of tRNA's for amino acids other than methionine. None were found. This raised an interesting question. In the proteins produced by E. coli cells the amino acid at the "front" end of the protein molecule is not always methionine; often it is alanine or serine. These amino acids are never found to be formylated. How, then, does either of them become the initial member of the protein chain?

Experiments with natural messenger RNA's (rather than synthetic polynucleotides) have suggested an explanation. Jerry Adams and Mario Capecchi, work-

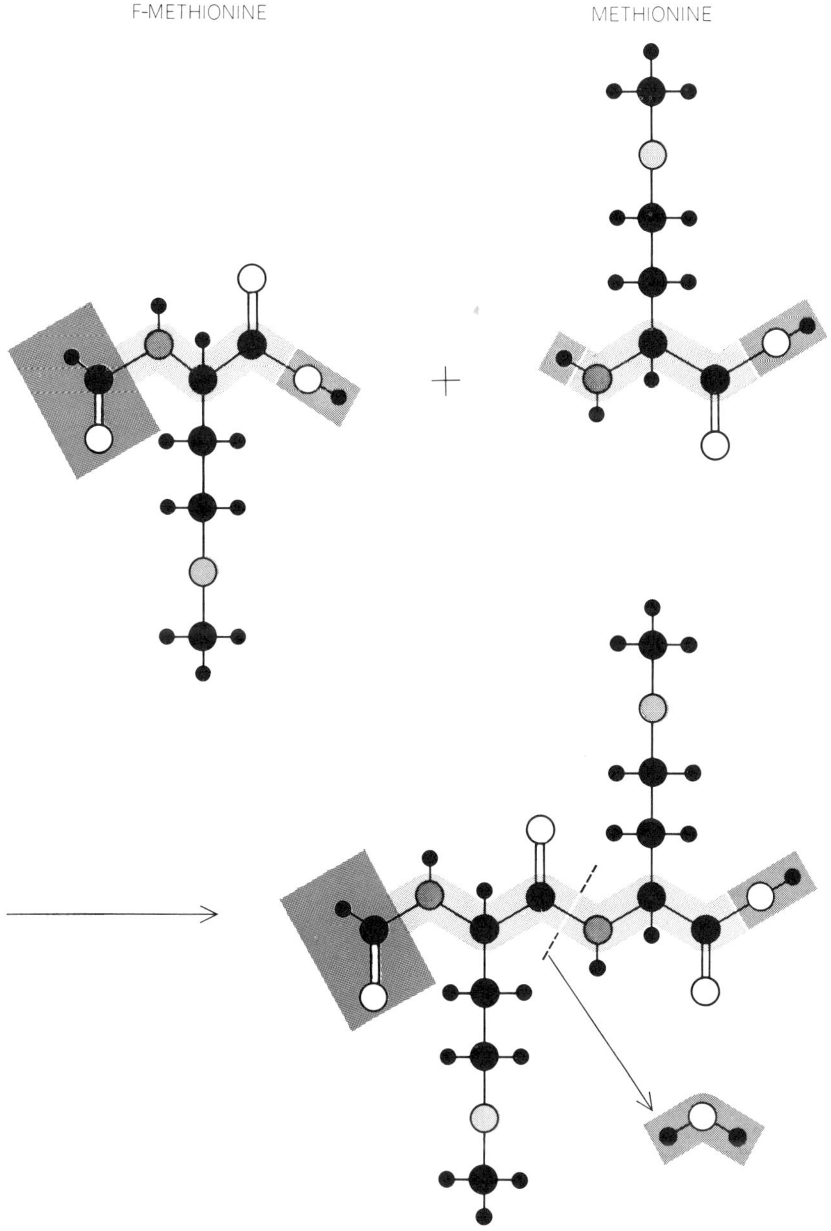

F-METHIONINE   METHIONINE

FORMYL METHIONINE, abbreviated F-Met, has a formyl group (CHO) where methionine (Met) has a hydrogen atom as part of a terminal amino (NH₂) group. When an amino acid enters a protein chain, one of the hydrogens from the amino end of one molecule combines with an OH group from the carboxyl (COOH) end of another molecule to form a molecule of water. The two molecules are then linked by a peptide bond. The formyl group prevents this reaction, hence F-Met can appear only at the beginning of a protein chain.

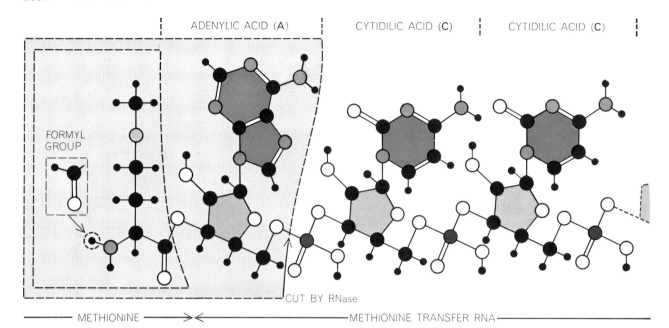

ADENYLIC ACID (A)   CYTIDILIC ACID (C)   CYTIDILIC ACID (C)

FORMYL GROUP

CUT BY RNase

—— METHIONINE ———>—<———————METHIONINE TRANSFER RNA ———————

TRANSFER OF AMINO ACID to the site of protein synthesis is accomplished by molecules of transfer RNA (tRNA). There is at least one species of transfer RNA for each amino acid. All transfer RNA molecules contain the base sequence CCA at the terminal that holds the amino acid. Such a terminal is diagrammed here and shown coupled to methionine. Methionine that subsequently can be converted to formyl methionine is transferred by a different tRNA. When treated with the enzyme ribonuclease (RNase), the final base (adenine) and its coupled amino acid are split off from the rest of the transfer RNA. The fragment is called an aminoacyl adenosine.

ing in the laboratory of James D. Watson at Harvard University, and Norton D. Zinder and his collaborators at Rockefeller University have used messenger RNA's extracted from bacterial viruses. These RNA's direct the synthesis of the proteins that form the coat of the virus. The experimenters in Watson's and Zin-

der's laboratories found that when such an RNA was added to cell-free materials in the test tube, formylated methionine turned up at the starting end of the coat proteins that were synthesized. This was most surprising, because normally in living systems the initial amino acid of the viruses' coat protein is alanine. A signifi-

cant clue was found, however, in the fact that the coat proteins synthesized in the cell-free systems invariably had an alanine in the second position, following the formyl methionine. From this it seems reasonable to deduce that in living systems, as in the cell-free system, the formation of the protein starts with formyl

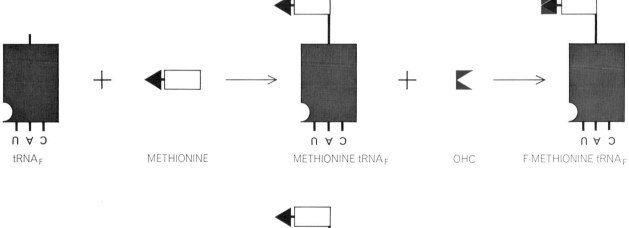

U A C   U A C   U A C
tRNA_F   METHIONINE   METHIONINE tRNA_F   OHC   F-METHIONINE tRNA_F

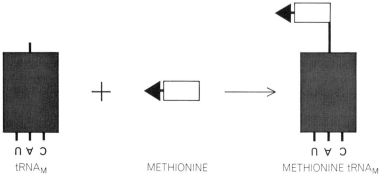

U A C   U A C
tRNA_M   METHIONINE   METHIONINE tRNA_M

TWO METHIONINE tRNA's, designated tRNA_F and tRNA_M, exhibit different characteristics. In the presence of an energy source (adenosine triphosphate) and a special coupling enzyme, both tRNA's combine with methionine. Only tRNA_F, however, forms a complex that is recognized by another enzyme that can convert methionine to formyl methionine. The formyl group is provided by 10-formyl tetrahydrofolic acid. The complex that results is F-Met-tRNA_F.

tRNA	CODONS
MET-tRNA$_M$	AUG
MET-tRNA$_F$	AUG
F-MET-tRNA$_F$	GUG

CODON ASSIGNMENTS show the bases in messenger RNA that cause the two Met-tRNA's to deliver methionine or formyl methionine for insertion in a protein chain.

methionine, and that the bacterial cells supply an enzyme that chops off the formyl methionine later, leaving alanine in the first position.

Experiments with *E. coli* RNA in our laboratory and others have produced similar results. Messenger RNA extracted from these bacteria, like that extracted from bacterial viruses, causes cell-free systems to synthesize proteins with formyl methionine in the first position. On the other hand, the proteins extracted from living *E. coli* cells usually have unformylated methionine or alanine or serine in the lead position. It therefore seems likely that the living cells remove the formyl group from methionine or split off the entire formyl methionine unit after synthesis of the protein chain has got under way. The significance of the frequent appearance of alanine and serine at the front end of *E. coli* proteins is not clear; no satisfactory explanation has yet been found for the cell's selection of alanine and serine to follow formyl methionine. At all events, what does seem plausible now is that in *E. coli* the synthesis of all proteins starts with formyl methionine as the first unit.

How does the messenger RNA convey the message calling for formyl methionine as the starting unit? Does it use a special codon addressed specifically to the formylatable variety of methionine tRNA? We tested various codons for their ability to bring about the delivery of formyl methionine to the protein-synthesizing ribosomes. A codon for methionine was already known: it is AUG. We found that AUG was "read" by both varieties of methionine tRNA—the formylatable and the unformylatable. Either variety of tRNA delivered and bound methionine to the ribosome in response to AUG. We found that the formylatable tRNA (but not the other variety) also recognized and responded to another codon: GUG.

These findings were consistent with our earlier observation that either poly-UAG or poly-UG could effect the incorporation of methionine into a polypeptide in a cell-free system. Poly-UAG, of course, can contain the codons AUG and GUG, depending on the sequence in which the bases happen to be arranged in this polynucleotide; poly-UG provides the codon GUG. That both AUG and GUG can initiate the synthesis of a methionine polypeptide was confirmed and clearly spelled out in detail by experiments in the laboratory of H. Gobind Khorana at the University of Wisconsin. Using synthetic messenger RNA's in which the bases were arranged only in these triplet sequences (AUG and GUG), Khorana's group showed that both codons led to the formation of a chain with formyl methionine in the starting position. AUG also placed methionine in internal positions in the chain, but GUG, which can code only for the formylatable version of the tRNA, incorporated methionine only at the starting end [see illustration below].

Investigators in the laboratories of Severo Ochoa at New York University and Paul M. Doty at Harvard obtained the same results. They also noted that both codons possess a certain versatility as signals, depending on their location in the messenger RNA. Located at or near the beginning of the messenger RNA chain, the AUG triplet is recognized by the formylatable variety of tRNA and leads to the placement of formylated methionine at the starting end of the polypeptide; farther on in the messenger RNA chain the same triplet is recognized by unformylatable tRNA and causes the placement of unformylated methionine in the internal part of the polypeptide. In short, at the "front" end of the RNA message the AUG codon says to the cell's synthesizing machinery, "Start the formation of a protein"; when it is located internally in the message, AUG simply says, "Place a methionine here." Similarly, the codon GUG was found to have two possible meanings: located at the beginning of the message, it orders the initiation of a protein with formylated methionine; in an internal position in the message it is the code word for placement not of methionine but of the amino acid valine.

How is it that each of these codons signifies a starting signal in one position and has a different meaning in another? Obviously this question will have to be answered in order to clarify the language of the protein-starting mechanism. Indeed, we cannot be sure that a codon in itself constitutes the entire message for the initiation of a protein. The signaling mechanism may be more complex than one might assume from the findings developed so far. Those findings are based almost entirely on work done with artificial messenger RNA's, and it is possible that the messages they provide are only approximations—meaningful enough to stimulate the cell machinery but not the full story.

When we consider how important the

SYNTHETIC MESSENGER	SOURCE OF METHIONINE		POSITION OF METHIONINE IN POLYPEPTIDE		CODONS USED
	MET-tRNA$_M$	MET-tRNA$_F$ / F-MET-tRNA$_F$	INTERNAL	N-TERMINAL	
RANDOM POLY-UG	−	+	−	+	GUG
RANDOM POLY-AUG	+	+	+	+	AUG GUG
POLY-(UG)$_n$	−	+	−	+	GUG
POLY-(AUG)$_n$	+	+	+	+	AUG

INCORPORATION OF METHIONINE in protein-like chains has been studied with synthetic messenger RNA's and the two species of methionine transfer RNA: tRNA$_M$ and tRNA$_F$. The plus sign indicates combinations that lead to incorporation. In random poly-UG and random poly-AUG the bases can occur in any sequence, but presumably the only effective sequences are GUG and AUG. Poly-(UG)$_n$ and poly-(AUG)$_n$ are synthetic chains of RNA consisting of 30 or more repetitions of the base sequences indicated.

codons AUG and GUG are in initiating the synthesis of polypeptides, it is certainly odd that a synthetic messenger RNA such as poly-U, which of course cannot supply those codons, nevertheless manages to cause the ribosomes to produce a polypeptide. We can only conclude that they do so by mistake, so to speak, that is, by acting in a way not entirely specified by the available information. (It is ironic that the genetic code was broken because artificial systems were able to make the right kind of mistake!) Are there circumstances that tend to assist these systems in accomplishing proper mistakes? One influential factor has been found. It is the concentration of magnesium in the cell-free system of building materials. A high magnesium concentration makes it possible for many kinds of synthetic messenger RNA to generate polypeptides; when the magnesium concentration is lowered, only the RNA's that contain AUG or GUG succeed in doing so. What magnesium may have to do with polypeptide initiation is still unclear.

Let us come back to the placement of the initial methionine as the normal first step in the construction of a protein. We have noted that the methionine-tRNA complex that places the amino acid in the initial position does not necessarily contain a formyl group. Evidently under conditions of a relatively high concentration of magnesium the formyl group of itself plays no essential role in the installation of the amino acid. What seems to be important is the character of the tRNA: only the formylatable variety of methionine tRNA can initiate the synthesis, and it can do so even when it is not formylated. What, then, are the specific properties that account for its role as an initiator?

A reasonable supposition is that this variety of tRNA has a special shape or configuration that helps it to fit into a particular site on the ribosome. As a matter of fact there is evidence that ribosomes possess two kinds of site for the attachment of tRNA's. One kind, called an amino acid site, simply receives and positions the tRNA when it arrives with its amino acid; the other kind, called a peptide site, holds the tRNA while a peptide bond is formed between its amino acid and an adjacent neighbor [see illustration at right]. It is therefore plausible to suppose that the formylatable variety of the tRNA for methionine may have a shape that helps it to fit into a peptide site on the ribosome and thus be in a position to start the linking together of amino acids.

Evidence in support of this hypothesis has been obtained in our laboratory by Mark S. Bretscher and one of us (Marcker) and by Philip Leder and his associates at the National Institutes of Health in experiments using the antibiotic puromycin. The structure of puromycin is similar to that of the end of a tRNA molecule that attaches to an amino acid [see illustration on next page]. Because it has an NH₂ group, puromycin can form a peptide bond with an amino acid, but since it lacks the free carboxyl (COOH) group of a normal amino acid it cannot form a second peptide bond.

Thus it cannot participate in chain elongation. Various experiments indicate that puromycin will add on to—and terminate—a growing polypeptide chain only when the tRNA holding the chain is bound in the peptide site.

In other experiments it has been found that the formylatable variety of methionine tRNA, when bound to a ribosome, will combine with puromycin; the unformylatable variety of the tRNA, on the other hand, will not react with puromycin. The experimental results therefore indicate that there are indeed two kinds of ribosomal site or state: one where a

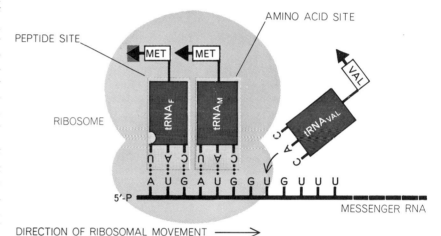

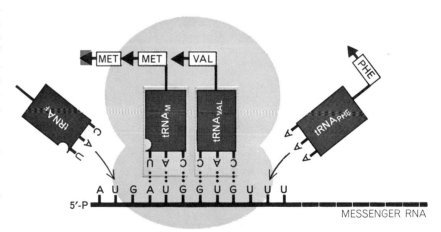

PROTEIN SYNTHESIS takes place on cellular particles called ribosomes, which travel along the "instruction tape" of messenger RNA, reading off the genetic message. The ribosome evidently has two sites for accommodating molecules of transfer RNA: a peptide site and an amino acid site. It appears that the structure of tRNA_F enables it to go directly to the peptide site, thereby initiating the protein chain. This special structure is symbolized by a notch in tRNA_F. Other tRNA's may acquire the configuration needed for the peptide site after first occupying the amino acid site. In step 1 (top) the codon AUG at the front end (5'-phosphate end) of messenger RNA pairs with the anticodon CAU that is believed to exist on tRNA_F, which delivers a molecule of formyl methionine to start the protein chain. The codon AUG in the second position is paired with the CAU anticodon of tRNA_M, which delivers a molecule of ordinary methionine. In step 2 (bottom) the tRNA_F molecule has moved away and the peptide site has been occupied by tRNA_M, which is now coupled to the growing protein chain. Valine transfer RNA has moved into the amino acid site.

peptide bond cannot be formed between the peptide chain and puromycin and one where it can. Most likely the latter is the peptide site. Furthermore, the experiments have strengthened the suspicion that the formylatable tRNA possesses a unique structure that somehow helps it to move into the peptide site on a ribosome. Apparently the structure of the formylatable tRNA has been particularly tailor-made for its function as a chain initiator.

The question therefore arises: What is the precise role of the formyl group? If

the formyl group per se has nothing to do with placing methionine in the starting position, what function does it have? Our earlier experiments, in which we used a relatively high magnesium concentration, suggested that the formyl group is involved somehow in the formation of the first peptide bond, which launches the building of the polypeptide chain. When the methionine tRNA complex is formylated, synthesis of the polypeptide proceeds much faster than when it is not. This effect can be ascribed to the fact that the presence of the formyl

group somehow facilitates the entry into the peptide site. It still remains to be determined just how the formyl group helps to promote such an effect.

Further light has been shed on the problem of protein-chain initiation in the past year by the work of several laboratories, including our own. Special protein agents, still poorly defined, have been implicated together with a cofactor in the formation of the initiation complex on the ribosome. When these new components are present and the supply of magnesium is low, the formyl group is necessary if the formylatable methionine tRNA is to be attached to the ribosomal peptide site by a messenger. Quite recently the cofactor has been identified as being a nucleotide derivative: guanosine triphosphate. Hence we are coming to the view that the conditions prevailing within the living cell are approached by these low-magnesium conditions, where there is strict specificity for forming the initiation complex and for unambiguous polypeptide formation. In our present state of knowledge, however, it is still unclear how these new components help to ensure the placement of the formylated methionine tRNA in the peptide site on the ribosome.

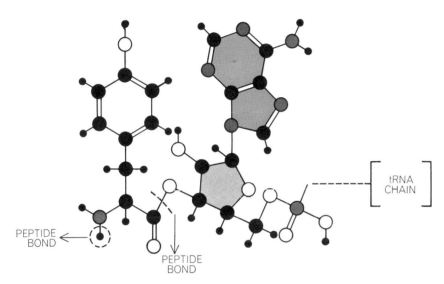

TYROSINE tRNA

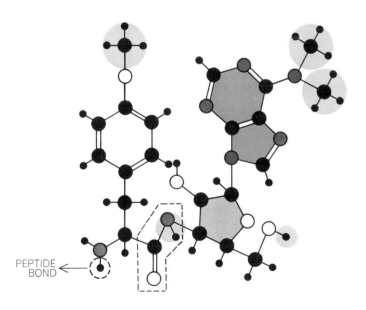

PUROMYCIN

PROTEIN-CHAIN TERMINATION can be induced by adding puromycin, an antibiotic, to a protein-synthesizing system. The structure of puromycin closely resembles the structure formed by the amino acid tyrosine and the terminal base of tRNA. Colored disks mark the atomic differences. Tyrosine can be inserted in a protein chain because it can form two peptide bonds. Puromycin can form only one peptide bond because the —CONH— linkage (*inside broken line*) is less reactive than the —COO— linkage in tyrosine tRNA.

The specific findings concerning the initiation of protein synthesis that we have discussed in this article apply only to bacterial cells. So far no such form of tRNA (containing the formyl group or any other blocking agent) has been found in the cells of mammals. Accordingly the mechanism of protein-chain initiation is possibly different in mammalian cells from the mechanism discussed here. The process of polypeptide initiation in the cells of higher organisms is currently under study in several laboratories.

Meanwhile the investigation of the *E. coli* system is being pursued with experiments that promise to yield further insights. The way in which the vaguely characterized protein agents and guanosine triphosphate are involved in the initiation of a polypeptide chain is being explored. Much work is under way on analyzing the sequence of nucleotides in natural messenger RNA's, with a view to determining whether or not AUG or GUG constitutes a complete coding signal for protein initiation. We are searching for differences between the formylatable and unformylatable varieties of methionine tRNA, in their nucleotide sequences and in their three-dimensional structures, that may throw light on their respective interactions with the ribosomes.

The Authors

BRIAN F. C. CLARK and KJELD A. MARCKER work together at the British Medical Research Council's Laboratory of Molecular Biology in Cambridge. Clark has a Ph.D. in organic chemistry from the University of Cambridge. He has also done research in biochemistry at the Massachusetts Institute of Technology and in biochemical genetics at the National Heart Institute in Bethesda, Md. Marcker has a Ph.D. in biochemistry from the University of Copenhagen. He is at Cambridge on a Carlsberg-Wellcome Fellowship.

Bibliography

A GTP REQUIREMENT FOR BINDING INITIATOR tRNA TO RIBOSOMES. John S. Anderson, Mark S. Bretscher, Brian F. C. Clark and Kjeld A. Marcker in *Nature*, Vol. 215, No. 5100, pages 490–492; July 29, 1967.

N-FORMYL-METHIONYL-S-RNA. K. Marcker and F. Sanger in *Journal of Molecular Biology*, Vol. 8, No. 6, pages 835–840; June, 1964.

THE ROLE OF N-FORMYL-METHIONYL-sRNA IN PROTEIN BIOSYNTHESIS. B. F. C. Clark and K. A. Marcker in *Journal of Molecular Biology*, Vol. 17, No. 2, pages 394–406; June, 1966.

STUDIES ON POLYNUCLEOTIDES, LXVII: INITIATION OF PROTEIN SYNTHESIS IN VITRO AS STUDIED BY USING RIBOPOLYNUCLEOTIDES WITH REPEATING NUCLEOTIDE SEQUENCES AS MESSENGERS. H. P. Ghosh, D. Söll and H. G. Khorana in *Journal of Molecular Biology*, Vol. 25, No. 2, pages 275–298; April 28, 1967.

SCIENTIFIC
AMERICAN   January 1968, Vol. 218, No. 1, pp. 86-96   OFFPRINT 1093

THE VENOUS SYSTEM

by J. Edwin Wood

The veins constitute a reservoir for the blood supply, not merely
a system of passive conduits. They constrict and dilate actively,
thus maintaining a satisfactory distribution of blood in the body.

The concept of the veins as passive conduits, fitted with valves to permit the flow of blood only toward the heart, was developed by William Harvey in the 17th century and persisted until fairly recently. In the past two decades, however, it has become clear that the veins have more subtle functions. Experiments showed first that the veins are capable of being distended far more than the arteries and then that at any given time they contain most of the blood in the body—perhaps 70 percent of the total. This suggested that they are not only conduits but also "capacity vessels," and that if this is true large quantities of blood must often accumulate in the lower parts of the body. This in turn made it seem likely that under such circumstances as exercise, blood loss or heart failure the veins must have to function actively in order to maintain venous blood pressure and perhaps to redistribute the blood supply.

The next line of investigation was therefore to establish whether or not the veins constrict actively in response to various stimuli, and if so how. It had long been known that changes in the tone, or degree of stiffness, of the arterial vessels occur in response to certain stimuli and that these changes exert a powerful and immediate influence on blood pressure and the well-being of the organism. Changes of tone in the veins were much less obvious, so that new methods had to be developed to measure them under various physiological conditions in human beings. This has been done over the past 10 years or so by several groups of workers, including those groups with which I have been associated at the Boston University School of Medicine, the Medical College of Georgia and the University of Virginia Medical School. It is now established

that the veins constrict actively, not only to preserve blood pressure in the highly distensible venous system but also to shift blood from the periphery toward the central circulation as required. The veins, in other words, comprise the reservoir of the circulatory system.

The successive components of the circulatory system can be described in terms of their characteristics as elements in a fluid-filled dynamic system. The heart is a pump with a power plant and valves that discharge oxygenated blood at a rate of flow of four quarts a minute. The arteries, which receive this blood at high pressure and velocity and conduct it throughout the body, are thickly walled with elastic fibrous tissue and a wrapping of muscle cells. The arterial tree terminates in short, narrow, muscular vessels called arterioles, from which the blood enters the capillary bed: a pervasive network of microscopic vessels whose tenuous walls act as a membrane across which nutrient and waste substances diffuse into and out of the tissues. From the capillaries the blood, now depleted of oxygen and burdened with waste products, moving more slowly and under low pressure, enters small vessels called venules and then the veins. These are generally larger and less thickly walled than comparable arteries; the layer of smooth muscle along their entire length is much thinner than that of the arteries, but then it has much lower pressures to cope with.

The entire vascular system, including the heart, is closely regulated by nerve impulses from the brain and spinal cord as well as by the intrinsic responsiveness of the vascular system itself. Nerve impulses reach the blood vessels and heart by way of the autonomic nervous system. The major source of nerve impulses

to the blood vessels is in the sympathetic ganglia, collections of nerve cells that control the sympathetic nerve fibers. The impulses ultimately result in the release of the hormone noradrenalin (also called norepinephrine) at the nerve endings on smooth muscle in the walls of the arteries and veins and in the heart.

An important concept of nerve function is the receptor theory, first proposed by Raymond P. Ahlquist of the Medical College of Georgia and later confirmed as an explanation of nerve and muscle function in the vascular system. It holds that small protein receptor complexes on the walls of smooth muscle cells have characteristics that determine the response the cell will make when the receptor is stimulated by a suitable substance. When noradrenalin is released by the nerve ending in the vicinity of the receptor, stimulation of an "alpha" receptor results in contraction and therefore in blood-vessel constriction, whereas stimulation of a "beta" receptor results in relaxation and therefore in dilatation of the vessel.

John W. Eckstein and his group at the University of Iowa College of Medicine found that arteries contain both alpha and beta receptors, so that certain stimuli result in the dilatation of arteries and other stimuli in their constriction. The preponderant receptor effect of noradrenalin is alpha stimulation (although some beta stimulation also occurs), and so the net effect of noradrenalin on arteries is constriction. In veins, on the other hand, only alpha receptors are present, so that sympathetic nerve stimulation results only in constriction. As the various responses of veins are described later in this article, it will be noted that an increase in tone of the smooth muscle—a constriction of the vessels—is the usual and expected response

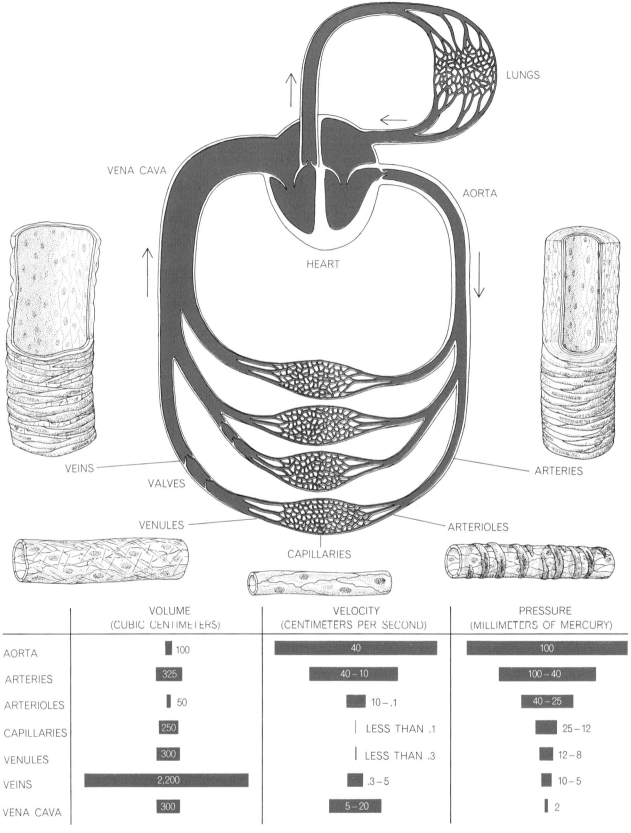

	VOLUME (CUBIC CENTIMETERS)	VELOCITY (CENTIMETERS PER SECOND)	PRESSURE (MILLIMETERS OF MERCURY)
AORTA	100	40	100
ARTERIES	325	40 – 10	100 – 40
ARTERIOLES	50	10 – .1	40 – 25
CAPILLARIES	250	LESS THAN .1	25 – 12
VENULES	300	LESS THAN .3	12 – 8
VEINS	2,200	.3 – 5	10 – 5
VENA CAVA	300	5 – 20	2

CIRCULATION is diagrammed schematically as a hydrodynamic system in which blood is forced into the arteries at high pressure by the heart and maintained at high pressure by the resistance effect of the arterioles, which supply the capillary bed that permeates all the tissues. Then, at low pressure, the blood enters the venules and finally the veins, many of which contain valves that keep the blood flowing toward the heart. The chart gives the total volume of all the vessels of each kind in man (note the great aggregate volume of the veins) and average values for the velocity and pressure in each kind of vessel. As shown by the drawings of segments of vessels, a single layer of endothelial cells lines the entire system, surrounded by fibrous tissue except in the case of the capillaries and by smooth-muscle cells in the case of arteries, arterioles and veins. The vessels are not drawn to scale.

of the veins, whereas arteries may dilate or constrict. One reason for this difference in the pattern of response of these two systems is the difference in receptor makeup. (In addition certain arteries have vasodilator nerves that apparently function differently from the sympathetic constrictor nerves.)

The nervous system functions so as to cause the heart to increase or decrease its pumping action in order to maintain the flow of blood needed by the entire body The arterial side of the circulation tends to respond to the need for flow to specialized tissues by maintaining arterial pressure. (For example, the arterial circulation to the brain tends to remain open while the circulation to less immediately essential parts of the body closes down during stressful circumstances.) On the venous side of the circulation nervous impulses impinge on the veins so as to maintain enough pressure and volume of blood in the central veins to form an adequate reservoir of blood for the heart to pump back into the lungs and ultimately into the arterial system.

If we were to understand the functioning of the veins as an active reservoir, we needed to learn how they constrict and dilate, in health and disease, in response to various environmental conditions and other stimuli. More specifically, we needed data on the distensibility of the veins, which would indicate the degree of contraction of the muscles controlling the walls. The distensibility of any hollow organ (or a similar inorganic element) is described by the way in which its volume varies with the pressure of the fluid in it.

There are standard methods of measuring changes in volume and pressure in veins. The volume is recorded by measuring, with the instrument called a plethysmograph, changes in the size of a patient's arm or leg. This is possible because so much of the blood in an extremity is contained in the veins, and because the arteries and capillaries change their volume relatively little; fluctuations in the volume of the extremity can therefore be ascribed almost completely to changes in the volume of the veins. The extremity (the forearm in most of our experiments) is placed in an airtight, watertight box that is partly filled with water. Any change in the volume of the forearm causes a proportionate rise in the water level, displacing air from the box into a measuring device, such as a simple bellows, that operates a recording pen [see top illustration on page 2612].

There are several ways to measure pressure within a vein, but the most graphic is to introduce into the vein a flexible tube attached to a glass column filled with a salt solution. The height of the solution above the tip of the tube is a measure of the excess of the local venous blood pressure over atmospheric pressure.

Clearly the mere measurement of volume and pressure is a simple affair. The problem is that in order to get reproducible pressure-volume curves and to compare results in different people under various conditions, one must start with a low and constant pressure and volume as a base line. Consider the problem of comparing the distensibility of a new and an old toy balloon. The distensibility of the new one is less than that of the old one; you have to blow harder to inflate it. When both of the balloons are inflated to the same pressure, say 30 millimeters of mercury, the volume of the old one is three times as great as the volume of the new one. That is a quantitative statement of the difference in distensibility of the two balloons and is valid because they began from the same base line: they had the same zero pressure and about the same low volume when they were deflated.

To accomplish the same result in the case of the veins we add water to the plethysmograph. This raises the venous pressure somewhat but, more important for our purpose, it reduces the "effective" venous pressure—the difference between the pressure within the vein and the pressure surrounding it—to less than a millimeter of mercury. Adding more water to the plethysmograph raises the venous pressure but does not change the effective pressure [see illustration on page 2612]. In other words, as long as water pressure in the plethysmograph exceeds the normal local venous blood pressure, the effective venous pressure— the pressure that actually tends to distend the vein—remains constant. The veins, like deflated balloons, are therefore in a state of low constant pressure and volume regardless of the tone of their walls.

The next step is to increase the effective pressure as one does by blowing up a balloon. This is done by inflating a blood-pressure cuff on the upper arm and thereby obstructing the flow of blood toward the heart. We inflate the cuff until the arm volume increases just a bit, showing that the cuff pressure barely exceeds the local venous blood pressure. Beyond this point, which we call an effective pressure of zero, any increment in cuff pressure causes an equal

SYMPATHETIC NERVES are one source of control of muscle in blood-vessel walls. According to one theory, noradrenalin released by the nerve endings affects "alpha" receptors, causing muscle contraction, or "beta" receptors, causing relaxation. Arteries have both receptors, veins only alpha receptors, and so sympathetic activity always constricts veins.

rise in effective venous pressure. Increasing the cuff pressure in five-millimeter-of-mercury increments, we note the change in volume recorded by the plethysmograph [*see illustration at upper left on page 2613*]. When these data are plotted, they yield the pressure-volume curve for the veins being studied [*see illustration at upper right on page 2613*].

The experiment can then be repeated under different conditions, in the presence of the hormone adrenalin, for example. Adrenalin (or epinephrine) is secreted by the adrenal gland and can also be synthesized. Whether it enters the bloodstream from the gland or is administered as a drug, it causes the arterioles to dilate but causes the veins to contract. (This is one of the cases in which the veins and the arterioles have opposite reactions to the same stimulus.) If we administer adrenalin to a subject, the muscle in the vein walls constricts and the veins become less distensible; the distensibility (or the change in volume at a pressure of 30 millimeters of mercury) is two cubic centimeters in the presence of adrenalin compared with 5.5 cubic centimeters under normal conditions.

One of the important functions of the circulation is to help maintain a constant body temperature, and we therefore studied the effect of a change in temperature on the distensibility of the veins. At a room temperature of 83 degrees Fahrenheit a lightly clothed male, lying down, is in approximate thermal equilibrium with his surroundings. In

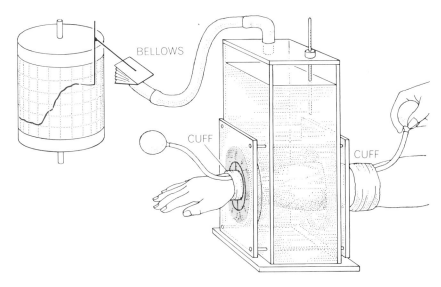

PLETHYSMOGRAPH records changes in forearm volume, thus measuring changes in the volume of the veins. The subject's arm is placed inside a rubber sleeve that forms a watertight seal with the sides of the plastic plethysmograph case. The pressure inside the veins is increased by inflating the pressure cuff (*right*), obstructing the flow of blood toward the heart; second cuff (*left*) temporarily cuts off circulation to the hand. Any increase in arm volume raises the water level, displacing air to operate the bellows and recording pen.

such an environment the venous distensibility averaged four cubic centimeters. When the environment was warmed to 95 degrees, the subject's venous distensibility remained the same; nevertheless, calculations (based on the rate of rise of the plethysmograph tracing) showed that the volume of venous blood flowing through the forearm increased. What happened was that the arterioles had dilated, allowing more blood to flow. Since the tone of the veins remained unchanged, the velocity of the blood in the veins increased. Then, when the temperature of the room was reduced to 68 degrees, the arterioles constricted almost at once; the blood flow diminished and the veins constricted 10 to 15 minutes later, thereby restoring blood velocity in the veins to about the normal level.

If the veins of the arm are considered as radiators, then the conservation of heat in cold weather could best be accomplished if a minimal amount of

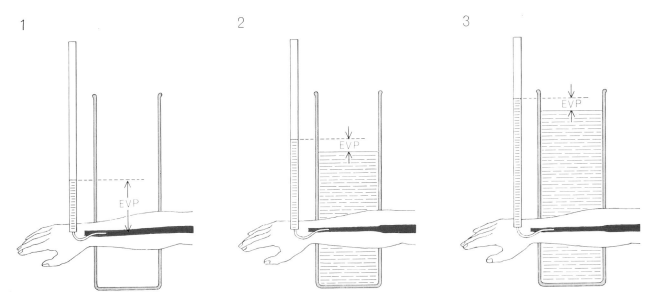

EFFECTIVE PRESSURE, the difference between the pressure inside and the pressure outside the veins, must be brought to a low, constant value for experiments on venous volume. This is done by adding water to the plethysmograph. In an empty plethysmograph the effective venous pressure (*EVP*) is equal to the blood pressure (*1*). If water is added, the pressure in the vein rises but the effective pressure becomes very small (*2*). More water further increases the internal pressure, but the effective pressure remains small (*3*).

blood flowed through them as rapidly as possible. That is what happened in the cold experimental situation. The removal of heat from the body in hot weather, on the other hand, could best be accomplished by a large quantity of blood exposed near the surface at a low velocity. Yet these conditions were not met in the warm experimental situation. The reason is that a certain pressure must be maintained in the veins in order to fulfill the system's functions as a reservoir, and this could not be done if the veins di-

lated enough to reduce the velocity.

One of the major physiological problems for man, the erect animal, is maintaining blood flow to all parts of the body. When a man stands up, the hydrostatic pressure in his leg veins approaches 100 millimeters of mercury; a large volume of blood tends to settle in the distensible vessels of the lower legs. This pooling of blood is counteracted, first of all, by simple flap valves in the veins, which act to prevent the return

flow of blood moving upward; they tend to hold the blood in a series of short, low-pressure columns. We have found that a second important mechanism counteracts pooling: a generalized constriction of the veins, not only in the legs but also elsewhere in the body. We demonstrated this effect by measuring venous distensibility in the forearm of a patient wearing inflatable legging-like pressure stockings. When the stockings were inflated, preventing the pooling of blood, the forearm venous distensibility in a number of sub-

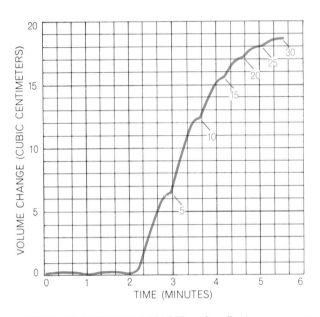

CHANGE IN FOREARM VOLUME as the effective venous pressure is raised is shown by a plethysmograph tracing. The effective pressure, from the base line to 30 millimeters of mercury, is shown by the numbers along the curve. The change in volume with each five-millimeter increment becomes less as the pressure rises.

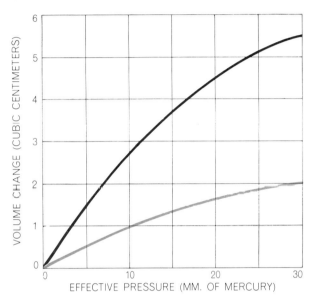

DATA FROM TRACING are plotted to yield a pressure-volume curve, with the volume changes corrected for the size of the subject's arm. Here the curve for a healthy subject under standard conditions (black) is compared with the curve obtained while he was being given adrenalin, which constricts the veins (color).

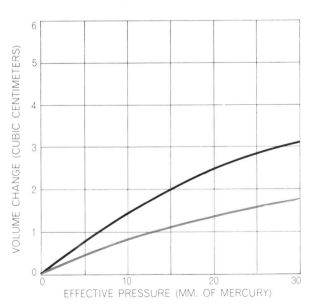

PRESSURE-VOLUME CURVES for a patient with heart failure show how the distensibility of the veins is reduced when the patient is at rest (black) and further reduced during exercise (color).

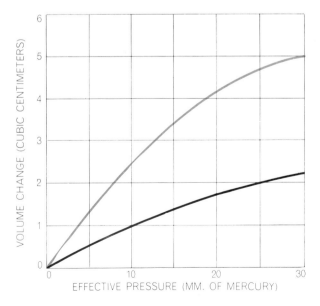

HEART-FAILURE PATIENT, exercising, has constricted veins (black). The administration of a drug that blocks the action of the sympathetic nervous system abolishes the venous response (color).

jects averaged 3.8 cubic centimeters; 15 minutes after the pressure was released, allowing blood to pool in the legs, distensibility in the arm averaged 2.6 cubic centimeters. In other words, although there was no change in the position of the forearm, the forearm veins constricted when blood pooled in the legs. This response has the ultimate effect of helping to maintain the pressure in the venous system and in particular near the heart. If this pressure is too low, there is inadequate blood flow into the heart and inadequate cardiac output; reduced blood flow to the brain often causes fainting, as in the case ot a soldier standing too long at attention.

When a person exercises, the tissues require an additional supply of blood. The heart rate increases to meet this need, but the heart may be limited in its ability to respond if not enough blood is being returned to it by the veins. An experiment that David E. Bass and I performed at the U.S. Army Natick Laboratories in Massachusetts, showed how the veins respond during exercise. The distensibility of veins in the forearm was measured in subjects walking a treadmill. The veins did constrict, showing a definite shift of blood away from the arm toward the heart to facilitate increased cardiac output. When the measurements were made at an elevated temperature, soldiers who were not conditioned to the heat showed evidence of inadequate cardiac output and specifically of inadequate venous response. Exercise for several days in the heat improved their responses, suggesting that the veins play a role in acclimatization to heat.

If the heart is subjected to some handicap that impairs its ability to function as a pump, such as high blood pressure, coronary artery disease or damage to the heart valves by rheumatic fever, then the condition known as heart failure may result. Some of the symptoms of heart failure are directly due to the fact that the heart is not pumping enough blood. Other symptoms, however, are caused by compensatory responses by the body to the lack of adequate blood flow, and one of the most interesting of these is a venous response. The normal heart is able to meet the total blood-flow needs of the body, even during mild exercise, without constriction of the veins. When the heart is handicapped, however, the veins are constricted even at rest. This chronic venous constriction, in association with the greater than normal total blood volume characteristic of heart failure, causes high pressure in the veins and capillaries, forcing large quantities of fluid through

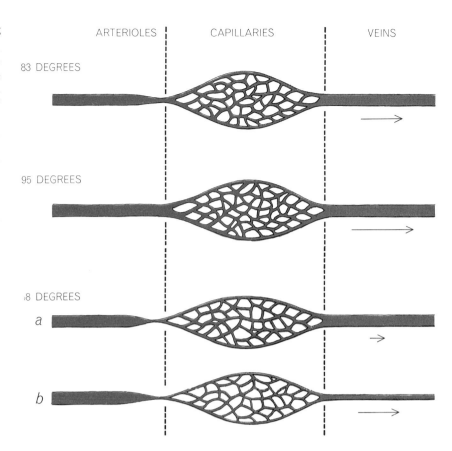

ARTERIOLES   CAPILLARIES   VEINS

83 DEGREES

95 DEGREES

.8 DEGREES
a

b

TEMPERATURE CHANGES prompt a venous response. At a room temperature of 83 degrees Fahrenheit arteriole resistance and venous volume are normal. At a temperature of 95 degrees the arterioles dilate but the veins do not, and so the blood velocity (*arrow*) in the veins increases. At 68 degrees the arterioles constrict and reduce blood flow, thus reducing venous velocity (*a*); then the veins constrict and restore the velocity to about normal (*b*)

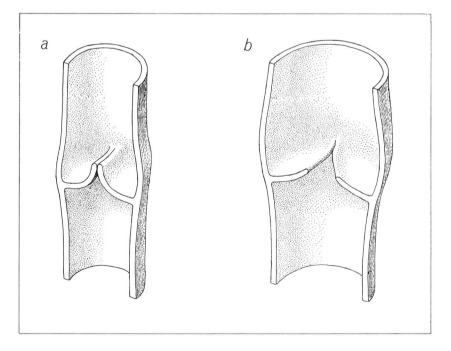

a   *b*

VALVE of a normal vein is composed of leaflets whose free edges meet intermittently to keep blood from flowing backward (*a*). If the valve functions badly, either because it has been damaged or because the vein walls are distended, blood pools in the vein, dilating it and further interfering with normal valve function. The result is a varicose vein (*b*).

the thin capillary walls into the tissues. The resulting dropsy, or edema, is a common symptom in heart failure. Constriction of the veins in heart failure is evident in a patient at rest and is accentuated during exercise [*see illustration at lower left on page 2613*]. Constriction of the veins in heart failure is induced by the autonomic nervous system. If a drug known to block sympathetic action is administered, the response is abolished, even during exercise [*see illustration at lower right on page 2613*].

What the tissues lack both in heart failure and during vigorous exercise is enough oxygen. We wondered if the veins would respond to the relative lack of oxygen at high altitudes, and in collaboration with Sujoy B. Roy of the All India Institute of Medical Sciences we carried a plethysmograph into the Himalayas last summer. We found that when a man is exposed to an altitude of more than 10,000 feet, the responses of the veins simulate those of the heart-failure patient in many respects.

When a person has high blood pressure that cannot be related to a cause such as kidney trouble, he is said to be suffering from essential hypertension. Although the origin of this condition is not known, the basic mechanical problem is clearly a constriction of the arterioles. Usually, as I have indicated, the arterioles and veins constrict together. It was therefore surprising to find, when we examined patients with essential hypertension, that their veins were not constricted. Other experiments suggest that the venous response may even be depressed in such patients. The importance of this finding is that it may narrow the field of possible causes of essential hypertension. For example, one suspected cause—increased activity of the autonomic nervous system—seems to be ruled out because it would almost surely have similar effects on both the veins and the arteries.

When blood flows too slowly, it has a tendency to clot. Clots that form in veins (thrombophlebitis) can break off and travel to the lungs, causing a pulmonary embolism, or stay in place and obstruct blood flow so badly as to cause serious edema. Thrombophlebitis can also damage the valves in a vein and is therefore one cause of varicose veins, a condition in which lack of valve function causes the blood to pool in the leg veins, which become chronically dilated. Varicose veins can arise from any injury to the valves or from the congenital absence of valves, but this is far from explaining the large number of cases. In 1966 we discovered that the venous distensibility is high in the forearm of patients who have varicose veins in the leg. Also in 1966, S. M. Zoster of the McGill University Faculty of Medicine, using different methods, found the same thing. This suggests that certain people inherit a venous disability that makes for lack of tone in the leg veins. When such a person stands up, abnormal dilatation of the veins keeps the valves from functioning properly, further dilating the vessels and leading to varicose veins.

Pregnant women have a predisposition to thrombophlebitis and to varicose veins, and some fatal cases of pulmonary embolism from thrombophlebitis have been ascribed to oral contraceptive drugs. We therefore measured venous distensibility in pregnant women and in women receiving oral contraceptives. In both cases we found that there was a generalized loss of venous tone com-

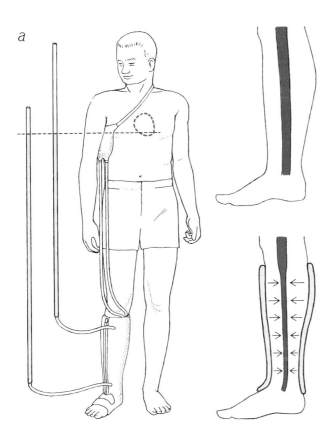
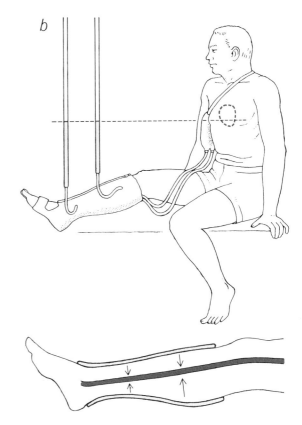

HYDROSTATIC STOCKING exerts a pressure that varies with the blood pressure in the leg. When the patient is standing, the pressure at the ankle (for a six-foot patient) would be about 100 millimeters of mercury (*a*). Without the stocking this would distend the vein (*top*); water pressure in the stocking keeps the vein from dilating (*bottom*). When the patient sits, the venous pressure at the ankle is about 50 millimeters (*b*); the stocking pressure also goes down. The stocking is used to treat ulcers from varicose veins.

pared with a control group. This may well explain the dilatation associated with poor valve function in varicose veins and with low velocity of blood flow in thrombophlebitis.

Usually varicose veins are superficial vessels that are not essential to the circulation; they can be tied off and removed by surgery. When a number of deep veins are involved, however, this is not possible. The constant high pressure of the pooled blood, unrelieved by valve action, can cause extreme edema that breaks down tissue and forms painful ulcers. Since dilatation of the vein wall depends on the difference between the pressure inside and the pressure outside the vein, the logical approach to therapy of advanced varicose veins would be to equalize those pressures. This would reduce the effective pressure to near zero and presumably force fluid

out of the tissues into the blood vessels. In addition, by bringing the valve leaflets closer together, it would enable the valves to function better. The trouble is that an external pressure high enough to counterbalance venous pressure when the patient was standing would be so high as to cut off all circulation in the legs when he sat down.

After a series of experiments we devised a way to exert just the right amount of pressure on the veins of the leg regardless of the patient's posture: a hydrostatic pressure stocking, in which the pressure would be supplied by a reservoir of water at about the level of the patient's heart. John E. Flagg of the David Clark Company designed the stocking, which is attached by two tubes to a water bag carried under the patient's arm. Any point on the surface of the patient's leg is therefore constantly sub-

jected to a pressure proportional to the vertical difference between that point and the level of the heart. The hydrostatic stocking has since proved beneficial to a number of patients suffering from ulcers due to varicose veins.

The stocking is one example of a practical therapy for malfunction of the veins that has been derived from experimental studies of normal vein function. As for our theoretical findings, I have a feeling that the most significant aspect may be the discovery of instances in which the veins constrict or dilate in a manner diametrically different from other components of the circulation. By pursuing these instances of apparently incongruous venous response we should be able to isolate mechanisms that combine to produce the complex responses of the vascular system in health and disease.

The Author

J. EDWIN WOOD is professor of medicine and Virginia Heart Association Professor of Cardiology at the University of Virginia. A graduate of Virginia, he received an M.D. from the Harvard Medical School in 1949. In 1950 he joined the faculty of Boston University, where he began his study of veins. He served in the U.S. Air Force from 1951 to 1953, during which time he investigated the effects of high-altitude pressure suits on the circulation. He returned to Boston, where he continued his work until 1958, when he went to the Medical College of Georgia to do cardiovascular research. He joined the Virginia faculty in 1964. Wood has also worked as a consultant to the U.S. Army on problems of excessive environmental heat and cold at its Natick, Mass., laboratories. He is currently chairman of the Council on Circulation of the American Heart Association.

Bibliography

HANDBOOK OF PHYSIOLOGY, SECTION 2: CIRCULATION, VOL. II. W. F. Hamilton and Philip Dow. American Physiological Society, 1963.

THE VEINS: NORMAL AND ABNORMAL FUNCTION. J. Edwin Wood. Little, Brown and Company, 1965.

SCIENTIFIC
AMERICAN July 1958, Vol. 199, No. 1, pp. 67-72

OFFPRINT **1094**

PREDATORY FUNGI

by Joseph J. Maio

Plants which trap insects are familiar in the folklore of biology.
Less well-known, but more important in the balance of nature, are
certain molds which ensnare and consume small animals in the soil.

Unless man succeeds in duplicating the process of photosynthesis, it appears that animals will always have to feed upon plants. But the plant world exacts its retribution. A number of plants have turned the tables on the animal kingdom, reversing the roles of predator and prey. These are the plant carnivores—plants that trap and consume living animals. Most famous are the pitcher plant, with its reservoir of digestive fluid in which to drown hapless insects; the sundew, with its flypaper-like leaves; and Venus's-flytrap, with its snapping jaws. But there are other carnivorous plants of larger significance in the balance of nature. We ought to know them better because they are to be found in great profusion and variety in any pinch of forest soil or garden compost. They are microscopic in size, but just as deadly to their animal prey as the sundew or Venus's-flytrap.

These tiny predators are members of the large group of fungi we call molds. They grow in richly branching networks of filaments visible to the naked eye as hairy or velvety mats. Molds do not engage in photosynthesis. Like most bacteria, they lack chlorophyll and so must

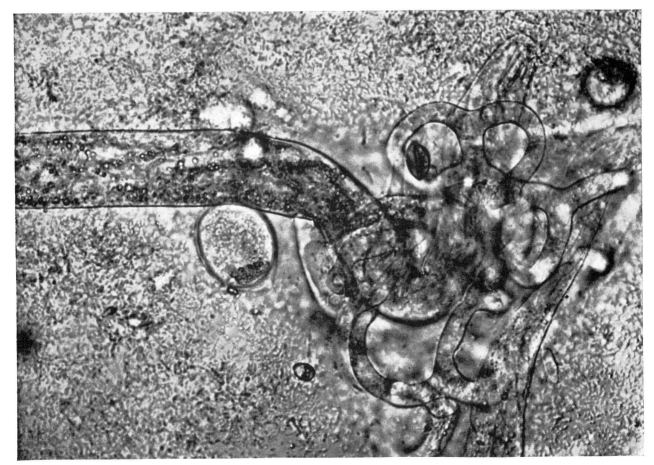

NEMATODE WORM IS TRAPPED by the adhesive fungus *Trichothecium cystosporium* in this photomicrograph by the British biologist C. L. Duddington. The entangling network of the fungus is at right; the body of the worm extends to the left. The oval object below the body of the worm is one of the spores by which the fungus reproduces. At lower right are the remains of another worm.

derive their food from other plants and from animals. Molds have long been familiar as scavengers of dead organisms, promoters of the process of decay. It was not until 1888 that a German mycologist, named Friedrich Wilhelm Zopf, beheld molds in the act of trapping and killing live animals—in this case the larvae of a tiny worm, the wheat-cockle nematode.

The nematodes (eelworms, hookworms and their like) are not the only prey of these animal-eating plants. Their victims run the gamut from the comparatively formidable nematodes down to small crustaceans, rotifers and the lowly amoeba. Charles Drechsler of the U. S. Department of Agriculture, a student of the subject for some 25 years, has identified a large number of carnivorous molds and matched them to their prey. Many are adapted to killing only one species of animal, and some are equipped with traps and snares which are marvels of genetic resourcefulness. How they evolved their predatory habits and organs remains an evolutionary mystery. These molds belong to quite different species and have in common only their behavior and some similarities of trapping technique. They present a challenging subject for investigation which may throw light on some fundamental questions in biology and may lead also to new methods for control of a number of crop-killing nematodes.

The simplest of the molds have no special organs with which to ensnare their victims. Their filaments, however, secrete a sticky substance which holds fast any small creature that has the misfortune to come in contact with it. The mold then injects daughter filaments into the body cavity of the victim and digests its contents. Most of the animals caught in this way are rhizopods—sluggish amoebae encased in minute hard shells. Sometimes, however, the big, vigorous soil nematodes are trapped by this elementary means.

More specialized is an unusual water mold, of the genus *Sommerstorffia*, which catches rotifers, its actively swimming prey, with little sticky pegs that branch from its filaments. When a rotifer, browsing among the algae on which this mold grows, takes one of these pegs in its ciliated mouth, it finds itself impaled like a fish on a hook.

Some molds do their trapping in the spore stage. The parent mold produces staggering numbers of sticky spores. When a spore is swallowed by or sticks

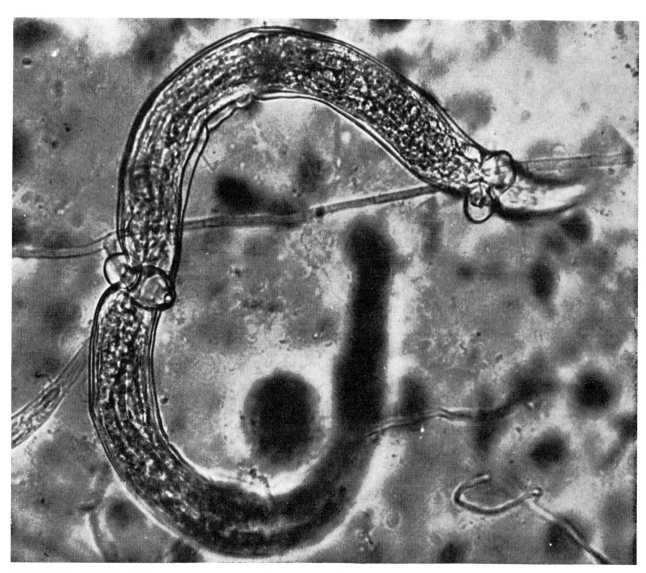

TWO CONSTRICTING RINGS of the fungus *Dactylaria gracilis* grasp another nematode in this photomicrograph by Duddington. The nematode was first caught by the head (*upper right*), and then flicked its body into another ring (*left center*). The rings deeply constrict the body of the worm. The horizontal line above the middle of the picture is a filament to which the rings are attached.

to a passing amoeba or nematode, it germinates in the body of its luckless host and sends forth from the shriveled corpse new filaments and new spores to intercept other victims.

The most remarkable of all killer molds are found among the so-called *Fungi Imperfecti*, or completely asexual fungi. The advanced specialization of these molds is particularly interesting because they are not killers by obligation but can live quite well on decaying organic matter when nematodes, their animal victims, are not available. If nematodes are present, these molds immediately develop highly specialized structures which re-adapt them to a carnivorous way of life. They will do so even if they are merely wetted with water in which nematodes have lived.

One of these molds is *Arthrobotrys oligospora*, the nematode-catching fungus that was first studied by Zopf. When nematodes are available, it develops networks of loops, fused together to form an elaborate nematode trap. An extremely sticky fluid secreted by the mold seems to play an important role in capturing the nematode, which need not even enter the network in order to be held fast. The fluid is so sticky that one-point contact with the network frequently is enough to doom the nematode. In its frenzied struggles to escape, the worm only becomes further entangled in the loops, and finally, after a few hours of exertion, weakens and dies. The destruction these molds can cause in a laboratory culture of nematodes is appalling, particularly from the point of view of the nematode!

Two French biologists, Jean Comandon and Pierre de Fonbrune, have made motion pictures which show that the fungus's secretion of this adhesive substance is accompanied by intense activity in its cells. Material in the cytoplasm of the cell streams toward the point of contact with the worm. The mold may be bringing up reserves of adhesive and digestive enzymes to subdue the nematode; it may also secrete a narcotic or an intoxicant to speed the process.

Even more artfully contrived are the "rabbit snares" employed by some molds. First fully described by Charles Drechsler, these are rings of filament which are attached by short branches to the main filaments, hundreds of them growing on one mold plant. The rings are always formed by three cells and have an inside diameter just about equal to the thickness of a nematode. When a nematode, in its blind wanderings

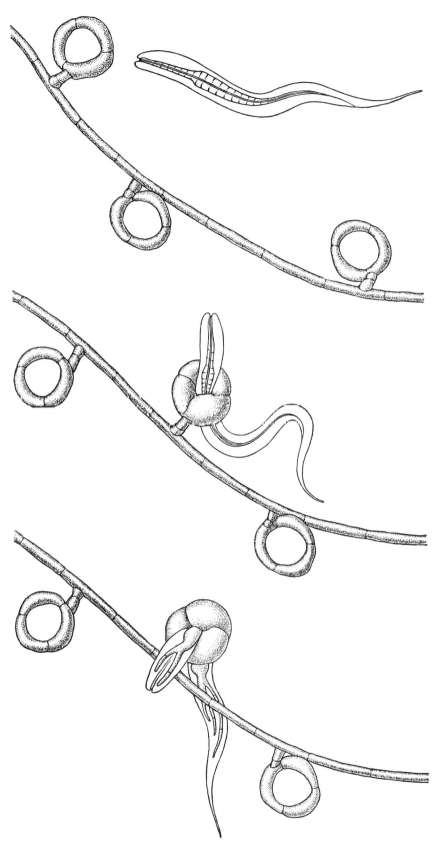

CAPTURE AND CONSUMPTION of a nematode worm by a fungus of the constricting-ring type is depicted in this series of three drawings. At the top the worm approaches a ring attached to a filament of the fungus. In middle the three cells of one ring have expanded to trap the worm. At bottom filaments have branched out of the fungus to digest the worm.

through the soil, has the ill luck to stick its head into one of these rings, the three cells suddenly inflate like a pneumatic tire, gripping it in a stranglehold from which there is no escape.

The rings respond almost instantaneously to the presence of a nematode; in less than one tenth of a second the three cells expand to two or three times their former volume, obliterating the opening of the ring. It is difficult to understand how the delicate filaments can hold the powerfully thrashing worm in so unyielding a grip. Occasionally a muscular worm does escape by breaking the ring off its stalk. But this victory only postpones the inevitable. The ring hangs on like a deadly collar and ultimately generates filaments which invade the worm, kill it and consume it.

We are not yet sure what cellular mechanisms activate these deadly nooses. We know that in the case of the constricting ring of one mold the activating stimulus is the sliding touch of the nematode as it enters the ring. A nematode that touches the outer surface of the ring will not trigger the mechanism. But if the worm passes inside the ring, its doom is certain. This mold, then, exhibits a sharply localized "paratonic" or touch response like that of the Venus's-flytrap.

Perhaps the inflation of the cells is caused by a change in osmotic pressure, resulting in an intake of water either from the environment or from neighboring cells. Or perhaps it results from changes in the colloidal structure of the cell protoplasm. There is a recent report from England that the constricting rings of one species react to acetylcholine—the substance associated with the transmission of impulses across synapses in the animal nervous system!

Some species of molds prey upon root eelworms that infest cereal crops, potatoes and pineapples. This has inspired experiments to use these fungi to control the pests. In one early experiment, conducted in Hawaii by M. B. Linford of the University of Illinois and his associates, a mulch of chopped pineapple tops was added to soil known to harbor the pineapple root-knot eelworm. This mulch produced an increase in the numbers of harmless, free-wandering nematodes which thrive in rich soil. The presence of these decay nematodes stimulated molds in the soil to develop nematode traps, which caught the eelworms as well as the harmless species. A recent experiment in England gives similar promise that the molds may be effective against the cereal root eelworm. Plants protected by stimulated molds showed slight damage compared to the eelworm-ravaged control plants.

Investigators in France have reported an experiment which suggests that molds may be used to control nematode parasites of animals as well as those of plants. Two sheep pens were heavily infested with larvae of a hookworm, closely related to the hookworms of man, which causes severe pulmonary and intestinal damage to sheep. One of the pens was sprinkled with the spores of three molds that employ snares or sticky nets to trap nematodes. Healthy lambs were placed in both pens. After 35 days of exposure the lambs in the pen inoculated with the molds were found free of infection, while those in the control pen showed signs of infestation with the worm.

The carnivorous molds offer many possibilities for future investigation. One subject that needs to be explored is their role in the complex biology of the soil. We would also like to know more about the physiological mechanism that underlies the extraordinary behavior of the nematode "snares." The results of experiments on mold control of nematodes are already encouraging. They suggest that one day these peculiar little plants may perform an even more important role in agriculture than they played in nature, silently and unobtrusively, throughout the millennia before their discovery.

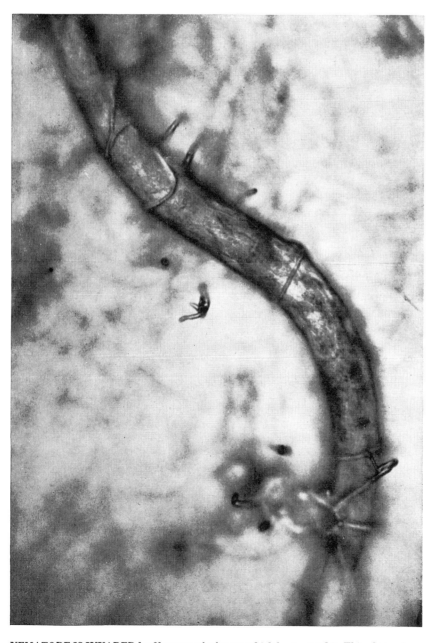

NEMATODE IS INVADED by filaments of a fungus which has trapped it. This photomicrograph was made by David E. Pramer of the New Jersey Agricultural Experiment Station.

The Author

JOSEPH J. MAIO originally intended to become a composer and still plays the piano and violin as a hobby. His interest in predatory plants stems from his high-school years, when he maintained a collection of Venus's-flytraps, sundews and pitcher plants. A three-year stay in a tuberculosis sanatorium made him especially aware of the activities of micro-organisms, and in 1955 he received his B.S. in microbiology from the University of Washington and his M.A. two years later. As a research assistant at the University of Washington School of Medicine he is at present working on bacterial viruses. In addition to his musical diversions he enjoys oil painting.

Bibliography

THE FRIENDLY FUNGI: A NEW APPROACH TO THE EELWORM PROBLEM. C. L. Duddington. Faber and Faber, 1957.

PREDACEOUS FUNGI. Charles Drechsler in *Biological Reviews*, Vol. 16, No. 4, pages 265-290; October, 1941.

THE PREDACIOUS FUNGI AND THEIR PLACE IN MICROBIAL ECOLOGY. C. L. Duddington in *Microbial Ecology*, pages 218-237; 1957.

THE PREDACIOUS FUNGI: ZOOPAGALES AND MONILIALES. C. L. Duddington in *Biological Reviews*, Vol. 31, No. 2, pages 152-193; May, 1956.

SOME HYPHOMYCETES THAT PREY ON FREE-LIVING TERRICOLOUS NEMATODES. Charles Drechsler in *Mycologia*, Vol. 29, No. 4, pages 447-552; July-August, 1937.

STIMULATED ACTIVITY OF NATURAL ENEMIES OF NEMATODES. M. B. Linford in *Science*, Vol. 85, No. 2,196, pages 123-124; January 29, 1937.

SCIENTIFIC
AMERICAN   October 1955, Vol. 193, No. 4, pp. 88-98

OFFPRINT 1095

NOCTURNAL ANIMALS

by H. N. Southern

Both predator and prey know their woodlands as intimately as an expert pianist knows the keyboard. The animals' blindness to red light makes it possible to observe them.

Animals avoid competing with one another by evolving more or less specialized ways of life. Such specialization may take the form of being bigger or smaller than other animals, of eating different foods, of living in different habitats or different niches of the same habitat, of foraging at different times of day. It is not surprising, therefore, that many animals concentrate their vital activity in the nighttime.

The study of night animals has always been a challenging one for the naturalist. Their dark world is as unknown and difficult to probe as life in the depths of the sea. Its investigation demands unusual methods, much labor and patience and a carefully governed imagination.

My own interest in nocturnal animals arose in a severely logical way. I wished to make a quantitative study of the predator-prey relationships among birds and mammals. Because both the predator and the prey populations would have to be laboriously censused, it was necessary to select a predator which did not range over too large a territory or prey on too many different kinds of animals. After considering various possibilities, I decided that the most promising was the tawny owl (*Strix aluco*). This species, slightly larger than the barn owl, feeds primarily upon small mice and voles. Although I was led to the choice of an owl by logic, it would be idle to pretend that the fascination of night work counted for nothing. I was soon absorbed in the problem of finding ways to learn the natural history of animals hidden by darkness.

First of all I had to examine closely the night animals' sensory equipment,

TAWNY OWL SWOOPS down on a wood mouse. The owl's flight is quite silent.

both to determine how they get about and to discover methods of observing them without detection. Obviously some nocturnal animals must have unusually acute vision and hearing. The wood mouse (*Apodemus sylvaticus*), a nightworker which is one of the chief foods of the tawny owl, has ultrasensitive eyes and greatly enlarged ear lobes. It responds quickly to the faintest sound, especially in the higher registers. But it is not itself a silent creature. On a still night about an hour after dark, the woodland floor can be heard rustling all over with the excursions of mice. It is the predator rather than the prey which must move in silence. The owl, flying softly on wings with frayed edges, is the most soundless of all night animals. One of the most unnerving hazards of field work in the woodland at night is the sudden, silent onslaught of the tawny owl, which may fiercely attack anyone approaching its chicks. The tawny owl apparently hunts both by sight and by ear, watching and pouncing from a perch. Most owls hunt mainly by ear, locating their prey with asymmetrical ears.

Many animals active by night have no obvious adaptations of sight or hearing to help them move in the darkness. We have learned in recent years that there is another sense which guides animals, including man, in moving about in a familiar territory. For want of a better term we call it the "kinesthetic" sense. It boils down to a conditioned, and therefore swift, repetition of set sequences of muscular movements. The trained muscles of a pianist's fingers produce almost miraculous sequences of movements. A person in his own home can walk down a flight of stairs in the dark and grasp a doorknob with uncanny precision. In night animals the kines-

thetic sense is all-important as a guide to movement and territory.

Let me cite two illustrations. If we put a house mouse into an unfamiliar cage, it will quickly explore its new home, traversing and retraversing it in every possible way. For ease of handling wild mice, I usually keep a refuge box in each cage, into which one can drive the mouse or mice to free the cage for cleaning and so forth. If a mouse is disturbed before it has made its preliminary explorations, it will panic and perhaps leap out of the cage. But if it has first had an opportunity to explore the cage, it will react to a disturbance by darting, swift as lightning, into the refuge box. It has achieved in a short time a familiarity with the environment which enables it to take the right path without "thinking," indeed almost without looking.

The second illustration is furnished by young owls. The chicks of the tawny owl are dependent upon their parents for an extremely long time—up to three months. This extended adolescence is devoted not to lessons in hunting but to learning the territory. A fortnight after learning to fly, the young owls have thoroughly explored a limited area. If one chases them with a flashing torch, they will fly only as far as the boundary of that area and then fly back to the middle of the territory. A week later this boundary has rippled outwards for 100 yards or so. Thus the young birds gradually extend their territory to the final range, which for an adult pair of tawny owls is some 25 to 80 acres of woodland. A territory is a thing which has to be known with an almost indecent intimacy. This appears to me the reason why animals find so little difficulty in maintaining their territorial rights.

Whether or not an animal is equipped with acute sight or hearing, it is likely to rely largely upon the kinesthetic sense in moving about its territory. We need only watch a wood mouse running through a dense tangle of undergrowth with truly fantastic speed and certitude to realize that it is not "seeing" its way. Kinesthetic sense is clearly the answer. As for shrews and moles, animals which have to forage both by day and by night because their appetites need perpetual appeasement, their journeyings are governed so completely by kinesthetic sense that they are equally efficient in the light and the dark. They have little need of eyes, and their eyes can hardly function at all. Indeed, it appears that certain animals have developed nocturnal vision almost, as it were, by chance. They are a sort of random sample of night creatures—a small percentage that has taken this line of specializa-

tion as an "extra." Among rodents the wood mouse is an outstanding example of adaptation to the faintest illumination; another, less outstanding, is the rabbit. Among predators, of course, the owls are pre-eminent.

All of these animals have eyes with similar characteristics: *viz.*, the eye is greatly enlarged and has many rods, which respond to dim light. Some owls can be trained to find even dead, motionless prey in the dark. But the owl's eye is especially fitted to detect movement across the field of view, because rods are sensitive principally to changes of light intensity; cones, which perceive patterns and colors, are absent, or nearly so, from the owl's retina. Less than a millionth of one candle power, about the amount of light that falls on the forest floor on a cloudy summer night, is suffi-

cient to reveal a mouse to the tawny owl. The bird, whose eyes are immobile, has a peculiar way of moving its head constantly up and down and from side to side when it concentrates on some object, presumably to make the object move across its field of view, even if it is motionless.

Having learned something about our subjects' adaptations to darkness, we could proceed to consider ways of outwitting them. To begin with, owls are noisy in communicating over their large territories. Their rather melancholy hooting, so characteristic of English woodlands at night, is a proclamation of territory. Vocal combats between neighboring tawny owls are not infrequent and can be recognized a quarter of a mile away by the hasty and excited way

HIDDEN IN BLIND in Wytham woods, the author spent many nights watching tawny owls. At first he tried to use an infrared Sniperscope, with indifferent success. Later he found he could flood the forest in rather strong red light without disturbing the birds.

in which the hooting rises to a screech or a wail. Furthermore, some males can be identified by a consistent aberration of their hooting, and thus their range of movements can be traced.

From intensive listening throughout the night, and especially in the few hours after dusk when activity is at a peak, a plan can be pieced together of the territories over quite a large area. I found it relatively easy, once I had learned the tricks, to census the tawny owls living on 1,000 acres of woodland.

This map of tawny owl territories was confirmed in an unexpected way. To take censuses of the populations of mice, I trapped large numbers of them, marked them with numbered metal leg-rings, and then released them. Simultaneously I was analyzing each month castings from the tawny owls. These contained the bones and fur of the prey they had eaten, and in the pellets I recovered many of the leg-rings with which I had marked the mice. From a single owl nest I might recover 20 to 30 rings. I knew, of course, at what points in the forest the mice marked with these rings had been released, and so I could test whether the area covered by the predations of the owls in question corresponded with the territory I had sketched in from listening. This correspondence was, in fact, almost perfect, so no room was left for doubting either the territoriality of the owls or the validity of the territory measurements.

When I turned to the task of finding a way to watch the animals directly, I first tried a wartime German invention: the infrared telescope popularly called the Sniperscope or, more flippantly, the Snooperscope. It converts an image formed from invisible, infrared rays into a visible, fluorescent image. I had beaten this sword into a plowshare by using it to watch the feeding behavior of wild Norway rats in the dark. My colleague, Dennis Chitty, had become interested in the problem of determining how it was that any alteration in the setting of bait, even so slight a change as placing it in a tin lid instead of on the bare floor, would cause rats to avoid an accustomed food. With the help of the Sniperscope we were able to watch the suspicious behavior of the rat in all its details.

Nevertheless, the Sniperscope had certain disadvantages. Among other things, the resolution of the image was coarse and the transformer supplying the current to the image-converter tube made a high-pitched whine. This last feature alone was enough to scare tawny owls. I therefore turned to another method of illumination: an automobile

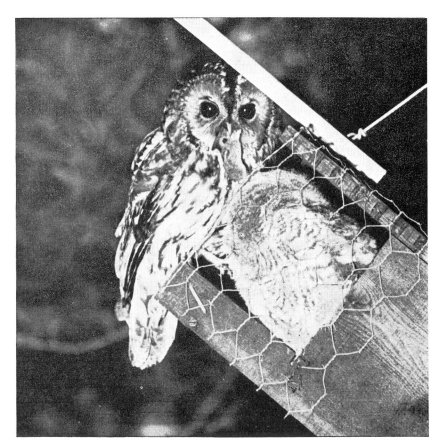

OWL FEEDS WOOD MOUSE to its owlet. Owl breeding closely follows the mouse population. In seasons when mice are abundant, tawny owls may attempt to rear two broods.

head lamp screened, like a darkroom lamp, to give visible red rays. At that time I had some 30 nest boxes for tawny owls distributed through the woodland, all fitted with electrical recording apparatus to show the number and frequency of visits by the parents throughout the night. It was urgent to evolve some method of overnight watching to check how many of the visits actually produced prey and what these prey were. The red light gave the perfect answer. It was invisible to the owls. Ensconced in a blind where I could watch an owl's nest through a large pair of 10 x 80 binoculars mounted on a tripod, I was able to see every detail of the owls' movements and the game brought home from their hunting, even the beetles which they occasionally brought to their young. Of some dozen families of owls that I watched intensively over two breeding seasons, no bird ever betrayed the slightest sign of nervousness of the red flood which illuminated its activities.

At one nest I managed to watch throughout eight complete nights. During this time the male brought home 20 prey, most of which were bank voles, then particularly abundant. This continuous watching was very fatiguing

even though the night was divided between two observers, because it was essential never to take one's eyes away from the binoculars.

During these watches much valuable information was gained on calls and the general behavior of the owls. It was most interesting to see the chicks, as they grew stronger, climbing up to the lip of the box to be fed. I remember one which made a somewhat premature attempt. It was so unsteady that it nearly fell from the box. Thereupon it fairly bolted back into the nest and refused to reappear until several more nights had passed.

The red-light watches disclosed that during the summer tawny owls eat large numbers of earthworms. For some time I had noticed that their castings often contained no fur or bones but had a matrix of vegetable fibers. Under the direct watch it developed that some tawny owl parents organized a regular ferry service of earthworms to their chicks. When the vegetable castings were examined under the microscope, they proved to be full of earthworm chaetae (bristles). Few of the large owls eat earthworms; the tawny owl's habit of doing so may well contribute to its success as a species.

When I applied the red light to watch-

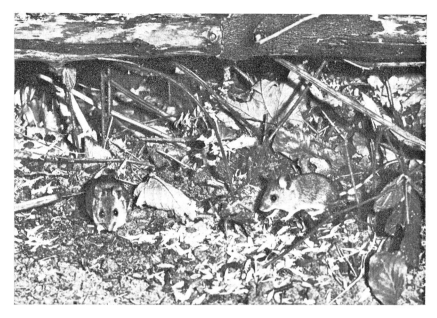

WOOD MICE NIBBLE BAIT near a live-trap. Dominant members of the mouse social hierarchy staked out a territorial claim to the trap and were caught in it repeatedly.

foraging animals at night. I have so far tried this only on the badger, which is a fairly noisy animal. By creeping up when the badger is stamping about and freezing when it is still, it is relatively easy to see closely what the badger is up to.

I believe that this method has great promise. If it can be used so successfully on the relatively few nocturnal animals of temperate latitudes, it should be of even greater value in exploring the richer faunas of the tropics. Its great merit is its simplicity. My apparatus has sometimes been cumbersome and costly, but a good flashlight with a concentrated spot and a red celluloid cap will serve for many purposes. To this one needs to add, however, plenty of perseverence.

The specific project which originally prompted me to test these various methods of night observation has been carried on now for eight years. The analysis and publication of the full results will take a long time. Nevertheless, we are able already to tell a number of things about the interaction between tawny owl and wood mouse populations.

In the first place, a continuous study of the contents of owl castings from all over the 1,000-acre woodland through eight years has shown very interesting seasonal variations in the diet. The owls' predations on wood mice and bank voles are greatest when ground cover is at its scantiest—from the fall of the leaves in late November until the beginning of May. During this period 70 to 80 per cent of the tawny owls' diet may consist of small mammals. Since the mouse population falls to its lowest ebb in early spring (just before the breeding season), it is clear that the preying of the owls must bear heavily on that population. During late spring, summer and early autumn, when the vegetation is thick, the owls turn to young moles and rabbits and to invertebrates such as cockchafers, ground beetles and earthworms.

The making of censuses and plotting of territories, achieved by the listening methods described earlier, revealed an astonishingly high density of owls. In 1947, after an exceptionally long and snowy winter, the breeding population on the 1,000 acres studied was 15 pairs. The following year it had increased to 20 pairs, and at the present this number has expanded further to 26 pairs, which means that each couple lives, feeds and sometimes rears young on only about 40 acres of woodland. It is unlikely that any other bird of prey which feeds principally on vertebrates can exhibit a steady density of this order.

ing the behavior of wood mice, I was especially interested in determining the answer to a perplexing question. Live-traps set out in the woodland catch mice already marked more frequently than they do unmarked mice. Obviously this must throw all askew any attempt to estimate the total mouse population. I therefore watched a trap for several nights, pinning up the door so that it would not close when mice visited the trap for food. I discovered that of half a dozen mice which came to the trap on those nights, one pair completely dominated the scene. The others were all younger ones and only came to feed when the dominant pair had had their

fill. Thus I learned that a social hierarchy existed among these wild mice, and that a trap would merely go on catching the same mouse over and over again.

I became interested to learn what other English mammals were red-blind. So far I have tested only badgers and foxes. Both of these animals appear to be quite unconscious of red illumination. Probably most carnivores that hunt at night have the same limitation. Their world is known to them more in terms of scent and kinesthetic conditioning than of vision; their nocturnal habit may be due partly to daytime persecution.

The red light can be useful not only for watching nests but also for tracking

NOISY BADGER at the entrance to its den pays no attention to red illumination. Southern crept up on the animal while it was stamping about and got close without being heard.

The population figures for the mice and voles are still very approximate. But the trends of the populations from year to year are quite clear. There is a very obvious link between the abundance of mice and the success of the owls in rearing young. In some years a tawny owl pair may raise two or three chicks; in others they make no attempt to breed at all. When the mice are neither abundant nor very scarce, the owls may lay eggs and allow them to chill (presumably because the hen must leave the nest to feed herself, if the male cannot bring her enough prey) or may lose chicks through starvation. If a clutch of eggs is lost, it is unusual for a second clutch to be laid. Second clutches were found on a widespread scale during only one of the years of the investigation—a year in which the populations of mice and voles reached the highest recorded peak.

The tawny owl's reluctance to replace lost clutches, as well as its early start in breeding (usually between March 18 and April 1), probably is due to the long period of the chicks' dependence on their parents. It is curious that, although the young hatchling in the nest is in great hazard of its life, once it has begun to fly it is extremely unlikely to be lost during the remainder of the dependence period.

August in woodland is a month of silence and inscrutability as far as tawny owls are concerned. The young have begun to fend for themselves, and their loud food cries cease. I have yet to evolve a method of studying the owls directly at that time. This is unfortunate, because an important part of the annual mortality falls just then. Indirect evidence suggests that the newly independent chicks suffer very high losses. For one thing, when the owls become territorially vociferous—in late September and October—their numbers have declined sharply. Returns from banded chicks show that the young owls rarely go outside the estate to establish new territories. The most important evidence, finally, is that a number of chicks have been found in a semistarved state during the latter part of August and early September.

With the exception of this period the life cycle of the tawny owl has been pieced together fairly coherently by now, and the history of this particular population is in a fair way to being depicted in quantitative terms. With certain reservations the same is true of the prey populations. In the time available very little information would have been obtainable without the methods of observation described above.

The Author

H. N. SOUTHERN is senior research officer in the Bureau of Animal Population at the Botanic Garden at Oxford, England. He writes that he cannot remember the time when he was not fascinated by birds, mammals and flowers. He studied classics at Oxford University and spent some time in a London publishing house. But the pull of natural history was so strong that he returned to Oxford, graduated a second time in zoology and joined the staff of the Bureau of Animal Population. During World War II he worked on the control of rabbits, rats and mice. Since then he has turned to examining the role of birds of prey in controlling populations of voles and mice.

Bibliography

THE SENSE ORGANS OF BIRDS. R. J. Pumphrey in *The Annual Report of the Smithsonian Institution for 1948,* pages 305-330; 1949.

TAWNY OWLS AND THEIR PREY. H. N. Southern in *The Ibis,* Vol. 96, No. 3, pages 384-410; July 1, 1954.

THE VERTEBRATE EYE AND ITS ADAPTIVE RADIATION. Gordon Lynn Walls. Cranbook Institute of Science, 1942.

SCIENTIFIC
AMERICAN   December 1959, Vol. 201, No. 6, pp. 140-151   OFFPRINT **1096**

THE PHYSIOLOGY OF THE CAMEL

by Knut Schmidt-Nielsen

How does the camel survive for weeks without drinking? Studies
in the Sahara Desert have exploded some old legends and have
elucidated the animal's remarkably parsimonious water economy.

The camel's ability to go without water for long periods is proverbial. Naturalists since Pliny the Elder have ascribed this talent to a built-in reservoir on which the animal can draw in time of need. Thus the English zoologist George Shaw wrote in 1801: "Independent of the four stomachs which are common to the ruminating animals, the Camels have a fifth bag which serves them as a reservoir for water. . . . This particularity is known to Oriental travelers, who have sometimes found it necessary to kill a Camel in order to obtain a supply of water."

Such tales appear even in modern zoology textbooks (though some writers have shifted the camel's reservoir from its stomach to its hump). There is remarkably little evidence for them. A few years ago I became interested in camel physiology and leafed through every book on the subject I could find. I accumulated a large amount of information on the camel's anatomy, its diseases and its evolution, but discovered that scientific knowledge of its water metabolism was almost nonexistent.

The camel's tolerance for drought is real enough. It can travel across stretches of desert where a man on foot and without water would quickly die of thirst. This fact poses some interesting questions for the physiologist. No animal can get along completely without water. Land animals in particular lose water steadily through their kidneys and from the moist surfaces of their lungs; mammals cool themselves in hot surroundings by evaporating water from their skins or oral membranes.

Desert animals have evolved a variety of mechanisms to minimize these losses. The kangaroo rat, for example, produces urine which contains so little water that it solidifies almost as soon as it is excreted. The animal avoids the heat and thus decreases evaporation by passing the daylight hours in a relatively cool burrow [see "The Desert Rat," by Knut and Bodil Schmidt-Nielsen; SCIENTIFIC AMERICAN Offprint 1050].

The camel can hardly escape the heat by burrowing. Does it emulate the desert rat by excreting a highly concentrated urine? Does it store water in its stomach, or anywhere else? For that matter, just how long can a camel go without water?

Our laboratory at Duke University was obviously no place to seek answers to these questions. Accordingly my wife and I, together with T. R. Houpt of the University of Pennsylvania and S. A. Jarnum of the University of Copenhagen, undertook to study camels in the Sahara Desert. The information we were seeking was not only of scientific interest but might also be of practical importance. In many arid lands the camel is the chief domestic animal. It serves not only as a beast of burden but also as a source of milk, meat, wool and leather. A better understanding of camel physiology might benefit the economy of these areas, which include some of the most poverty-stricken regions on earth.

The expedition required a year's planning, because we had to be completely independent of outside supplies in setting up our desert laboratory. Late in 1953 we arrived at the oasis of Béni Abbès, located in the desert south of the Atlas Mountains. To many Americans the word "oasis" suggests a few date palms clustered around a well. Béni Abbès, however, is a community of several thousand people. Its location over an underground river flowing down from the mountains ensures water for drinking and some irrigation.

Camels are valuable beasts in the Sa-

hara, and we had considerable difficulty in securing animals for experimental purposes. With the aid of our native assistant, Mohammed ben Fredj, I managed after considerable haggling to purchase one animal and later to rent two others. Meanwhile we sought to track down the source of some of the abundant camel folklore.

Explorers have often reported that their camels have gone without water for as long as several weeks. Is this possible? Under certain circumstances it is. Much desert exploration has been conducted in winter, because the desert summers are unbearably hot. During the winter the camel can often meet its need for water by browsing on bushes and succulent plants, which flourish after a rain and contain considerable water. Indeed, we found that in the Sahara grazing camels are not watered at all in winter. Some that had not drunk for as much as two months refused water when we offered it to them. When we examined such animals that had been butchered for meat, we found that all their organs contained the normal amount of water.

Thus the camel's drought tolerance in winter seems entirely ascribable to its diet. Even a man would have no difficulty in abstaining from drinking in cool weather if he fed largely on juicy fruits and vegetables. During the winter months the only way to dehydrate a camel is to restrict it to a fodder of dry grass and dried dates (not the soft, sweet dates of the grocery but a hard, fibrous and bitter variety). Even when we kept our camels on this completely dry diet, they could go for several weeks without drinking, though they lost water steadily through their lungs and skin and through the formation of urine and feces. We were able to measure these losses quite accurately simply by weighing the beasts, because unlike most animals deprived of water they continued to eat normally and thus did not lose weight in the usual sense. When we finally offered them water, they would in a few minutes drink enough to bring their weight back to normal. In no case did they drink more water than they had previously lost. Evidently they were not storing up a surplus but replacing a deficit.

Our studies of butchered camels supported this conclusion; we could find no evidence for special water-storing organs. The camel's first stomach, or rumen, does have pouches that are not found in the stomachs of other ruminants. These pouches have been called water sacs by some investigators, but their total fluid capacity is only about a gallon, and they contain coarsely masticated fodder rather than water. The main part of the rumen and the other stomachs contained considerable fluid, but less than the stomachs of other ruminating animals do. Chemical analysis revealed the similarity of the fluid to digestive juices; in salt content it resembled blood rather than water. Though it could furnish no significant reserve of water for the camel, it might well save the life of a thirsty man. The stories of travelers who killed their camels for water may thus be true. However, a man would have to be terribly thirsty to resort to such an expedient, because the fluid is usually a foul-smelling greenish soup. As for other possible sites of water storage, we found that neither the hump nor any other part of the animal's body contained an unusual quantity of fluid.

The tale that the camel stores water has probably gained color from the animal's remarkable capacity for rapid and copious drinking. During a later experiment one of our camels drank more than 27 gallons of water in 10 minutes. Camels are always watered before a long desert journey, and if they have not been watered for some days they will drink greedily. The uncritical observer could easily conclude that the animals are storing water in anticipation of future needs.

Some writers have alleged that the camel's fatty hump, though it contains no water reservoir, nonetheless provides the animal with a reserve of water as well as of food. Their logic is simple enough. All foods contain hydrogen and therefore produce water when they are oxidized in the body; the desert rat obtains all its water from this source. Fat contains a greater proportion of hydrogen than any other foodstuff and yields

CAMEL CAN LOSE WATER amounting to more than 25 per cent of its body weight. A dehydrated camel (*top*) appears emaciated but can still move about; a similarly dehydrated man would be unconscious or dead. The animal can quickly drink back its water losses and resume its normal appearance (*bottom*). Photographs were made about 10 minutes apart.

a correspondingly larger amount of water: about 1.1 pounds of water for each pound of fat. Thus a camel with 100 pounds of fat on its back might seem to be carrying 110 pounds—more than 13 gallons—of water. But in order to turn the fat into water the camel must take in oxygen through its lungs; in the process it loses water by evaporation from the lung surfaces. Calculation of the rate of loss indicates that the animal will lose more water by evaporation than it can gain through oxidation.

If the camel possesses no special reserves of water, it must have evolved methods for economizing on water expenditure. Its kidney functions in particular seemed worth investigating. All vertebrates rely on the kidney to rid their bodies of nitrogenous wastes (in mammals primarily urea). Few animals can tolerate any accumulation of these substances in the body. The kangaroo rat conserves water by eliminating a urine very rich in urea, and we were not surprised to find the camel following its example. When we kept our camels on dry fodder, their urine output declined and its urea content rose. In the case of one young animal, however, the urea output dropped sharply along with the volume of urine. The animal did not retain the urea in its blood, because the concentration of urea in the blood plasma also fell. Even the injection of urea into the bloodstream did not raise its concentration in the urine.

Further work revealed that the low level of urea had to do with nutrition rather than water conservation. The dry fodder contained too little protein to meet the nutritional needs of a young and growing camel. Under these circumstances the animal was apparently able to reprocess "waste" urea into new protein. From recent studies of sheep we have found that they share the camel's ability to use nitrogenous compounds over and over again. Experiments with cows by other investigators suggest that these ruminants may husband nitrogen in the same manner [see "The Metabolism of Ruminants," by Terence A. Rogers; SCIENTIFIC AMERICAN, February, 1958].

To determine the maximum efficiency with which the camel can excrete urea, we tried to put one of our camels on a high-protein diet. Unfortunately the camel's taste in food is exceedingly conservative. When we sought to trick the animal into increasing its protein intake by feeding it dates stuffed with raw peanuts, it balked at the first date and refused to accept any more for the rest of the day. Over a period of weeks we gradually accustomed it to the adulterated food, but we were forced to abandon the experiment when we found that we had exhausted the entire stock of peanuts in the oasis!

The kidney of the desert rat has a remarkable capacity for concentrating not only urea but also salt, and one would expect the camel's kidney to operate with similar efficiency. We were not able to test this hypothesis because

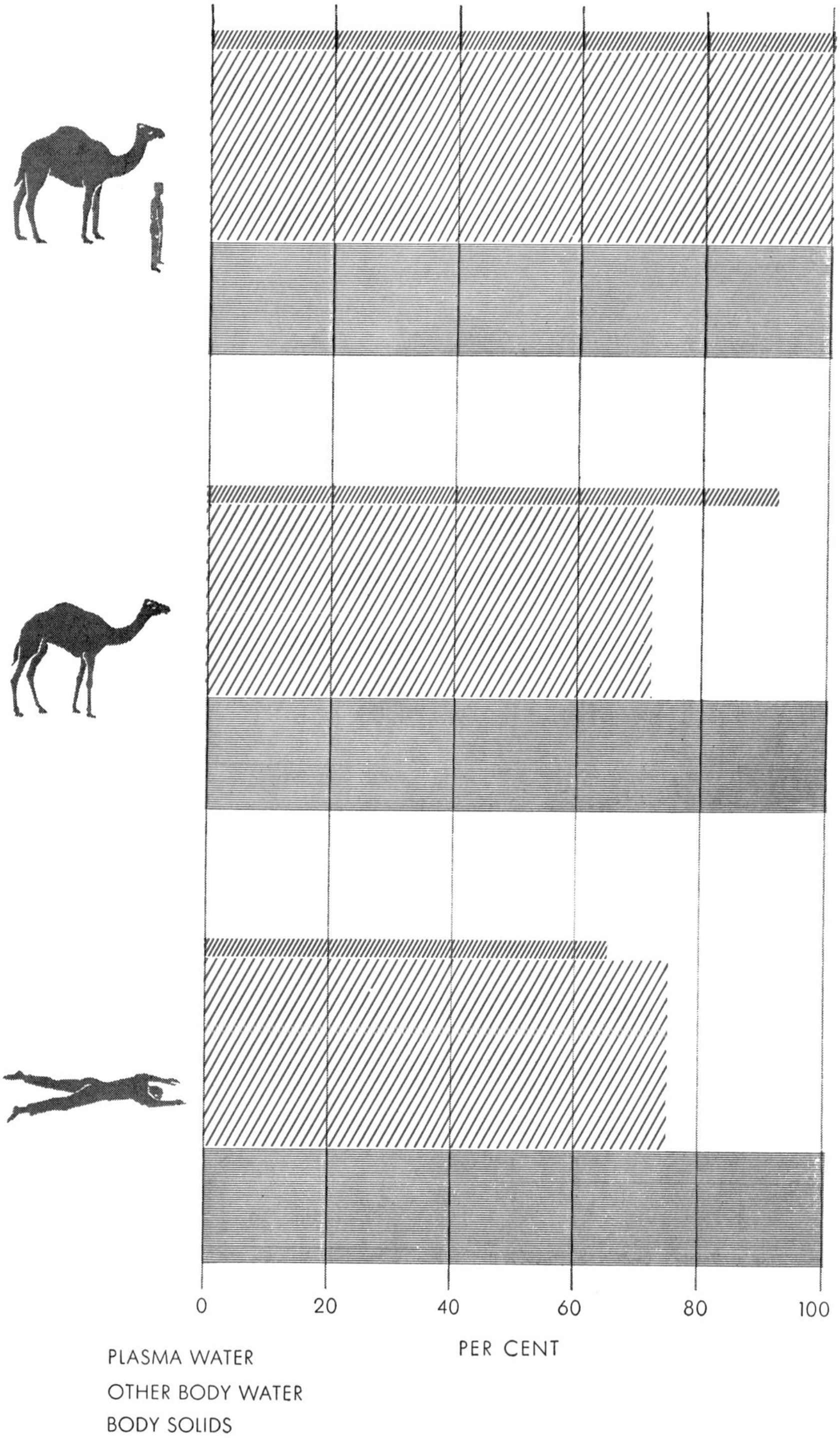

PLASMA WATER

OTHER BODY WATER

BODY SOLIDS

PER CENT

CAMEL SURVIVES DEHYDRATION partly by maintaining the volume of its blood. Under normal conditions (top) plasma water in both man and camel accounts for about a 12th of total body water. In a camel that has lost about a fourth of its body water (center) the blood volume will drop by less than a 10th. Under the same conditions a man's blood volume will drop by a third (bottom). The viscous blood circulates too slowly to carry the man's body heat outward to the skin, so that his temperature soon rises to a fatal level.

neither the fodder nor the local wells contained much salt. We were told, however, that camels in other areas could and did drink "bitter waters" containing enough magnesium sulfate (epsom salt) to sicken a man. Camels in coastal regions were said to browse on seaweed along the shore; these plants would of course have a salt content equal to that of sea water. It seems quite possible that the camel can approach the performance of the desert rat, whose efficient kidney permits it to drink sea water without harm.

Even the most efficient kidney cannot explain the camel's ability to get along on low water-rations during the desert summer. In this season the temperature soars to 120 degrees Fahrenheit or more. Since higher animals cannot survive body temperatures much over 100 degrees, they must either seek shelter, as do many small desert creatures, or cool themselves by evaporation of water from the mouth (panting) or from the skin (sweating). On a hot desert day a man may produce more than a quart of sweat an hour.

A man losing water at this rate rapidly becomes intensely thirsty. If the loss passes 5 per cent of his body weight (about a gallon), his physical condition rapidly deteriorates, his perceptions become distorted and his judgment falters. A loss of 10 per cent brings on delirium, deafness and insensitivity to pain.

In cool surroundings a man may cling to life until he has lost water up to 20 per cent of his body weight. In the desert heat, however, a loss of no more than 12 per cent will result in "explosive heat death." As the blood loses water it becomes denser and more viscous. The thickened blood taxes the pumping capacity of the heart and slows the circulation to the point where metabolic heat is no longer carried outward to the skin and dissipated. The internal temperature of the body suddenly increases, and death quickly follows.

How does the camel fare under similar conditions? We kept one camel without water for eight days in the heat of the desert summer. It had then lost 220 pounds—about 22 per cent—of its body weight. It looked emaciated; its abdomen was drawn in against its vertebrae, its muscles were shrunken and its legs were scrawny and appeared even longer than usual. But even though the animal would not have been able to do heavy work or travel a long distance, it was not in serious condition. It had no difficulty in emptying bucket after bucket of water and quickly recovered its normal appearance. Later experiments showed that camels can lose more than 25 per cent of their body weight without being seriously weakened. The lethal limit of dehydration is probably considerably higher, but for obvious humanitarian reasons we did not attempt to determine it.

How does a camel that has lost water amounting to a quarter of its body weight—more than a third of the water in its system—manage to avoid the explosive heat death that would overtake a man long before this point? It can do so because it maintains its blood volume despite the water loss. We were able to

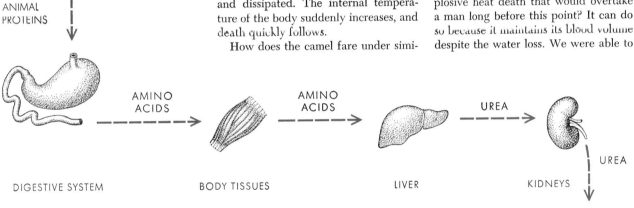

PLANT AND ANIMAL PROTEINS

DIGESTIVE SYSTEM — AMINO ACIDS → BODY TISSUES — AMINO ACIDS → LIVER — UREA → KIDNEYS — UREA

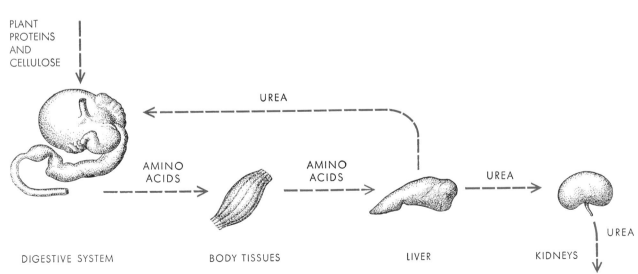

PLANT PROTEINS AND CELLULOSE

UREA

DIGESTIVE SYSTEM — AMINO ACIDS → BODY TISSUES — AMINO ACIDS → LIVER — UREA → KIDNEYS — UREA

UNUSUAL NITROGEN METABOLISM helps the camel to get along on low-grade fodder. In man (top) nitrogenous metabolic wastes are constantly excreted in the form of urea. In the camel (bottom) urea may return via bloodstream to the stomach, where in combination with broken-down cellulose it is reprocessed into new protein. Sheep and perhaps cattle husband nitrogen similarly.

calculate the blood volume before and after dehydration by injecting a non-poisonous dye into the bloodstream and measuring its concentration when it had become evenly distributed. In the case of a young camel that had lost about 11 gallons of water the reduction in blood volume was less than a quart. The water had been lost not from the blood but from the other body fluids and from the tissues.

One might suppose that the camel has some special physiological mechanism for maintaining its blood volume. In theory, however, a man should be able to do the same thing. When he lacks water, the osmotic pressure of the proteins in his blood plasma should draw it from the rest of his body. Since this does not happen, the real question is not how the dehydrated camel maintains its

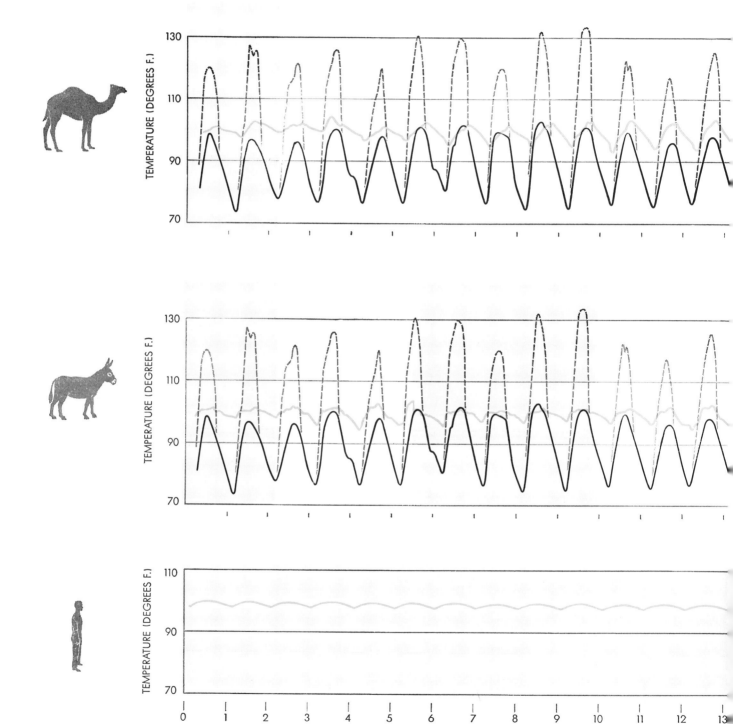

"FLEXIBLE" BODY TEMPERATURE partly accounts for the camel's ability to conserve water. Colored curves show variations in body temperature for a camel and a donkey over a 20-day period during the desert summer; a typical temperature curve for a man is shown for comparison. Solid gray curves show variations in air temperature during the same period; the broken curves indicate the additional heat load due to radiation from the sun and ground. The camel's temperature can rise to 105 degrees before the animal begins to sweat freely; during the night its temperature drops as low as 93 degrees, thereby delaying the next day's rise. The camel

blood volume but why a dehydrated man does not.

The camel not only tolerates dehydration much better than a man but also loses water much more slowly. One reason is that it excretes only a small volume

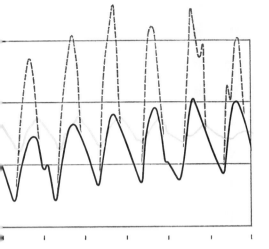

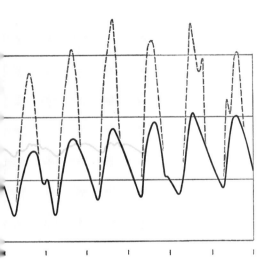

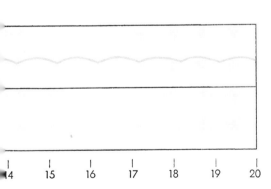

survived though watered (*shading*) only at the beginning and end of the period. Donkey, with a less flexible temperature, sweated more and drank more often. In the desert summer man must drink every day.

of urine—in summer often as little as a quart a day. The animal saves much more water, however, by economizing on sweat.

In the burning heat of the desert an inanimate object such as a rock may reach a temperature of more than 150 degrees F. A camel in such an environment, like a man, maintains a tolerable body temperature by sweating. But where the temperature of the man remains virtually constant as the day grows hotter, the temperature of the camel increases slowly to about 105 degrees. As the temperature of the camel rises, the animal sweats very little; only when its temperature reaches 105 degrees does it sweat freely. The camel's elevated temperature also lessens its absorption of heat, which of course depends on the difference between the temperature of its body and that of the environment.

The camel lowers the heat load on its body still further by letting its temperature fall below normal during the cool desert night. At dawn its temperature may have dropped as low as 93 degrees. Thus much of the day will elapse before the animal's body heats up to 105 degrees and sweating must set in. As a result of its flexible body temperature the camel sweats little except during the hottest hours of the day, where a man in the same environment perspires almost from sunrise to sunset.

These remarkable temperature fluctuations—about 12 degrees for the camel as against two degrees for a man—might seem to indicate a failure of normal heat regulation. This, however, is not so. The camel's body temperature never rises above 105 degrees. Moreover, if the camel has free access to water, the fluctuations are much smaller: about four degrees, comparable to the fluctuations during the cool winter months.

These findings may have implications for animal husbandry in the tropics. Cattle breeders have long sought ways of adapting the highly productive European dairy and beef animals to hot climates. In the process they have generally assumed that animals whose body temperature rises in hot weather will adapt poorly to heat. Although this assumption is often correct, the camel's example suggests that the reverse may sometimes be true.

People in desert areas and outside them have long appreciated the excellent insulating properties of camel hair. The camel employs camel-hair insulation to lower its heat load still fur-

ther. Even during the summer, when the camel sheds much of its wool, it retains a layer several inches thick on its back where the sun beats down. When we sheared the wool from one of our camels, we found that the shorn animal produced 60 per cent more sweat than an unshorn one. To Americans, who wear light clothing during the summer, the idea that a thick layer of wool is advantageous in the desert may seem unreasonable. The Arabs, however, typically dress in several layers of loose clothing, frequently made of wool. I have seen a nomad returning from the desert shed one thick wool burnous after another. We ourselves quickly learned that Arab dress was more comfortable than "civilized" garments.

The camel's hump also helps indirectly to lessen the heat load on the animal. Nearly all mammals possess a food reserve in the form of fat, but in most of them the fat is distributed fairly uniformly over the body just beneath the skin. In having its fat concentrated in one place the camel lacks insulation between its body and its skin, where evaporative cooling takes place. The absence of insulation facilitates the flow of heat outward, just as the insulating wool slows the flow of heat inward.

To clarify the relative importance of insulation and temperature fluctuation in lowering the camel's need for water, we repeated several of our experiments on a donkey. Because the donkey, like the camel, is a native of the arid lands of Asia, it should presumably have developed similar devices for conserving water.

We found that the donkey could tolerate as much water loss as the camel—up to 25 per cent. The donkey, however, lost water three times more rapidly than the camel does; thus it could not go as long without drinking. In one experiment a camel went for 17 days without drinking, while the donkey had to be watered at least once every four days. The difference in water loss seems partly due to the fact that the body temperature of the donkey is more stable than that of the camel. The donkey begins to sweat when its temperature has risen only slightly. Furthermore, the donkey's coat is quite thin in comparison with the camel's and thus does not provide so effective a barrier against heat from the environment.

The donkey outdoes the camel in one respect: its drinking capacity. A camel that has lost 25 per cent of its body weight can drink back the loss in about

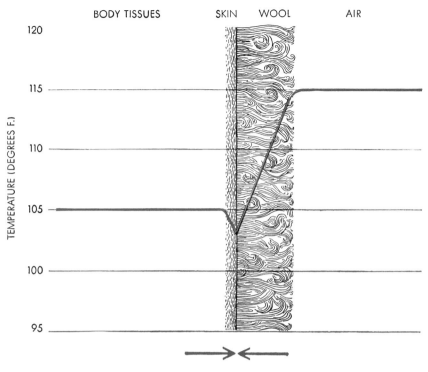

BODY TISSUES   SKIN   WOOL   AIR

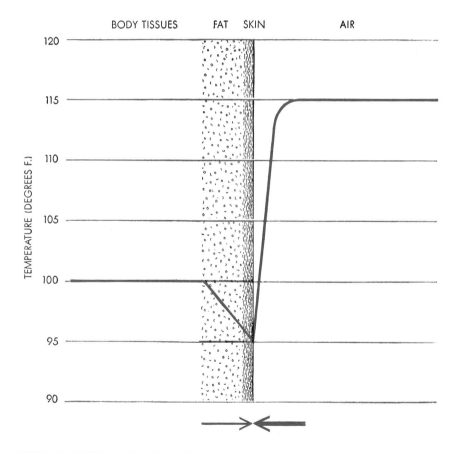

BODY TISSUES   FAT   SKIN   AIR

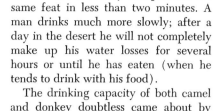

10 minutes; a donkey can perform the same feat in less than two minutes. A man drinks much more slowly; after a day in the desert he will not completely make up his water losses for several hours or until he has eaten (when he tends to drink with his food).

The drinking capacity of both camel and donkey doubtless came about by evolutionary adaptation in the wild state. Water holes are rare in an arid land, and predators often lie in wait near them. The animal that can quickly replenish its water losses and depart is likely to live longer than one that must linger by the water to satisfy its needs.

HEAT FLOW IN CAMEL AND MAN is compared in these schematic cross sections of a small portion of body surface. Colored curves show temperature; heat flow, suggested by arrows, increases with the slope of the curves. Camel (*top*) can eliminate heat from its body with an average skin temperature of about 103 degrees; its wool slows the heat flow from the environment. Man (*bottom*), with a lower body temperature and an insulating layer of fat beneath the skin, must maintain a lower skin temperature (that is, must evaporate more sweat) to obtain the same flow of heat from his interior. The lack of insulation between his skin and the air raises heat flow from the air, necessitating still more evaporation.

The Author

KNUT SCHMIDT-NIELSEN is professor of physiology in the department of zoology at Duke University. He studied the camel when he led an expedition of four scientists to the Sahara Desert in 1953 and 1954, while on a Guggenheim fellowship. He has long been interested in the water metabolism of animals, as indicated by two previous articles in SCIENTIFIC AMERICAN: one ("The Desert Rat") in collaboration with his wife in July, 1953, and the second ("Salt Glands") in January, 1959. He was born in Norway and holds degrees from the University of Copenhagen. In addition to his work in physiology, he has done research in biochemistry and analytical chemistry. He studied for four years in Copenhagen under August Krogh, Nobel laureate in physiology, and married Krogh's daughter. The Schmidt-Nielsens came to this country in 1946. They worked at Swarthmore College, Stanford University and the University of Cincinnati before they went to Duke in 1952.

Bibliography

BODY TEMPERATURE OF THE CAMEL AND ITS RELATION TO WATER ECONOMY. Knut Schmidt-Nielsen, Bodil Schmidt-Nielsen, S. A. Jarnum and T. R. Houpt. Department of Zoology, Duke University, 1959.

THE CAMEL, ITS USES AND MANAGEMENT. Major Arthur Glyn Leonard. Longmans, Green and Co., 1894.

PHYSIOLOGY OF MAN IN THE DESERT. E. F. Adolph and associates. Interscience Publishers, 1947.

SCIENTIFIC
AMERICAN   August 1952, Vol. 187, No. 2, pp. 20-23

OFFPRINT 1097

THE SPIDER AND THE WASP

by Alexander Petrunkevitch

About the odd relationship between the tarantula and the giant wasp Pepsis. Although the tarantula can easily kill Pepsis, one species permits the wasp to sting it and lay an egg in its body.

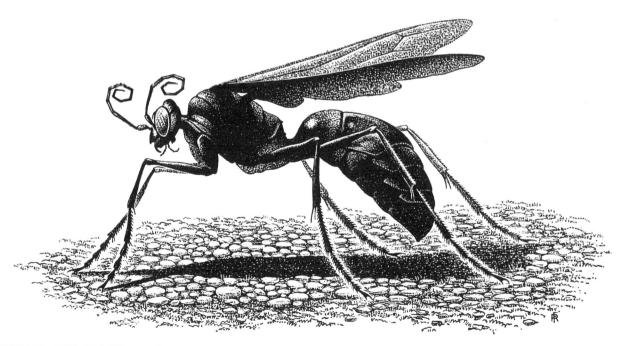

SPIDER AND WASP are the tarantula *Cyrtopholis portoricae* (*top*) and the digger wasp *Pepsis marginata* (*bottom*). The tarantula is shown in an attitude of de-fense. The wasps of the genus Pepsis are either a deep blue or blue with rust-colored wings. The largest species of the genus have a wingspread of about four inches.

TO HOLD ITS OWN in the struggle for existence, every species of animal must have a regular source of food, and if it happens to live on other animals, its survival may be very delicately balanced. The hunter cannot exist without the hunted; if the latter should perish from the earth, the former would, too. When the hunted also prey on some of the hunters, the matter may become complicated.

This is nowhere better illustrated than in the insect world. Think of the complexity of a situation such as the following: There is a certain wasp, *Pimpla inquisitor,* whose larvae feed on the larvae of the tussock moth. *Pimpla* larvae in turn serve as food for the larvae of a second wasp, and the latter in their turn nourish still a third wasp. What subtle balance between fertility and mortality must exist in the case of each of these four species to prevent the extinction of all of them! An excess of mortality over fertility in a single member of the group would ultimately wipe out all four.

This is not a unique case. The two great orders of insects, Hymenoptera and Diptera, are full of such examples of interrelationship. And the spiders (which are not insects but members of a separate order of arthropods) also are killers and victims of insects.

The picture is complicated by the fact that those species which are carnivorous in the larval stage have to be provided with animal food by a vegetarian mother. The survival of the young depends on the mother's correct choice of a food which she does not eat herself.

In the feeding and safeguarding of their progeny the insects and spiders exhibit some interesting analogies to reasoning and some crass examples of blind instinct. The case I propose to describe here is that of the tarantula spiders and their arch-enemy, the digger wasps of the genus Pepsis. It is a classic example of what looks like intelligence pitted against instinct—a strange situation in which the victim, though fully able to defend itself, submits unwittingly to its destruction.

MOST tarantulas live in the Tropics, but several species occur in the temperate zone and a few are common in the southern U. S. Some varieties are large and have powerful fangs with which they can inflict a deep wound. These formidable looking spiders do not, however, attack man; you can hold one in your hand, if you are gentle, without being bitten. Their bite is dangerous only to insects and small mammals such as mice; for a man it is no worse than a hornet's sting.

Tarantulas customarily live in deep cylindrical burrows, from which they emerge at dusk and into which they retire at dawn. Mature males wander about after dark in search of females and

NEST OF THE MUD DAUBER WASP illustrates an intricate predatory relationship. A single cell of the nest, enlarged 10 times, contains one pupa of a secondary predator and five smaller pupae of a tertiary predator.

DEATH OF THE SPIDER is shown in these drawings. In the first drawing the wasp digs a grave, occasional-

occasionally stray into houses. After mating, the male dies in a few weeks, but a female lives much longer and can mate several years in succession. In a Paris museum is a tropical specimen which is said to have been living in captivity for 25 years.

A fertilized female tarantula lays from 200 to 400 eggs at a time; thus it is possible for a single tarantula to produce several thousand young. She takes no care of them beyond weaving a cocoon of silk to enclose the eggs. After they hatch, the young walk away, find convenient places in which to dig their burrows and spend the rest of their lives in solitude. Tarantulas feed mostly on insects and millepedes. Once their appetite is appeased, they digest the food for several days before eating again. Their sight is poor, being limited to sensing a change in the intensity of light and to the perception of moving objects. They apparently have little or no sense of hearing, for a hungry tarantula will pay no attention to a loudly chirping cricket placed in its cage unless the insect happens to touch one of its legs.

But all spiders, and especially hairy ones, have an extremely delicate sense of touch. Laboratory experiments prove that tarantulas can distinguish three types of touch: pressure against the body wall, stroking of the body hair and riffling of certain very fine hairs on the legs called trichobothria. Pressure against the body, by a finger or the end of a pencil, causes the tarantula to move off slowly for a short distance. The touch excites no defensive response unless the approach is from above where the spider can see the motion, in which case it rises on its hind legs, lifts its front legs, opens its fangs and holds

this threatening posture as long as the object continues to move. When the motion stops, the spider drops back to the ground, remains quiet for a few seconds and then moves slowly away.

The entire body of a tarantula, especially its legs, is thickly clothed with hair. Some of it is short and woolly, some long and stiff. Touching this body hair produces one of two distinct reactions. When the spider is hungry, it responds with an immediate and swift attack. At the touch of a cricket's antennae the tarantula seizes the insect so swiftly that a motion picture taken at the rate of 64 frames per second shows only the result and not the process of capture. But when the spider is not hungry, the stimulation of its hairs merely causes it to shake the touched limb. An insect can walk under its hairy belly unharmed.

The trichobothria, very fine hairs growing from disklike membranes on the legs, were once thought to be the spider's hearing organs, but we now know that they have nothing to do with sound. They are sensitive only to air movement. A light breeze makes them vibrate slowly without disturbing the common hair. When one blows gently on the trichobothria, the tarantula reacts with a quick jerk of its four front legs. If the front and hind legs are stimulated at the same time, the spider makes a sudden jump. This reaction is quite independent of the state of its appetite.

These three tactile responses—to pressure on the body wall, to moving of the common hair and to flexing of the trichobothria—are so different from one another that there is no possibility of confusing them. They serve the tarantula adequately for most of its needs and enable it to avoid most annoyances and

dangers. But they fail the spider completely when it meets its deadly enemy, the digger wasp Pepsis.

These solitary wasps are beautiful and formidable creatures. Most species are either a deep shiny blue all over, or deep blue with rusty wings. The largest have a wing span of about four inches. They live on nectar. When excited, they give off a pungent odor—a warning that they are ready to attack. The sting is much worse than that of a bee or common wasp, and the pain and swelling last longer. In the adult stage the wasp lives only a few months. The female produces but a few eggs, one at a time at intervals of two or three days. For each egg the mother must provide one adult tarantula, alive but paralyzed. The tarantula must be of the correct species to nourish the larva. The mother wasp attaches the egg to the paralyzed spider's abdomen. Upon hatching from the egg, the larva is many hundreds of times smaller than its living but helpless victim. It eats no other food and drinks no water. By the time it has finished its single gargantuan meal and become ready for wasphood, nothing remains of the tarantula but its indigestible chitinous skeleton.

The mother wasp goes tarantula-hunting when the egg in her ovary is almost ready to be laid. Flying low over the ground late on a sunny afternoon, the wasp looks for its victim or for the mouth of a tarantula burrow, a round hole edged by a bit of silk. The sex of the spider makes no difference, but the mother is highly discriminating as to species. Each species of Pepsis requires a certain species of tarantula, and the wasp will not attack the wrong species. In a cage with a tarantula which is not its normal prey the wasp avoids the spider, and is usually killed by it in the night.

Yet when a wasp finds the correct species, it is the other way about. To

ly looking out. The spider stands with its legs extended after raising its body so the wasp could pass under it. In the second drawing the wasp stings the spider, which falls on its back. In the third the wasp licks a drop of blood from the wound. In the final drawing the spider lies in its grave with the egg of the wasp on its abdomen.

identify the species the wasp apparently must explore the spider with her antennae. The tarantula shows an amazing tolerance to this exploration. The wasp crawls under it and walks over it without evoking any hostile response. The molestation is so great and so persistent that the tarantula often rises on all eight legs, as if it were on stilts. It may stand this way for several minutes. Meanwhile the wasp, having satisfied itself that the victim is of the right species, moves off a few inches to dig the spider's grave. Working vigorously with legs and jaws, it excavates a hole 8 to 10 inches deep with a diameter slightly larger than the spider's girth. Now and again the wasp pops out of the hole to make sure that the spider is still there.

When the grave is finished, the wasp returns to the tarantula to complete her ghastly enterprise. First she feels it all over once more with her antennae. Then her behavior becomes more aggressive. She bends her abdomen, protruding her sting, and searches for the soft membrane at the point where the spider's leg joins its body—the only spot where she can penetrate the horny skeleton. From time to time, as the exasperated spider slowly shifts ground, the wasp turns on her back and slides along with the aid of her wings, trying to get under the tarantula for a shot at the vital spot. During all this maneuvering, which can last for several minutes, the tarantula makes no move to save itself. Finally the wasp corners it against some obstruction and grasps one of its legs in her powerful jaws. Now at last the harassed spider tries a desperate but vain defense. The two contestants roll over and over on the ground. It is a terrifying sight and the outcome is always the same. The wasp finally manages to thrust her sting into the soft spot and holds it there for a few seconds while she pumps in the poison. Almost immediately the tarantula falls paralyzed on its back. Its legs stop twitching; its heart stops beating. Yet it is not dead, as is shown by the fact that if taken from the wasp it can be restored to some sensitivity by being kept in a moist chamber for several months.

After paralyzing the tarantula, the wasp cleans herself by dragging her body along the ground and rubbing her feet, sucks the drop of blood oozing from the wound in the spider's abdomen, then grabs a leg of the flabby, helpless animal in her jaws and drags it down to the bottom of the grave. She stays there for many minutes, sometimes for several hours, and what she does all that time in the dark we do not know. Eventually she lays her egg and attaches it to the side of the spider's abdomen with a sticky secretion. Then she emerges, fills the grave with soil carried bit by bit in her jaws, and finally tramples the ground all around to hide any trace of the grave from prowlers. Then she flies away, leaving her descendant safely started in life.

IN ALL THIS the behavior of the wasp evidently is qualitatively different from that of the spider. The wasp acts like an intelligent animal. This is not to say that instinct plays no part or that she reasons as man does. But her actions are to the point; they are not automatic and can be modified to fit the situation. We do not know for certain how she identifies the tarantula—probably it is by some olfactory or chemo-tactile sense—but she does it purposefully and does not blindly tackle a wrong species.

On the other hand, the tarantula's behavior shows only confusion. Evidently the wasp's pawing gives it no pleasure, for it tries to move away. That the wasp is not simulating sexual stimulation is certain, because male and female tarantulas react in the same way to its advances. That the spider is not anesthetized by some odorless secretion is easily shown by blowing lightly at the tarantula and making it jump suddenly.

What, then, makes the tarantula behave as stupidly as it does?

No clear, simple answer is available. Possibly the stimulation by the wasp's antennae is masked by a heavier pressure on the spider's body, so that it reacts as when prodded by a pencil. But the explanation may be much more complex. Initiative in attack is not in the nature of tarantulas; most species fight only when cornered so that escape is impossible. Their inherited patterns of behavior apparently prompt them to avoid problems rather than attack them. For example, spiders always weave their webs in three dimensions, and when a spider finds that there is insufficient space to attach certain threads in the third dimension, it leaves the place and seeks another, instead of finishing the web in a single plane. This urge to escape seems to arise under all circumstances, in all phases of life and to take the place of reasoning. For a spider to change the pattern of its web is as impossible as for an inexperienced man to build a bridge across a chasm obstructing his way.

In a way the instinctive urge to escape is not only easier but often more efficient than reasoning. The tarantula does exactly what is most efficient in all cases except in an encounter with a ruthless and determined attacker dependent for the existence of her own species on killing as many tarantulas as she can lay eggs. Perhaps in this case the spider follows its usual pattern of trying to escape, instead of seizing and killing the wasp, because it is not aware of its danger. In any case, the survival of the tarantula species as a whole is protected by the fact that the spider is much more fertile than the wasp.

The Author

ALEXANDER PETRUNKEVITCH was professor of zoology at Yale University.

Bibliography

TARANTULA VERSUS TARANTULA-HAWK: A STUDY IN INSTINCT. Alexander Petrunkevitch in *The Journal of Experimental Zoology,* Vol. 45, No. 2, pages 367-397; July 5, 1926.

SCIENTIFIC
AMERICAN August 1963, Vol. 209, No. 2, pp. 38-46

OFFPRINT 1098

THE EVOLUTION OF BOWERBIRDS

by E. Thomas Gilliard

It now seems that the complex mating stages these birds maintain
and their specialized sexual displays constitute an advanced form
of avian behavior that tends to speed up evolutionary processes.

A 19th-century naturalist once suggested that just as mammals were commonly divided into two groups—man and the lower forms—all birds should be split into two categories: bowerbirds and other birds. No one who has observed the behavior of these remarkable creatures of Australia and New Guinea and examined their artifacts can scoff at this proposal. The males of some species build elaborate walled bowers of sticks and decorate them with bright objects and even with paint. Others construct towers up to nine feet high, some with tepee-like roofs and internal chambers, on circular lawns that they tend carefully and embellish with golden resins, garishly colored berries, iridescent insect skeletons and fresh flowers that are replaced as they wither. The bowers are stages set by the males on which to perform intricate routines of sexual display and to mate with the females of their species. The bowerbirds' architectural, engineering and decorating skills and their courtship displays constitute behavior that, as G. Evelyn Hutchinson of Yale University has said, "in its complexity and refinement is unique in the nonhuman part of the animal kingdom."

The student of evolution inevitably asks how such extremely specialized behavior came about. The answer, I suspect, can be unmasked if one steps back to survey all the birds with behavioral affinities to the bowerbirds, that is, those birds that practice the pattern of courtship behavior known as arena behavior. There are only 18 species called bowerbirds, at least 12 of which actually build bowers, but there are in all some 85 species that have been described as arena birds. This is still a small proportion—about 1 per cent—of the avian species of the world. But arena birds are a world-wide assemblage including species in such disparate families as sandpipers, grouse, bustards, blackbirds, small tropical manakins and the bizarrely beautiful birds of paradise.

It has fallen to my lot to be able to make comparative ethological investigations of many of these species in the tropics of New Guinea and South America. As a result I have been able to reach some conclusions that seem to be new. I believe that arena behavior, wherever it appears, probably has a common origin and that it represents an advanced stage in avian development. Once set in motion, I think, it has a predictable evolution leading rather quickly to the development of the highly specialized combinations of structure and behavior found in all the far-flung arena species. The bowerbirds are at the pinnacle of arena evolution. They have gone a step beyond the most richly ornamented arena birds, substituting fancy houses and jewelry for colorful plumage.

Arena behavior was defined by the ornithologist and student of evolution Ernst Mayr as a pattern of territorial behavior in which the males establish a mating station that has no connection with feeding or nesting. I would add that it is a rather rare form of courtship behavior involving a group of males usually living in an organized band on or about a long-established mating space: the arena. Each arena is composed of a number of courts, the private display territories of individual males. To establish their right to a territory the males go through ritualistic combat routines, fighting, charging, displaying their plumage or brandishing twigs, singing or producing "mechanical" sounds. Once territories are established there is little fighting for mates because the females do the choosing. The sexes live apart for long periods of the year and are often

so dissimilarly dressed as to look like different species. Since there is no true pair bond, the males play no part whatever in building or defending the nest or in rearing the young.

This advanced courtship pattern is in sharp contrast to the less advanced behavior of the other 99 per cent of the world's birds. For them the central event is the establishment of a pair bond between a male and a female, with the pair proceeding to share the work of raising the young. (The word "advanced" is not intended to imply a value judgment on the state of matrimony. Ornithologists simply assume that pair-bonding and work-sharing habits represent the less advanced evolutionary condition in birds because these habits are so nearly universal.) The pair-bond pattern is found regularly not only in the phylogenetically recent passerine (perching) order of songbirds, which is currently the most numerous and highly differentiated avian group, but also in the older non-passerine birds; it is a "conservative" behavioral pattern that has resisted modification. Yet the breakthrough to arena behavior seems to occur, apparently at random, just about anywhere in the world and at scattered points on the family tree of birds [see illustration on page 2642].

The characteristics that define arena behavior and argue for its common origin and line of evolution emerge from the study of a fairly large number of arena birds. The pattern is most evident when the arena is small, as in the case of the ruff, a sandpiper of northern Europe and Asia whose behavior has been described in detail by C. R. Stonor. The males and females apparently live apart except for a few minutes in the breeding season. Each spring the males gather in isolated clans, each of which populates

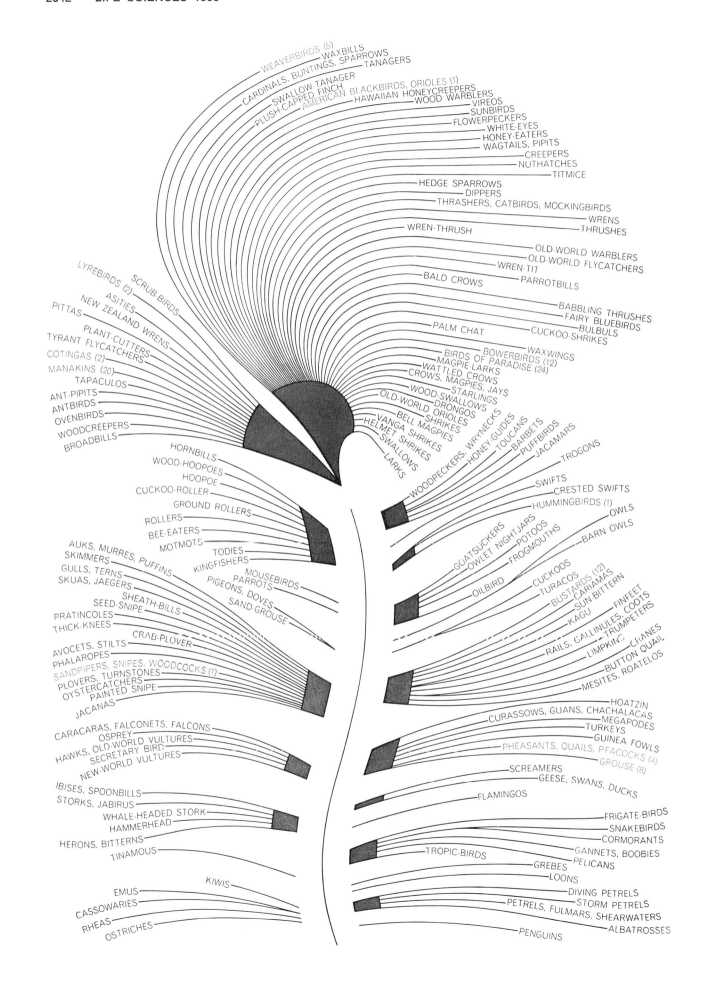

a small, grassy hillock in rolling meadow-land. After a period of fighting and display among themselves the males learn to recognize one another as individuals, and arrange themselves on the mound in a social order that presumably remains fairly fixed throughout the breeding season. Each male's territory is a private court about two feet in diameter that he defends vigorously against other males. The clan waits day after day for the visits of occasional females in search of mates. When one appears, the males go to their courts and assume strangely stiff postures, extending the colorful plumage of their neck ruffs. Displaying in this manner they reminded Stonor of a bed of flowers. The female wanders through this cluster and pecks at the neck feathers of the bird she prefers. Mating occurs immediately—whereupon the rejected males immediately collapse on their courts as if in a fainting spell.

Arena behavior of a similar sort but on a larger scale is practiced by the sage grouse and prairie chicken of North America. The grouse's arena may be half a mile long and 200 yards wide, with 400 males within its boundaries, each standing 25 to 40 feet apart on its private court. The zoologist John W. Scott was able to study the breeding hierarchy in a clan of these grouse. He found that the great majority of matings went to four "master" and a few "submaster" cocks with courts located along the center line of the long, narrow arena. Of 114 observed matings involving males whose place in the hierarchy had been determined, 74 per cent went to the four master cocks. Only after these birds had become satiated did 13 per cent of the matings go to the submasters, and the few remaining matings went to scattered owners of peripheral courts. These and other observations make it clear that arena matings are not random: the co-ordinated clan activities that serve to establish the territorial hierarchy, and thus the breeding rights, are of primary evolutionary importance.

Some years ago on an expedition to South America I studied arena be-

MALE RUFFS display on small private courts close to one another in a tight arena. When reeves (female ruffs) are in the vicinity, the males posture stiffly on their courts, extending their colorful plumage (top). Two reeves approach (middle); one selects a mate by pecking at its neck feathers (bottom). The photographs were made by Arthur Christiansen.

havior in a very different bird, a cotinga called cock of the rock (Rupicola rupicola). I found a clan of these brilliant orange birds, which wear a great semicircular crest resembling that of a Roman helmet, in the Kanuku Mountains of British Guiana. The males held and defended an arena some 40 by 80 feet in extent including about 40 small courts—cleared areas on the ground under saplings and vines that provided convenient perches. For 20 consecutive days I watched three members of the clan, readily recognizable as individuals, that held adjacent courts in one part of the arena. I was struck by the silence and deliberateness of movement that characterized their behavior on the courts; it was reminiscent of the behavior of a pair-bonded male at its nest. During the period of

observation females visited the arena several times. Whenever a female arrived, the three males, if they were not already on the ground, would fall almost like stones from their perches to their courts. There, with bodies flattened and heads tilted so that the crests were silhouetted against the bare ground, they would posture stiffly for many minutes. Again there was a resemblance to the attitude of a male attending a nest. The three birds jealously defended from one another their own courts and a cone-shaped space above them. But when a wandering nonclan male visited them, they would fly up and attack him as a team with violent chasing displays, wing-buffeting, strange cries and whinnying sounds.

The 24 species of birds of paradise

CRESTED BOWERBIRD (*Amblyornis macgregoriae*) of New Guinea is a member of a remarkable genus that builds high towers of sticks surrounded by courtyards. *A. macgregoriae*, the most colorfully plumed of the genus, builds the least complex bower.

GREATER BIRD OF PARADISE (*Paradisaea apoda*) is native to the New Guinea region. This bird is a member of a breeding colony established in 1909 on Little Tobago in the Caribbean Sea. **Photographs on this page and page 3 were made by the author.**

in which arena behavior is seen vary widely in physical characteristics and in the details of their displays. Some clear courts on or near the ground, some inhabit the middle levels of tropical forests and some display high in the treetops. In many of them the arena is so large that it has not usually been considered an arena at all. The distances between the individual courts can mislead one into believing that each male is operating in solitude, but this is almost certainly not the case. Apparently these species have "exploded" arenas; the birds' calls and mechanisms for the production of other sounds are always highly developed and powerful, so that the males can interact in spite of their seeming isolation. Strong evidence in favor of the exploded-arena hypothesis is the fact that in many of these species the courts have been found to be concentrated in certain areas of the forest year after year.

I have studied two of the species that clear courts on or near the ground beneath low branches, vines and saplings: the magnificent bird of paradise (*Diphyllodes magnificus*) and Queen Carola's bird of paradise (*Parotia carolae*). For both species the court is the property of a single male that remains in attendance many hours a day, probably for several months a year. Austin L. Rand of the Chicago Natural History Museum has described how the magnificent bird of paradise spends countless hours trimming away the forest leaves above its court, thereby enabling a shaft of sky light to enhance the bird's iridescent coloring. Similar but less well developed court-clearing is practiced by other arena birds. In an arena of blue-backed manakins on Tobago in the West Indies I saw that much of the foliage had been cut around the arboreal courts. Frank M. Chapman of the American Museum of Natural History studied clans of Gould's manakins located miles apart in Panama that had cleared many small, platelike clearings in 200-foot-long strips of the rain-forest floor. The cock of the rock clears its court with violent wing-thrashing, whereas most other arena birds use their bills for this purpose. However it is done, the court of many arena birds is swept clean of fallen debris if it is terrestrial, and stripped of many leaves and twigs if it is arboreal.

It seems not too big a step from the court-clearers to such elementary bower-builders as Archbold's bowerbird (*Archboldia papuensis*). These are clearly arena birds: the males and females apparently live apart most of the year. The males spend the breeding season on or close to table-sized stages on the ground in high mountain forests of New Guinea. Each stage is owned and defended by a single male who carpets it with ferns, decorates it with shafts of bamboo, piles of resin, beetle skeletons, snail shells and lumps of charcoal. Each is within audible range of other stage-tending males. On one slope of Mount Hagen I found five stages concentrated in a zone about two miles in diameter, and my native hunters reported others I did not see. Although the species seemed rather common, a number of expeditions failed to find any of these birds elsewhere on Mount Hagen; however, a similar group was discovered on another mountain some 20 miles away. In my opinion each of these groups represents a clan, and its gathering place is the clan's arena. The males within each clan maintain contact with one another by uttering mighty whistles and harsh, rasping notes, and it seems likely that they all know when a female is in the arena and act in concert as many other arena birds do.

I watched one male receiving a visit from a female. As soon as the female arrived near the court the male dropped to its colorful stage and began to act in a manner resembling that of a young bird begging food. With its wings outstretched and its tail spread, it crawled tortuously toward the female, which perched at the edge of the court and kept moving around its periphery. The male held its head up like a turtle, made gasping movements with its bill and kept up a deep, penetrating "churr" song. In spite of the vigor of this display the ceremony was apparently not consummated by mating. After 22 minutes something disturbed the birds and the female flew off. Soon the male began rearranging the piles of ornaments and resumed its long, solitary wait.

Another New Guinea bowerbird with an exploded arena is the extraordinary gardener bowerbird (*Amblyornis*). A male of this genus builds its bower by piling sticks against a sapling on the floor of a mountain rain forest and clearing a mossy saucer around the tower. Some species build large towers with roofs and internal chambers and decorate the moss court with snail shells, insect and spider silk and fresh flowers changed daily for months on end. Others build only a small roof and use fewer ornaments, and still others merely maintain a clearing around a modest tower of intertwined sticks.

Some years ago I noticed that there is an inverse ratio in the three known *Amblyornis* species between the complexity of the bower and the plumage of the male bird (the three females are virtually indistinguishable). In the species *A. macgregoriae*, which builds the simple bower, the adult male wears a long golden-orange crest [*see top illustration on page 2644*]. In *A. subalaris*, which builds the somewhat more complex bower, the male wears a shorter crest. And in the aptly named species *A. inornatus*, which builds the most elaborate bower (with a broad roof overhanging a court decorated with berries, shells and piles of flowers), the male wears no crest at all and cannot be distinguished from any of the females!

I believe that in these birds the forces of sexual selection have been transferred from morphological characteristics—the male plumage—to external objects and that this "transferral effect" may be the key factor in the evolution of the more complex bowerbirds. This would explain the extraordinary development and proliferation of the bowers and their ornaments: these objects have in effect become externalized bundles of secondary sexual characteristics that are psychologically but not physically connected with the males. The transfer also has an important morphological effect: once colorful plumage is rendered unimportant, natural selection operates in the direction of protective coloration and the male tends more and more to resemble the female.

Further evidence of this sort came from observations of Lauterbach's bowerbird (*Chlamydera lauterbachi*), a grassland and forest-edge species of New Guinea. In an area several miles in diameter I once found 16 bowers of this "avenue-building" species hundreds to many thousands of feet apart. One bower I examined contained almost 1,000 pale pebbles weighing nearly 10 pounds. More than 3,000 sticks and 1,000 hair-like strands of grass had gone into the four-walled structure. The sticks were interlocked to form a rigid structure and the grass was used to line the vertical walls facing the inner court. Three times during the many days I watched a female entered a bower. The male became highly excited and began to dance. The female jumped quickly within the walls and then stood still and alert. Almost as soon as she was in the bower the male picked up with its bill a marble-sized red berry, held it high and displayed it to the female much as it would have displayed its bright crest feathers—if it had had any. *C. lauterbachi*, like *A. inornatus*, is the most advanced builder of its genus. It is also a species in which the male and

INVERSE RATIO was noted by the author between the complexity of gardener bowerbirds' bowers and the brilliance of their plumage. The most complex bower, seen at left in a photograph made by S. Dillon Ripley of Yale University, is built by the crestless *Amblyornis inornatus*. The simplest bower (*right*) is that of the orange-crested *A. macgregoriae* [*see illustration on page 2644*].

female cannot be told apart except by dissection. The transferral effect seems to be operating in this case too.

It appears that once the female has selected a bower-owner she stays for several days. (This has also been reported in some arena birds, such as the argus pheasant.) These stays may be responsible for the assumption made by many investigators that there is a pair bond in bowerbirds. Pair-bonding cannot be proved or disproved except by marking and observing females; my investigations indicate that at least most of the bower-building bowerbirds are polygynous, with exploded arenas like those of their close relatives, the birds of paradise.

One further observation bearing on the transferral effect should be mentioned. In the Finisterre Mountains of New Guinea I watched and filmed the courtship behavior of the fawn-breasted bowerbird (*Chlamydera cerviniventris*), in which both sexes are an identical drab brown. When a female entered the two-walled avenue bower and squatted on the floor, the male immediately approached. On the ground several feet from the bower the male suddenly appeared to be overcome by a spasm. Its head seemed to turn involuntarily away from the female again and again. Finally the bird appeared to regain control, seized a sprig of green berries in its bill, faced the female and waved the berries up and down as it slowly approached the bower. I saw several more such visits by a female, and each time the male went through the curious twisting motions that presented the back of its head to the female.

Later, watching the films I had made of these movements, I was struck by the thought that the head-screwing might constitute crest display—except for the fact that *C. cerviniventris* has no crest! But many males closely related to this species do have glittering violet-to-pink crests at the nape of the neck and the Australian ornithologist John Warham has described how they twist their necks to display the crest to a female in the bower. I concluded that the head-twisting of *C. cerviniventris* is a relict movement dating from the time when the species had such a crest. With the later incorporation of the berries as ornaments in the courtship ceremony, I postulated, the crest became unimportant. Since it was now simply a liability in terms of protective coloration, it was lost through natural selection—but the movement associated with it persists. This I consider a strong second line of evidence for the transferral effect.

Again it must be emphasized that the courtship behavior of this species and probably that of all other ground-displaying bowerbirds, even though complicated and camouflaged by refinements of ornamentation and stick architecture, follows the basic pattern of

COURTSHIP BEHAVIOR of avenue-building bowerbirds is shown in this sequence of drawings. The male builds a walled bower by inserting thousands of sticks into a foundation mat (*left*) and decorates it with pebbles and berries (*second from left*).

arena behavior the world around. It is the behavior of a clan of males interacting in an arena, each on its own territory and competing with the other males for itinerant females. Many arena species clear courts and some do it more effectively than others; some build stages or erect walls, towers or houses. All these actions, I believe, are merely levels of refinement of the same basic behavior.

Is the history of the bower, then, the same as the history of the arena bird's court? I think so. Arena behavior, I suggest, can develop fortuitously at any period in the history of any bird group as a result of a shift in the work load shared by a pair-bonded male and female. The division of labor in nest construction and care and the rearing of the young varies from species to species. In extreme cases the males may be completely released from all nesting duties—perhaps because natural selection favors a stock in which brightly colored males stay away from the nest. Emancipated from the pair bond, the males can live apart from the females in bachelor clans. Now sexual selection can operate freely, tending in the direction of brighter plumage and more complex display behavior that will attract more females.

The next step, from elementary arena behavior to bower-building, may not be so great as it seems at first. I have pointed out that most arena birds clear some sort of display space for themselves. In the species that have come down from the trees to the ground, such as the cock of the rock and some birds of paradise, the males spend much of their time clearing away twigs and leaves and perhaps berries, stones and shells, if there are any about. A. J. Marshall of Monash University in Australia and Erwin Stresemann of the Berlin Natural History Museum have speculated that

the handling of these objects may accidentally have become incorporated in and important to the courtship ceremony for which the court is maintained, and so have led to bower-building. I think it likely that both court-clearing and bower-building are deeply rooted in the nesting impulses of the male birds. Nest-building and the actions associated with it by each species constitute fixed behavioral patterns that are not easily abandoned and are more likely to be diverted into new directions. Other investigators have noted actions in arena birds, and particularly in bowerbirds, that reminded them of nesting behavior. V. G. L. van Someren remarked some years ago that in shaping its court the male of the weaverbird species known as Jackson's dancing whydah "creates recesses resembling the early stages of a nest, butting into the grass and smoothing it down with his breast." Edward A. Armstrong commented in his classic book on bird behavior that "this performance would seem to be due to the survival of the nest-building impulse."

Marshall, a leading student of the bowerbirds, has called attention to many activities he believes stem from displaced nesting habits. Certainly as one looks at a New Guinea stick bower, particularly that of Lauterbach's bowerbird, one cannot but feel it is some sort of monstrous nest. The wall of sticks, the lining of grass, even the way the male places egg-sized berries or pebbles near the center of the basket-like structure—all suggest aspects of nest-building that still survive in males that have had no nesting responsibilities for tens of thousands of years and probably much longer. In other bowerbirds and arena birds this impression of a physical nest is, to be sure, not so vivid. But, as noted in the case of the cock of the rock, I have often been impressed by the male's strangely

quiet and attentive manner when it visits its court or bower, a manner that reminds an ornithologist of a parent bird arriving at its nest.

To sum up, I would define arena behavior as courtship behavior reshaped by emancipated males to include their nondiscardable nesting tendencies. I would further suggest that bower behavior has developed in certain arena birds under the influence of natural and sexual selection, that some of the ground-clearing arena birds are even now on the way to becoming builders of bowers, and that the dully dressed bowerbirds that build the most complex and ornamented structures are at the leading edge of avian evolution.

This hypothesis does not in itself explain the great variety and variability of bowers or the complexities of behavior and plumage in arena birds, all of which seem to imply that these birds are evolving at an accelerated rate compared with other birds. The biological advantage of arena behavior may be precisely that it does speed up evolution. Because of promiscuous polygyny a few males in each generation are enough to propagate a species. Losses by predation can be very acute (and indeed must be in the case of terrestrially displaying males), and both natural and sexual selection can operate more severely than usual.

An indication that some such process may be at work is the fact that "intergeneric hybrids," although extremely rare in all animals, are rather more common in arena birds. Since a species—a limb-tip on the avian tree of evolution—is identified as such by its "reproductive isolation" from the other limb-tips, it is difficult to explain even one case of interfertility between genera, the main limbs of the tree. Yet in our collection at the American Museum of Natural History

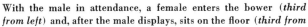

With the male in attendance, a female enters the bower (*third from left*) and, after the male displays, sits on the floor (*third from* *right*). The two birds mate (*second from right*). Then the female leaves to build a nest and rear her young by herself (*right*).

MALE COCK OF THE ROCK (*Rupicola rupicola*) perches above its court in a British Guiana forest. Like the bowerbirds, this mem- ber of the cotinga family is an arena bird: the males live apart from the females in clans and establish individual breeding stations.

THREE COCKS occupying adjacent courts in a small arena were watched by the author for 20 days as they defended their terri- tories and displayed to visiting females. The cock at left is postur- ing on its terrestrial court; the others perch above their courts.

"TRANSFERRAL EFFECT" is illustrated by a relict head-turning movement in a bowerbird. *Chlamydera nuchalis* (*left*) displays to the female a bright pink crest at the nape of its neck. *C. cer-* *viniventris* (*right*) makes a similar movement although it has no crest. This bird's use of berries as ornaments made the crest unnecessary and it disappeared, but the turning motion persists.

we have no less than 11 adult male intergeneric hybrid offspring of the magnificent bird of paradise (*Diphyllodes magnificus*) and the king bird of paradise (*Cicinnurus regius*), which were long ago classified in different genera. The number of hybrids occurring between these birds leads me to suspect that *Diphyllodes* and *Cicinnurus* may not be nearly so distantly related as their fundamental structures seem to indicate. Perhaps arena behavior, once it takes hold of a species, fashions structural changes (body form) more rapidly than it does genetic changes (reproductive barriers). Such uneven radiation

might explain the hybrids between arena birds so different in size, shape and color that any taxonomist would accept them as distinct genera. Is there perhaps a correlation between such birds and the many varieties of domestic dogs, in the case of which man has acted as the agent of rapid selection and has bred such different but interfertile forms as the Pekingese and the great Dane?

This idea is probably premature and may be fanciful. Keeping to firmer ground, it is safe to say that the highly specialized combinations of structure and behavior seen in arena birds argue most eloquently that these birds are

evolving at a faster rate than most birds and that this accelerated evolution is due to their behavior. One has only to consider the magnificent plumage of the argus pheasant, the great inflatable bibs of the bustard, the radiant orange paraphernalia of the cock of the rock and the lacy plumage of the birds of paradise to be seized by the notion that some rapidly operating mechanism is directing the evolution of these birds. The same holds true for the even more wonderful arena birds called bowerbirds, with their houses and ornamented gardens and their courtship displays that replace plumage with glittering natural jewelry.

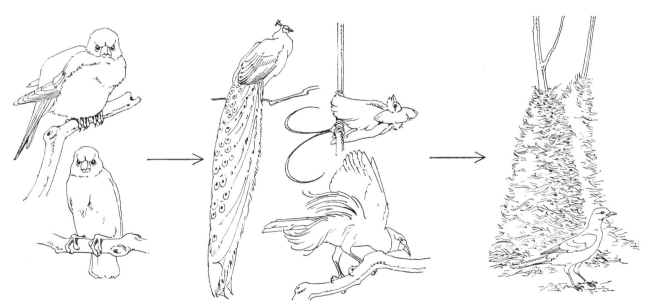

EVOLUTION OF BOWERBIRDS is diagramed here according to the author's hypothesis in highly simplified form. "Ordinary" birds (*left*) develop a pair bond, with a male and female mating and then tending the nest. If that pair bond is broken, there is a breakthrough to arena behavior (*center*) and a consequent proliferation of specialized plumage and courtship behavior. A few arena birds go one step further, to bower-building (*right*). With sexual selection transferred to objects, males may become dully colored.

The Author

E. THOMAS GILLIARD is curator of the Department of Birds at the American Museum of Natural History. Since he began his association with the American Museum in 1932 he has led numerous ornithological expeditions into remote regions of North and South America, Central Asia and the East and West Indies. He has had a part in the discovery and collection of many new species and subspecies of birds, mammals, fishes and reptiles. In 1956 Gilliard made his fifth trip into the interior of New Guinea, where he has been studying the behavior of bowerbirds and birds of paradise.

Bibliography

BIRD DISPLAY AND BEHAVIOR: AN INTRODUCTION TO THE STUDY OF BIRD PSYCHOLOGY. Edward A. Armstrong. Lindsay Drummond Ltd., 1942.

BIRDS OF CENTRAL NEW GUINEA. Ernst Mayr and E. Thomas Gilliard in *Bulletin of the American Museum of Natural History*, Vol. 103, Article 4, pages 315-374; April, 1954.

BOWER ORNAMENTATION VERSUS PLUMAGE CHARACTERS IN BOWER-BIRDS. E. Thomas Gilliard in *The Auk*, Vol. 73, No. 3, pages 450-451; July, 1956.

COURTSHIP AND DISPLAY AMONG BIRDS. C. R. Stonor. Country Life Limited, 1940.

ON THE BREEDING BEHAVIOR OF THE COCK-OF-THE-ROCK (AVES, RUPICOLA RUPICOLA). E. Thomas Gilliard in *Bulletin of the American Museum of Natural History*, Vol. 124, Article 2, pages 37-68; July, 1962.

SCIENTIFIC
AMERICAN   April 1961, Vol. 204, No. 4, pp. 150-160   OFFPRINT 1099

THE ECOLOGY OF FIRE

by Charles F. Cooper

Fire has played a major role in shaping the world's grasslands
and forests. Attempts to eliminate it have introduced problems
fully as serious as those created by accidental conflagrations.

Before Europeans came to North America, fires periodically swept over virtually every acre on the continent that had anything to burn. Along with climate, soil, topography and animal life, these conflagrations helped shape the pattern of vegetation that covered the land.

Civilization brought a tendency to regard fire as pure disaster, together with massive efforts to exclude fire completely from forest and grassland. The attempts frequently succeeded all too well. Over wide regions the pattern of plant life has changed, but not always in a way that users of the land could wish. Paradoxically, in some forest areas fire prevention has greatly increased the destructiveness of subsequent fires.

There is evidence that natural fires have occurred over most of the earth for thousands of years. Buried layers of charcoal testify to prehistoric fires. Historical writings mention great conflagrations witnessed by men. In the narratives of the explorers of North America are numerous accounts of traveling for days through smoke from distant fires, and of passing through burned-over prairies and woodlands.

Tree trunks in forested areas contain a record of past fires. A moderately intense fire often kills an area on one side of a tree, leaving the rest of the tree unharmed. As new layers of tissue grow over the dead spot, they count off the years since the fire. Examining freshly cut stumps of large redwoods, the California forester Emanuel Fritz found evidence of about four fires a century during the 1,100-year history of the stand. The figure is probably conservative, because there must have been many fires not severe enough to leave scars. In the ponderosa pine forests of California and Arizona, fire scars indicate an average of one burning every eight years.

Many forest fires are started by lightning; on the prairies rain immediately extinguishes lightning-set grass fires. Most prehistoric fires were undoubtedly the work of man.

Notwithstanding the popular conception, American Indians were not cautious in using fire. They did not conscientiously put out camp fires nor, unless their villages were threatened, did they try to keep fires from spreading. Often they burned intentionally—to drive game in hunting, as an offensive or defensive measure in warfare, or merely to keep the forest open to travel. A contemporary history of the Massachusetts Bay Colony, dated 1632, relates that "the Salvages are accustomed to set fire of the country in all places where they come; and to burn it twize a year, vize, at the Spring, and at the fall of the leafe. The reason that moves them to do so, is because it would be otherwise so overgrown with underweedes that it would all be a copice wood, and the people could not be able in any wise to passe through the country out of a beaten path."

In open country fire favors grass over shrubs. Grasses are better adapted to withstand fire than are woody plants. The growing point of dormant grasses, from which issues the following year's growth, lies near or beneath the ground, protected from all but the severest heat. A grass fire removes only one year's growth, and usually much of this is dried and dead. The living tissue of shrubs, on the other hand, stands well above the ground, fully exposed to fire. When it is burned, the growth of several years is destroyed. Even though many shrubs sprout vigorously after burning, repeated loss of their top growth keeps them small. Perennial grasses, moreover, produce seeds in abundance one or two years after germination; most woody plants require several years to reach seed-bearing age. Fires that are frequent enough to inhibit seed production in woody plants usually restrict the shrubs to a relatively minor part of the grassland area.

Most ecologists believe that a substantial portion of North American grasslands owe their origin and maintenance to fire. Some disagree, arguing that climate is the deciding factor and fire has had little influence. To be sure, some areas, such as the Great Plains of North America, are too dry for most woody plants, and grasses persist there without fire. In other places, for example the grass-covered Palouse Hills of the southeastern part of the state of Washington, the soil is apparently unsuited to shrub growth, although the climate is favorable. But elsewhere—in the desert grasslands of the Southwest and the prairies of the Midwest—periodic fires must have tipped the vegetation equilibrium toward grasses.

Large parts of these grasslands are now being usurped by such shrubs as mesquite, juniper, sagebrush and scrub oak. Mesquite alone has spread from its former place along stream channels and on a few upland areas until now it occupies about 70 million acres of former grassland. Many ecologists and land managers blame the shrub invasion entirely on domestic livestock; they argue that overgrazing has selectively weakened the grasses and allowed the less palatable shrubs to increase. These explanations do not suffice; even on plots fenced off from animals shrubs continue to increase. A decrease in the frequency of fires is almost surely an essential part of the answer.

Fire has played an equally decisive role in many forests. A good example is found in the forests of jack pine that now spread in a broad band across Michigan, Wisconsin and Minnesota. When lumbermen first entered this region, they found little jack pine; the forests consisted chiefly of hardwood trees and white pines that towered above the general forest canopy. The loggers singled out the white pines for cutting, considering the other species worthless. Their activities were usually followed by fires, accidental or intentional. Supported by the dry debris of logging, the fires became holocausts that killed practically all the remaining vegetation. The mixed forest had little chance to regenerate; even the seeds of most trees were destroyed. But those of the jack pine survived. Unlike most pine cones, which drop off and release their seeds in the fall, jack pine cones stay closed and remain attached to the tree, sometimes so firmly that

FIRE MAINTAINS GRASSLAND by holding back the spread of mesquite *(shown here)* and other shrubs, which originally constitute a small part of the vegetation *(a)* but which soon proliferate and reduce areas that are available to grass *(b)*. Fire *(c)* reduces grasses and shrubs alike, but while the growing point of grass lies near or beneath the ground *(root system at right in "c")* and is left unharmed, the buds and growing tissue of shrub stand fully exposed and are destroyed. The balance is further tipped toward grasses *(d)* because they produce abundant seed a year or two after germination; as a result of this they lose only one or two years' growth in fire. Shrubs lose several years' growth.

branches grow right around them. Inside the cones the seeds remain viable for years. When the cones are heated, as in a forest fire, they slowly open and release their seeds. Thus the fires simultaneously eliminated the seeds of competing species and provided an abundant supply of jack pine seed together with a bed of ash that is ideal for germination. The result of the process is a pure stand of jack pine.

The valuable Douglas fir forests of the Pacific Northwest also owe their origin to fire. This species requires full sunlight; it cannot grow in the dense shade cast by a mature fir forest. When old Douglas firs die, their place is taken not by new Douglas firs but by cedars and hemlocks, more tolerant of shade, which therefore constitute the "climax" vegetation of the region. Forest fires, however, arrest the succession by creating openings in the forest into which the light, winged seeds of Douglas firs can fly from adjacent stands. The seedlings take advantage of the sunlight in the openings: they flourish and top competing vegetation; ultimately they grow into pure stands of uniform age.

Jack pine and Douglas fir are dependent on fire for their establishment but cannot endure frequent burning thereafter. In other forests fire is a normal part of the environment during the whole life of the stand. The longleaf pine of the southeastern U. S. is a striking example. This species is almost ideally adapted to recurring fires.

Unlike most pines, the young longleaf does not grow uniformly after germination. The seedling reaches a height of a few inches in a few weeks. Then it stops growing upward and sprouts a grasslike ring of long drooping needles that surrounds the stem and terminal bud. During this so-called grass stage, which usually lasts from three to seven years, the plant's growth processes are concentrated in forming a deep and extensive root system and in storing food reserves.

Longleaf pine is easily shaded out by competing hardwoods and is susceptible to a serious blight known as brown spot. The brown spot fungus multiplies during the dry summer, and the autumn rains splash its spores onto the needles of the low seedlings. Unless overtopping vegetation is cleared away and brown spot is controlled, the young pines may remain in the grass stage indefinitely.

One of America's first professional foresters, H. H. Chapman of Yale University, perceived in the early 1920's that periodic fires were essential to the life of longleaf pine. Protected by its canopy of needles, the longleaf seedling can withstand heat that kills the aboveground portions of competing hardwoods and grasses. At the same time the flames consume dry needles infected with brown spot, destroying the principal source of fungus spores. After the young pine emerges from the grass stage, its phenomenal growth—often four to six feet a year for the first two or three years—quickly carries the buds beyond the reach of surface fires. The thick, corky bark of the sapling protects its sensitive growing tissue. As the tree

SEROTINOUS JACK PINE CONES resist fire (*top*). Unlike other pine cones, which open and release their seeds in the fall, jack pine cones open after being heated (*bottom*).

ORIGINAL FORESTS of Great Lakes region were of mixed hardwoods, some jack pines (*at right in "a"*) and white pines (*middle distance and background*). Early loggers cut white pines and left other species standing (*jack pine cone is in foreground of*

grows and the bark thickens, it becomes resistant to any but the most intense fires.

Largely at Chapman's urging, prescribed fire has become an accepted management tool in the southeastern longleaf pine forests. Before a stand is harvested a fire is run through during the dry summer, when it will burn fairly hot. This clears out most of the undergrowth without killing the trees and prepares a good seedbed. The old trees are cut during the following winter, after the seeds have fallen. About three winters later, when burning conditions permit only a relatively cool fire and the seedlings have entered the grass stage, the area is burned again. Fires at regular three-year intervals thereafter keep down the worthless scrub oaks and help control brown spot. They hold back the normal succession, which would lead to a climax oak-hickory forest. Moreover, a regime of periodic fires reduces the accumulation of dry fuel on the ground that might otherwise lead to an uncontrollable holocaust.

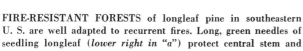

FIRE-RESISTANT FORESTS of longleaf pine in southeastern U. S. are well adapted to recurrent fires. Long, green needles of seedling longleaf (*lower right in "a"*) protect central stem and bud against surface fires that burn out forest debris and saplings of competing hardwoods (*middle distance and background in "b"*). Rapid vertical growth of tree after seedling stage carries

"b"). Debris of logging supported holocausts that consumed remaining vegetation. Although jack pines were destroyed, their
cones survived and released seeds (c). Seedling jack pines (d)
grew in fertile ashes, giving rise to pure jack pine stands today (e).

My own work has dealt with the ponderosa pine forests of the Southwest. As
19th-century chronicles attest, they used
to be open, parklike forests arranged in a
mosaic of discrete groups, each containing 10 to 30 trees of a common age.
Small numbers of saplings were dispersed among the mature pines, and
luxuriant grasses carpeted the forest
floor. Fires, when they occurred, were
easily controlled and seldom killed a
whole stand. Foresters in other regions
envied the men assigned to the "asbestos
forests" of the Southwest.

Today dense thickets of young trees
have sprung up everywhere in the
forests. The grass has been reduced, and
dry branches and needles have accumulated to such an extent that any fire is
likely to blow up into an inferno that
will destroy everything in its path. Foresters have generally blamed the over-
production of trees on a period of unusually favorable weather conditions, or
on removal of competing grasses and
exposure of bare soil through past

vulnerable bud beyond reach of bigger fires (c); thickening bark
affords increasing insulation for delicate cambium against hotter
fires. At the same time the tree drops more needles, supporting
hotter fires that clear out larger saplings (d). Self-governing mechanism keeps forest open (e). Illustrations follow single tree from
seedling ("a" and "b") to sapling ("c" and "d") to maturity (e).

trampling and grazing by domestic animals. But it is becoming increasingly apparent that a vigorous policy of fire exclusion, too long followed, is at least partly responsible.

Lightning is frequent in the ponderosa pine region, and the Indians set many fires there. Tree rings show that the forests used to burn regularly at intervals of three to 10 years. The mosaic pattern of the forest has developed under the influence of recurrent light fires. Each even-aged group springs up in an opening left by the death of a predecessor. (After remaining intact for 300 years or more, groups break up quite suddenly— often in less than 20 years.) The first fire that passes through consumes the dead trees, and leaves a good seedbed of ash and mineral soil, into which seed drifts from surrounding trees. Young ponderosa seedlings cannot withstand even a light surface fire, but in the newly seeded opening they are protected by

PRIMEVAL DOUGLAS FIR FORESTS of western U. S. have origin in fires of previous centuries. Young Douglas firs, which are intolerant of shade, cannot grow beneath mature Douglas fir forest (a), yield to cedars and hemlocks, which make up the climax

REPLACEMENT OF DOUGLAS FIRS by hardwoods is in part attributable to exclusion of fire. Cedar and hemlock saplings (a), unlike young Douglas firs, grow well in shade of mature Douglas fir forest, take over as older firs die (b). As cedar and hemlock

the lack of dry pine needles to fuel such fires. Consequently the young stand escapes burning for the first few years. Eventually the saplings drop enough needles to support a light surface fire, which kills many smaller saplings but leaves most of the larger ones alive. The roots of the survivors quickly appropri-

ate the soil made vacant, and their growth is stimulated.

The degree of thinning accomplished by a fire depends upon the quantity of fuel on the ground. The denser the sapling stand, the more needles it drops and the hotter the fire it will support. The process is thus a sort of self-regulating

feedback mechanism governed by the density of the stand. Thinning by fire is less efficient than the forester might wish, but it does help to prevent the stagnation resulting from extreme overcrowding.

As a group of trees grows toward maturity, new seedlings germinate beneath

vegetation of the region. Succession was interrupted by frequent small fires that burned out cedar and hemlock trees (b). Douglas

fir seeds from adjacent stands blew into new openings (c), grew well in seedbed of ashes and became pure stands of Douglas fir (d).

trees grow (c), they diminish remaining opportunities for Douglas fir seedlings to survive. Climax vegetation that results is composed

of cedar and hemlock (d). These illustrations, like those on preceding and following pages, are drawn from same point of view.

it. The volume of dry fuel dropped by the older trees, however, supports fires hot enough to eradicate the seedlings entirely. Fire and shade together prevent younger trees from developing; the even-aged character of the group is maintained throughout its life.

Wild-land managers have historically, and properly, concentrated on sup-pressing accidental fires in forests and grasslands and on discouraging deliberate overburning by man. While fire may favor the establishment of jack pine forests, the annual burning long practiced in the South will prevent the growth of any forest at all. By the same token, occasional fires may be needed to maintain African grasslands, but deliberately set-ting fire to the country every few months has unquestionably damaged them seriously. It is time to relax many of the ingrained prejudices against fire and to utilize it, judiciously, as a tool in the management of both forests and grasslands.

PARKLIKE PONDEROSA PINE FORESTS of Southwest were typically a mosaic of even-aged groups (*mature stand in middle distance of "a"; young stand in background*). Frequent fires kept forest debris from accumulating (*b*); thus the fires were mild and created openings for seedlings (*c*), which cannot grow in shade. As new trees matured, they dropped more needles, providing more fuel for hotter fires, which killed new seedlings (*d*). Mosaic and parklike character of the forest was thereby maintained.

CLUTTERED PONDEROSA PINE FORESTS (*a*), in contrast to those illustrated on page 2658, result from the elimination of the periodic fires that occurred naturally. Saplings that would have been thinned out by fire now vie for space in the formerly open avenues between trees the grass cover is reduced and forest debris and undergrowth have accumulated to the point (*b*) that fires that formerly would have been mild and easily controlled often explode into holocausts (*c*) that destroy the entire stand of trees (*d*)

The Author

CHARLES F. COOPER works for the U.S. Agricultural Service in Boise, Idaho. He acquired a B.S. in forestry at the University of Minnesota in 1950, an M.S. at the University of Arizona and a Ph.D. in plant ecology at Duke University in 1958. "In the course of a rather short career," he writes, "I have been variously occupied as an aircraft mechanic, forester, pulpwood logger, manufacturer of creosoted fence posts, range conservationist and college teacher."

Bibliography

CHANGES IN VEGETATION, STRUCTURE AND GROWTH OF SOUTHWESTERN PINE FORESTS SINCE WHITE SETTLEMENT. Charles F. Cooper in *Ecological Monographs*, Vol. 30, No. 2, pages 129-164; April, 1960.

THE DESERT GRASSLAND: A HISTORY OF VEGETATIONAL CHANGE AND AN ANALYSIS OF CAUSES. Robert R. Humphrey in *The Botanical Review*, Vol. 24, No. 4, pages 193-252; April, 1958.

FIRE AS THE FIRST GREAT FORCE EMPLOYED BY MAN. Omer C. Stewart in *Man's Role in Changing the Face of the Earth*, edited by William L. Thomas, Jr., pages 115-133. University of Chicago Press, 1956.

SCIENTIFIC
AMERICAN   August 1962, Vol. 207, No. 2, pp. 29-35   OFFPRINT **1100**

THE THALIDOMIDE SYNDROME

by Helen B. Taussig

A mild and supposedly safe sedative taken by pregnant women
has deformed the limbs and other organs of several thousand
infants in West Germany, England, Canada and other countries.

Two grossly deformed infants were the subject of an exhibit at the annual meeting of the pediatricians of the Federal Republic of Germany held in October, 1960, in the city of Kassel. Photographs and X-ray pictures showed that the long bones of the infants' arms had almost completely failed to grow; their arms were so short that their hands extended almost directly from their shoulders. Their legs were less affected but showed signs of a similar distortion of growth. Both infants were also marked by a large hemangioma (strawberry mark) extending from the forehead down the nose and across the upper lip; one of them was also found to have a duodenal stenosis, that is, a constriction of the beginning of the small intestine. The physicians who presented these cases, W. Kosenow and R. A. Pfeiffer, members of the staff of the Institute of Human Genetics in Münster, had never seen quite this combination of anomalies in a single infant. They regarded it as a new clinical entity.

The deformity of the limbs was characteristic of a malformation known as phocomelia, from the Greek words *phoke*, meaning seal, and *melos*, meaning limb. Phocomelia is so rare that most physicians never see it in a lifetime; moreover, it usually affects only one limb. Kosenow and Pfeiffer reported that they could find no hereditary indication for the condition in the history of either family, no incompatibility in the blood types of the parents and no

abnormality in the chromosomes of the tissue cells of either child. Guido Fanconi, a Swiss pediatrician who has long been interested in congenital deformities, declared that he too had never seen infants afflicted this way. Otherwise little note was taken of the exhibit. I missed it myself, although I was at the meeting.

In retrospect it is surprising that the exhibit did not attract a great deal of attention. During 1960 almost every pediatric clinic in West Germany had seen infants suffering such defects. In Münster there had been 27, in Hamburg 30 and in Bonn 19. There had been perhaps a dozen cases of phocomelia in 1959, whereas in the preceding decade there had been perhaps 15 in all of West Germany. During 1961 the incidence of phocomelia increased rapidly; hundreds of afflicted infants were born.

When the West German pediatricians gathered for their 1961 meeting in November at Düsseldorf, almost all of them were aware of the mysterious outbreak of phocomelia. At the meeting Widukind Lenz of Hamburg made the disclosure that he had tentatively traced the disease to a new drug that had come into wide use in sedatives and sleeping tablets. The generic name of the drug was thalidomide. Under the trade name Contergan, it had been marketed as freely as aspirin in West Germany from 1959 into the spring of 1961. Lenz had found that many mothers of "seal limb" infants admitted to the Hamburg clinic had taken this drug early in pregnancy. Contergan

and other preparations containing thalidomide have now been withdrawn from sale. But infants injured by the drug are still in gestation. When the last of them has been born by the end of this summer or early in the autumn, thalidomide will have produced deformities in 4,000 or even as many as 6,000 infants in West Germany alone, and probably more than 1,000 in other countries where it has been marketed. The one-third who are so deformed that they die may be the luckier ones.

It happens that thalidomide-containing drugs did not reach the market in the U.S. This was because of a lucky combination of circumstances and the alertness of a staff physician at the Food and Drug Administration—not because of the existence of any legal requirement that the drug might have failed to meet. If thalidomide had been developed in this country, I am convinced that it would easily have found wide distribution before its terrible power to cause deformity had become apparent. The marketing techniques of the pharmaceutical industry, which can saturate the country with a new drug almost as soon as it leaves the laboratory, would have enabled thalidomide to produce thousands of deformed infants in the U.S. I believe that it is essential to improve both the techniques for testing and the legal controls over the release of new drugs.

The news that a large number of malformed infants had been born in West

Germany and that a sleeping tablet was suspected as the cause first came to me in late January of this year. I was particularly concerned because of my lifelong interest in malformations. That a drug was implicated was of especial interest, because little is known about the cause of the various congenital anomalies that arise in the course of gestation. I immediately went to West Germany to investigate the situation, and I have also conferred and corresponded with physicians in other countries where thalidomide, under various names, has been sold.

In West Germany I was told that a Swiss pharmaceutical house, interested in producing a new sedative, had first synthesized thalidomide in 1954. Because it showed no effects on laboratory animals the company discarded it. Then the West German firm Chemie Grünenthal undertook the development of the compound. Once again thalidomide showed no effect on laboratory animals. Since the structure of the molecule suggested that it should work as a sedative, Grünenthal tried it as an anticonvulsant for epileptics. It did not prevent convulsions, but it worked as a hypnotic, acting promptly to give a deep, "natural" all-night sleep without a hangover. Given the trade name Contergan, it became during 1960 the favorite sleeping tablet of West Germany, inexpensively available without a prescription and widely used in homes, hospitals and mental institutions. It turned out to be as safe for humans as for animals. Would-be suicides who tried it after it came on the market survived large doses of it without harm.

Grünenthal combined thalidomide with aspirin and other medicines. Germans consumed these compounds—Algosediv, Peracon Expectorans, Grippex and Polygrippan—for such conditions

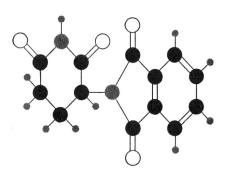

THALIDOMIDE is a synthetic drug. In this diagram of its molecule, carbon atoms are represented by black balls, hydrogen by small gray balls, oxygen by white balls and the two nitrogen atoms by large gray balls.

as colds, coughs, grippe, nervousness, neuralgia, migraine and other headaches and asthma. A liquid form made especially for children became West Germany's baby sitter. Hospitals employed it to quiet children for electroencephalographic studies. As an antiemetic, it helped to combat the nausea of pregnancy, and of course Contergan gave many a pregnant woman a good night's sleep. Grünenthal was manufacturing it almost by the ton.

Soon pharmaceutical companies in other countries began to make or market thalidomide under license from Grünenthal. Distillers (Biochemicals) Ltd. sold it as Distaval in the British Isles, Australia and New Zealand. Combinations received the trade names of Valgis, Tensival (a tranquilizer), Valgraine and Asmaval. An advertisement in Great Britain emphasized the safety of the drug with a picture of a small child taking a bottle from a medicine shelf. From Portugal it went into local and international channels of distribution as Softenon. In Canada Frank W. Horner Ltd. of Montreal marketed it as Talimol and the Canadian branch of the Wm. S. Merrell Company of Cincinnati as Kevadon. In September, 1960, the Merrell Company applied to the Food and Drug Administration for clearance to sell Kevadon in the U.S.

At that time no one had reported any untoward side effects from thalidomide. During the next few months, however, German medical journals carried reports of a new polyneuritis associated with long-term use of the drug. Patients complained of tingling hands, sensory disturbances and, later, motor disturbances and atrophy of the thumb. By April, 1961, there was a sufficient number of ill effects reported in West Germany following the use of the drug to place the thalidomide compounds on the list of drugs for which prescriptions were required. (It was under prescription from the beginning in most other countries.) Nevertheless, thalidomide remained popular and continued in widespread use in the home and in hospitals.

By the summer of 1961 physicians all over West Germany were realizing with alarm that an increasing number of babies were being born with disastrous deformities of their arms and legs. In Kiel, Münster, Bonn and Hamburg four different investigations were under way. From a study of 32 cases in Kiel and its environs H. R. Wiedemann found that the malformations followed a specific pattern, although they varied in severity.

Abnormality of the long bones of the arms characterizes the great majority of the cases, with the legs involved in half of these. The radius or ulna (the forearm bones) or both may be absent or defective. In extreme cases the humerus (upper-arm bone) also fails to appear. Typically both arms are affected, although not necessarily equally. When the legs are involved, the hip girdle is not fully developed. Dislocation of the hip and outward rotation of the stub of the femur turns the deformed feet outward. The worst cases have neither arms nor legs; since they cannot turn over in the crib or exercise they usually succumb to pneumonia.

The hemangioma of the face, as Pfeiffer and Kosenow pointed out, is possibly the most characteristic feature of the syndrome. It is, however, neither harmful nor permanent. A saddle-shaped or flattened nose is common. In some cases the external ear is missing and the internal auditory canal is situated abnormally low in the head. In spite of this deformity hearing tends to be fairly good if not normal. Many of the children display paralysis of one side of the face. Many suffered from a variety of malformations of their internal organs, involving the alimentary tract and also the heart and circulatory system. Most of the children seem to be normally intelligent.

Pfeiffer and Kosenow in Münster had found no evidence that the phocomelia in their first two cases was hereditary. Eventually they completed detailed studies of 34 cases, with the same result. This was surprising because many of the previous cases of phocomelia could be traced back in the family. These two investigators concluded that an unknown agent from the environment, affecting the embryo at some time between the third and sixth week of pregnancy, had caused the damage. During this period, when most women do not yet know they are pregnant, the embryo goes through the principal stages of development.

Was the unknown agent a virus? An infection by rubella, or German measles, during this critical period of gestation results in severe malformations but not in phocomelia. That it might be some other virus seemed to be ruled out by the fact that the increase in the incidence of phocomelia had been steady, not abrupt, and by the fact that the cases were confined within the boundaries of West Germany. By the time of last year's pediatric meeting at Düsseldorf in late

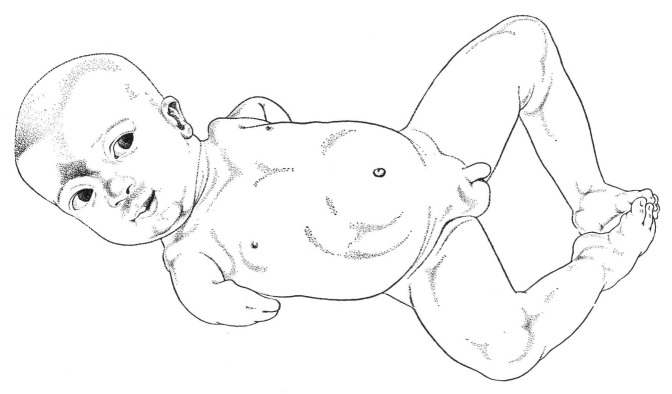

VICTIM OF THALIDOMIDE SYNDROME typically has short, deformed and useless arms and hands. The actual case shown in this drawing displays the hemangioma, or strawberry mark, on the forehead, nose and upper lip, which is the most characteristic (al-though harmless) feature of syndrome. Other abnormalities that may occur include deformed legs and feet and a wide variety of deformations of the ears, digestive tract, heart and large blood vessels. Most of the afflicted children have normal intelligence.

November the range of speculation included radioactive fallout.

Lenz meanwhile had formed a new suspicion. Like the other investigators, he had been sending out lengthy questionnaires to the parents of deformed infants and to the physicians who attended them, asking about X-ray exposure, drugs, hormones, detergents, foods and food preservatives, contraceptive measures and tests for pregnancy. In his initial returns he noted that approximately 20 per cent of the mothers reported taking Contergan during pregnancy. On November 8, he recalls, it occurred to him that Contergan might be the cause. He now asked all the parents specifically about Contergan, and 50 per cent reported use of the drug. Many of the mothers said that they had considered the drug too innocuous to mention on the questionnaire.

On November 15 Lenz warned Grünenthal that he suspected Contergan of causing the catastrophic outbreak of phocomelia and he urged the firm to withdraw it from sale. On November 20, at the pediatric meeting, he announced that he suspected a specific but unnamed drug as the cause of the "Wiedemann syndrome" and said that he had warned the manufacturer. That night a physician came up to Lenz and said: "Will you tell me confidentially, is the drug Contergan? I ask because we have such a child and my wife took Contergan." Before the meeting was over the doctors generally knew that Lenz suspected Contergan.

On November 26 Grünenthal withdrew the drug and all compounds containing it from the market. Two days later the West German Ministry of Health issued a firm but cautious statement that Contergan was suspected as the major factor in causing phocomelia. Radio and television stations and the front pages of newspapers promptly carried announcements warning women not to take the drug.

On the other side of the world W. G. McBride, a physician in New South Wales, Australia, saw three newborn babies with severe phocomelia during April, 1961. In October and November he saw three more. From the histories of the mothers he found that all six had taken Distaval in early pregnancy. McBride notified the Australian branch of Distillers Ltd. and it cabled its findings to the London headquarters on Novem-ber 27. This and the news from Germany caused the firm to withdraw the drug on December 3. Because of the demand by physicians it has been returned to limited sale in England, but in Germany it is now illegal to possess thalidomide.

The news of the Australian experience prompted A. L. Speirs, a physician of Stirlingshire, Scotland, to review 10 cases of phocomelia that he had seen in his practice during the preceding months. By checking prescription records and medicine cabinets in the victims' homes, he obtained positive proof that eight of the mothers had taken Distaval in early pregnancy.

Thus in the last weeks of 1961 circumstantial evidence accumulating in various parts of the world indicated that thalidomide played an important role in the causation of phocomelia. Physicians now began asking women who were still pregnant about their experience with the drug. One obstetrician in Germany asked 65 pregnant women if they had taken Contergan in early pregnancy. Only one said that she had. The physician declared that if she had an abnormal baby, he would believe Lenz. She did!

Among 350 pregnant women in Lübeck, W. von Massenbach found that 13 had taken Contergan, six during the second half of pregnancy and seven in the first four and a half months. Of the seven, two had babies with phocomelia, one had a baby with a closed anus and four had normal infants. By March, 1961, clinical records in Düsseldorf showed that 300 women who had not taken Contergan had given birth to healthy infants, whereas half of those who reported taking Contergan bore deformed infants. At Hamburg, meanwhile, Lenz had set out to investigate the exact connection between phocomelia and the drug and to fix reliably the date or dates of exposure to the drug in each case. He considered the use of Contergan to be proved only by a photostatic copy of a prescription or by a hospital record. This was difficult because the drug had been sold without prescription before April, 1961, and because nurses in German hospitals dispense sleeping tablets

as freely as nurses in the U.S. give laxatives. In one case, Lenz told me, the attending physician swore that the mother had not received Contergan, although her baby had been born with phocomelia. The physician insisted that he had prescribed a different sedative. At the pharmacy Lenz found the prescription marked by the druggist: "Drug not in stock. Contergan given instead."

By the middle of March of this year Lenz had assembled histories on 50 cases in which he had established documentary evidence for use of the drug and had determined the date of the last menstrual period before pregnancy. He had proof in each case of the date or dates on which Contergan had been taken. All but five of the women had taken the drug between the 30th and 50th day after the last menstrual period and the five had taken it between the 50th and 60th day. In the 21 instances in which Lenz managed to ascertain the date of conception,

the mother had taken Contergan between the 28th and 42nd day after conception. Thus the exact time when thalidomide can damage the embryo varies somewhat, but the period in which the embryo is especially vulnerable to the drug appears to be relatively brief.

In the human embryo the first signs of future limbs can be discerned with a microscope when the embryo is only 10 days old. By 42 days the tiny limbs are visible to the naked eye, although the embryo is only a little more than an inch long. The fact that the arm buds develop slightly earlier than those of the legs may be of significance in accounting for the greater frequency of arm damage. As the malformations indicate, the drug arrests and deranges those processes of development that are in progress when the embryo is exposed to it. Just how thalidomide interferes with growth remains to be determined. Some German doctors still doubt that Contergan is the sole

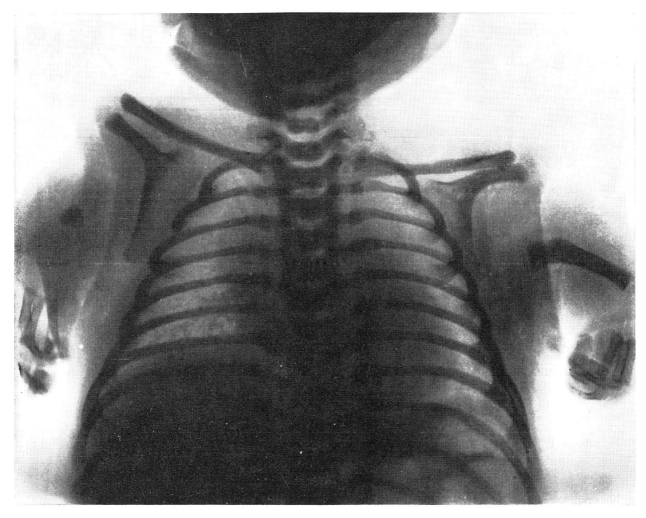

TYPICAL PHOCOMELIA, or "seal limb," is readily apparent in this X ray of chest, shoulders and arms of West German infant. In "classic" phocomelia usually only one arm was affected. Phocomelia caused by thalidomide almost always deforms both arms.

cause of the phocomelia syndrome, and a number of English physicians hold that some other substance or factor also causes phocomelia because they cannot get a history of Distaval or compounds containing it in every case. Furthermore, there is apparently no relation between the amount of the drug ingested and the severity of the malformation. A single dose of 100 milligrams appears to be enough to cause severe phocomelia, yet in other instances the same doses may produce only a mild abnormality. This must be due to a lack of susceptibility or to the fact that the drug was not taken in the sensitive period.

A drug with a molecular structure similar to that of thalidomide is Doriden, also used as a sedative. Although in a few cases of phocomelia the mother says she took Doriden, not Contergan, Doriden has been widely used in Switzerland since 1955, and phocomelia did not appear there until 1961. Almost all the few Swiss cases have been traced to Contergan from Germany.

Little is known about the metabolism of thalidomide, how the body excretes it or how long the deformity-producing factor persists in the body. About all that is certain is that it is insoluble in water and in fat. Obviously the usual laboratory animals metabolize it differently from human beings; it does not induce sleep in the animals. Investigators at the Grünenthal laboratories have tried unsuccessfully to produce phocomelia in rats, mice and rabbits. They have shown that the drug passes through the placenta of rabbits, but the offspring were normal in these experiments. G. F. Somers of the Distillers Ltd. laboratories has fed massive doses to pregnant rabbits. The rabbits did not sleep; they did, however, produce offspring with abnormalities remarkably similar to those in human infants. Since thalidomide makes a horse sleep, it may be that the horse will react in other ways as man does. Experiments with monkeys and apes will also be of interest. When the proper experimental animal is found, thalidomide does offer the possibility of studying the origin of malformations.

It is not yet possible to determine the exact number of infants born with phocomelia in West Germany, but the outbreak was devastating. The records of the Institute of Human Genetics in Münster show three cases of bilateral phocomelia in 1959, 26 cases in 1960 and 96 in 1961. Up to this spring 13 pairs of twins afflicted with phocomelia had been

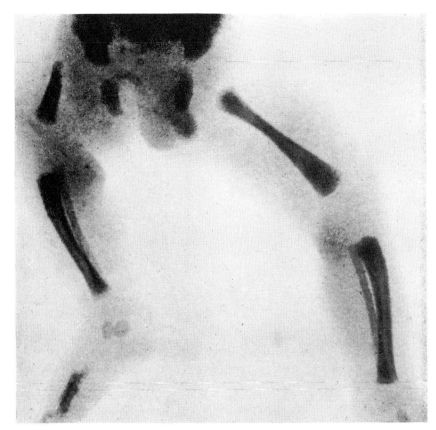

DEFORMITY OF LEG, here a very short femur, or thighbone (left), characterizes quite a few cases of the thalidomide syndrome. The bones of the hip girdle are also abnormal.

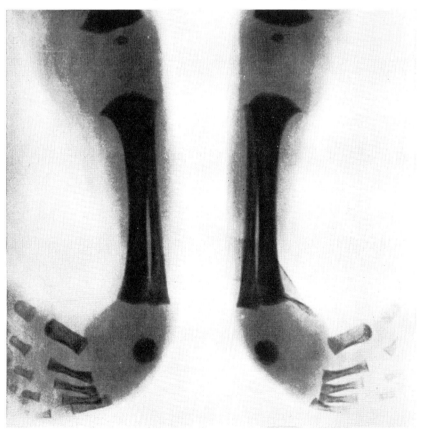

FEET ROTATE OUTWARD in this case. Even though badly deformed, this child may be able to learn to walk. Widukind Lenz, Hamburg pediatrician, supplied these three X rays.

registered. Since twins occur once in every 100 births, the institute estimates that there will be 1,300 cases in the state of North Rhine–Westphalia, where it is located. It is an indication of the prevalence of phocomelia that the state's Ministry of Health has set up a registry for all children with defective hands and arms who will need orthopedic help. As of January, 800 had been registered, 80 per cent suffering from phocomelia, and reports were in from only half the state. By now the total may have reached 2,000. Applying this experience to the population of West Germany as a whole, the country anticipates a minimum of 4,000 cases. I should not be surprised by a total of 6,000. There is every reason to believe that two-thirds of the infants will live for many years; indeed, the children appear to have a normal life expectancy.

In England, alas, the incidence is also high. Reports of phocomelia associated with Distaval appear regularly in *The Lancet,* the British weekly medical journal. Clifford G. Parsons of Birmingham has advised me that almost every physician at a medical meeting in England last spring had seen at least one case. The total for the country will probably be in the hundreds, however, not in the thousands.

Reports are still coming in from all over the world showing that phocomelia has occurred wherever thalidomide has been used. Sweden has had 25 cases, from Contergan purchased in Germany. Switzerland has had four cases. The Portuguese preparation, Softenon, has caused seven cases in Lebanon. Distaval has produced a case in Israel. In Peru, Contergan obtained by the father in Germany caused a case. Lenz has written me of an outbreak of phocomelia in Brazil. As yet I have received no figures for Portugal.

In September, 1960, when the Merrell Company applied to the Food and Drug Administration for permission to distribute the thalidomide compound, none of these untoward developments could have been anticipated. Clearance was delayed because the initial submission of papers was found to be "incomplete." Over the next few months, while the manufacturer gathered and filed additional material in support of the application, the first indications of the drug's neuropathic side effects were reported in the German medical press. Frances Oldham Kelsey, a physician and pharmacologist at the agency, took note of these reports. She also noted that the proposed label for the drug recommended its use against the nausea of pregnancy. From her work with quinine in connection with the malaria project during World War II, Mrs. Kelsey had become "particularly conscious of the fact that the fetus or newborn may be, pharmacologically, an entirely different organism from the adult." She therefore requested more data from the manufacturer to show that the drug was safe in pregnancy. Before her questions were answered the outbreak of phocomelia in Germany had brought withdrawal of the drug from the market in that country.

If thalidomide had been developed in this country, the story would have been quite different. Almost everyone agrees that with no knowledge of the delayed neuropathic effects of the drug and no appreciation of its dangers in pregnancy, the thought would not have occurred to anyone that it might injure the unborn child. Therefore permission for sale of the drug as a sedative would have been granted; it was an excellent sedative and appeared to be safe.

In the U.S. there have been only a few cases of the syndrome—two of them the twin offspring of a German woman who had married an American and brought Contergan with her to the U.S. Even the families of U.S. personnel stationed abroad have escaped—with one exception. At the U.S. Army headquarters in Heidelberg, in March, Thomas W. Immon was able to assure me that not one of the 16,000 babies born in U.S. hospitals in Germany during 1961 had phocomelia. More recently, however, he has had to report the birth in a U.S.

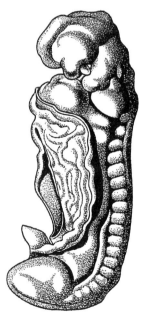
THIRD WEEK

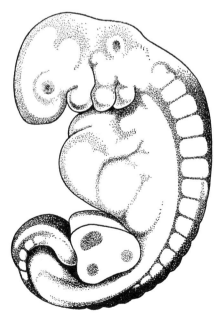
FOURTH WEEK

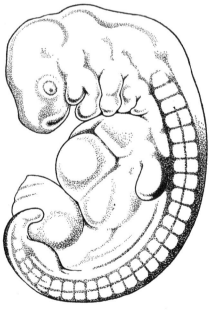
FIFTH WEEK

DEVELOPMENT OF HUMAN EMBRYO from third week after conception (*far left*) through eighth week (*far right*) is crucial. The embryo grows from about a quarter of an inch in length at end of third week to one and a quarter inches at end of eighth. Thalid-

Army hospital of one infant with phocomelia. The mother, a German, reported that she had taken Contergan in the early weeks of her pregnancy.

Unfortunately the people of Canada have had a different experience, even though the Dominion Government has a drug-regulating agency like that of the U.S. With two thalidomide preparations on the market in 1961, many pregnant women were exposed to the drug. At least 12 have delivered offspring afflicted with deformed arms and legs. The manufacturers issued a warning to physicians in December, advising them not to prescribe thalidomide for pregnant women. It was not until March, however, that governmental authorities asked the manufacturers to withdraw the drug entirely. Between now and the fall there will undoubtedly be additional casualties.

A generation ago new drugs, particularly those for relatively minor complaints such as insomnia, only gradually achieved widespread popularity. The rather small number of people using them in the first few years provided, albeit unwittingly, test cases not only for the efficacy but also for the long-term safety of the drug. Today "educational" representatives of drug houses visit each physician regularly. Pounds of lavish and expensive drug brochures assault the physician by mail. Most medical journals are crowded with handsome advertisements, many printed in full color on heavy cardboard or metallic paper, extolling the virtues of this year's model or modification of some recently invented tranquilizer, diuretic or antihypertensive compound. New drugs thus find huge markets within a few months.

In most countries, with the exception of Canada, governmental regulation of the pharmaceutical trade is less stringent than it is in the U.S. The Food and Drug Administration, however, is limited to considering only the safety and not the efficacy of a drug, and it exercises no control until the drug is ready for sale. During testing, conducted by and for the drug houses, a new compound may be distributed for clinical trial to many physicians. They are supposed to warn patients that the drug is experimental and to obtain a release signed by the patient. Not all physicians keep careful records of the cases in which they have distributed such test drugs. Clearance by the Food and Drug Administration, which rests on evidence of safety submitted by drug companies, must often be based in part on reports from observations made under clinical conditions that are, to say the least, not ideal. Certainly the procedure needs strengthening here.

Until recently no thought had been given to the need for the testing of drugs for potential harmfulness to the human embryo. In my laboratory at the Johns Hopkins School of Medicine I have not been able to obtain abnormalities in baby rabbits with thalidomide primarily because the massive doses I have used bring on so many abortions. This illustrates one of the problems of testing new drugs: what size dose in animals makes for a fair test? As thalidomide shows, animals may not react at all like humans.

Of course, no drug can ever be certified as completely safe. But all the hazards of a given drug should be established before it is marketed. In dealing with cancer and other serious diseases there is some justification for taking chances with new drugs. The less serious the illness is, the more certain it should be that the drug is harmless as well as effective. In the case of thalidomide, I wonder how long it would have taken to determine the cause of the malformations if the drug had produced some more common but less spectacular congenital defect. Any drug labeled safe should be relatively harmless for all people of all ages, including the unborn. Married women of childbearing age should avoid drugs as much as possible, particularly new ones.

For most people the story of thalidomide has ended. The tragedy will go on, however, for the infant victims of the "harmless" sedative and their families for the rest of their lives.

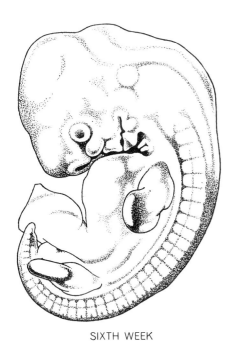 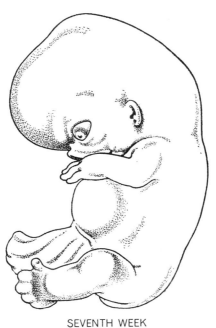 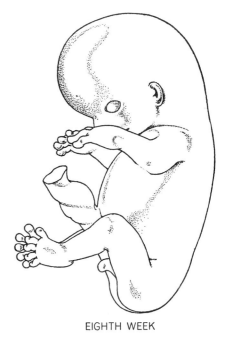

SIXTH WEEK   SEVENTH WEEK   EIGHTH WEEK

omide seems to cause almost all its deformities when the mother takes the drug during the fourth, fifth or sixth week of pregnancy, as limb buds, ears, intestinal tract, heart and blood vessels are forming and going through first growth. These embryos are normal.

The Author

HELEN B. TAUSSIG is professor of pediatrics at the Johns Hopkins School of Medicine, where she received her M.D. in 1927. With Alfred Blalock, a surgeon at Johns Hopkins, she conceived the famous "blue baby" operation, first performed on a human patient in 1945. The operation alleviates an interrelated group of congenital heart defects that lead to inadequate oxygenation of the blood, producing a characteristic blueness of the skin. Miss Taussig has received honorary degrees from more than a dozen institutions, including Columbia University, Harvard University and the University of Athens. Her other honors include the Lasker Award and the Feltrinelli prize. She has also been decorated by the French as Chevalier of the Legion of Honor.

Bibliography

KINDLICHE MISSBILDUNGEN NACH MEDI-KAMENT-EINNAHME WÄHREND DER GRAVIDITÄT? W. Lenz in *Deutsche Medizinische Wochenschrift*, Vol. 86, No. 52, pages 2555-2556; December 29, 1961.

THALIDOMIDE AND CONGENITAL ABNOR-MALITIES. A. L. Spiers in *The Lancet*, Vol. 1, No. 7224, pages 303-305; February 10, 1962.

THALIDOMIDE AND CONGENITAL ABNOR-MALITIES. G. F. Somers in *The Lancet*, Vol. 1, No. 7235, pages 912-913; April 28, 1962.

ZUR FRAGE EINER EXOGENEN VERURSA-CHUNG VON SCHWEREN EXTREMITÄ-TENMISSBILDUNGEN. R. A. Pfeiffer and W. Kosenow in *Münchener Medizinische Wochenschrift*, Vol. 104, No. 2, pages 68-74; January 12, 1962.

SCIENTIFIC
AMERICAN   February 1968, Vol. 218, No. 2, pp. 32–39   OFFPRINT **1101**

THE MEMBRANE OF
THE MITOCHONDRION

by Efraim Racker

The folded inner membrane of this intracellular body is the site
of the major process of energy metabolism in the living cell. It is
studied by taking it apart and attempting to put it together again.

The seat of oxidative phosphorylation, the process by which most
plant and animal cells produce the
energy required to sustain life, is the inner membrane of the intracellular particles called mitochondria [*see bottom
illustration on page 2670*]. Associated
with this membrane are enzymes of oxidative phosphorylation, embedded in a
complex matrix that binds them tena-
ciously in an ordered array. The mechanism of energy production in mitochondria has long defied analysis, since
a complex chemical pathway in a living
organism cannot really be understood
until its intermediate products have been
identified and the enzymes that catalyze
each step of the process have been in-
dividually resolved as soluble components. A decade ago my colleagues and
I set out to attack the problem by trying
to take the inner membrane of the mitochondrion apart and put it back together. We have been partially successful in
the attempt, and along the way we have
made some exciting discoveries and developed new methods of studying enzymes bound in membranes.

The universal energy carrier of the
cell is adenosine triphosphate (ATP).
This molecule functions by transferring
its energetic terminal phosphate group
to another molecule. In so doing it
is converted to adenosine diphosphate
(ADP), which in turn can be transformed
into ATP by energy-generating systems
in the cell. This regeneration of ATP
occurs at several stages in the course of
the breakdown and oxidation of foodstuffs. Some ATP is formed during glycolysis, a well-understood metabolic
pathway utilizing soluble enzymes that
break carbohydrates down to simpler
compounds.

Most of the ATP is formed, however,

during the course of oxidative phosphorylation in mitochondria. Pyruvate,
the end product of glycolysis, is delivered to the mitochondria, where it is
oxidized to carbon dioxide and water
by enzymes of the Krebs cycle [*see top
illustration on pages 2670 and 2671*]. As
hydrogen is removed from the successive
intermediate products, it is captured by
the coenzyme diphosphopyridine nucleotide (DPN), which contains the vitamin
nicotinamide. The electrons of hydrogen
are passed along a series of respiratory
enzymes, notably yellow flavoproteins
and red cytochromes, ultimately combining with protons and oxygen to form water. The energy of this oxidation process
is utilized at three sites to regenerate
ATP from ADP and inorganic phosphate. Under physiological conditions
such "coupling" of oxidation with phosphorylation is compulsory, and respiration takes place only when ADP and
phosphate are available, which is to say
when ATP is being utilized. This "tight
coupling" represents an ingenious control mechanism through which energy
production is regulated by the rate of
energy consumption.

Two chemicals that affect oxidative
phosphorylation serve as tools with
which to analyze the process. One is
dinitrophenol (DNP), which uncouples
oxidation from phosphorylation so that
respiration proceeds but produces heat
instead of ATP. The other is the antibiotic oligomycin, which acts differently.
It interferes with the production of ATP,
thereby inhibiting respiration as long as
the system is tightly coupled. When dinitrophenol is added, the inhibition by
oligomycin is overcome and respiration
returns to its original rate, although it
produces no ATP.

The enzymes that catalyze electron
transport had been isolated and char-

acterized, but only after the disruption
of mitochondria with detergents. This
process left the oxidation enzymes able
to function but damaged the phosphorylation system severely. It was accordingly
believed for a long time that the intact
mitochondrial structure was essential for
oxidative phosphorylation and that the
component parts could not be separated
without destroying them.

In 1956 the system did begin to yield
to fractionation. Independent experiments reported almost simultaneously
from the laboratories of Albert L. Lehninger, Henry A. Lardy, David E. Green
and W. W. Kielley showed that chemicals such as digitonin and physical methods such as sonic oscillation would break
mitochondria into "submitochondrial
particles" much smaller than mitochondria and yet able to catalyze oxidative
phosphorylation. This accomplishment
was an important step forward, and yet
there was still no indication that a true
resolution—a separation of soluble components—would ever be possible. Such
resolution was the task we undertook at
the Public Health Research Institute of
the City of New York.

In 1957 the first successful resolution of
the system of oxidative phosphorylation was achieved when Harvey Penef-
sky, Maynard E. Pullman and I fragmented beef-heart mitochondria by
agitating them with glass beads in a
powerful device called a Nossal shaker.
We removed the heavier unbroken mitochondria by centrifuging the mixture at
low speed and then respun the lighter
fraction at high speed [*see top illustration on page 2672*]. Resulting sediment—the submitochondrial particles—
still contained the respiratory enzymes
but could not produce much ATP; the
remaining fraction contained a soluble

component that was necessary for the coupling of phosphorylation to oxidation. We called this soluble component a coupling factor, F_1. In time various treatments of mitochondria separated other coupling factors that were also required for phosphorylation, and these we called F_2, F_3 and F_4.

The experiments demonstrating the resolution of F_1 were difficult to reproduce. In laboratory jargon our data were "in the right direction"—and that is always a sign of trouble. F_1 as much as doubled phosphorylation, but that was not enough stimulation to provide a reliable assay of its coupling activity. And a reliable assay was required if we were to purify F_1 and characterize it. Now, it was known that mitochondria could catalyze the splitting of ATP into ADP and inorganic phosphorus. In fact, as early as 1945 Lardy, working at the University of Wisconsin, had suggested that this enzymatic ("ATP-ase") activity might be the inverse of some step in oxidative phosphorylation. We discovered that partially purified F_1 did in fact exhibit ATP-ase activity.

Since this was the first time ATP-ase had been extracted as a soluble component from mitochondria, we decided to go after this enzyme. We realized it was a gamble that might not shed any light on oxidative phosphorylation, but the ATP-ase assay was simple and accurate and at least we had had experience in purifying soluble enzymes. We felt that it should not take long to establish whether or not the ATP-ase and the coupling-factor activity were related.

Yet sometimes experience gets in one's way. Working in a "cold room," as one ordinarily does in enzyme research, we found the ATP-ase to be quite unstable (in contrast to the ATP-ase activity of submitochondrial particles, which was quite stable) and we made little progress.

One day we discovered that this enzyme was "cold labile": at 0 degrees centigrade it lost all activity in a few hours, but at room temperature it was stable for days. That was a turning point in our investigations; from then on purification was simple. Furthermore, we had a decisive tool for determining the relationship between ATP-ase and coupling factor. The fact that both activities decayed at the same rate at 0 degrees indicated that the same protein was responsible for both.

Chemical fractionation of F_1 gave us a pure enzyme in good yield. In fact, at first the yield seemed to be too good: often our final preparation had more units of ATP-ase than we had estimated were present in the crude preparation. Moreover, the ratio of coupling activity to ATP-ase activity was not constant during purification. An examination of these discrepancies by Pullman revealed that the crude mitochondrial extract contained a protein that inhibited ATP-ase activity but not coupling activity, and that the removal of this inhibitor during purification explained the unexpected increase in total ATP-ase activity.

The purified F_1 had one puzzling property. Whereas Lardy and his collaborators had shown that both oxidative phosphorylation and the ATP-ase activity of mitochondria were very sensitive to oligomycin, our soluble enzyme was completely insensitive. This apparent discrepancy caused some of our colleagues to challenge the significance of our observations with the soluble enzyme. I had heard that the late Oswald T. Avery had once said: "It doesn't matter if you fall down, as long as you pick up something from the floor when you get up." And so we accepted the challenge and embarked on a project to find

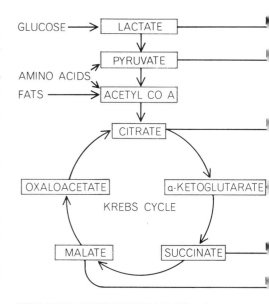

OXIDATIVE PHOSPHORYLATION is the process whereby energy from the oxidation of foodstuffs is harnessed to produce ATP, the energy carrier of the cell. Sugars, fats

out why oligomycin inhibited the enzyme in mitochondrial particles but not the soluble enzyme.

We started with the working hypothesis that there must be a component in mitochondria that confers oligomycin sensitivity on the enzyme. To show this we first had to prepare submitochondrial particles from which all the bound, oligomycin-sensitive ATP-ase had been removed, then add F_1 to them and observe what happened. We were able to eliminate the ATP-ase activity from particles by treating them with urea at 0 degrees, but to our surprise oligomycin-sensitive ATP-ase activity kept reappearing on dilution or aging. It developed that most of the ATP-ase in submitochondrial particles was latent—masked, apparently, by Pullman's inhibitor—and was more resistant to urea than the manifest enzyme was. We had to learn how

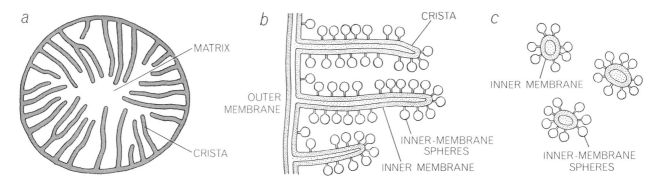

MITOCHONDRION, seen in a schematic cross section (a), has two membranes, each about 60 angstrom units (six millionths of a millimeter) thick. The inner membrane is deeply folded into "cristae" covered with the inner-membrane spheres, each about 85 ang- stroms in diameter (b). The inner membrane, with its spheres, is the site of oxidative phosphorylation. Mitochondria exposed to sonic oscillation become fragmented into small submitochondrial particles (c), which are still capable of oxidative phosphorylation.

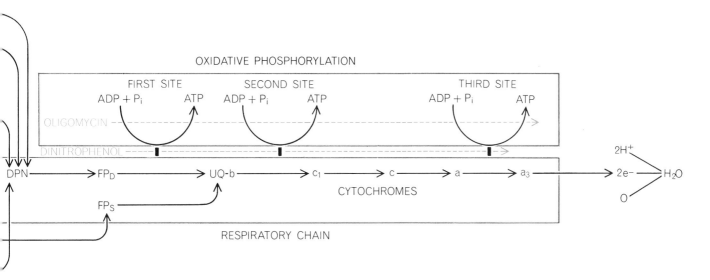

OXIDATIVE PHOSPHORYLATION AND RESPIRATORY CHAIN begin with foodstuffs, a key step in the generation of the ATP that powers the living cell. Carbohydrates, fats

and proteins are partially metabolized and then, in mitochondria, enter the Krebs cycle, in which they are broken down to carbon dioxide. In the process hydrogen atoms are accepted by the coenzyme diphosphopyridine nucleotide (*DPN*). The chain of respiratory enzymes, including flavoproteins (*FP*) and cytochromes b, c_1, c, a and a_3, catalyze a stepwise transfer of electrons to form water. At three sites phosphorylation is "coupled" to electron transfer. It can be uncoupled by dinitrophenol and inhibited by oligomycin.

to unmask this ATP-ase by removing the inhibitor before urea treatment. We found that if submitochondrial particles were first treated with trypsin, a digestive enzyme, and only then with urea, the resulting "*TU* particles" were depleted of virtually all ATP-ase activity [*see middle illustration on next page*]. More recently, when we found that the trypsin was damaging the mitochondrial membrane, Lawrence Horstman of our laboratory discovered that the inhibitor could be removed more gently by passing the submitochondrial particles through a column of Sephadex, a molecular sieve that separates small bodies from large ones. When this procedure is followed by treatment with urea, the resulting "*SU* particles" are analogous to *TU* particles but are much more effective in reconstitution experiments.

When we added F_1 to *TU* or *SU* particles, the enzyme was bound to the particles and the ATP-ase activity became not only sensitive to oligomycin but also stable at 0 degrees. Thus our working hypothesis was confirmed: mitochondria contain a component or components that alter the properties of F_1. We have become increasingly aware that this phenomenon is not unusual. Enzymes bound to membranes almost invariably have some properties that are different from those of the same enzymes in solution. Gottfried Schatz of our laboratory suggested the word "allotopy" (from the Greek for "other" and "position") to designate this phenomenon. We observed, furthermore, that the properties not only of the enzyme but also of the

membrane to which it is attached are changed depending on whether they are separate or bound to one another [*see top illustration on page 2673*]. An allotopic property of an enzyme can be used to devise a quantitative assay to serve during the purification of the membrane, since one can test successively purer membrane preparations to see if they are still capable of changing the properties of the added enzyme.

The *TU* particles that conferred oligomycin sensitivity on F_1 still contained the entire electron-transport chain, and we went on, with the allotopic property of F_1 as the tool, in an attempt to further resolve this membrane system. One day I subjected *TU* particles to sonic oscillation without including the usual salt buffer. Centrifugation of the resulting mixture at high speed yielded a soluble extract that conferred oligomycin sensitivity on F_1. We called the factor responsible for this property F_0. The discovery seemed even more exciting when the soluble preparation turned out to contain the entire electron-transport chain and even some residual phosphorylating activity: it appeared that we had actually rendered the entire system soluble. Then the addition of salt solution made the preparation turbid, which meant that particles had formed from the soluble system. In other words, in the presence of salt buffer—which must be added in biological experiments to keep the medium constant—F_0 was still particulate. At the time this was disappointing, but the observation led us

into new investigations of the relation between membrane structure and function.

In collaboration with Donald F. Parsons of the University of Toronto Faculty of Medicine and Britton Chance of the University of Pennsylvania School of Medicine, we examined all our membranous preparations of F_0 by negative staining in the electron microscope. We saw, first, that the submitochondrial particles we had started with were similar to those prepared by earlier investigators: sac-shaped structures outlined by a membrane that was covered with the characteristic "inner-membrane spheres" that had been discovered by Humberto Fernandez-Moran of the University of Chicago. The treatment with trypsin caused little change in structure. Subsequent treatment with urea, however, had a dramatic effect: although it left the membrane intact, it removed the inner-membrane spheres [*see bottom illustration on next page*]. This was unexpected, since David Green had once maintained that these spheres, which he called "elementary particles," represented groups of enzymes of the electron-transport chain [see "The Mitochondrion," by David E. Green; SCIENTIFIC AMERICAN, January, 1964]. We had found, on the contrary, that the *TU* particles (which lacked spheres) contained the entire electron-transport chain!

If the spheres did not contain respiratory enzymes, what did they contain? We calculated that most of the protein removed by urea treatment could be accounted for by the removal of ATP-ase,

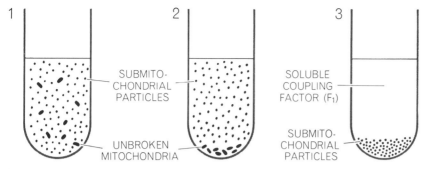

COUPLING FACTOR F_1 is separated by centrifugation. Mitochondria subjected to sonic oscillation (1) are centrifuged at low speed to separate particles from intact mitochondria (2). Then the particles are spun at high speed. The resulting light fraction (3) contains a soluble component (F_1) that is required for ATP production in the course of oxidation.

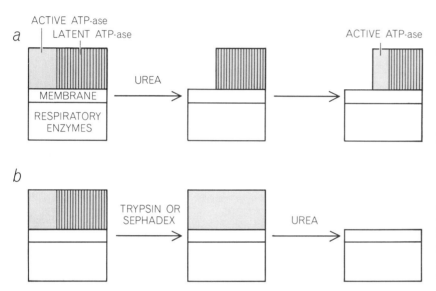

ENZYMATIC ACTIVITY (ATP-ase activity) of soluble F_1 was found resistant to oligomycin, unlike that of the intact membrane. To see if the membrane conferred this sensitivity on soluble F_1, it was first necessary to remove all native F_1 from the membrane. Most of the F_1 is masked by an inhibitor (bars), however; treatment with urea removed only exposed F_1, and ATP-ase activity reappeared (a). The destruction of inhibitor by trypsin (b) exposed the latent F_1 to removal by urea. Later Sephadex was substituted for trypsin.

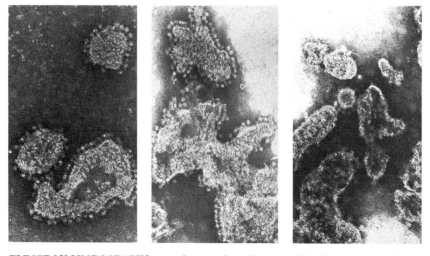

ELECTRON MICROGRAPHS trace the procedure diagrammed in the preceding illustration. The membrane of submitochondrial particles, enlarged about 100,000 diameters, is lined with inner-membrane spheres (left). Trypsin has little effect on appearance (center). Urea removes the spheres from the particles, leaving "TU particles" without spheres (right).

and so we suspected that the spheres were identical with F_1. We were encouraged in this belief when a preparation of pure F_1 turned out to have the characteristic appearance of the 85-angstrom-unit inner-membrane spheres [see micrograph at bottom left on page 2676].

One further experiment was needed to identify F_1 unambiguously with the spheres: the reconstitution of a depleted particle by the addition of F_1, resulting in the restoration of the submitochondrial particle's typical shape and function. This was accomplished only recently, after the development of the SU particles. The addition of F_1 to these particles yielded a preparation that was indistinguishable in structure from fully functional submitochondrial particles [see illustration on page 2676] and confirmed that coupling factor F_1 is identical with the inner-membrane spheres.

The morphological reconstitution was not paralleled by restoration of function, however. In an effort to regain oxidative phosphorylation we added three more coupling factors, F_2, F_3 and F_4—proteins that had been obtained from mitochondria by various extraction procedures. SU particles reconstituted with all four coupling factors oxidized succinate, a compound of the Krebs cycle, with a high efficiency of ATP synthesis: for each molecule of oxygen consumed, up to 1.8 molecules of ATP were formed. That is very close to the best value—two molecules—that can be achieved with intact mitochondria.

With these experiments one of the aims of our investigation had been achieved: a resolution of soluble components and a reconstitution of structure and function. Another aim has been to get some insight into the mode of action of the coupling factors. How do they fit into the mechanism of oxidative phosphorylation?

There are currently two views of the general nature of that mechanism. One is a chemically oriented hypothesis originally suggested by E. C. Slater of the University of Amsterdam in 1953, in analogy to the mechanism of ATP formation in glycolysis. It proposes that during electron transport high-energy intermediate compounds ($A{\sim}x$, $B{\sim}x$, $C{\sim}x$) are formed at each coupling site, composed of a member of the respiratory chain (A, B, C) and an unknown (x). These compounds are transformed into a common intermediate by interaction with another unknown (y) to form $x{\sim}y$. This intermediate in turn combines with inorganic phosphate to yield $x{\sim}P$, which

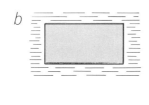
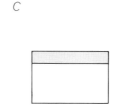

ALLOTOPIC PROPERTIES of purified F_1 and the mitochondrial membrane are indicated. The ATP-ase activity of particles (*a*) was known to be sensitive to oligomycin (*light color*). When soluble F_1 was discovered (*b*), it was found to be resistant to oligomycin (*dark color*), and membrane from which F_1 had been removed (*c*) was sensitive to trypsin (*light gray*). When the enzyme and membrane were bound (*d*), each was changed: the F_1 became sensitive to oligomycin, the membrane resistant to trypsin (*dark gray*).

ultimately transfers its energetic phosphate group to ADP to form ATP.

Recently Peter Mitchell of the Glynn Research Ltd. laboratories in England has challenged this chemical hypothesis with some new and provocative ideas. Instead of a high-energy intermediate compound of the respiratory chain, he proposes that an electrical potential develops during respiration that provides the energy for ATP production: The positively charged hydrogen ions (protons) are moved to one side of the membrane while the negatively charged electrons are channeled to the other side. The separation of charges is utilized by a complex mechanism to give rise to the high-energy intermediate $x{\sim}y$, which powers the formation of ATP. At the core of this hypothesis is an ATP-ase located in the inner membrane.

In some respects the two hypotheses are not much different: both include a high-energy intermediate, $x{\sim}y$, to generate ATP from ADP and phosphate. In the Mitchell hypothesis, however, $x{\sim}y$ is formed by means of an electrical membrane potential. This requires a much higher integrity of the membrane structure than is required by the chemical hypothesis. Indeed, Mitchell considers that uncouplers such as dinitrophenol act by making the membrane "leaky" to protons, thus preventing a separation of charges. It is apparent, therefore, that further studies of the inner membrane are of utmost importance for the evaluation of the two hypotheses.

What is the role of coupling factor according to these two formulations? Mitchell proposes that F_1, together with F_O, represents the reversible ATP-ase that utilizes the electrical potential to generate ATP. According to the chemical hypothesis, F_1 catalyzes the last step in ATP formation, the "transphosphorylation" from $x{\sim}P$ to ADP. Indeed, every reaction associated with oxidative phosphorylation that requires ATP can be shown to be dependent on F_1. June Fessenden-Raden in our laboratory has prepared an antibody against F_1 and has found that these ATP-dependent reactions are inhibited by the antibody.

In collaboration with Mrs. Fessenden-Raden, Richard McCarty and Gottfried Schatz, I have recently found that a coupling factor may have, in addition to a catalytic function that is inhibited by its antibody, a second, "structural" function that is not impaired by the antibody. Our first example was the stimulation by F_1 of a reaction that is catalyzed by mitochondria but does not involve ATP. The second example was the observation that in chloroplasts, the energy-generating particles of plant cells, a coupling factor (chloroplast F_1) is required not only for all reactions that involve ADP or ATP but also for a "proton pump" that is driven by light energy without ATP. In contrast to the ATP-dependent reactions, however, this proton pump was not inhibited by an antibody against the chloroplast coupling factor. The factor therefore appears to contribute to the integrity of the chloroplast membrane, which is required for proton transport.

A third example of the "structural" role of a coupling factor was observed with a preparation of F_1 from yeast mitochondria, which stimulated phosphorylation in beef-heart particles that still contained some residual beef F_1. This stimulation was apparently due to a structural effect of yeast F_1, since it was not inhibited by an antibody against yeast F_1. In beef-heart particles (*SU* particles) that were completely devoid of native F_1, the yeast factor had no effect; apparently it could not fulfill the catalytic functions of native coupling factor.

We are beginning to suspect that the dual role played by F_1 is representative of a common occurrence in the interaction of enzymes and membranes and is an important expression of the allotopy phenomenon. The contribution of F_1 to the integrity of the mitochondrial membrane and the fact that it is required for the operation of the proton pump of the chloroplast have obvious bearing on the Mitchell hypothesis, and may lead to a clarification of the role played by the membrane in oxidative phosphorylation.

While the work with F_1 was going forward as described above, we therefore

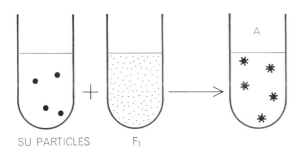
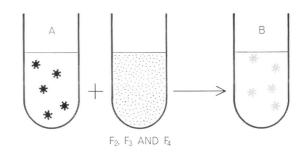

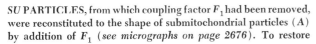

SU PARTICLES, from which coupling factor F_1 had been removed, were reconstituted to the shape of submitochondrial particles (*A*) by addition of F_1 (*see micrographs on page 2676*). To restore function as well as structure it was necessary also to add F_2, F_3 and F_4. When this was done, the reconstituted particles (*B*) were capable of generating ATP almost as well as intact mitochondria.

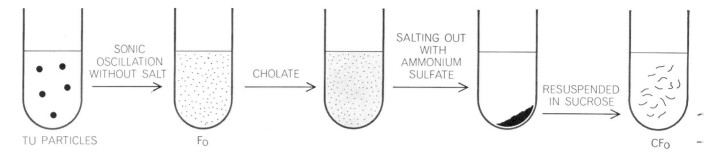

MEMBRANE OF MITOCHONDRION was isolated as shown here. TU particles were subjected to sonic oscillation, producing F_O, which had the capacity to bind F_1. When F_O was dissolved in cho- late and fractionated by "salting out," the colorless precipitate CF_O was obtained. It lacked respiratory enzymes and lipids; added to F_1, it inhibited ATP-ase activity. Addition of phospholipid

also pursued the problem of resolving the inner mitochondrial membrane. F_O had turned out to be a yellow-brown complex of many components including the entire electron-transport chain. By chemical fractionation in the presence of a bile salt, Yasuo Kagawa isolated a vir- tually colorless fraction (CF_O) with some interesting properties. It lacked respira- tory activity, having lost almost all the flavoproteins and cytochromes present in the original submitochondrial particles, and it contained only traces of phospho- lipids, the fat constituents of the mem- brane.

When F_1 was added to CF_O, the ATP- ase activity of F_1 was almost completely inhibited. The subsequent addition of phospholipid to this inactive complex fully restored ATP-ase activity, which

was now sensitive to oligomycin! Equally striking results were observed in the elec- tron microscope. CF_O was amorphous. After the addition of F_1 numerous inner- membrane spheres became attached to CF_O, which remained amorphous. Then, with the addition of phospholipid, the characteristic saclike membranous struc- tures covered with spheres became ap- parent [see illustration below]. They could not be distinguished from func- tional submitochondrial particles, even though they lacked major components of such particles, the respiratory en- zymes. These enzymes had always been assumed to be an integral part of the inner membrane—and now they were found not to be present in what appears to be the isolated membrane. How, then, are the respiratory enzymes associated

with the membrane? What are the con- stituents of the inner membrane itself? How are they organized?

To answer some of these questions we proceeded to disrupt the membrane fur- ther to see which constituents were nec- essary for the interaction with F_1. Sev- eral years ago Thomas E. Conover and Richard L. Prairie in our laboratory had separated a coupling factor, F_4, that was necessary for phosphorylation in particles obtained by sonic oscillation of mitochondria under highly alkaline con- ditions. When Kagawa exposed TU par- ticles to sonic oscillation under alkaline conditions, after high-speed centrifuga- tion he obtained a sediment ("TUA par- ticles") that no longer made F_1 sensitive to oligomycin. The addition of F_4 to TUA particles restored oligomycin sen-

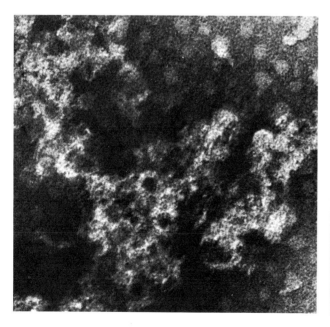

CF_O FRACTION, enlarged about 250,000 diameters in an electron micrograph, appears amorphous (left). When F_1 is added to the CF_O preparation, it appears that the F_1 spheres attach themselves to the CF_O, but no distinct structure is seen (center). When phospho- lipids are added, the distinct structure that emerges (right) re- sembles that of submitochondrial particles. In other words, CF_O

Racker / THE MEMBRANE OF THE MITOCHONDRION 2675

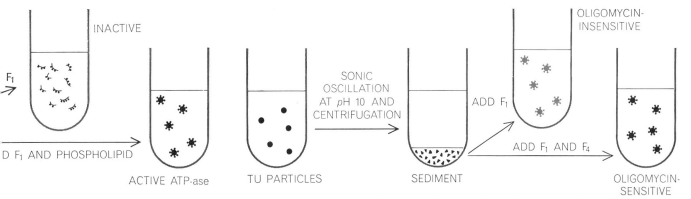

restored ATP-ase activity, which was now oligomycin-sensitive. CF_O and phospholipid may thus comprise the membrane proper.

COUPLING FACTOR F_4 is apparently required for oligomycin sensitivity. When *TU* particles are broken down under alkaline conditions, the sedimented particles cannot confer oligomycin sensitivity on F_1. Addition of F_4 in the presence of salt restores this capacity.

sitivity to the complex. Recent experiments by Bernard Bulos in our laboratory at Cornell University revealed that F_1 is bound by *TUA* particles in the presence of salt but is nevertheless not inhibited by oligomycin. On addition of small amounts of a highly purified preparation of F_4, he observed a time-dependent restoration of oligomycin sensitivity, suggesting that an enzymatic process may be taking place. This is the first clue to the mode of action of F_4.

Several years ago Richard S. Criddle, Stephen H. Richardson and their collaborators at the University of Wisconsin isolated an insoluble "structural protein" from mitochondria with the help of detergents and solvents. Our crude preparation of F_4 was similar to that protein in its capacity to combine with some flavo-

proteins and cytochromes of the respiratory chain but, not having been exposed to damaging chemicals, it remained soluble. In the electron microscope it appeared quite amorphous. When phospholipids were added to soluble F_4, however, a precipitate formed that appeared to be membranous and shaped into sacs [*see illustration at right below*]. We have therefore proposed that F_4 may have a function as an organizational protein in the mitochondrial membrane—a kind of backbone for the association of the respiratory enzymes and the coupling factors involved in the transformation of oxidative energy into ATP.

With this concept as a working hypothesis we have embarked on what promises to be a long and venturesome

journey. Taking the membrane-like complex of F_4 and phospholipid as starting material, we are adding isolated soluble flavoproteins and cytochromes of the respiratory chain step by step, checking at each stage to see if some of the allotopic properties of the respiratory chain are restored. In our laboratory Alessandro Bruni and Satoshi Yamashita have constituted sections of the respiratory chain with the appropriate allotopic properties (such as sensitivity to the respiratory poison antimycin). These experiments have given us confidence that we shall eventually achieve a complete reconstitution of the respiratory chain from soluble components. Then we shall turn to the final task: the reconstitution of the system of oxidative phosphorylation from its individual components.

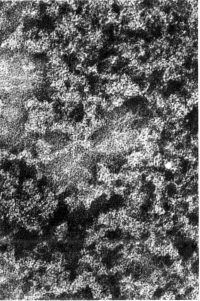
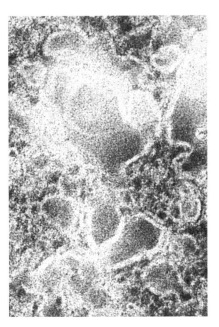

and phospholipid seem to suffice, without respiratory enzymes, to bind F_1 and reconstitute the shape of submitochondrial particles.

F_4 PREPARATION, seen in an electron micrograph at a magnification of 250,000 diameters, appears to be amorphous (*left*). The addition of phospholipid to soluble F_4 yields particles with a sac-shaped structure similar to that of submitochondrial particles (*right*).

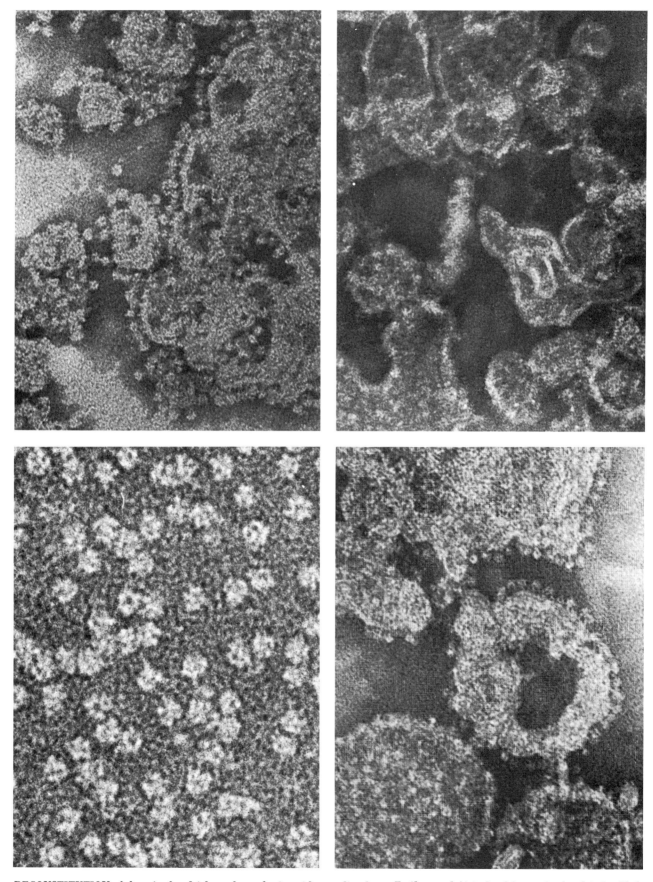

RECONSTITUTION of the mitochondrial membrane begins with submitochondrial particles lined with inner-membrane spheres (*top left*). Treatment with a molecular sieve (Sephadex) and urea produces "SU particles" without spheres (*top right*). When cou-pling factor F_1 (*bottom left*) isolated from mitochondria is added, the characteristic shape of submitochondrial particles is restored (*bottom right*). F_1 spheres are enlarged about 600,000 diameters, other preparations about 300,000 in these electron micrographs.

The Author

EFRAIM RACKER is Albert Einstein Professor of Biochemistry at Cornell University and chairman of the Section of Biochemistry and Molecular Biology in the university's Division of Biological Sciences. Born in Poland, Racker received an M.D. degree from the University of Vienna in 1938. From 1938 to 1940 he was research assistant in biochemistry at the Cardiff Mental Hospital in Wales. He then spent a year as research associate in the department of physiology at the University of Minnesota and two years as an intern and research fellow in pneumonia at Harlem Hospital in New York City. From 1943 to 1952 he taught microbiology at the New York University College of Medicine. In 1947 he became a U.S. citizen. From 1952 to 1954 he was associate professor of biochemistry at the Yale University School of Medicine. He then served on the staff of the Public Health Research Institute of the City of New York until he went to Cornell in 1966.

Bibliography

THE ENERGY-LINKED FUNCTION OF MITOCHONDRIA. Edited by Britton Chance. Academic Press Inc., 1963.

MITOCHONDRIAL OXIDATIONS AND ENERGY COUPLING. Maynard E. Pullman and Gottfried Schatz in *Annual Review of Biochemistry*, Vol. 36, pages 539–610; 1967.

PARTIAL RESOLUTION OF THE ENZYMES CATALIZING OXIDATIVE PHOSPHORYLATION, X: CORRELATION OF MORPHOLOGY AND FUNCTION IN SUBMITOCHONDRIAL PARTICLES. Yasuo Kagawa and Efraim Racker with Rolf E. Hauser in *The Journal of Biological Chemistry*, Vol. 241, No. 10, pages 2475–2484; May 25, 1966.

SYMPOSIUM ON ENERGY COUPLING IN ELECTRON TRANSPORT. Albert L. Lehninger, Efraim Racker, Britton Chance, Chuan-Pu Lee, Leena Mela, S. N. Grazen, S. Estrada-O, Henry A. Lardy, A. T. Jagendorf and P. Mitchell in *Federation Proceedings*, Vol. 26, No. 5, pages 1333–1334; September–October, 1967.

SCIENTIFIC
AMERICAN   February 1968, Vol. 218, No. 2, pp. 108-116

OFFPRINT 1102

THE MIGRATION OF POLAR BEARS

by Vagn Flyger and Marjorie R. Townsend

These large carnivores travel widely in pursuit of their prey. Exactly how widely is unknown, but steps are being taken to find out with the help of artificial satellites.

The polar bear, the largest non-aquatic carnivore, has to roam widely over its Arctic habitat in quest of its principal food, which is seal meat. Few human investigators are inclined to spend much time following polar bears in such a forbidding environment. Even if they were, their presence in an otherwise virtually barren landscape would be likely to disturb the bears and alter their normal pattern of behavior.

As a result remarkably little is known about these lumberingly graceful animals. Eskimos have provided some information, which is partly fact and partly folklore. Scattered accounts have come from hunters and members of Arctic expeditions. By tracking and other means it has been ascertained that an individual bear may move hundreds of miles, but this kind of information reveals little about the activities of the animal during the journey.

Now, however, a technique is at hand that promises to fill some of the gaps in our knowledge of what kind of life the polar bear leads. The technique involves following a number of bears over an extended period by means of an artificial satellite in a polar orbit. If present plans are realized, a program of tracking polar bears will be incorporated in one of the Nimbus satellites to be launched by the National Aeronautics and Space Administration within the next two years.

The information that the satellites can be expected to yield will be of both scientific and economic value. Polar bears, which apparently have evolved from the brown bears of the Temperate Zone forests, represent a remarkable instance of adaptation to a singular environment. They also represent a significant element in the subsistence economy of Arctic regions, not only as a source of meat and

skins for Eskimos but also as an attraction to hunters, who pay substantial amounts of money for guide fees and air travel in pursuit of the bear. The U.S. Department of the Interior has estimated that "each polar bear harvested in Alaska at the present time contributes at least $1,500 to the economy of the state in one way or another," so that the harvest of about 300 bears in Alaska in 1965 meant an income of about $450,000, much of which was "expended in relatively small Arctic villages."

A harvested polar bear is of course a dead one. No one knows with certainty how many of the bears are killed by hunters from various nations, but in the light of increasingly efficient techniques for hunting bears there is concern about the survival of the species. The hunting techniques include the use of small, ski-equipped airplanes that fly out over the ice in pairs. In Norway hunters can charter a yacht and travel along the edge of the ice pack near the Svalbard Islands shooting bears from the deck of the ship. As a result of pressures of this kind the International Union for the Conservation of Nature and Natural Resources has put polar bears on its list of animals in danger of extinction.

For both economic and scientific reasons, therefore, it is important to establish the trend of population among polar bears. Present estimates of the bear population range from 5,000 to 20,000 (another indication of the state of knowledge about the bears). Satellites will not count bears, but by providing accurate data about their migratory habits they will make counts more meaningful.

A vivid description of the environment to which polar bears have adapted was provided in 1856 by one of the early Arctic explorers, Elisha

Kent Kane of the U.S. He wrote of a scene during the long polar night: "Intense moonlight, glittering on every crag and spire, tracing the outline of the background with contrasted brightness, and printing its fantastic profiles on the snowfields. It is a landscape such as Milton or Dante might imagine—inorganic, desolate, mysterious. I have come down from deck with the feelings of a man who has looked upon a world unfinished by the hand of its creator."

Polar bears are abroad in this formidable environment during the entire year—even in the dead of winter, when the temperature can drop as low as 65 degrees below zero Fahrenheit. Unlike other mammals, the polar bear appears to have no home territory. The apparent reason is that the bear does not find its food in one place for very long.

The bears do occasionally seek shelter. Females in particular excavate dens in the snow to give birth and rear their young. A female and her cubs (usually two) will stay in the den for as long as five months. The cubs survive by nursing, and the female lives off her reserve of fat.

Emerging from the den in March or April, the mother leads her cubs down to the sea ice. During the journey the cubs play constantly, tumbling, wrestling with one another and sliding on the ice. C. R. Harington of the National Museum of Canada has described what he has seen such a group do after it reaches the sea ice:

"If we watch the group closely for a few hours during early April, we will observe the mother prowling, head down, along the drifted leeward margin of some hummocky ice. Catching the scent of a snow-covered seal den, she crouches motionless before it—the cubs behind following her example. With lightning-

IMMOBILIZED POLAR BEAR is marked by one of the authors (Flyger) with a long-lasting dye. The dye helps in the identification of the bear when the animal is seen in subsequent months and so provides an indication of the extent of the animal's migration. The bear was on the Arctic ice cap north of Point Barrow, Alaska. Flyger and his colleagues immobilized the animal by shooting it with a syringe filled with succinylcholine chloride, a drug that acts quickly on the voluntary muscles and wears off in about 30 minutes.

RECOVERY OF BEAR shown in the top illustration on this page took place approximately 30 minutes after the immobilizing drug had taken effect. The bear was marked not only with the dye but also with a tag on its left ear. Drug appears to leave no aftereffects.

like blows of her paws she shatters the hard upper layer of snow, rises on her hind legs and drives both forelegs down with the entire weight of her body. The den collapses and the breathing hole is stopped with snow. She scoops out the young 'whitecoat' seal within—almost simultaneously dispatching it."

Except when a mother is traveling with her young and during the mating season, the polar bear is a solitary animal, seldom seen in the company of other bears unless a large supply of food is available. In such a case a number of bears may appear; once as many as 42 were counted in the vicinity of a dead whale. Bears that are really hungry—and it is sometimes a long while between meals—have been known to kill and eat other bears, which is perhaps one reason the bears tend to avoid one another's company.

A mature male polar bear is likely to be from eight to nine feet long and to weigh between 800 and 1,000 pounds. Bigger bears have been reported, not always reliably. A mature female will be perhaps six feet long and will weigh about 700 pounds. The normal color of a mature bear is a creamy or yellowish white; in some kinds of light the bear looks more yellow than white.

The normal gait of a polar bear is a somewhat ponderous shuffle that covers ground much more rapidly than one might expect. When pressed, a bear will break into a lope or a gallop; at top speed the animal can travel over rough ice at a speed estimated as 20 to 25 miles

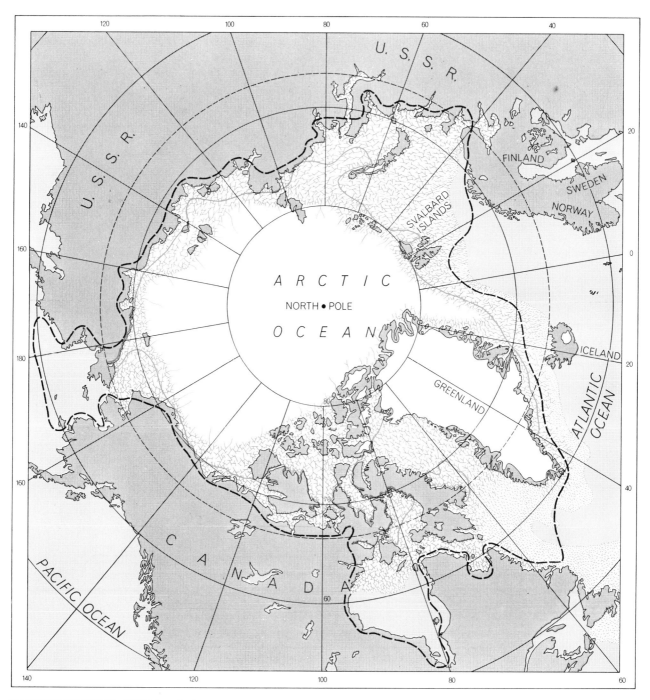

HABITAT OF POLAR BEARS is the large ice cap surrounding the Arctic pole. The map shows the extent of the winter ice pack and, with a colored line, the normal limits of the summer ice. The broken line shows the approximate maximum range of polar bears.

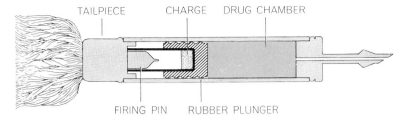

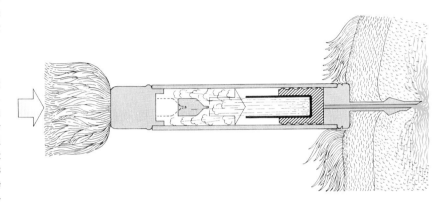

EXPLOSIVE SYRINGE is used to immobilize bears. At top the syringe appears as it looks before it is fired from a special rifle. At bottom the syringe has struck a bear. The resulting inertia drives the firing pin forward into the charge, which explodes and forces the drug into the bear. The amount of drug that is used depends on the size of the bear, so that an appropriately loaded syringe cannot be put into a rifle until a bear has been spotted.

an hour. The bear also spends a considerable amount of time swimming, particularly in summer when the ice is broken over wide areas. A bear moving from ice into water seems hardly to break its stride, entering the water with a great belly flop and moving off rapidly with a strong swimming stroke. It became known only recently, through underwater photographs made under the supervision of Martin W. Schein of Pennsylvania State University, that a swimming bear normally uses only its front legs for propulsion; the hind legs appear to serve as rudders [*see illustration on page 2683*].

Such is the kind of information about polar bears that has been accumulated through occasional (and seldom very lengthy) observations. If, as is now proposed, some bears are to be followed more systematically by satellite, the first task to be mastered is equipping bears so that the satellite can communicate with them. The task requires immobilizing a bear long enough so that a unit for receiving and transmitting radio signals can be fastened to its neck. This must be done in such a way that the bear recovers quickly and is able to continue its normal activities while it is being followed by the satellite.

One of us (Flyger) has been experimenting with techniques for immobilizing polar bears with drugs. The drug is injected by means of a syringe that is shot from close range. When the syringe strikes the bear, the inertia of the projectile activates a firing mechanism that injects the drug into the animal [*see illustration on this page*]. The work presents several problems, not the least of which are the hostility of the environment and the potential hostility of the bear.

The kind of drug used to immobilize a bear must have several characteristics besides those of taking effect rapidly and wearing off within a matter of hours. First, there must be a wide margin of safety between the effective dose and the lethal dose. Second, the drug must be so potent that it can be contained in no more than 10 milliliters of fluid; 10 milliliters is the capacity of the largest projectile syringe. Third, the drug must produce no serious aftereffects. Fourth, the drug should be one for which an antidote can be given if necessary.

Few drugs have these characteristics. The ones most commonly used for immobilizing various animals have been nicotine alkaloid, succinylcholine chloride, phencyclidine and the synthetic opiate M. 99. Nicotine alkaloid acts on nerve ganglia and causes immobilization lasting from 10 minutes to an hour. During that time the animal is completely anesthetized.

Succinylcholine chloride causes immobilization by depolarizing voluntary muscles so that they are unable to function. The animal is immobile for between five and 30 minutes but retains such faculties as hearing, sight and smell. Phencyclidine affects the ganglia and the central nervous system, immobilizing an animal for periods of one to 12 hours. M. 99 takes effect slowly, sometimes requiring 20 minutes to immobilize an animal, but then the animal is completely anesthetized and immobile for as long as eight hours.

In March, 1966, Schein and one of us (Flyger) went to Point Barrow, Alaska, to develop methods of capturing and marking polar bears. The Arctic Research Laboratory of the University of Alaska put its facilities, including two aircraft, at our disposal, and on days when the weather permitted we flew over the polar ice cap searching for bears. One plane flew at an altitude of 100 feet, looking for polar-bear tracks. We were in the other plane, flying at an altitude of 500 feet and slightly behind the lower plane so that it could maneuver freely.

When we found tracks, we followed them to the bear. Our plane then flew on ahead in the direction of the bear's travel while the pilot of the other plane attempted to judge the size of the bear and notified us of it by radio so that we could load the appropriate charge into the syringe. Our plane landed about two miles beyond the bear. We got out and walked to a ridge of ice while our plane took off again and joined the other plane in herding the bear toward us. When the animal was within range, we shot it in the hip or shoulder with the syringe.

The next minute or so always gave us some concern. Our syringe guns, which we had expected to have a range of about 70 yards, actually had a maximum range of about 40 yards, apparently because the intense cold had congealed the lubricant in the gun barrel. We had several narrow escapes from charging bears. Although both of us carried rifles, the man who had fired the syringe gun had very little time to unlimber his rifle against the charging bear, and the other man often could not shoot without endangering his companion.

We were using succinylcholine chlo-

SWIMMING ABILITY of polar bears is exhibited by a bear diving from an ice floe as a research vessel draws near in Diskobukta Bay of the Svalbard Islands of Norway. The vessel was used as a base for temporarily immobilizing bears with drugs and marking them.

YOUNG BEAR is almost immobilized after being injected with a drug from the syringe visible in the bear's left flank. The rope is to keep the bear from drowning while drugged.

ride, and the results were disappointing. Some bears were unaffected. Others died, apparently from a combination of factors. One factor was that it is difficult to estimate the size of a bear in the absence of reference points on the ice, so that we often had an overcharge of the drug. Another factor was that the bears, having been chased for some distance, were rather out of breath when we shot them. As a result the drug seemed to have an exaggerated effect on their respiratory muscles, causing death from suffocation.

The difficulties we encountered both in hunting the bears and in immobilizing them led us to the conclusion that the ideal platform from which to shoot the syringe would be a helicopter. A helicopter could hover over a bear long enough for the hunters to make an accurate estimate of the bear's size, so that they could load the syringe with the proper dose. Helicopters would make it unnecessary to drive the bears, with the result that they would not become winded. The helicopter would also have the obvious advantage of keeping the men out of the bear's reach until the animal was immobilized.

Our experience in Alaska was by no means a total failure. We did manage to capture, mark and release some bears. We learned that the bears can be successfully marked with ear tags and long-lasting dyes. We also established that a collar can be attached firmly to a polar bear, which is a prerequisite to equipping the animal with a radio transmitter. The point had been in doubt because of the bear's tapering neck, which might have meant that the bear could slip out of a collar in a short time.

Later in 1966 one of us (Flyger) accompanied Albert W. Erickson of the University of Minnesota and Thor Larsen of the University of Oslo on an expedition to capture bears in the Svalbard Islands. This time we operated under summer conditions, working from a ship in the pack ice. With the drug M. 99 we immobilized, marked, measured and released four bears. The drug did leave the bears sleepy for a long time, but that is not a serious handicap for the animal in the summer.

In various expeditions by various investigators, about 100 bears have thus been captured, marked and released. The work seems to establish that it will be feasible to put a radio transmitter on a bear. Indeed, it has already been done with grizzly bears and other Temperate Zone animals, which have then

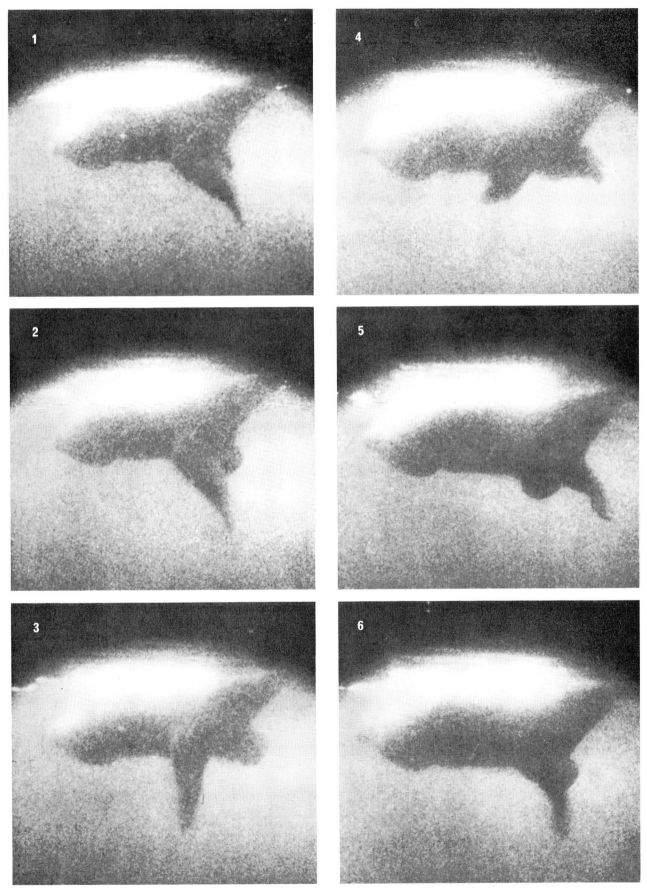

SWIMMING MOTION of a polar bear normally involves only the front legs, as seen in this series of photographs from a motion-picture film made underwater in the Svalbard Islands region. The photographer was in a cage that was lowered over the side of a ship. The rear legs of a swimming bear apparently function as rudders. In these photographs only the bear's head is above water.

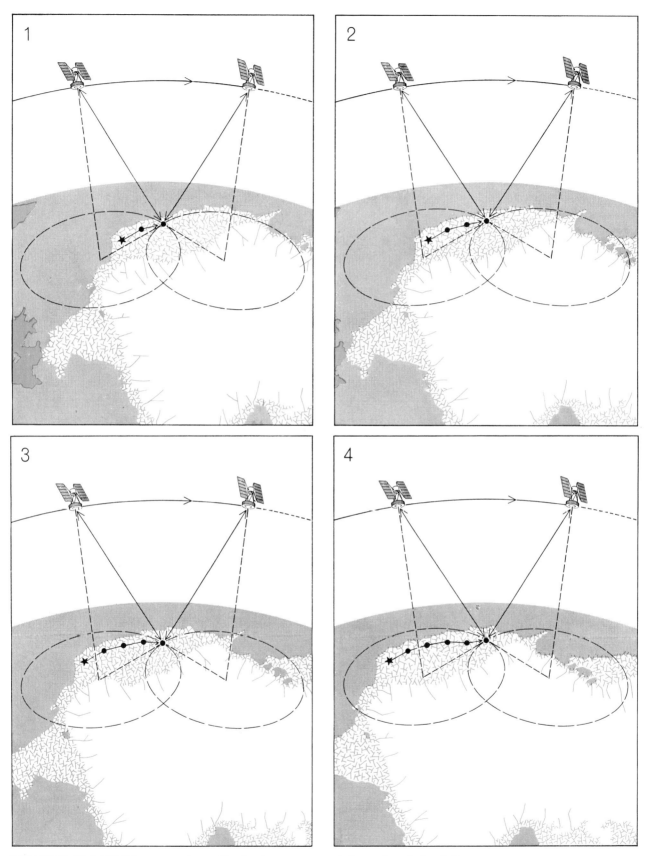

LOCATION OF A BEAR by means of a satellite involves an exchange of radio signals between the satellite and the bear. The position of the satellite is known and the distance between the satellite and the bear is established by the time required for the exchange of radio signals. Thus two sides of a right triangle are known and the third side, which forms the radius of an imaginary circle, can be calculated (1). The bear is on the circumference of the circle. Another exchange of signals a few minutes later establishes another circle. Bear is at one of the two intersections of the circles. Since one knows where it was when caught (star) or last reached by the satellite (dots), the proper intersection can be identified. The bear can thus be followed for many days (2–4).

been followed by ground-based radio receivers [*see illustration on next page*].

No polar bear has actually been fitted with a transmitter, however, because some technological problems remain to be overcome. Most electric batteries cannot produce enough current to operate a transmitter at the low temperatures of the polar bear's environment. Moreover, the equipment must be able to withstand frequent immersion as the bear goes in and out of the water. It nonetheless seems probable that these difficulties will be overcome before long, so that polar bears can be equipped to communicate with a satellite.

The Nimbus satellite, which will be used primarily for oceanography and meteorology, is being designed to orbit the earth from pole to pole at an altitude of about 600 miles. During a single 24-hour rotation of the earth the satellite will pass over every point on the earth twice—once as it moves from south to north and again, 12 hours later, as it moves from north to south and the point has rotated to that side of the orbit. At the higher latitudes the satellite will actually be within radio range of a given point on several sequential orbits. As a result the satellite could provide the location of a polar bear every two hours for half a day. For the other half the bear would be out of communication.

The system being developed by NASA to follow bears and other moving objects is called the Interrogation, Recording and Location System. In its ultimate form it will enable the satellite to interrogate some 32,000 "platforms" (bears, ocean buoys, balloons and a variety of other units), accurately fixing their position at least twice a day and recording other data they transmit.

Let us consider the case of a single polar bear that has been instrumented to communicate with a satellite. At a specified time, when the satellite is expected to be over the bear, the satellite broadcasts a signal that will turn on the bear's transmitter if the animal is within the signal beam. Once the transmitter is on, it returns a signal to the satellite. Because the position of the satellite is known at all times, and the distance between the satellite and the bear can be computed from the length of time it takes the radio signals to travel between them, it is possible to calculate the position of the bear by triangulation after two communications with the animal about three minutes apart. The location data and any other information the bear's communication system has been programmed to

DUMMY RADIO is fastened to a collar on an immobilized bear in Alaska. The dummy matches size and weight of a radio that could be used for communication between a bear and a satellite. Experiment helped to show that a collar will stay on a bear's tapering neck.

MATERNITY DEN of a polar bear was photographed at Southampton Island of Canada's Northwest Territories by C. R. Harington of the National Museum of Canada. The bear that excavated the den in snow used it to give birth and to rear her cubs during their first winter.

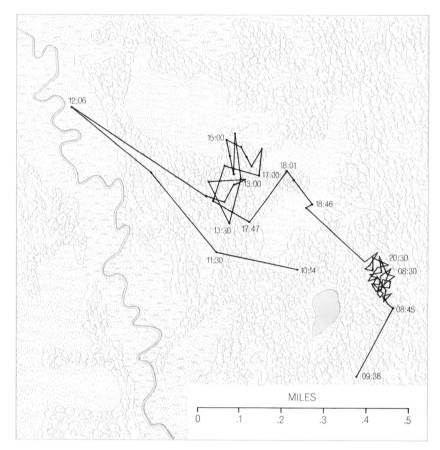

MILES

| 0 | .1 | .2 | .3 | .4 | .5 |

TRACKING OF DEER and other animals has been accomplished by means of radio communication between the animals and ground-based receivers, indicating the feasibility of following polar bears with a satellite. The map, which is based on the work of John R. Tester of the University of Minnesota and Keith L. Heezen of the Michigan Conservation Department, shows the movements of a single deer in the Cedar Creek Natural History Area of Minnesota from 10:14 A.M. on a January day to 9:38 A.M. the following day.

provide are stored in the satellite and periodically sent by it to ground stations, from which they are transmitted to a computer for processing.

Step by step the sequence proceeds as follows. The set of interrogations that the satellite will conduct on a given orbit is worked out by the computer, and a punched tape is prepared so that the entire set can be transmitted to the satellite when it passes over its command station. A command for a single interrogation has two parts: (1) the address, or radio code, that has been assigned to the platform (in this case a bear) and (2) the exact orbital time at which the interrogation is to be conducted.

At the stipulated time the satellite begins transmitting the address. If the bear is within the range of the signal, the radio unit on the bear receives the signal, decodes it for comparison with the address assigned to the bear and, if the signal corresponds to the address, verifies the platform's identity to the satellite. The satellite then signals the platform to begin transmitting whatever data it is programmed to provide. In the case of a bear the data would presumably be physiological. The entire exchange takes about three seconds.

The results of all the interrogations are stored in the satellite. On completion of an orbit the ground station commands the satellite to transmit the data it has stored. When the satellite has sent all the data, it automatically signals the ground station to that effect. The satellite is then ready to accept a new set of interrogations.

The data received at the ground station are given some further processing there and then transmitted to a central data-processing unit at NASA's Goddard Space Flight Center in Maryland. There the location of each platform is computed and the information obtained from the platform is distributed to the people who want it. The Goddard computer also generates the next set of interrogations for the satellite.

So much for the interrogation and recording functions of the Nimbus system. The location function is predicated on the fact that the location of the satellite is precisely known at all times. Hence the distance from the satellite to the point on the ground directly below the satellite is known at any instant. That distance represents one side of a right triangle.

The hypotenuse of the triangle is the line from the satellite to the platform. Its length can be computed from the time required for the special signal generated in the satellite to reach the platform, be regenerated and return to the satellite—traveling at the speed of light. With the length of two sides of the right triangle known, the length of the third side can be calculated. The third side is the hypothetical line on the ground from the platform to the point directly below the satellite. That line constitutes the radius of a circle. The complete calculation establishes that the bear is somewhere on the circumference of the circle [see illustration on page 2684].

In order to establish exactly where the animal is, another exchange of signals between the satellite and the bear is required. It takes place about three minutes after the first exchange. By means of the second exchange another imaginary circle, with the bear on the circumference, is described on the ground. The two circles intersect at two places. The bear is at one of these places. Since one knows where the bear was during the satellite's previous orbit (or where the animal was captured, in the case of the first orbit in which its location was determined), it is not hard to decide which intersection represents the location of the bear; the two intersections are normally hundreds of miles apart. The precision of the technique is sufficient to determine the location of the bear to within about a mile.

It will be well within the capacity of a satellite system to provide information not only about the migration of polar bears but also on such matters as their respiration and heart action under various circumstances. With such data it would be possible to relate the activity of polar bears to weather conditions on the polar ice cap and thus to come to a better understanding of how an animal can survive in the harsh conditions regularly encountered by the polar bear. Perhaps some of this information will be useful to men who are obliged to adapt themselves to living under Arctic conditions.

The Authors

VAGN FLYGER and MARJORIE R. TOWNSEND are respectively at the University of Maryland and the Goddard Space Flight Center of the National Aeronautics and Space Administration. Flyger is a research professor and head of the Inland Resources Division of the university's Natural Resources Institute. Born in Denmark, he was graduated from Cornell University in 1948 with a bachelor's degree in zoology and entomology. In 1952 he received a master's degree in zoology and forestry at Pennsylvania State University. Four years later he obtained an Sc.D. in vertebrate ecology from Johns Hopkins University. He has worked for the state of Maryland in various capacities since 1948, going to the Natural Resources Institute in 1962. Besides his work on polar bears he has studied intensively the habits of the gray squirrel and, on a visit to Antarctica in 1963, the habits of penguins and seals. Mrs. Townsend is manager of the small-astronomy-satellite project at Goddard. The project is developing a series of satellites that will map the sky for X-ray sources and make studies in other regions of the electromagnetic spectrum. Earlier she was involved in the development of the satellite communication system described in the present article. Mrs. Townsend was graduated from George Washington University in 1951 with a degree in electrical engineering. She is the wife of an obstetrician in Washington, D.C., and the mother of four boys.

Bibliography

THE IMMOBILIZATION OF CAPTIVE WILD ANIMALS WITH SUCCINYLCHOLINE: II. Warren R. Pistey and James F. Wright in *Canadian Journal of Comparative Medicine and Veterinary Science*, Vol. 25, No. 3, pages 59–68; March, 1961.

THE POLAR BEAR: A MATTER FOR INTERNATIONAL CONCERN. Vagn Flyger in *Arctic*, Vol. 20, No. 3, pages 147–153; September, 1967.

PROCEEDINGS OF THE FIRST INTERNATIONAL SCIENTIFIC MEETING ON THE POLAR BEAR. U.S. Department of the Interior and The University of Alaska, 1966.

THE WORLD OF THE POLAR BEAR. Richard Perry. Cassell & Company Ltd; 1966.

SCIENTIFIC
AMERICAN   March 1968, Vol. 218, No. 3, pp. 32-37   OFFPRINT **1103**

HUMAN CELLS AND AGING

by Leonard Hayflick

When normal cells are grown outside the body, their capacity to survive dwindles after a period of time. This deterioration may well represent aging and an ultimate limit to the span of life.

The common impression that modern medicine has lengthened the human life-span is not supported by either vital statistics or biological evidence. To be sure, the 20th-century advances in control of infectious diseases and of certain other causes of death have improved the longevity of the human population as a whole. These accomplishments in medicine and public health, however, have merely extended the *average* life expectancy by allowing more people to reach the upper limit, which for the general run of mankind still seems to be approximately the Biblical fourscore years. Aging and a limited life-span apparently are characteristic of all animals that stop growing after reaching a fixed, mature size. In the case of man, after the age of 30 there is a steady, inexorable increase in the probability of death from one cause or another; the probability doubles about every eight years as one grows older. This general probability is such that even if the major causes of death in old age—heart disease, stroke and cancer—were eliminated, the average life expectancy would not be lengthened by much more than 10 years. It would then be about 80 years instead of the expectancy of about 70 years that now prevails in advanced countries.

Could man's life-span be extended— or is there an inescapable aging mechanism that restricts human longevity to the present apparent limit? Until recently few biologists ventured to attempt to explore the basic processes of aging; obviously the subject does not easily lend itself to detailed study. It is now receiving considerable attention, however, in a number of laboratories [see "The Physiology of Aging," by Nathan W. Shock; SCIENTIFIC AMERICAN, January, 1962]. In this article I shall discuss some new findings at the cellular level.

No doubt many mechanisms are involved in the aging of the body. At the cell level at least three aging processes are under investigation. One is a possible decline in the functional efficiency of nondividing, highly specialized cells, such as nerve and muscle cells. Another is the progressive stiffening with age of the structural protein collagen, which constitutes more than a third of all the body protein and serves as the general binding substance of the skin, muscular and vascular systems [see "The Aging of Collagen," by Frederic Verzár; SCIENTIFIC AMERICAN Offprint 155]. In our own laboratory at the Wistar Institute we have addressed ourselves to a third question: the limitation on cell division. Our studies have focused particularly on the structural cells called fibroblasts, which produce collagen and fibrin. These cells, like certain other "blast" cells, go on dividing in the adult body. We set out to determine whether human fibroblasts in a cell culture could divide indefinitely or had only a finite capacity for doing so.

Alexis Carrel's famous experiments more than a generation ago suggested that animal cells per se (that is, cells removed from the body's regulatory mechanisms) might be immortal. He apparently succeeded in keeping chick fibroblasts growing and multiplying in glass vessels for more than 30 years—a great deal longer than a chicken's life expectancy. Later experimenters reported similar successes with embryonic cells from laboratory mice. It has since been learned, however, through improved techniques and a better understanding of cell cultures, that the conclusions drawn from those early experiments were erroneous.

In the case of chick fibroblasts it has been repeatedly demonstrated that, if care is taken not to add any living cells to the initial population in the glass vessel, the cell colony will not survive long. The early cultures, including Carrel's, were fed a crude extract taken from chick embryos, and it is now believed these feedings must have contained some living chick cells. That is to say, in all probability the reason the cultures continued to grow indefinitely was that new, viable fibroblasts were introduced into the culture at each feeding.

Restudy of the experiments in culturing mouse cells has brought to light a highly interesting fact. It has been found that when normal cells from a laboratory mouse are cultured in a glass vessel, they frequently undergo a spontaneous transformation that enables them to divide and multiply indefinitely. This type of transformation takes place regularly in cultures of mouse cells but only rarely in cultures of the fibroblasts of man or other animals. These transformed cell populations have several abnormal properties, but they are truly immortal: many of the mouse-derived cultures have survived for decades. Similarly, the famous line of transformed human cells called HeLa, originally derived from cervical tissue in 1952 by George O. Gey of the Johns Hopkins University School of Medicine, is still growing and multiplying in glass cultures.

On microscopic examination the transformed cells show themselves to be indeed abnormal. Instead of the normal number of 46 chromosomes in a human diploid cell, the "mixoploid" HeLa cells may have anywhere from 50 to 350 chromosomes per cell. They differ from normal chromosomes in size and shape and also stain differently. Moreover, they often behave like cancer cells: inoculated into a suitable laboratory animal, they can grow as tumors. This property of transformed cells has become an important tool in investigations of the

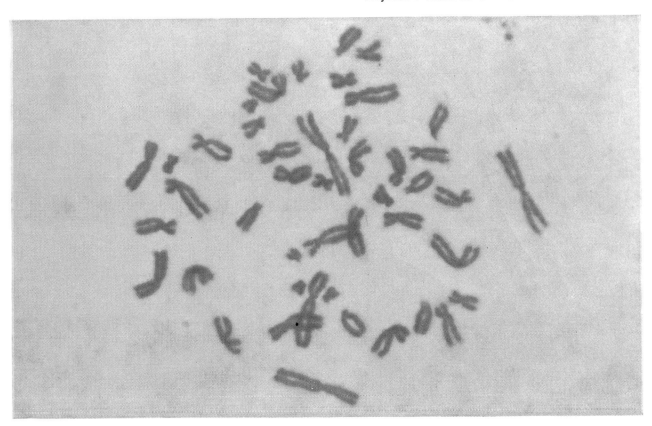

HUMAN CHROMOSOMES are seen magnified 3,000 diameters in a normal diploid cell grown in culture. The chromosomes assume this compact form during mitotic cell division. When grown in cul- ture, human cells display a limited capacity to divide; their finite longevity may be related to the finite span of human life. The chromosomes in the picture are stained with a dye called Giemsa.

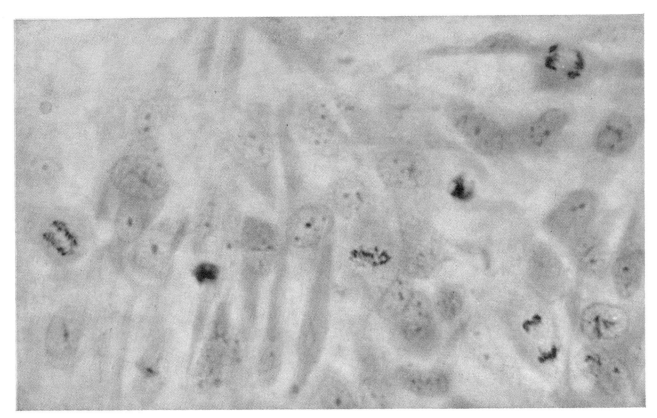

STAGES OF CELL DIVISION appear in this photomicrograph of human cells. The chromosomes are seen as dark clusters. Those in the nucleus at the center have not begun to divide. At left is a nucleus in which the chromosome cluster has separated into two parts; at right, top and bottom, are two nuclei in a later stage of the division cycle. The culture is stained with aceto-orcein and green counterstain. The magnification is 750 diameters. Both photomicrographs were made by Paul S. Moorhead of the Wistar Institute.

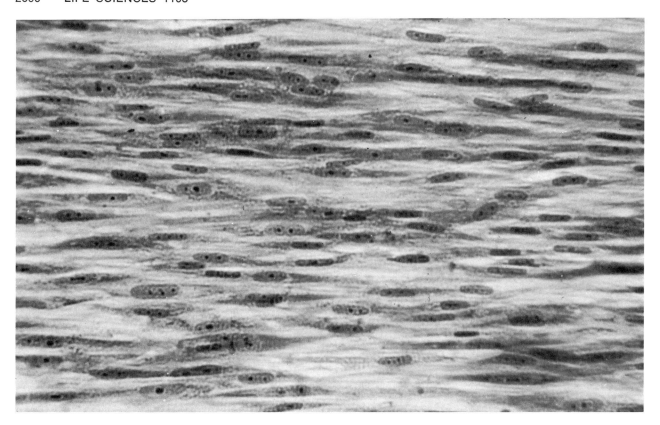

GROWING HUMAN CELLS cover the surface of the glass vessel in which they were planted. They form a layer one cell deep. Under proper conditions this normal cell population will continue to proliferate; it will not, except in unusual cases, multiply indefinitely.

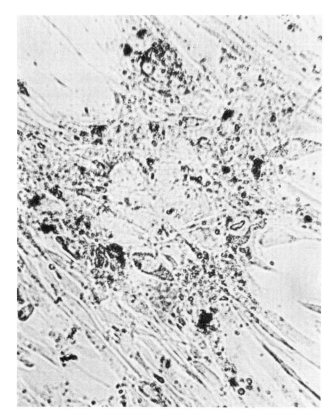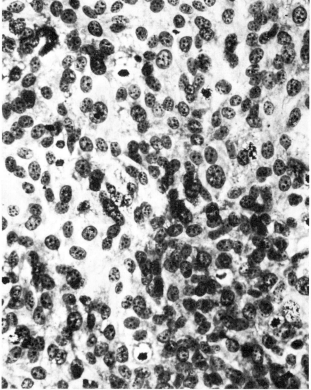

AGED HUMAN CELLS are irregular in appearance; they no longer divide. The aging of such normal cell populations is apparently due to an intrinsic process, not a deficiency in growing conditions. The cells were taken from lung tissue and grown in glass. This photomicrograph and the one at top were made by the author.

IMMORTAL HUMAN CELLS appear in this photomicrograph made by Fred C. Jensen of the Wistar Institute. The cells, once normal like those at top, have been treated with the monkey virus SV-40; thus transformed, they can apparently multiply indefinitely. The magnification in these photomicrographs is 300 diameters.

genesis of cancer. Although the spontaneous transformation of human cells is rare, investigators of cancer are making use of the recent discovery that normal human cells can be routinely transformed into cancer cells by exposing them to the monkey virus known as SV-40.

A crucial consideration for the relevance of cells in culture to aging, of course, is that they are normal cells. Our interest is therefore directed not to abnormal cells but to the observation that normal cells do not divide indefinitely. Whereas a population of transformed cells will proliferate and survive for decades in cell culture, no one has succeeded in perpetuating a culture of normal animal cells. The same is true of cells implanted in a living animal. Transformed cells will go on growing indefinitely in a series of tissue transplants from animal to animal, but normal cells will not. Peter L. Krohn of the University of Birmingham has shown, for example, that normal mouse skin can survive only a limited number of serial grafts from one mouse to another in the same inbred strain.

Over the past seven years in our laboratory Paul S. Moorhead and I have been studying cell cultures of normal human fibroblasts. Unlike highly specialized cells, fibroblasts (which serve as the structural bricks for most body tissues) will grow and multiply in a nutrient medium in glass bottles. We have used lung tissue as our principal source of the cells. We break down the tissue into separated cells by means of the digestive enzyme trypsin, then remove the trypsin by centrifugation and seed the cells in a bottle containing a suitable growing medium. After a few days of incubation at 99 degrees Fahrenheit the fibroblasts have spread out on the glass surface and begun to divide. In a week or so they cover the entire available surface with a layer one cell deep. Since they will proliferate only in a single layer, we strip off the layer that has covered the bottle surface, again separate the cells with trypsin and plant half of the cells in each of two new bottles with fresh medium. In three or four days the inoculated fibroblasts grow over all the available surface in each bottle, thus doubling the number of cells taken from the original bottle. As this procedure is repeated at four-day intervals, the fibroblasts continue to proliferate in new bottles, doubling in number each time.

We found that fibroblasts taken from four-month-old human embryos doubled in this way about 50 times (the limit ranged between 40 and 60 doublings).

After reaching this limit of capacity for division the cell population died. It could therefore be concluded that human fibroblasts derived from embryonic tissue and grown in cell culture have a finite lifetime amounting to approximately 50 population doublings (which in our culture covered a span of six to eight months).

Further study reinforced this conclusion. It turned out, for example, that if cell division was interrupted and then resumed, the total number of population doublings was not altered. In our experiments we did not, of course, double the number of bottles at each step; after only 10 doubling passages we would have had 1,024 bottles, and 50 doublings of the original seeding of fibroblasts could have produced about 20 million tons of cells! To keep the yield within reasonable bounds we set aside most of the cells from the subcultivations and put them in cold storage. We found they could be kept in suspended animation for apparently unlimited periods; even after six years in storage they proved to be capable of resuming division when they were thawed and placed in a culture medium. They "remembered" the doubling level they had reached before storage and completed the course from that point. For example, cells that had been stored at the 30th doubling went on to divide about 20 more times.

The geometric rate of increase of the cells in culture has made it possible to provide essentially unlimited supplies of the cells for experiments. Samples of one of our strains of normal human fibroblasts (WI-38), "banked" after the eighth doubling in liquid nitrogen at 190 degrees below zero centigrade, have been distributed to hundreds of research laboratories around the world. The stored cells presumably would be available for study far in the future. For instance, well-protected capsules containing frozen cells might be buried in Antarctica or deposited in orbit in the cold of outer space for retrieval many generations hence. Investigators might then be able to use them to study, among other things, whether or not time had brought about any evolutionary change in the aging of man or other animals at the cellular level.

We found further confirmation of the finite lifetime of human fibroblasts when we cultured such cells from adult donors. Samples of lung fibroblasts taken at the time of death from eight adults, ranging in age from 20 to 87, underwent 14 to 29 doublings in cell culture afterward. The number of doublings in these

tests did not show a clear correlation with the age of the donor, but presumably this was because our method of measuring the doubling of cell populations in bottles is not sufficiently precise to disclose such a correlation in detail. Our current experiments do suggest, however, that consistent differences can be found between broad age groups. It appears that fibroblasts from human embryos will divide in cell culture 50 ± 10 times, those from persons between birth and the age of about 20 will divide 30 ± 10 times and those from donors over 20 will divide 20 ± 10 times.

We have tested fibroblasts from several human embryonic tissues besides lung tissue and found that they too are limited to a total of about 50 divisions in culture. The doubling lifetimes of fibroblasts from animals other than man have also been studied. As one would expect, the cells of the shorter-lived vertebrates show less capacity for division. For example, normal fibroblasts from embryos of chickens, rats, mice, hamsters and guinea pigs usually double no more than 15 times in cell culture, and cells that have been taken from adults of the same species undergo considerably fewer than 15 divisions.

Early in our experiments it became evident that we had to examine the possibility that a lethal factor in the culture medium or a defect in the culture technique might be responsible for the limitation of division and ultimate death of the cells. Did the cells stop dividing because of some lack in the nutrient mixture or the presence of contaminating microorganisms such as viruses? We explored these possibilities by various experiments, one of which consisted in culturing a mixture of normal fibroblasts taken from male and female donors. Female cells can be distinguished from male cells either by the presence of special chromatin bodies (found only in female cells) or by the visible difference between the XX female sex chromosomes and the XY male chromosomes. As a consequence we were able to use these cell "markers" as a label for following the progeny of the respective original parents.

We seeded a bottle with a certain number of fibroblasts from a male population that had undergone 40 doublings and with an equal number of fibroblasts from a female population that had gone through only 10 doublings. If some inadequacy of the culture or the accumulation of a killing factor in the medium were the primary cause of cell death in our cultures, then in this experiment the number of doublings in culture should have been the same for both the male and the female cells after we had mixed them together; there is no reason to suppose that a nutritional inadequacy or a lethal factor such as a virus would act preferentially on the cells of a particular sex. Actually the male and female cells composing the mixture, presumably because of their difference in age, proved to have sharply different survival rates. Most of the male population, having already undergone 40 doublings before the mixture was made, died off after 10 more doublings of the mixed culture, whereas the "young" female population of cells in the same mixed culture, with only 10 previous doublings, went on dividing for many more than the male. After 25 doublings all the male cells had disappeared and the culture contained only female cells. These results appear to confirm in an unambiguous way that the life-span of fibroblast populations in our cell cultures is determined by intrinsic aging of the cells rather than by external agencies.

Does this aging result from depletion or dilution of the cells' own chemical resources? We considered the possibility that the eventual death of the cells might be attributable to the exhaustion of some essential metabolite the cells could not synthesize from the culture medium. If that were the case, however, the original

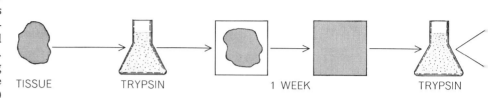

CULTIVATION OF HUMAN CELLS in the author's laboratory begins with the breakdown of lung tissue into separate cells. This is accomplished by means of the digestive enzyme

store of this substance must be very large indeed to enable the cells to multiply for 50 generations. Simple mathematics showed that in order to provide at least one molecule of the hypothetical substance for each cell by the 50th doubling, the original cell would have to have at least three times its known weight even if the substance in question were the lightest element (hydrogen) and

the original parent cell were composed entirely of that element!

By isolating individual cells and developing clones (colonies) from them we have been able to establish that each human fibroblast from an embryo is capable of giving rise to about 50 doubling generations in cell culture. As the cell proliferation proceeds there is a gradual decline in the capacity for reproduction.

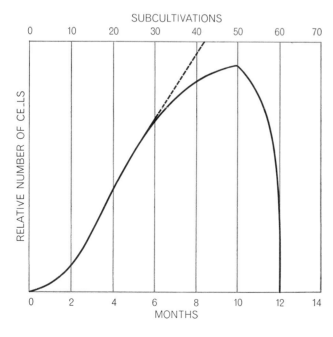

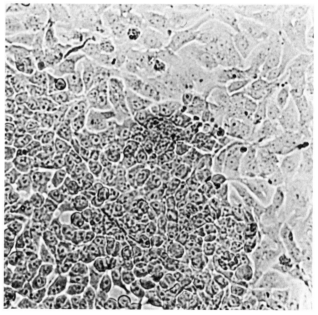

LIFETIME OF HUMAN CELLS was determined by allowing a population to multiply until it had doubled in size. After a culture of cells from embryonic tissue had grown to a particular point, it was divided in two (see illustration at top of these two pages). Cell division ceased after about 50 such subcultivations had doubled. It is possible at any time (although it is rare) for a spontaneous change to occur after which the cells multiply indefinitely (broken line).

TRANSFORMED HUMAN CELLS are distinguished by their morphology and reaction to staining. The darker amnion cells, forming a large island at lower left, have undergone a spontaneous transformation. They will continue to divide after the neighboring cells have died. Transformed, or "mixoploid," cells have more chromosomes than diploid cells do; they are utilized in cancer research. The magnification of this photomicrograph is about 180 diameters.

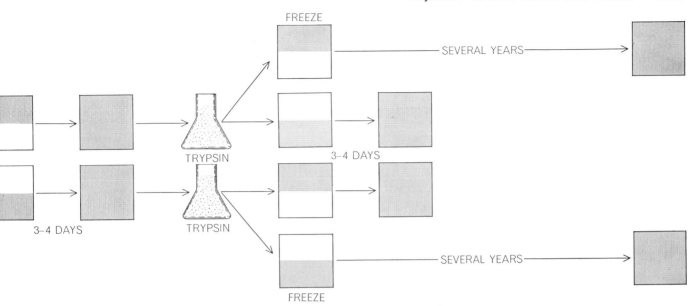

trypsin. After the cells, seeded in a bottle, have multiplied to cover its surface, they are again treated with trypsin, then divided into two halves and replanted. Most cells thus grown are placed in cold storage. Thawed and planted years later, they resume division.

Investigators in several laboratories have found that in successive generations of the multiplying cells a larger and larger fraction of the progeny becomes incapable of dividing until, by the 45th or 50th doubling generation, the entire population has lost this ability.

Curiously, as the population of human fibroblasts approaches the end of its lifetime, aberrations often crop up in the chromosomes. Chromosome aberrations and cell-division peculiarities related to age have also been observed in human leucocytes and in the liver tissue of living mice. The question of whether or not cell abnormalities are a common accompaniment of human aging remains a moot point, however; there is no clear clinical or laboratory evidence on the matter.

Our information on the aging of cells so far is limited to what we can observe in cell cultures in glassware. It has not yet been established that fibroblasts behave the same way in living animals as they do in an artificial culture. However, in man several organs lose weight after middle age, and this is directly attributable to cell loss. The human brain weighs considerably less in old age than it does in the middle years, the kidney also shows a large reduction in nephrons accompanying cell loss and the number of taste buds per papilla of the tongue drops from an average of 245 in young adults to 88 in the aged. We cannot be certain that fibroblasts stop dividing or divide at a lower rate as an animal ages or that the bodily signs of aging can be explained on that basis. It is well known, however, that certain cell systems in animals do stop dividing and die in the normal course of development. Familiar examples are the larval tissues of insects, the tail and gills of the tadpole and some embryonic kidney tissues (the pronephros and mesonephros) in the higher vertebrates.

Reviewing these phenomena, John W. Saunders, Jr., of the School of Veterinary Medicine of the University of Pennsylvania has suggested that the death of cells resulting in the demise of specific tissues is a normal, programmed event in the development of multicellular animals. By the same reasoning we can surmise that the aging and finite lifetime of normal cells constitute a programmed mechanism that sets an overall limit on an organism's length of life. This would suggest that, even if we were able to checkmate all the incidental causes of human aging, human beings would inevitably still succumb to the ultimate failure of the normal cells to divide or function.

In this connection it is interesting to consider what engineers call the "mean time to failure" in the lifetime of machines. Every machine embodies built-in obsolescence (intentional or unintentional) in the sense that its useful lifetime is limited and more or less predictable from consideration of the durability of its parts. By the repair or replacement of elements of a machine as they fail its lifetime can be extended, but barring total replacement of all the elements eventual "death" of the machine as a functioning system is inevitable.

What might determine the "mean time to failure" of an animal organism? I suggest that animal aging may result from deterioration of the genetic program that orchestrates the development of cells. As time goes on, the DNA of dividing cells may become clouded with an accumulation of copying errors (analogous to the "noise" that develops in the serial copying of a photograph). The coding and decoding system that governs the replication of DNA operates with a high degree of accuracy, but the accuracy is not absolute. Moreover, there is some experimental evidence that, as Leslie E. Orgel of the Salk Institute for Biological Studies has suggested, certain enzymes involved in the transcription of information from DNA for the synthesis of proteins may deteriorate with age. At all events, since the ability of cells to divide or to function is controlled by the inherited information-containing molecules, it seems likely that some inherent degeneration of these molecules may hold the key to the aging and eventual death of cells.

Pursuing the machine analogy, we might surmise that man is endowed with a longer life-span than other mammals because human cells have evolved a more effective system for correcting or repairing errors as they arise. Such an evolution would account for the generally progressive lengthening of the fixed life-span from the lower to the higher animals; presumably the march of evolution has developed improvements in the cells' error-repairing mechanisms. It is clear, however, that even in man this system is far from perfect. In the idiom of computer engineers we might say that man, like all other animals, has a "mean time to failure" because his normal cells eventually run out of accurate program and capacity for repair.

The Author

LEONARD HAYFLICK is professor of medical microbiology at the Stanford University School of Medicine. Until he went to Stanford recently he had spent most of his life in Philadelphia, where he was born and educated. He received all his university degrees from the University of Pennsylvania: a bachelor's degree in microbiology in 1951, a master's degree in medical microbiology in 1953 and a Ph.D. in medical microbiology and biochemistry in 1956. For several years he was a member of the faculty at the university and then was associated with the Wistar Institute of Anatomy and Biology in Philadelphia. In addition to his studies of cultured human cells Hayflick has worked with the smallest free-living microorganisms, the myco-plasmas. In 1961 he identified the agent causing primary atypical pneumonia in man as a mycoplasma; previously the organism had been thought to be a virus.

Bibliography

SENESCENCE AND CULTURED CELLS. Leonard Hayflick in *Perspectives in Experimental Gerontology: A Festschrift for Doctor F. Verzár*. Charles C Thomas, Inc., 1966.

THE SERIAL CULTIVATION OF HUMAN DIPLOID CELL STRAINS. L. Hayflick and P. S. Moorhead in *Experimental Cell Research*, Vol. 25, No. 3, pages 585–621; December, 1961.

TOPICS IN THE BIOLOGY OF AGING. Edited by Peter L. Krohn. Interscience Publishers, 1966.

SCIENTIFIC
AMERICAN   November 1948, Vol. 179, No. 5, pp. 46-51   OFFPRINT **1104**

BACTERIAL VIRUSES AND SEX

by Max and Mary Delbrück

Some fascinating experiments have demonstrated
that the tiny organisms which prey on bacteria
employ a primitive kind of sexual reproduction.

TWO YEARS AGO, at a summer symposium in Cold Spring Harbor, N.Y., experiments were presented which showed that bacteria, and even some viruses that live on bacteria, apparently have a method of sexual reproduction. This finding was a considerable surprise. Up to that time it had been generally supposed that the simple one-celled bacteria had no sex and that they multiplied simply by splitting in two; the method of reproduction of the still more rudimentary bacterial viruses was entirely unknown. The simplest organisms previously known to have a sexual mode of reproduction were the molds, yeasts and paramecia. Indeed, the recognition of sex even in those organisms was less than 20 years old.

Sex was once thought to be the exclusive possession of life's higher forms. Yet as biologists have looked more carefully down the line, simpler and simpler forms have been found to be possessed of it. Now, among the viruses, we are searching for it at the lowest known level of life.

Since the Cold Spring Harbor symposium, this research has been pushed further, and some rather remarkable facts have been uncovered. This article will discuss a group of the viruses which are parasites of bacteria, and particularly will go into what has been learned recently concerning their reproduction.

Sexual reproduction is the coming together and exchanging of character factors of two parents in the making of a new individual. Aside from its other aspects, sex has a special interest for biologists as a highly useful and indeed almost necessary device for an organism to survive in the competitive evolutionary scheme of life. Plant and animal species, to avoid extinction in the changing environments of geologic time, evolve by utilizing mutations (changes in the basic hereditary material) which enable them better to adapt themselves to their environment. These mutations turn up spontaneously and spread through the population by the convenient means of sexual reproduction.

Mutations are assorted and combined anew in every generation. Thus species that reproduce sexually always have in store a vast array of new types, some of which may be adapted to a changed environment and can become the parents of the next link in the evolutionary chain. This is the evolutionary advantage of sex.

It is logical, therefore, to look for sex in every known form of life. It was with great caution, however, that the discovery of sex in the simplest organisms was reported two years ago. E. L. Tatum and J.

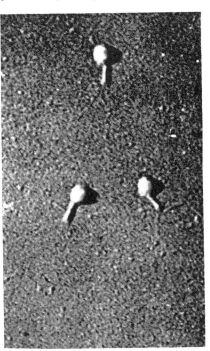

VIRUS T4, shadowed with gold to make a specimen for the electron microscope, has shape of a tadpole.

Lederberg of Yale University told of experiments in which they had found bacteria which seemed to combine certain traits of two parental strains that had been mixed. Similar findings with respect to the viruses that attack bacteria were reported by A. D. Hershey of Washington University, and by W. T. Bailey, Jr. and M. Delbrück at Vanderbilt University.

The bacterial virus is a very small organism which enters a bacterium, reproduces itself and eventually destroys its host. From the latter a generation of new viruses then emerges. The virus thus "infects" a bacterium, even as plant and animal viruses infect plants and animals. Bacterial viruses were first discovered 30 years ago by the French bacteriologist F. D'Herelle, who noticed that the bacteria growing in some of his test tubes mysteriously dissolved. After experimentation D'Herelle concluded that their dissolution was due to some agent much smaller than a bacterium, and that this agent grew at the expense of bacteria. He called the agents that had destroyed the bacteria "bacteriophages" (bacteria-eaters); the same organisms are now often called bacterial viruses.

For many years thereafter bacteriologists and medical men were sure that bacterial viruses existed. The viruses were even measured, isolated and grouped, although they were never actually seen. Bacterial viruses are too small to be seen under the most powerful microscope of the conventional type; they have been made visible only recently by new types of microscopes.

D'Herelle's discovery raised the great expectation that bacterial viruses might be used as "agents of infectious health" to destroy the bacteria that caused human and animal diseases. It was the hope of early research workers that a population infected by a bacterial epidemic could be cured by infecting it with the virus inimical to that bacterium. Their hope has not been realized, but the bacterial viruses remain a subject of keen interest—for good and sufficient reasons.

Viruses seem to lie on that uncertain and perhaps unreal borderline between life and non-life. The uncertainty about this boundary line is both very old and very new. In ancient times all nature was supposed to be animate. Spirits dwelt in stones as well as in animals, and as recently as a few centuries ago the spontaneous generation of complex living organisms from mud was a matter of universal belief. The advance of the scientific method has taught us that there is an enormous difference between the living and

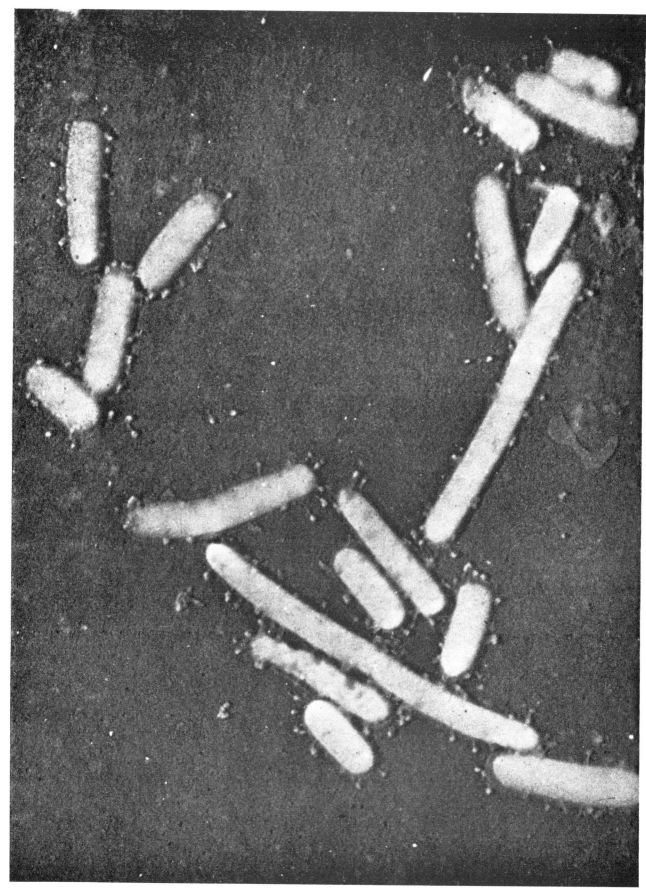

BACTERIA UNDER ATTACK by a swarm of bacterial viruses are shown by the electron microscope. The viruses, which are of the strain T4 described in this article, attach themselves to bacteria and sometimes push inside them. There the viruses reproduce until the bacterium bursts, liberating an entire new generation of viruses.

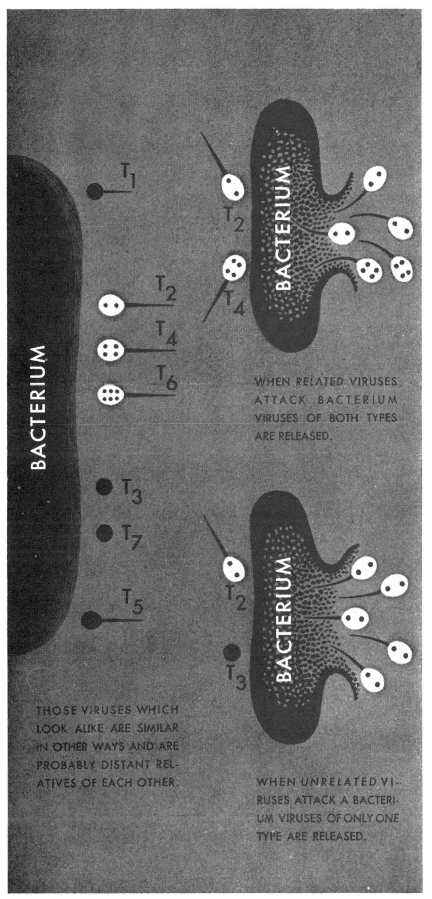

WHEN RELATED VIRUSES ATTACK BACTERIUM VIRUSES OF BOTH TYPES ARE RELEASED.

THOSE VIRUSES WHICH LOOK ALIKE ARE SIMILAR IN OTHER WAYS AND ARE PROBABLY DISTANT REL- ATIVES OF EACH OTHER.

WHEN UNRELATED VI- RUSES ATTACK A BACTERI- UM VIRUSES OF ONLY ONE TYPE ARE RELEASED.

EXPERIMENTAL ORGANISMS of the research discussed in this article are seven bacterial viruses that attack the same species of bacterium.

inorganic worlds. During the development of classical biology in the 19th century, there arose two great generalizations: 1) the theory of evolution, and 2) the cell theory. The theory of evolution proclaims the relatedness of all living things; the cell theory sets forth a universal principle of construction for them. Both of these generalizations unified biology and distinguished it from the study of the inorganic world.

IN our generation, however, the pendulum has begun to swing the other way. The great refinement of scientific technique has pushed the limit of observation beyond the point where it had stood for about 100 years, namely, at the resolving power of the light microscope. With this advance has come the recognition of the existence of many things below cellular size which do not fit into the established categories of life or non-life. Of these the viruses have become the most controversial. To learn all we can about them becomes, then, even more intriguing than the original idea that bacterial viruses might be useful in medicine.

Bacteria-eating viruses are common, and where bacteria exist in the natural state, viruses capable of destroying them almost always can be found. Outside of the bacterium, the virus seems dead. But it does not die; it lies quiescent and functionless until a bacterium presents itself. The virus then attaches itself firmly to the bacterium. Many viruses may cling to a single bacterium, but only one needs to enter the cell to begin a cycle of viral reproduction. Once within the host, the virus quickly comes to life and multiplies prodigiously. How it grows from the one or more particles that are known to enter the cell to the several hundred that burst from the suddenly ruptured host is a secret still closely guarded within the walls of the bacterium.

The guinea pigs of bacterial virus genetics have been seven different viruses which all attack the same bacterium. Some of these viruses, which we shall speak of as T1, T2 and so forth, are surprisingly complex in form and behavior.

The viruses that were first made visible by the electron microscope in 1941 were revealed to be spermlike forms. Some were seen lying free, others were clinging to the exterior of a bacterium. Other pictures revealed the bacterium with new viruses streaming from a hole ripped in its cell wall.

The seven viruses do not all look alike. In appearance they fall into four categories. The members of one family, consisting of T2, T4 and T6, look like tadpoles, with dark forms visible within their bodies. T5 has a round, solid body and a tail. T1 is similar to T5, but smaller. T3 and T7 are the smallest, with spherical bodies and no visible tail.

The viruses which look alike are related in several other respects, and the way they behave as a family is illustrated by a very curious phenomenon. When two viruses which are not related happen to attach themselves to the same bacterium, one successfully enters the bacterium and multiplies, but the other perishes without leaving any offspring. If the two viruses seeking the same bacterial home are related, however, both enter and reproduce. This rule has certain exceptions and certain special modifications, depending on the degree of relatedness between the contending viruses; as among human beings, the restriction of real estate among the viruses has subtle points.

One might wonder how the biologist can learn anything about the behavior of organisms so small that he generally cannot see them. The answer is that bacterial viruses make themselves known by the bacteria they destroy, as a small boy announces his presence when a piece of cake disappears. Much of what we know about the viruses is based on the following experiment, which requires only modest equipment and can be completed in less than a day.

Bacteria first are grown in a test tube of liquid meat broth. Enough viruses of one type are added to the test tube so that at least one virus is attached to each bacterium. After a certain period (between 13 and 40 minutes, depending on the virus, but strictly on the dot for any particular type), the bacterium bursts, liberating large numbers of viruses. At the moment when the bacteria are destroyed, the test tube, which was cloudy while the bacteria were growing, becomes limpid. Observed under the microscope, the bacteria suddenly fade out.

Before the bacteria burst, however, part of the liquid is taken from the test tube and diluted. From this diluted liquid the experimenter takes a small sample expected to contain only a single infected bacterium. When this bacterium has liberated its several hundred viruses into a liquid medium in a test tube, the liquid is poured on a plate covered with a layer of live bacteria. Each virus deposited on the plate will start attacking the bacterium on which it rests. Each of the offspring of the virus, in turn, will attack the nearest bacterium. Successive generations of offspring from the one original virus will spread out in a circle, attacking bacteria until after a few hours a small round clearing becomes visible to the naked eye. The number of such clearings, or "colonies," formed on the plate is a count, therefore, of the number of viruses liberated by the original infected bacterium.

The union of the virus and its bacterium takes place under rather complicated and specific conditions which are not well understood. Of the life of the virus inside the bacterial host, still less can be divined. Does the virus multiply by one individual producing another, by simple splitting, or by some other process? What specific elements of nutrition are necessary for virus

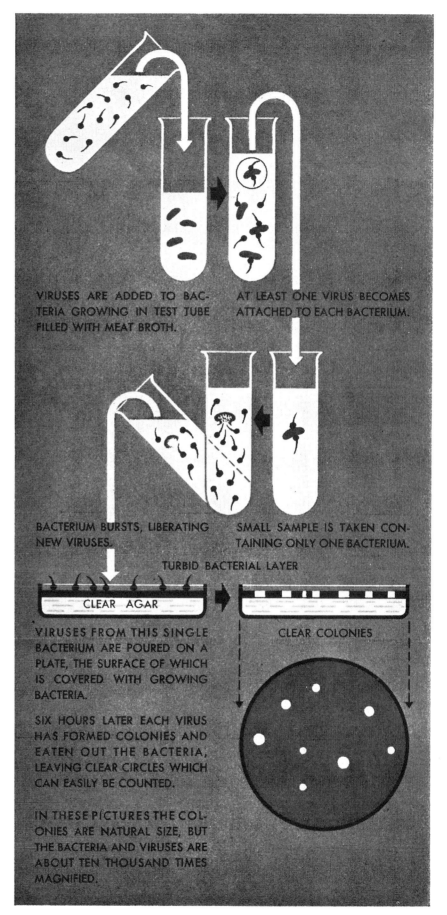

VIRUSES ARE ADDED TO BACTERIA GROWING IN TEST TUBE FILLED WITH MEAT BROTH.

AT LEAST ONE VIRUS BECOMES ATTACHED TO EACH BACTERIUM.

BACTERIUM BURSTS, LIBERATING NEW VIRUSES.

SMALL SAMPLE IS TAKEN CONTAINING ONLY ONE BACTERIUM.

TURBID BACTERIAL LAYER

CLEAR AGAR

VIRUSES FROM THIS SINGLE BACTERIUM ARE POURED ON A PLATE, THE SURFACE OF WHICH IS COVERED WITH GROWING BACTERIA.

SIX HOURS LATER EACH VIRUS HAS FORMED COLONIES AND EATEN OUT THE BACTERIA, LEAVING CLEAR CIRCLES WHICH CAN EASILY BE COUNTED.

IN THESE PICTURES THE COLONIES ARE NATURAL SIZE, BUT THE BACTERIA AND VIRUSES ARE ABOUT TEN THOUSAND TIMES MAGNIFIED.

CLEAR COLONIES

EXPERIMENTAL TECHNIQUE that is used in bacterial virus research is outlined in this drawing. The equipment required is remarkably simple.

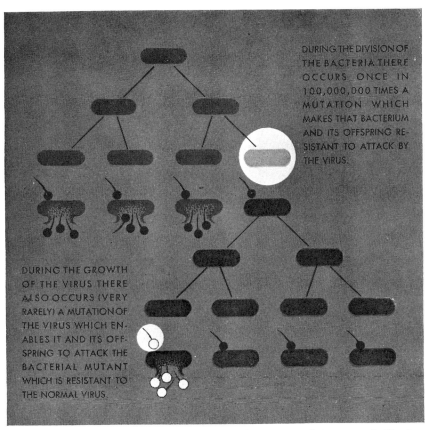

During the division of the bacteria there occurs once in 100,000,000 times a mutation which makes that bacterium and its offspring resistant to attack by the virus.

During the growth of the virus there also occurs (very rarely) a mutation of the virus which enables it and its offspring to attack the bacterial mutant which is resistant to the normal virus.

MUTATION is the mechanism that enables bacteria and viruses and all other living things to adapt themselves to changing environmental conditions.

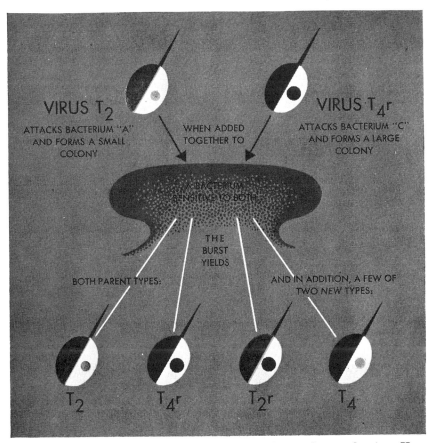

VIRUS T₂

ATTACKS BACTERIUM "A" AND FORMS A SMALL COLONY

WHEN ADDED TOGETHER TO

VIRUS T₄ʳ

ATTACKS BACTERIUM "C" AND FORMS A LARGE COLONY

BACTERIUM SENSITIVE TO BOTH

THE BURST YIELDS

BOTH PARENT TYPES:

AND IN ADDITION, A FEW OF TWO NEW TYPES:

T₂   T₄ʳ   T₂ʳ   T₄

EXCHANGE of virus characteristics is form of sexual reproduction. Here characteristics are type of bacterium attacked and the size of the colony.

reproduction? Is it possible to break open the cell before it would normally burst and from the contents at this intermediate stage learn something of the process of multiplication? What causes the violent disruption and dissolution of a bacterium?

Although we cannot fully answer these questions, it has nevertheless been possible to wrest some remarkable secrets from the viruses. We have learned something about the way in which they transmit characteristics and survive from generation to generation. One method of investigation has been to study the fashion in which viruses are able to meet emergencies in their environment. For example, when viruses are mixed with bacteria, most of the bacteria are destroyed. One in perhaps 100 million bacteria, however, will mutate to a form that is resistant to the virus, thus establishing a line of defense for its species. The virus, on the other hand, is capable of launching a new attack by mutating to a form which can destroy the resistant bacteria.

ALL kinds of mutations, many of them easy to recognize, turn up among the viruses. Some produce variant types of colonies on the bacterial plate; they may create fuzzy clearings instead of sharp-edged ones, or large clearings instead of small round ones. This kind of virus mutation was discovered in 1933 by I. N. Asheshov (who now heads a research project on bacterial viruses at the New York Botanical Garden) during his studies of anti-cholera vibris viruses in India. Other breeds of viruses have been found which need some particular substance, such as a vitamin or calcium, to become capable of attaching themselves to the bacterium. T. F. Anderson, of the Johnson Foundation for Medical Research at the University of Pennsylvania, opened up a totally unexpected new angle in viral research when he discovered that viruses T4 and T6 will not attack a bacterium in a medium lacking a simple organic compound called l-tryptophane.

The discovery that bacterial viruses have a sexual form of reproduction came about in the following way. M. Delbrück and W. J. Bailey, Jr. were working with viruses T2 and T4r (a mutant of T4), which are relatives that can reproduce in the same bacterium of strain B. Each has two distinguishing characteristics: T2 produces a small colony and can destroy a mutant strain of bacteria called A; T4r produces a large colony and can destroy a mutant strain of bacteria called C. When T2 and T4r were added to a bacterium, viruses of both these parent types were released upon burst, as expected. But in addition two new types of virus came out, with' their characteristics switched! One of the new types produced a large colony and destroyed bacterium A; the other produced a small colony and destroyed bacterium C. Obviously the parents had got together and exchanged something. The number of individuals of these new forms coming from a single bacterium varied, but the maximum number found was about 30 per cent of the total yield.

The most surprising discovery of all, however, was made by S. E. Luria of Indiana University. It came about as a sequel to an accidental observation in our laboratory at Vanderbilt University. When a virus that has been "killed" by exposure to ultraviolet light is added to a bacterium, the bacterium is destroyed but no new viruses issue from it. In one such experiment, Bailey irradiated viruses long enough so that most, but not quite all, of them were killed. He then transferred some samples, as usual, to a bacteria-covered plate. He wanted to determine the number of survivors, and expected to find less than 100 virus colonies on the plate. Instead, the next morning he found thousands of colonies! Puzzled, Bailey repeated the experiment, with the same result. The supposedly dead viruses had in some way come to life.

Later, at Indiana, Luria took this problem up seriously. He discovered a curious fact: although a bacterium infected with only one "killed" virus dies and yields nothing, a bacterium infected with two or more "killed" viruses bursts and yields several hundred new viruses. Luria therefore assumed that inside a bacterium two or more "killed" viruses (or perhaps we had now better call them mortally damaged) can pool their undamaged parts to make whole individuals capable of reproducing themselves and of escaping from the bacterium. He estimated that each virus of the T2, T4 or T6 type has about 20 vital units. Assume that each time a virus is shot at, or exposed to ultraviolet light, one vital unit is knocked out. If there are two viruses, each of which has been shot at four times, there is a good chance that the same vital unit has not been hit in both. The remaining units then seem to have a way of combining and forming effective individuals.

This "revival of the dead," as we might call it, which indicates some substitution of vital material, is interrelated with the previously mentioned exchange of character traits in viruses, a phenomenon which has been explored very successfully by A. D. Hershey at Washington University.

Gradually the study of these two phenomena should reveal something more of the way in which one virus produces another and—the most ambitious hope—even something of the simple facts of life. Here, as far as mind and imagination and skill can reach, is a vast region of the very small that is open for exploration.

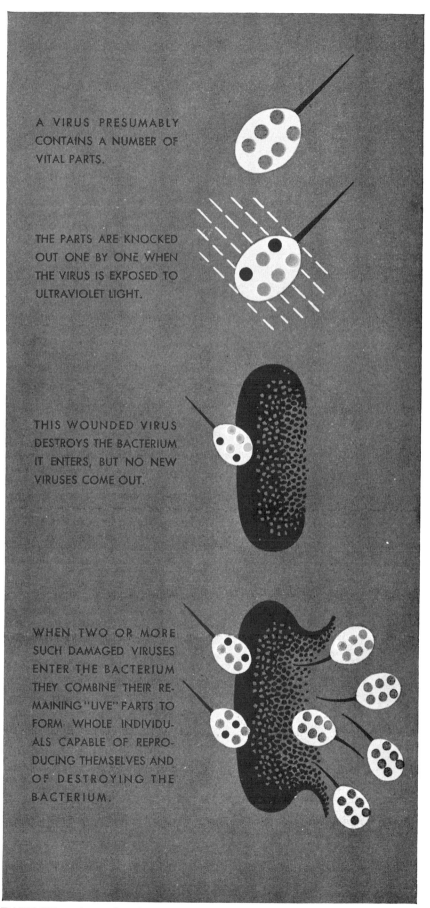

A VIRUS PRESUMABLY CONTAINS A NUMBER OF VITAL PARTS.

THE PARTS ARE KNOCKED OUT ONE BY ONE WHEN THE VIRUS IS EXPOSED TO ULTRAVIOLET LIGHT.

THIS WOUNDED VIRUS DESTROYS THE BACTERIUM IT ENTERS, BUT NO NEW VIRUSES COME OUT.

WHEN TWO OR MORE SUCH DAMAGED VIRUSES ENTER THE BACTERIUM THEY COMBINE THEIR REMAINING "LIVE" PARTS TO FORM WHOLE INDIVIDUALS CAPABLE OF REPRODUCING THEMSELVES AND OF DESTROYING THE BACTERIUM.

DAMAGE to viruses by ultraviolet rays (*second drawing from top*) is added proof that they exchange characteristics. Damaged viruses pool resources.

The Authors

MAX DELBRÜCK is professor of biology at the California Institute of Technology. Mary Bruce Delbrück is his wife.

Bibliography

EXPERIMENTS WITH BACTERIAL VIRUSES. M. Delbrück in the *Harvey Lecture Series,* Vol. 41, pages 161-187; 1945-46.

RECENT ADVANCES IN BACTERIAL GENETICS. S. E. Luria in the *Bacteriological Reviews,* Vol. 2; March, 1947.

HEREDITY AND VARIATION IN MICROORGANISMS. *Cold Spring Harbor Symposia on Quantitative Biology,* Vol. 11; 1946.

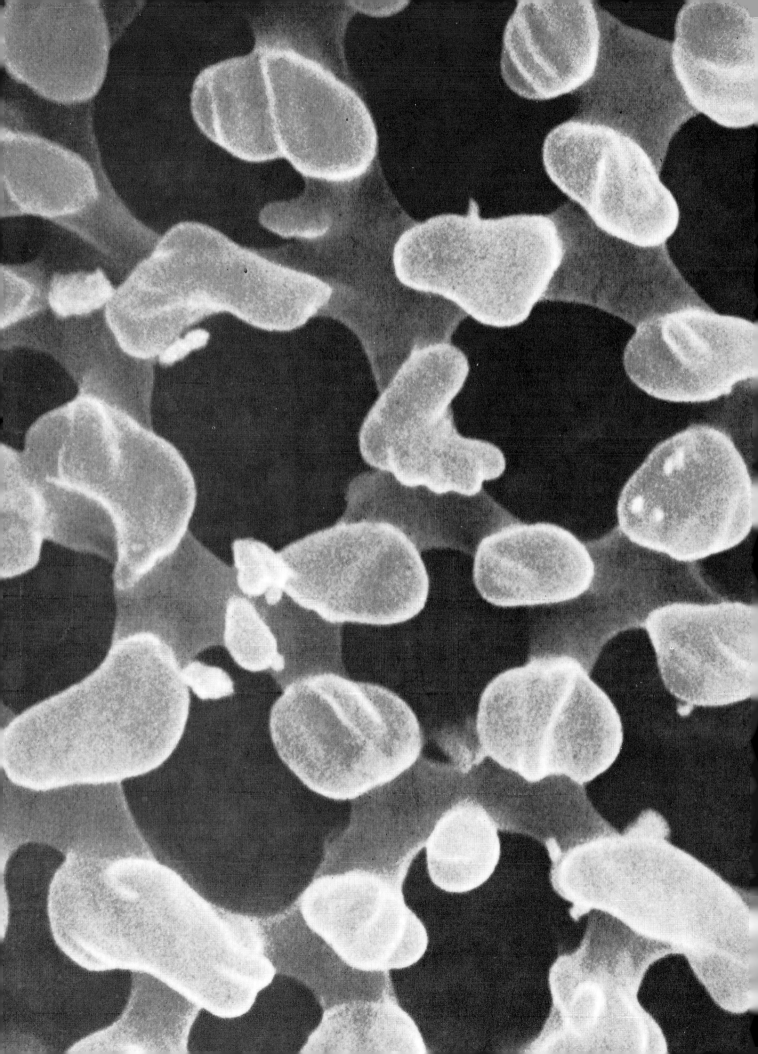

SCIENTIFIC
AMERICAN   April 1968, Vol. 218, No. 4, pp. 80-90   OFFPRINT **1105**

POLLEN

by Patrick Echlin

The tiny grains that carry the male genetic material of plants are closely studied with, among other instruments, the scanning electron miscroscope. Their baroque architecture raises fundamental questions.

Pollen grains, the dustlike cells that transfer the male genetic material in the reproduction of flowering plants, are among the shortest-lived independent bodies in nature, and yet a major component of pollen grains is one of the most enduring natural materials. Few pollen grains remain alive for more than a few days after they have been dispersed; some live only a few hours. With the exception of the pollen of aquatic plants such as *Zostera* (eelgrass), however, all pollen grains are covered with an extremely tough substance called sporopollenin. The sporopollenin outer wall of a pollen grain remains intact in concentrated acids and alkalis at temperatures as high as 500 degrees Fahrenheit. Samples taken from the depths of ancient bogs contain grains of pollen that are clearly recognizable in spite of having been buried for hundreds of thousands of years. (Fossilized pollen grains have been isolated from Cretaceous deposits some 100 million years old.) The durability of pollen, together with the fact that the pollen grains of each plant genus (and even some species) have their own distinctive form, have made pollen analysis an important tool for investigating the vegetation and the climate of the past.

In the Botany School at the University of Cambridge we have been studying the form of pollen grains for a number of years. We have been concerned not only with the differences among grains but also with changes in the grains as they grow to maturity. With the recent development of the reflection scanning electron microscope we have been able to examine pollen grains in unprecedented detail. This kind of microscope differs from the conventional transmission electron microscope in that the electron beam, instead of passing through an ultrathin specimen, scans the surface of an opaque specimen. The scattered electrons, together with secondary electrons emitted by the specimen itself, are then amplified and form an image of the surface in strong relief on the face of a cathode-ray tube [*see illustration on page 2706*]. Although resolution of the scanning microscope is not as high as that of the transmission microscope, its usable depth of focus is much greater and its specimens are simpler to prepare. It is therefore possible to examine a large number of specimens in a relatively short time. Such an instrument is of immense value in the study of specimens as complex as the surfaces of pollen grains.

Before describing what we have learned about the maturation of pollen grains it will be useful to review the role of pollen in plant reproduction and to describe the morphology of the mature pollen grain. Let us begin with the stamen, the male organ of the flower where the pollen is formed [*see top illustration on next page*]. The stamen usually consists of a short stalk or filament, and at its tip is the structure known as the anther. The anther consists of four elongated sacs within which the pollen grows. At maturity the anther bursts open along a predetermined line of cells that are weaker than their neighbors, thereby releasing the pollen.

The pollen eventually travels to the female organ of the same flower or of another flower of the same species, a journey that can be accomplished in a variety of ways. The female organ, the pistil, consists of an ovary at the base of a stalk called the style and, at the top of the style, the sticky structure known as the stigma. If the pollination has been successful, the grain adheres to the stigma and fertilization begins. The pollen grain puts forth an extension, the pollen tube, which by the secretion of special enzymes is able to penetrate the surface of the stigma. Then the pollen tube grows downward between the cells of the style until it reaches the ovary, providing a path along which the pollen grain's two nuclei travel. One nucleus fuses with the egg nucleus in the ovule to form the embryo; the other fuses with two "polar" nuclei to give rise to the nutritive endosperm tissue.

Fertilization is only one of two independent functions served by pollen. The other is setting in motion the physiological processes that form the plant's fruit. The chemical constituents of pollen, although they are mainly protein and fat, also include vitamins, free amino acids, pigments and small amounts of two growth hormones: indoleacetic acid and gibberellin. The last two substances induce the production of hormones in the plant's female organ, which in turn stimulate the growth of the ovary wall and the formation of the fruit. The fact that this function of pollen is separate from fertilization is indicated by experiments in which the formation of fruit was induced by dead pollen, pollen extracts and even pollen from other species. Such fruit, however, lacks male nuclei and is therefore sterile.

The amount of pollen produced by a single plant varies greatly among species, but even in the most pollen-poor

SURFACE OF POLLEN from the geranium is enlarged 11,000 times in a micrograph made with the scanning electron microscope (*opposite page*). The instrument produces a contoured image which shows that the surface of the grain is made up of tiny rods that project upward from an openwork floor.

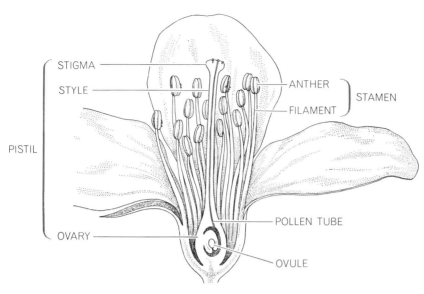

FLOWER CROSS SECTION shows a cluster of stamens, each with a pollen-filled anther at its tip, surrounding a central pistil with a sticky stigma at the tip and an ovary in the base. One pollen grain adhering to the stigma has grown a long tube (*color*) reaching the ovary.

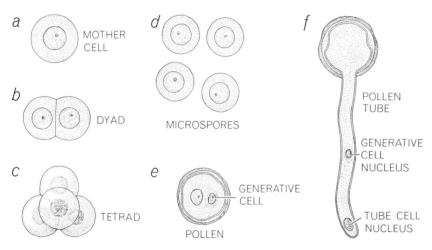

GRAIN OF POLLEN, beginning as a single cell (*a*), undergoes meiosis (*b, c*) to become a cluster of four microspores called a tetrad. When mature, the tetrad separates into individual microspores (*d*), each of which has two cell nuclei (*e*). If a free pollen grain comes in contact with a stigma, the tube cell becomes involved with the growth of the pollen tube toward the ovary (*f*) and the generative cell then travels down the tube to fertilize the ovule.

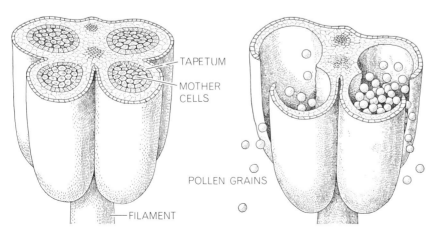

ANTHER CROSS SECTIONS show the two kinds of tissue (*left*) that respectively give rise to mature grains of pollen and to the tapetum, or inner wall, of the anther cavity. When mature (*right*), the anther splits open and the loose pollen in its cavities is dispersed.

flowers many thousands of pollen grains are shed for each ovule that is eventually fertilized. In the hazel tree (*Corylus*), for example, the average is 2.5 million grains per fertilized ovule. A single shoot of the hemp plant (*Cannabis*) may produce more than 500 million grains. At the other extreme, even a large flax plant (*Linum*) may produce no more than 20,000 grains.

Forest trees, many of which are pollinated by the wind, manufacture large quantities of pollen, the number of grains per acre running into many billions. So much pollen is given off by evergreens that the discharge is sometimes visible as a cloud over the forest, and pollen can be scooped up by the handful from the surface of a forest lake. It has been estimated that the spruce forests of southern and central Sweden produce 75,000 tons of pollen a year.

Pollen grains display a wide range of sizes. Among the largest are the grains of the pumpkin (*Cucurbita*); they may be 250 microns (.25 millimeter) in diameter. At the other end of the scale are the grains of forget-me-not (*Myosotis*), only two to five microns in diameter. Once they are shed the pollen grains of most plants travel separately, but the grains of some species—orchids, for example—remain in clusters.

Speaking generally, the size of the pollen grain is related to the means of its dispersal. Grains between 20 and 60 microns in diameter are usually carried from one plant to another by the wind. Grains that are larger or smaller than that are usually transported by insects. Some plants do not depend on either wind or insects: the pollen from each flower's anthers simply falls onto its own stigma.

Plants that are pollinated by insects generally possess some mechanism for attracting the pollinators. One is odor, including not only odors that are pleasant to man but also many that are unpleasant. Another, often combined with odor, is conspicuous color, usually on the petals of the flower but occasionally on other parts of the plant. Visits by pollinators are also encouraged by edible pollen and by nectar. The pollen of insect-pollinated plants cannot easily be blown away by the wind, and it is often located within the flower in such a way that it can be easily reached only by the appropriate insect.

The flowers of wind-pollinated plants are characterized by simplicity of structure. They are not scented or showy, do not produce nectar and usually produce large amounts of powdery pollen. The

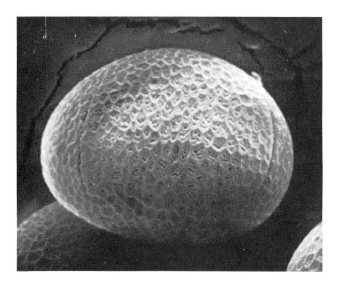

TOUCH-ME-NOT (*Impatiens grandiflora*) pollen is seen magnified 3,500 times. The furrow lines of the grain are faintly visible.

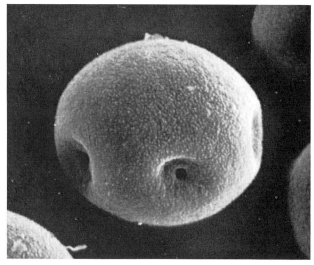

HORNBEAM (*Carpinus*) pollen grain is magnified 2,200 times. The characteristic deep-sunken pores have even deeper center openings.

WALLFLOWER (*Cheiranthus*) pollen grains are magnified 2,300 times. They are marked by prominent furrows and many tiny pits.

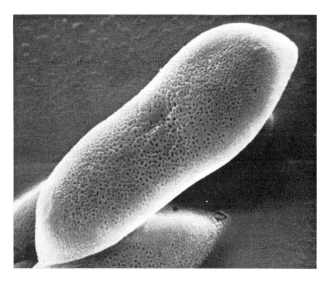

BLUE-EYED GRASS (*Sisyrinchium bermudiana*) pollen is magnified 1,250 times. The furrow in the oblong grain lies out of sight.

THRIFT (*Armeria maritima*) pollen, magnified 1,400 times, shows a deeply sculptured surface and a furrow marked by broken ridges.

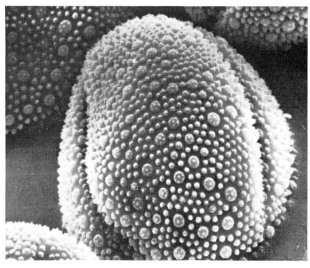

FLAX (*Linum austriacum*) pollen grain is magnified 1,250 times. It displays deep furrows and a surface dotted with small blunt spines.

flowers of many wind-pollinated trees open and release their pollen before the leaves appear, so that pollen transport is not obstructed by the dense mass of foliage that develops later. The wind can carry small grains of pollen a surprisingly long way. Air samples collected over the North Atlantic, more than 400 miles from the nearest land, contained pollen grains of the alder *Alnus viridis*. Peat deposits on the isolated South Atlantic islands of Tristan da Cunha contain pollen grains of the evergreen beech (*Nothofagus*) and the joint fir (*Ephedra*). It is unlikely that either of these plants ever grew on the islands, and the evergreen beech is not found in southern Africa, the closest land. It therefore seems likely that the evergreen beech pollen, and probably the pollen of both plants, was borne to Tristan da Cunha by the prevailing westerly winds

from South America, nearly 2,500 miles away. This, of course, is an extreme case; pollen is usually not carried very far by the wind because it is collected and washed out of the air by rain.

Pollen grains are divided into two major classes on the basis of gross morphology. One class is characterized by a single germinal furrow, the other by three germinal furrows. The grains also have a wide range of other morphological characteristics, such as the form and location of the apertures in the grain wall known as pores and other features of wall structure. When pollen grains are arrayed in order of size, it becomes apparent that the walls of the smallest grains are comparatively featureless. The grains of middle size, mostly the windborne pollens, are also generally rounded and smooth. The larger pollen grains typical of insect-pollinated plants,

however, are often elaborately sculptured. An example is the pollen grain of the mallow (*Malva*), the surface of which appears in the illustration at top right on page 2708. The grain walls of the larger pollens are also sometimes coated with an oily-adhesive substance.

How do these variations in surface morphology arise? Is there a point during the growth and maturation of the grain at which its characteristic form appears? What factors are primarily responsible: heredity, influences from the microenvironment or a mixture of both? In an effort to find answers to such questions we have examined the ultrastructure of pollen grains from a number of plants at various stages on the way to maturity. The description of pollen-grain growth that follows is based on our study of pollen from a species of hellebore (*Helleborus foetidus*). With

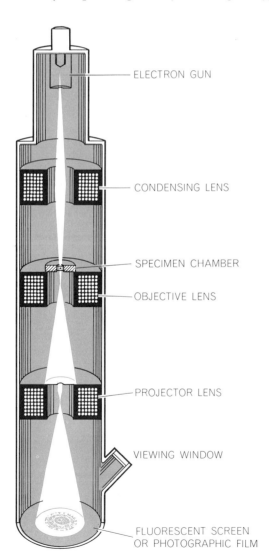

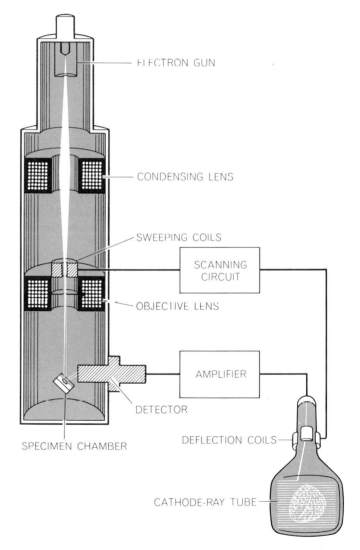

SCANNING MICROSCOPE (*right*) differs from the conventional transmission electron microscope (*left*) in its use of whole specimens rather than thin sections. In both instruments a beam of electrons is used to bombard the object being studied. In the scan-

ning microscope some electrons that are scattered by impact with the specimen and others that are emitted by the specimen during bombardment are collected to form a cathode-tube image characterized by great depth of focus and a three-dimensional appearance.

certain minor differences, the process is the same as the one that has been observed in the pollen grains of other plants.

Pollen grains resemble most other living plant cells in that the living cytoplasm of the grain is surrounded by a thin wall of cellulose; in pollen this wall is called the intine. Immediately outside the intine is another layer known as the exine. The principal component of the exine is sporopollenin, the tough substance that gives pollen its remarkable durability. The chemical nature of sporopollenin is somewhat obscure. Gas chromatography indicates, however, that it is primarily a polymer of monocarboxylic or dicarboxylic fatty acids with a fairly high molecular weight.

In a few groups of plants the exine is a fairly uniform sheath. More commonly it is complex and is itself divided into an outer component (the ektexine) and an inner one (the endexine). The inner component, which completely covers the intine, is usually a smooth layer. It is the outer component that forms the structures that give the grain walls their rich detail. These structures are made up of tiny rods called bacula. The rods differ widely in size and may be either isolated or clustered in groups. In some plant genera the tips of the bacula are fused to form a tectum, or roof, that is perforated or sculptured in characteristic ways [see *illustrations at top of next two pages*]. In other genera the bacula have a spinelike appearance, and the tips of the spines can be rounded, pointed or shaped in other ways. Long and short spines on the surface of the tectum may surround crater-like pores, as they do in morning-glory pollen, looking rather like a high-speed photograph of a splashing water droplet. Or the spines may cover the surface of the pollen grain in symmetrical arrays of peaks, as they do in woodsmallow pollen, like some surrealist portrait of a lunar landscape.

The early stages of pollen growth are intimately associated with the development of the anther, the pollen-producing organ at the tip of the stamen. Precursor cells in the young anther, comprising what is called archesporial tissue, give rise to two distinctively different components. One of them, the primary parietal layer, forms both the outer wall of the anther and its specialized inner wall, called the tapetum. The tapetum, consisting of one or more layers of cells, plays an essential role in the formation of pollen grains: abnormal development of the tapetum invariably

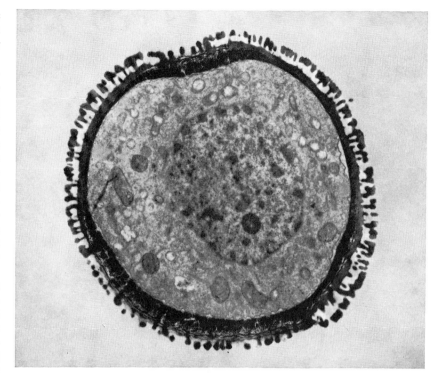

TRANSMISSION MICROGRAPH shows many details of grain structure visible in a transverse section through a mature pollen grain. This is pollen from a hellebore (*Helleborus foetidus*), a member of the buttercup family. Structures visible in the micrograph include the main components of the grain's tough outer coat. The inner layer of the coat, the endexine, is darker than the outer layer, the ektexine, and is quite thick in the vicinity of a furrow (*lower left*). The rodlike elements whose fused ends form both the roof and the floor of the pollen grain's outermost covering are visible as dark upright structures along the entire circumference of the pollen grain. The electron micrograph enlarges the grain 6,200 times.

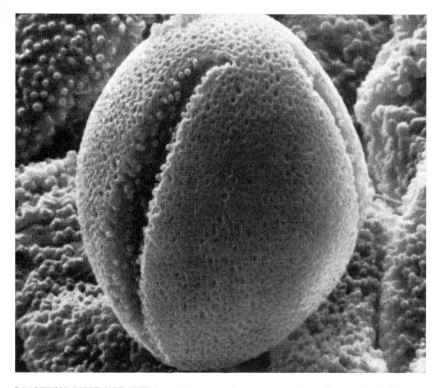

SCANNING MICROGRAPH shows the external appearance of a pollen grain of the same hellebore species. The smoothness of the outermost grain covering is broken by two deep furrows. The tiny, crater-like openings in the surface are areas in which the tips of rods have not fused completely in forming a roof. The pollen grain is seen magnified 3,400 times.

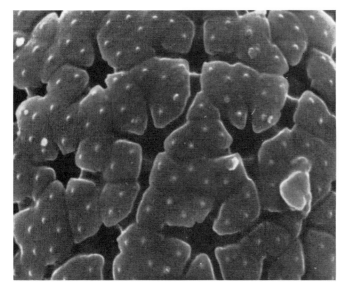

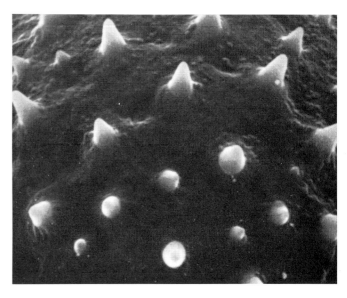

POLLEN SURFACE DETAILS are visible in four scanning micrographs. Magnified 4,500 times (*left*), pollen of the Asiatic shrub *Pimelea*

has a roof of large plates that bear many tiny spines. Pollen of the mallow (*Malva*) is magnified 2,500 times (*second from*

disrupts the maturation of the pollen.

The second archesporial component is the primary sporogenous tissue. This tissue produces the many pollen mother cells from which the pollen grains later arise. Each mother cell divides by the process of meiosis, in which the usual diploid number of chromosomes in the cell nucleus is reduced by half. At the end of this process the original cell has been transformed into four microspores, each with a haploid number of chromo-

somes. The four microspores, collectively called a tetrad, then mature into four pollen grains within the anther cavity.

The tetrad stage is an important one in the development of the pollen grain. The surface features of the mature grain are definitely related to the original orientation of the microspore within the tetrad. One such feature is the single furrow that characterizes one of the two major classes of pollen grains; the furrow develops on the side of the grain that is

not in contact with the other three grains during the tetrad stage. It is also during the tetrad stage that the outer layer of the grain wall develops the more complex configurations characteristic of each genus.

In our study of the development of pollen within the hellebore anther the earliest stage we have recognized so far comes after the differentiation of archesporial tissue into its parietal and sporogenous components. Transmission elec-

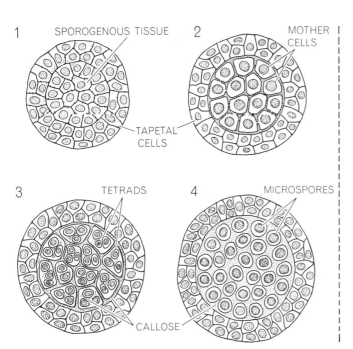

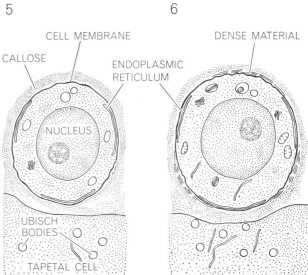

POLLEN-GRAIN DEVELOPMENT is shown in schematic cross sections. The first four stages (*left*) are not at the same scale as the next five. When the two kinds of tissue that comprise the anther interior can first be distinguished (*1*), cells within each kind are linked by bridges but no links exist between the two kinds. The

inner, sporogenous tissue gives rise to pollen mother cells that share a common pool of protoplasm (*2*). Next the mother cells undergo meiotic division until each consists of a tetrad of microspores (*3*). Each tetrad is isolated by a layer of callose tissue. The tetrads then separate (*4*). Meanwhile aggregates that will become Ubisch

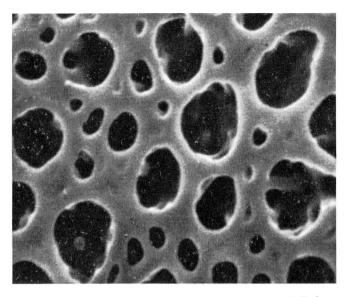

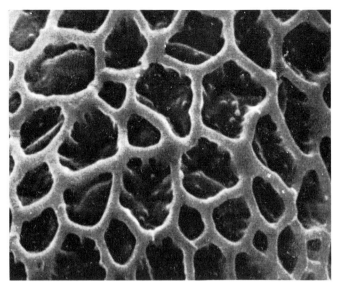

left); it is covered with spines. Pollen of the bluebell (*Endymion*) is magnified 16,000 times (*third from left*); it has many

large and small roof openings. Pollen of the Australian shrub *Callistemon*, magnified 16,700 times (*right*), has only a lacelike trace of a roof.

tron micrographs show sporogenous cells surrounded by tapetal cells. Bridges of protoplasm called plasmodesmata are visible between tapetal cells and between sporogenous cells but not between the two kinds of cell.

The next event we can recognize is the transformation of sporogenous cells into a mass of mother cells. This is the result of mitosis, the process of cell division in which the number of chromosomes is not reduced by half but remains

the same. The pollen mother cells are interconnected by canals through which the cytoplasm passes freely. Accordingly the entire mass of pollen mother cells is a kind of large single cell that shares a common pool of cytoplasm.

Each pollen mother cell continues to grow in size and is progressively enveloped in a layer of callose, an amorphous cell-wall substance. Callose has been shown to be resistant to the diffusion of relatively small molecules; it seems clear

that its function at this stage is to act as a barrier between mother cells. Thus far the sharing of cytoplasm has presumably presented no problems. With the microspores about to emerge, however, the requirements of genetic identity probably demand that each nucleus act within its own independent unit of cytoplasm. In any case it is not until each mother cell is successfully isolated from the others by a layer of callose that the mother cell undergoes meiosis and is

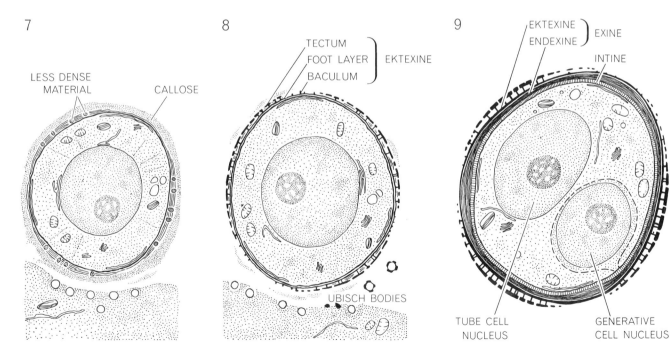

bodies have formed in the surrounding anther tissue, the tapetum. Networks of dense endoplasmic reticulum now begin to form in the cytoplasm within each microspore (*5, only one microspore shown*). In areas where these networks are absent the microspore cell membrane becomes coated with a dense fibrous material (*6*), containing

gaps in which a less dense material now appears (*7*). Callose diminishes (*8*), the maturing grain's layer of ektexine develops and Ubisch bodies erupt from the tapetum. As the endexine and the intine appear (*9*) the pollen grain reaches the state of full maturity and its nucleus divides into a pollen tube cell and a generative cell.

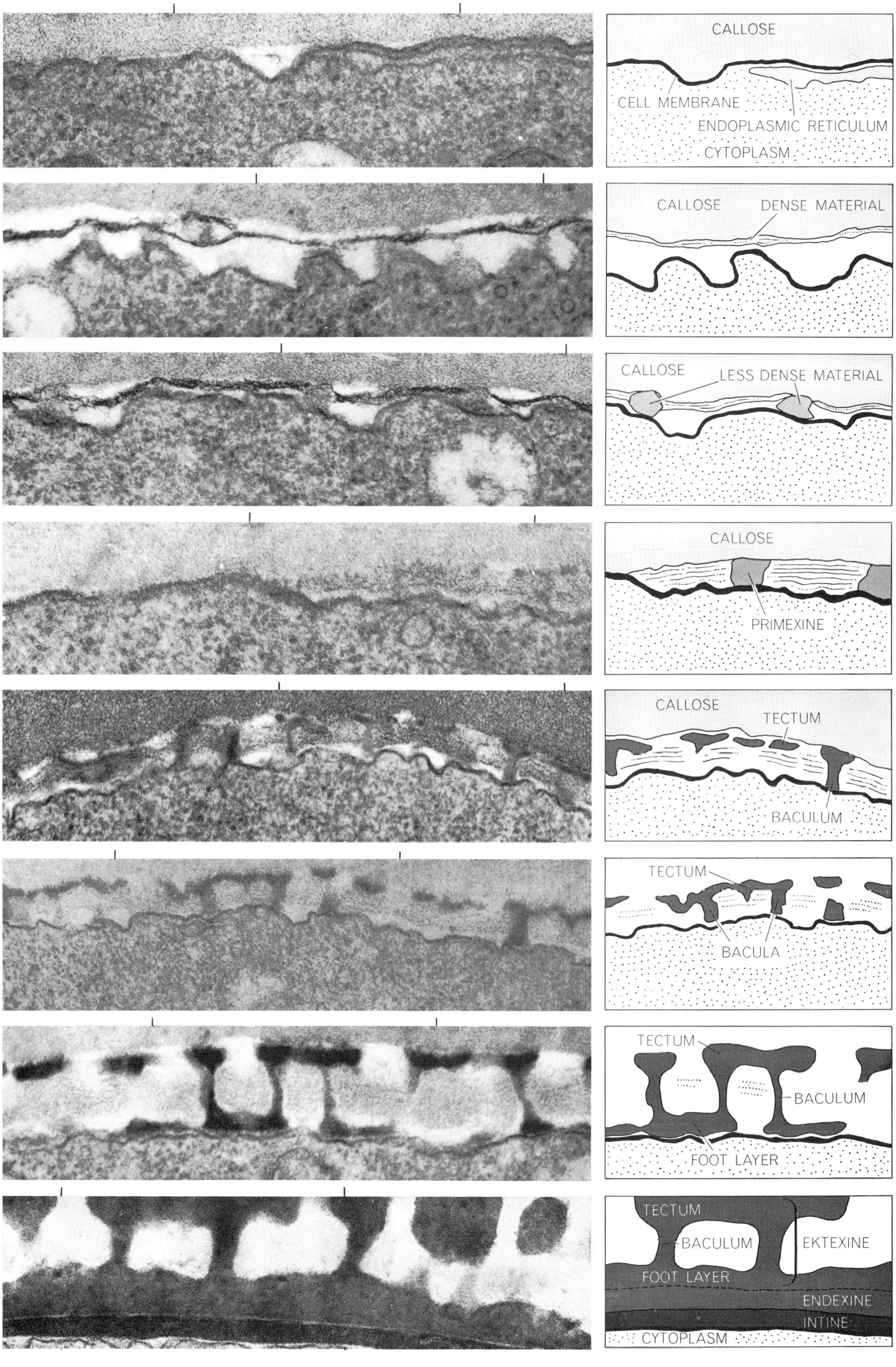
CALLOSE
CELL MEMBRANE
ENDOPLASMIC RETICULUM
CYTOPLASM
CALLOSE
DENSE MATERIAL
CALLOSE
LESS DENSE MATERIAL
CALLOSE
PRIMEXINE
CALLOSE
TECTUM
BACULUM
TECTUM
BACULA
TECTUM
BACULUM
FOOT LAYER
TECTUM
BACULUM
EKTEXINE
FOOT LAYER
ENDEXINE
INTINE
CYTOPLASM

transformed into a tetrad of microspores. Each of the four microspores is similarly sheathed in callose and isolated from its siblings.

In the next stage the first precursor of the mature pollen grain's exine outer coat appears. The microspore is still enclosed within its wall of callose, but the cell membrane that forms a boundary between the callose and the microspore cytoplasm begins to increase in thickness. A network of denser substances in the microspore cytoplasm, consisting of elements in the endoplasmic reticulum, appears in some regions immediately below the cell membrane. We believe the region in which this endoplasmic reticulum develops corresponds to the area on the surface of the mature pollen grain where the furrow (or furrows) will eventually form.

In the regions not associated with endoplasmic reticulum the thickened cell membrane now has a convoluted appearance. A thin layer of fibrous material, which is dark in transmission electron micrographs, begins to appear along the top of the convolutions outside the cell membrane. When the fibrous material has reached a certain thickness, gaps appear in it, and material that has a somewhat lighter appearance in transmission electron micrographs is deposited in the gaps. The regions where the deposition of this lighter (and more uniform) material takes place correspond to the areas where the first elements of the precursor to exine, known as primexine, are subsequently deposited. This phase of maturation thus appears to be the one in which the surface pattern of the mature pollen grain is determined.

The progression from primexine to exine is not difficult to follow. Elements of the primexine give rise to precursors of the rodlike bacula of the mature exine. Hellebore pollen grains have a distinctive roofed appearance; our studies show how the roof develops in a series of steps. The first is the positioning of the bacula, which in cross section look like pillars. The upper end of each baculum now spreads sideways in all directions. When the spreading tips meet, they form a perforated roof on top of the pillars. While this is going on the lower ends of the bacula also spread sideways and coalesce to form a floor at the base of the pillars. Roof, pillars and floor together comprise the outermost, or ektexine, layer of the pollen-grain wall. As the ektexine structure is completed it is impregnated with sporopollenin. As this feat of construction and coating is being accomplished the layer of callose that surrounds the maturing pollen grain gradually becomes thinner. Finally the callose layer disappears, the tetrads break apart and the individual grains of pollen lie free within the anther.

During the microspores' growth into mature pollen grains structural changes also occur in the cells of the tapetum, the inner wall of the anther. At the same time that meiosis is transforming mother cells of the hellebore into tetrads, the tapetal cells of the anther enlarge and begin to form curious spherical bodies. These bodies, which in their eventual free state are called Ubisch bodies (after their discoverer, G. von Ubisch of the University of Heidelberg), are at first closely associated in the tapetal cells with rows of the cytoplasmic particles known as ribosomes. Later they are associated with the tapetal cells' network of endoplasmic reticulum.

The next change to occur in the tapetum coincides with the microspores' development of their first structural wall elements. Some of the spherical bodies within each tapetal cell now begin to migrate to the cell surface, where they break through into the anther cavity. As soon as a sphere breaks through it is encased in sporopollenin and can be considered a mature Ubisch body.

It must be more than coincidence that the Ubisch bodies receive their coat of sporopollenin at the same time the ektexine elements of the pollen-grain wall do. We believe that most tapetal cells and pollen-grain cells alike have the capacity to secrete sporopollenin but that only in developing pollen grains is the substance laid down in an elaborately organized manner.

During the period when the spherical precursors of the Ubisch bodies appear and begin to migrate the cells of the tapetum show increasing signs of dissolution. Their cellulose walls diminish in thickness, their cytoplasm becomes less dense and many of the subcellular bodies in it become unrecognizable. By the time the Ubisch bodies break free into the anther cavity, the tapetal-cell walls have disappeared and only a thin cytoplasmic membrane surrounds each tapetal cell. What function the Ubisch bodies serve is not known.

As the pollen grain's protective ektexine covering is forming, the second component of the outer coat—the endexine—is deposited below it. The process involves a number of thin white membranes that appear to arise from the cytoplasm and provide a locus around which sporopollenin is deposited. As the deposition proceeds the membranes grow into plates that finally merge to form the endexine zone. No sign of either the membranes or the plates can be seen in the mature pollen-grain wall.

Formation of the endexine zone signals that the pollen grain has almost reached maturity. All that remains now is for the last component of the grain wall—the intine—to form inside the endexine. This takes place quite rapidly. At about the same time the nucleus of the pollen grain also takes a last step toward maturity: the simple mitotic division that gives rise to the male sexual nucleus and the nucleus of the future pollen tube. Recently Roger Angold, working in our laboratory, has shown that in the development of the pollen grain of the bluebell (*Endymion*) the male generative, or sexual, nucleus becomes surrounded by a thin wall of cellulose. This wall forms in close association with the intine, then surrounds the generative nucleus and eventually pinches off and becomes completely separate from the intine. Our own studies have shown that a similar wall is formed at this stage in hellebore pollen.

As these climactic events occur in the pollen grain the tapetal cells of the anther wall also make their final contribution. Within each cell the spherical bodies that have not broken free to become Ubisch bodies gradually coalesce, and when the tapetal-cell membrane finally dissolves, the aggregates enter the anther cavity. It may be that these coalesced bodies give rise to the sticky material—an oleaginous substance called tryphine—that covers the mature pollen grains of most insect-pollinated plants.

Almost all we know about the details of pollen growth and development comes from the analysis of electron micrographs. Returning to the questions raised earlier—whether the surface patterns of pollen grains are created by nature or nurture, something within the pollen grain itself or some element in the microenvironment—it is interesting to note that the first responses to these questions were given long before electron microscopy was available as a research tool.

It is quite clear to us that, in hellebore at least, the patterning of the pollen

DEVELOPMENT of the outer coating of a hellebore pollen grain is seen in a series of transmission micrographs magnified 50,000 times (*opposite page*). Marks above each photograph indicate area in the drawing at right, in which structures are identified.

grain's surface is a consequence of information inherited by the microspore when meiosis transforms each mother cell into four microspores. We have selected as the key point in time the moment when locations for the later deposition of primexine appear in the fibrous material between the microspore membrane and its covering of callose. Are there any factors within the microspore cytoplasm that determine these locations? As long ago as 1911 Rudolf Beer observed thin threads radiating from the nucleus of the pollen grain of the morning glory (*Ipomoea*). Beer hinted that in some mysterious way these threads determined the organization of the exine. With the aid of the electron microscope we have located in the cytoplasm of hellebore pollen grains that appear to be microtubules radiating from the nucleus in the manner of Beer's morning-glory threads. We believe the microtubules, together with other cytoplasmic organelles called dictyosomes, may be the factors that determine the pollen grain's exine pattern. This view, based at least in part on a duplication of Beer's finding of nearly 60 years ago, is not unchallenged. John Heslop-Harrison of the University of Wisconsin, together with John Skvarla and Donald A. Larson of the University of Texas, believe surface patterns in the pollen grains they have investigated are determined by the endoplasmic reticulum instead.

Can the microenvironment be dismissed as an insignificant factor in the patterning of pollen-grain surfaces? Again we can turn to an earlier answer. In 1935 Roger P. Wodehouse, then with the Arlington Chemical Company, produced convincing evidence that in many types of pollen the pattern of pores and furrows bore a close relation to the contact geometry of the microspores during their association in the tetrad. Our hellebore studies have produced evidence in general support of Wodehouse's conclusion: One of the three furrows that mark each hellebore pollen grain may well be initiated at the point of common contact within the tetrad.

It would thus appear that both heredity and environment play a part in determining the complex surface patterns of pollen grains. If one or the other factor must be declared the more significant, heredity far outweighs environment. Granting that some gross features such as furrowing may be attributed to tetrad contact geometry, all the rest—including the myriad variations in wall structure and patterning—is the product of inheritance alone.

The Author

PATRICK ECHLIN is head of the electron microscopy laboratory in the department of botany at the University of Cambridge and a fellow of Clare Hall. He obtained a teacher's certificate at Goldsmiths' College of the University of London in 1954 and a bachelor's degree in botany at University College London in 1957. For the next five years he was a Fulbright Scholar at the School of Medicine of the University of Pennsylvania, where he received a Ph.D. in medical microbiology in 1961. In addition to pollen his research interests include the biology of blue-green algae (he was the author of the article "The Blue-Green Algae" in the June 1966 issue of SCIENTIFIC AMERICAN) and the origins of photosynthesis. He writes that outside of his work his interests include "photography, gourmet cooking and the camping variety of worldwide travel."

Bibliography

HALF A CENTURY OF MODERN PALYNOLOGY. A. A. Manten in *Earth-Science Reviews*, Vol. 2, No. 4, pages 277–316; December, 1966.

POLLEN. H. F. Linskens in *Handbuch der Pflanzenphysiologie/Encyclopedia of Plant Physiology, Vol. XVIII: Sexuality, Reproduction, Alternation of Generations*, edited by W. Ruhland. Springer-Verlag, 1967.

POLLEN GRAINS. Roger P. Wodehouse. McGraw-Hill Book Company, 1935.

POLLEN MORPHOLOGY AND PLANT TAXONOMY. Gunnar Erdtman. Hafner Publishing Company, 1966.

THE PRINCIPLES OF POLLINATION ECOLOGY. Knut Fægri and L. van der Pijl. Pergamon Press, 1966.

TEXTBOOK OF POLLEN ANALYSIS. Knut Fægri and Johannes Iversen. Blackwell Scientific Publications, 1964.

SCIENTIFIC
AMERICAN   April 1968, Vol. 218, No. 4, pp. 108-116

OFFPRINT 1106

THE SEXUAL LIFE OF A MOSQUITO

by Jack Colvard Jones

Modern methods of insect control call for detailed knowledge of an insect's physiology and behavior. Reproduction in *Aedes aegypti,* the yellow-fever mosquito, is surprisingly elaborate.

The spectacular success achieved in recent years in eradicating the screwworm fly by blocking its reproduction has encouraged hopes that a similar strategy might be effective against a much more serious insect pest—the mosquito. Mosquitoes, transmitting malaria, yellow fever, encephalitis and other grave diseases, have probably caused more human deaths than any other group of insects. The control or elimination of the mosquito, like the control or elimination of any insect pest, may call for an entire arsenal of judiciously chosen methods, and new methods are constantly being sought. Among the most attractive methods are biological ones: they are aimed at a specific pest and do not involve the use of chemicals that may be hazardous to other organisms. Biological methods of control, however, require an intimate knowledge of the target species' way of life.

The surest method of eradicating a species is to destroy its ability to reproduce. In the case of the screwworm fly this was accomplished by sterilizing males and releasing them into the wild population in saturating numbers. This tactic is now being studied with various species of mosquito, and it has proved highly effective in the laboratory. Unfortunately it has not met with much success under natural conditions. Clearly there is a need for more information about the sexual life and reproductive mechanisms of mosquitoes. The relatively small amount of information already available has suggested a number of approaches to preventing the reproduction of the insect and bringing about the autocide of disease-carrying species.

The sexual behavior of mosquitoes varies remarkably from species to species. The species for which the most de-

tailed information has been gathered is the yellow-fever mosquito *Aedes aegypti,* which has probably caused as much human illness and as many human deaths as any other man-biting mosquito. Here I shall describe what has been learned about the sexual life of this species by many investigators, including my colleagues and myself at the University of Maryland.

Every mosquito begins its life with the fertilization of the egg. Within the protective shell of the egg the embryo develops rapidly. When eggs containing fully developed embryos are placed in water that is poor in oxygen, they give rise to larvae. In about a week the larvae are mature. All mosquito larvae go through four distinct stages of growth, shedding their old skin before each new stage. Microscopic examination of the insect in its last larval stage shows that it has already formed the rudiments of the antennae, eyes, legs and wings of an adult mosquito. The antennae are enclosed within two small sacs in the larva's head. These sacs disclose the individual's sex: if they are large and well developed, the larva will give rise to a male; if they are small and poorly developed, the adult will emerge as a female.

After the transformation from the larval to the pupal state, the pupa (which to the unaided eye looks something like a comma) completes the development of the anatomical structures that will characterize the adult. In this development the alimentary tract is completely remodeled, the female forms its internal reproductive organs, the male develops mature sperm in its testes and many other changes hidden from surface view take place. The decaying larval tissues provide food for the synthesis of the adult tissues. In about two days the pupa sheds its skin and the insect emerges

as an adult mosquito. The female then differs from the male not only anatomically but also in behavior. For example, whereas the *Aedes aegypti* male feeds only on plant nectar and water, the female has a thirst, in addition, for animal blood—a thirst the male mosquito does not share because its proboscis lacks the necessary cutting tools.

The male is relatively small and is distinguished by large, hairy antennae. Shortly after the male's emergence into the winged state its rear end undergoes a remarkable rotation. The last two segments of the abdomen, pivoting on the membrane between the seventh and the eighth segment, begin to rotate (either clockwise or counterclockwise). In the first three hours this end portion turns 90 degrees, and by the 20th hour it has made a full turn of 180 degrees, so that the male's rear end is upside down from its original position. The change is permanent. Were it not for the 180-degree reversal of the abdominal tip, the male would be unable to copulate with the female—whose abdomen always retains its original position.

Similar rotations of the male abdomen are characteristic of all Diptera (two-winged insects). What kind of mechanism is responsible for this curious twist? The process is not fully understood. There is no indication that the muscles of the body wall cause the rotation. It appears that the twisting force may be applied by powerful rotational contractions of the hindgut, operating rather like a screwdriver. It has also been suggested that the membrane on which the end segments pivot may originally be plastic and then slowly harden during the rotation, thus fixing the rear end in the new position. The membrane itself shows no external

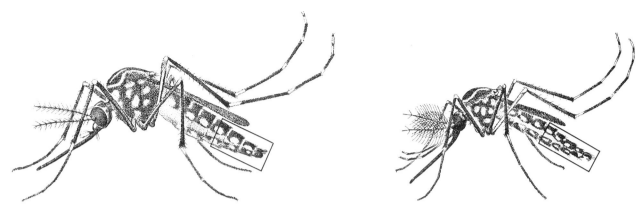

MALE AND FEMALE YELLOW-FEVER MOSQUITOES are shown 10 times actual size. The adult female (*left*) is larger than the adult male (*right*), and her antennae are less elaborate. The two rectangles outline the parts depicted in the cross sections below.

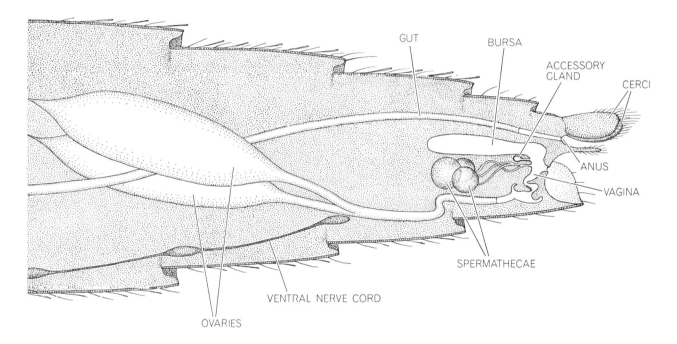

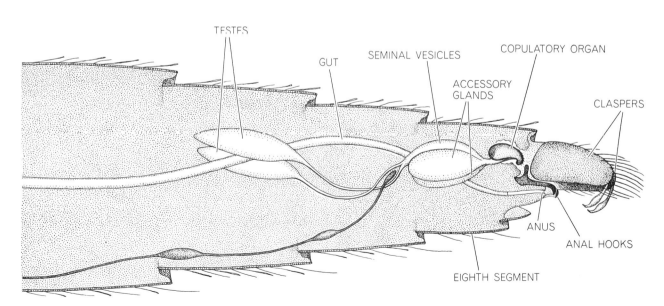

REPRODUCTIVE ORGANS of the female mosquito (*top*) and the male (*bottom*) appear in longitudinal section. The external struc-ture of the male's genital apparatus is more complex than that of the female; her internal structure is more complex than the male's.

sign of having been twisted, but if the insect is dissected, the twist is clearly visible in internal organs such as the tracheae (air tubes), the nerve cord and the sperm ducts.

During the rotation of the male's rear end, spermatozoa from its two testes pass down thin-walled sperm ducts to fill two seminal vesicles at the end of the abdomen. The sperm, consisting of needle-like heads and long tails, all become precisely aligned inside the vesicles with the heads pointing toward the abdominal tip. On both sides of the seminal vesicles are large, pear-shaped accessory glands that secrete a major component of the seminal fluid. A small ejaculatory duct opens into the male's copulatory organ.

The genital apparatus is extremely complex [see bottom illustration on opposite page]. The copulatory organ itself, deeply retracted in a fleshy pocket, is composed of a hinged pair of tiny, curved plates; these are intricately attached to a pair of large claspers positioned externally on the abdominal tip. A large anal cone between the two claspers obscures the copulatory organ. On both sides of the cone are hooks that serve as grasping accessories. The external genital structures of male mosquitoes are so intricate and distinctive that taxonomists use them to identify many mosquito species.

Whereas the male mosquito's reproductive system is complex on the outside and relatively simple on the inside, the female's is relatively simple on the outside and complex on the inside. Externally it consists of two paddle-like plates (called cerci) with long sensory hairs above the anus and a long tonguelike structure next to the retracted vagina below the anus. The vagina is S-shaped and contains three interlocking valves of different shapes, one of them with teeth along its tip. Four distinct internal structures open into the vagina: a sac called the bursa, which first receives sperm from the male; a tiny, globular accessory gland; three spherical organs called spermathecae to which the sperm migrate by way of a tiny funnel and long, twisted ducts, and a long oviduct through which the eggs pass from the ovaries.

In general female mosquitoes must feed on blood to develop eggs. Some species can develop the first batch of eggs without having had a blood meal, but even these require blood in order to lay subsequent batches. Some species must have more than one blood meal before

ROTATION of the tip of the male mosquito's abdomen takes place early in adult life. Before the 180-degree turn (which may be clockwise or counterclockwise) the clasper claw points up (left); it points down when the turn is completed about 20 hours later (right). With the tip reversed, copulation can occur (see illustrations at bottom of next two pages).

their eggs can mature. The blood meal initiates a chain of essential physiological events in the female. It is believed that her stomach, greatly distended by the drink of blood, presses on the nerve cord and causes it to send electrical signals to the brain. Within an hour the message excites certain cells in the upper part of the brain to secrete a hormone into the insect's circulating body fluid, and this in turn results within a few hours in the secretion of a secondary hormone from a pair of small glands in the female's neck. The latter hormone triggers a spectacular series of events in the ovaries. Submicroscopic pits (as many as 300,000 of them) appear on the surface of the egg. The eggs then begin to imbibe protein from the fluid in the ovary that collects in the surface pits; droplets of yolk form rapidly and soon fill the egg. The egg cell enlarges enormously and the egg nucleus, originally large, shrinks to a tiny mass of genetic material. A thin shell forms over the egg within the ovary.

A female mosquito that has fed on blood will produce and lay eggs even if she has not been fertilized. Fertilization, however, strikingly increases her egg production. Immediately after her blood meal the female is not very attractive to males, but at any other time she attracts them instantly merely by flying about. The males are drawn by the buzzing sound of the wings. The attractive

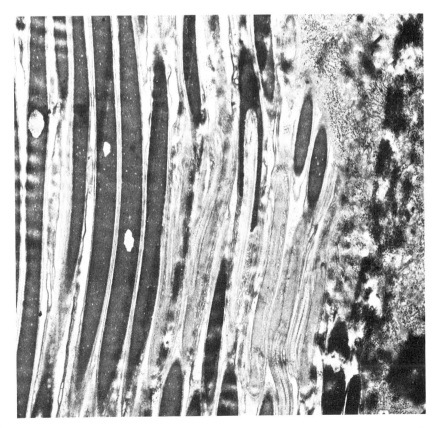

SPERM inside the seminal vesicle, the male's storage sac, are precisely aligned. The headpieces (dark cigar-shaped areas) all point in the same direction. The tails are wormlike. At right is the wall of the vesicle (mottled area). Magnification in this electron micrograph, made by Victor H. Zeve of the National Institutes of Health, is about 15,000 diameters.

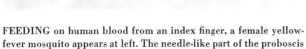

FEEDING on human blood from an index finger, a female yellow-fever mosquito appears at left. The needle-like part of the proboscis is piercing the skin. At right the mosquito is seen with abdomen swollen from the blood meal. Only the female insect drinks blood.

sounds are in the range between 300 and 800 vibrations per second. Experiments conducted by Louis M. Roth, then at the U.S. Army Quartermaster Research and Development Center, show that a male mosquito will pursue a sound in this range regardless of the source—whether it is male or female, of its own or another mosquito species or even simply a tuning fork. (Mature females and males just emerged from the pupal state beat their wings at about the same rate and are pursued by older males; young, newly emerged females have a different beat and therefore are not pursued.) According to Roth, the male's hearing range is only about a foot. Roth has also demonstrated that the male is deaf to the flying female if his antennae are removed or prevented from vibrating.

The stimulating sound causes the male not only to pursue the source but also to seize it with his claspers. He will clasp the cloth walls of a cage if a tuning fork

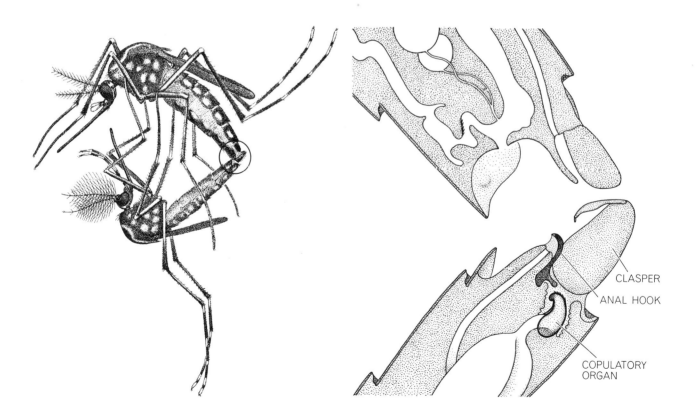

CLASPER

ANAL HOOK

COPULATORY
ORGAN

COPULATORY POSITION, bringing the male yellow-fever mosquito below the female, is shown at left. In the schematic diagram at right of the anatomy of the male and female insects, the organs are drawn as they appear before coital contact. The male's copulatory organ (here retracted), the anal hook and the clasper are rigid elements; in coitus they stretch and distort the female's more flexible tissues.

is sounded outside the walls, and indeed will seize anything that comes within his reach in the direction of a suitable vibrating source, including a male mosquito.

The process of copulation with a female is rapid but complex. On reaching the flying female the male first seizes her back with his legs, the tips of which are equipped with little grappling hooks. Then, with remarkable agility, he swiftly swings around until he is hanging face-to-face below his partner. The female, exhibiting no obvious response, may continue her flight, or the pair may fall to the floor of the cage. In either case the male quickly proceeds to bring his genitalia into contact with those of the female [see illustrations at bottom of opposite page and below]. His claspers grasp her cerci; this causes the tongue-like plate under her anus to move upward and expose the edges of the vagina. The male then uses his anal hooks to pull the female's genitalia toward him. He rapidly extends his previously retracted copulatory organ so that teeth at its tip mesh with the teeth on the dorsal valve in the vagina. The forceful entry everts the valve, and the hinged plates of the male's organ then spread out and widen

the vagina, enlarging the opening into the bursa. At that instant the male organ discharges a large quantity of seminal fluid, containing about 2,000 sperm, into the bursa. The pair then quickly separate, frequently with a parting kick against the male by the female's hind legs. The entire copulatory act takes from 14 to 20 seconds under natural circumstances, and about 30 seconds when a nonflying pair of mosquitoes are artificially induced to mate in the laboratory under a microscope.

Experiments in our laboratory have established that an *Aedes aegypti* male can be induced to mate mechanically if its rear end is placed in contact with the genitalia of an unfertilized female. The technique is quite simple. A male and a female are lightly anesthetized, a drop of glue is put on the head of a pin and the glue is applied to the abdomen of each insect. The mosquitoes are then rubbed together while being observed under the microscope. The technique makes possible a number of experiments that otherwise could not be done. With it we have demonstrated that the rear end of the male alone, immediately after being severed from the rest

of the body, is capable of copulating with and inseminating a female. A noteworthy and perhaps critically important finding is the fact that in such experiments only the rear end of an unfertilized female will cause the male to extend his copulatory organ and ejaculate. When a fertilized female is offered mechanically to a male under the microscope, he somehow recognizes instantly that she has been mated and he makes no attempt to copulate with her; indeed, he frequently draws away. A flying female, however, may induce the male to copulate even if she has been fertilized. Andrew Spielman and his colleagues at the Harvard School of Public Health have recently found that if these females are reinseminated, the mass of semen is rapidly ejected. Presumably the enticement of the female's buzzing overrides the male's recognition of her fertilized condition.

A male mosquito can inseminate five or six females in rapid succession. He may copulate with many more—as many as 30 within 30 minutes—but his supply of seminal fluid and sperm is exhausted after five or six matings. It takes about two days for the male to refill the seminal vesicles with a fresh supply of sperm and

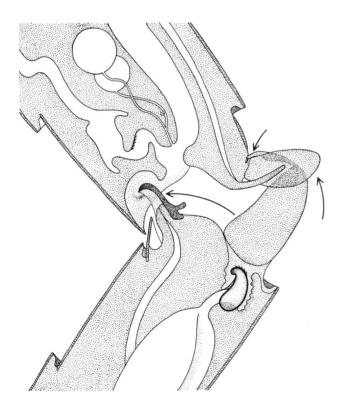
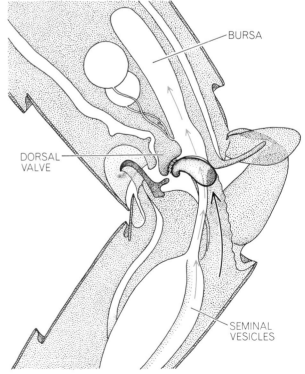

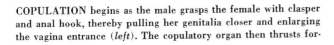
COPULATION begins as the male grasps the female with clasper and anal hook, thereby pulling her genitalia closer and enlarging the vagina entrance (*left*). The copulatory organ then thrusts forward, engages the teeth of the dorsal valve and spreads to widen the entrance into the bursa (*right*). In this position the male ejaculates into the bursa sperm from the seminal vesicles (*colored arrows*).

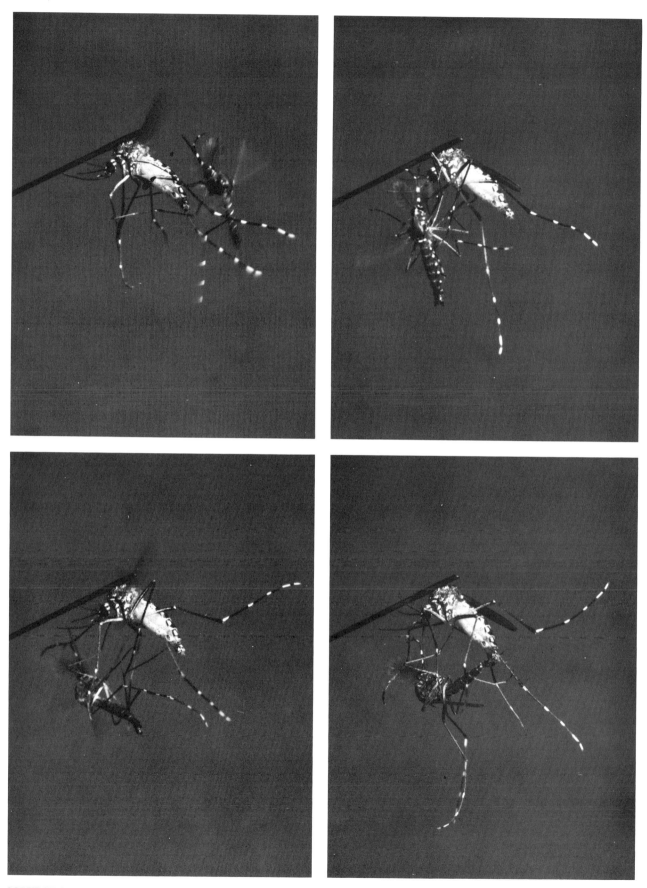

COPULATING MOSQUITOES belong to the yellow-fever species *Aedes aegypti*. The male mosquito is attracted to the female (tethered to a fine steel wire by rubber cement) by the sound of her moving wings (*top left*). With tiny hooks on his leg tips he grasps the female while swinging around to hang below her (*top right*). In this position he brings his genital organs into contact with those of the female (*bottom left and right*). These photographs and the others in this article were made by Thomas Eisner of Cornell University.

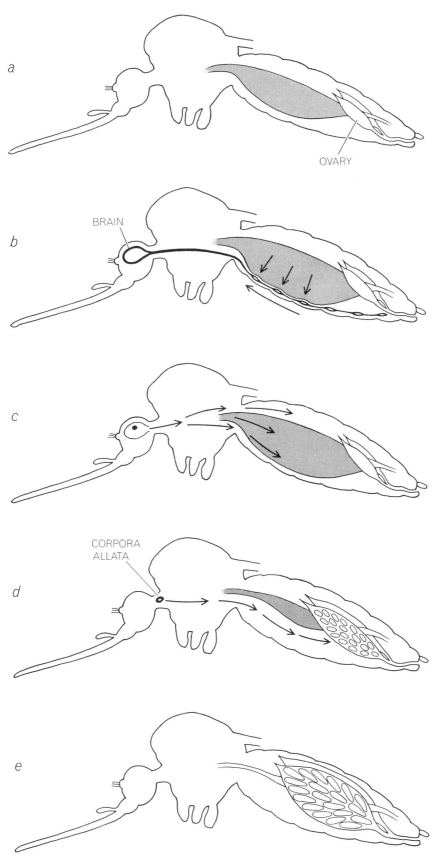

DEVELOPMENT OF EGGS begins after the female mosquito imbibes blood (*color*) and her gut distends (*a*). The gut then presses on the ventral nerve cord and signals are sent to the brain (*b*). In response certain brain cells secrete a hormone into the insect's circulating fluid (*c*), giving rise to the secretion of a second hormone by the corpora allata (glands in the neck). With the arrival of this hormone at the ovary (*d*), protein from the blood enters the ovary fluid and is taken up by the eggs. Yolk droplets form and eggs enlarge (*e*).

regenerate the accessory-gland material.

Let us turn our attention now to the female. After her bursa has been filled with semen from the male the long, threadlike sperm swim toward the opening of the bursa and line up in dense bundles with their tails undulating rapidly. They remain there for 30 to 40 seconds. Then groups of the sperm abruptly make a sharp U-turn, swim to the funnel leading to the spermathecae and make their way up long, twisted ducts to those receiving organs. Generally they reach only two of the three spermathecae. Within less than five minutes the migrating sperm (perhaps 1,000 of them, probably together with seminal fluid) have filled the two reservoirs, and there the sperm swim rapidly in circles and remain active as long as the female lives. In short, once the female has been mated, she is fertilized for the rest of her lifetime.

The behavior of the sperm in the female's reproductive tract presents several puzzles. Why do the highly active sperm wait in the bursa for 30 to 40 seconds before starting their journey to the spermathecae? How do they make their way there without error, although neither the bursa nor the ducts provide contractions that might give them direction? Why do no sperm or so few sperm enter the third spermatheca? Curiously, our experiments have failed to develop any definite proof that the sperm can actually travel to the spermathecae without assistance. In fact, our efforts to fertilize females by artificially inseminating their bursae with highly active sperm and accessory gland material have generally been unsuccessful. We deduce from these facts that in some unknown way the female exercises a control of her own over fertilization.

At all events, the consequences of her fertilization are quite clear. The sperm that reach her spermathecae enable her to produce fertile eggs. And a fertilized female lays a great many more eggs than an unfertilized one, as a result of the slow absorption into her body of the seminal material (sperm and granular material from the male's accessory glands) that remains behind in her bursa after the migration of sperm to the spermathecae stops. It is not clear how the spermatozoa get out of the spermathecae to reach the egg. The sperm cell does, however, enter the egg through a specialized tiny opening at its front end just as it is laid.

The female lays her eggs one at a time. After finding a suitable water site for depositing them, she first lifts her hind legs and fans them up and down. Then she wipes the tip of her abdomen with them and vigorously scrubs them together. The tip of the abdomen almost touches the water surface. Her cerci and the tonguelike plate below the anus are pressed close together and point down. As an egg descends from the ovary through the oviduct there is considerable contraction and twitching of the genital structures, particularly the edges of the vagina. In emerging from the vagina the large white egg pushes the cerci and subanal plate upward, everts the vaginal opening and then drops into the water. One egg follows another until the female has delivered her batch. Some species of mosquito array the eggs in rafts on the water. Some species lay their eggs while hovering over the water, the eggs dropping like little bombs. Others eject their eggs into water that has collected in the hollow of a tree. There is even a mosquito that lays her eggs on her hind legs and then dips her legs into the water to set the eggs free.

The female mosquito has a remarkable capacity for retaining eggs in her ovaries after the eggs have fully ripened. The majority of mosquitoes will not lay eggs unless they have located an appropriate site for them. We have been able to force such females to deposit their eggs under the microscope, however, by a shock treatment such as crushing the head or thorax or cutting off the head or abdomen. Within less than a minute after the injury eggs begin to emerge from the vaginal opening, and occasionally a considerable number will be laid.

We have also investigated the action of portions of the egg-delivering system.

If the ovary itself is cut away from the body when it is greatly distended with ripe eggs and placed in a drop of saline solution on a glass slide, eggs will slide out through the open end. It appears, however, that in normal circumstances the main force for drawing eggs out of the ovary under natural conditions comes from vigorous rhythmic contractions of the internal oviducts (the exit ducts of the ovaries). Ordinarily the lateral oviducts are constricted, and this constriction may be responsible for the ovary's retention of ripe eggs until the female has found a suitable site in which to lay them.

The female mosquito invariably lays her eggs in daylight, principally in midafternoon between 2:00 and 3:00 P.M. It has been demonstrated by A. J. Haddow and J. D. Gillett of the East African Virus Research Institute in Uganda that light controls the egg-laying rhythm (perhaps by activating some hormone); these workers showed that the mosquito will lay her eggs in the nighttime hours and refrain in the daytime hours if the light cycle in her cage is reversed.

A female that is ready to lay eggs explores possible sites primarily with her feet. She flies about and alights on various water surfaces, setting all six legs down on the water. The hairs on her legs evidently possess a chemical sensitivity, particularly to salinity. It has been shown by Robert C. Wallis, who was then at the Johns Hopkins School of Medicine, that they can distinguish between distilled water and weak saline solutions. Most species of mosquito prefer fresh water for laying their eggs, but some favor brackish waters. On finding a favorable site the female walks to the edge of

the water and lays her eggs there.

Soon after they are laid the white eggs swell and turn black. Normally the egg hatches into a larva in two days. If the conditions are not suitable, the larva, even though they are fully developed within the shell of the egg, can remain alive for a considerable time. Charles L. Judson of the University of California at Davis has shown that when such eggs are placed in water with a low oxygen content, the pharynx of the larvae suddenly becomes active. It makes rapid swallowing motions, apparently without actually imbibing fluid, and proceeds to break out of the shell. A small spine on top of its head presses against the shell along a preformed line and the top of the shell neatly snaps off.

The larva wriggles out quickly, air suddenly enters its breathing tubes, which take on the appearance of silvery capillaries, and its heart begins to beat. The insect at once starts to feed, and so a new cycle of mosquito life gets under way.

There are definite periods in the life cycle of mosquitoes in which reproduction can be profoundly affected by some specific treatment. William R. Horsfall and J. R. Anderson of the University of Illinois have demonstrated, for example, that the sex of larvae in certain strains of mosquito can be completely reversed by subjecting the larvae to a certain temperature at a certain period. Many other entomologists are examining similar phenomena in the sexual life of mosquitoes in the hope of finding some way of eliminating the insect when it is not wanted.

The Author

JACK COLVARD JONES is professor of entomology at the University of Maryland. He was graduated from Auburn University in 1942 and received a master's degree there in 1947. In 1950 he obtained a Ph.D. from Iowa State University. From 1950 to 1958 he was at the National Institutes of Health, investigating the anatomy, histology and physiology of *Anopheles quadrimaculatus*, the southeastern malaria mosquito, and also studying the mode of action of various insecticides on the hearts of those insects. He has been at the University of Maryland since 1958. Of his work there he writes: "Teach general and advanced insect physiology. Conduct research on the anatomy and physiology of the reproductive system of mosquitoes and on the structure and function of insect blood cells in general. Interest in comparative hematology extends from 1946 to present. Interest in the circulatory system of insects in general extends from 1950 to present." Jones plans to begin studying the alimentary canal of mosquitoes in the near future.

Bibliography

A STUDY OF MOSQUITO BEHAVIOR. Louis M. Roth in *The American Midland Naturalist*, Vol. 40, No. 2, pages 265–352; September, 1948.

A STUDY OF OVIPOSITION ACTIVITY OF MOSQUITOES. Robert Charles Wallis in *American Journal of Hygiene*, Vol. 60, No. 2, pages 135–168; September, 1954.

SCIENTIFIC
AMERICAN   May 1968, Vol. 218, No. 5, pp. 83-90

OFFPRINT **1107**

THE FLIGHT-CONTROL SYSTEM
OF THE LOCUST

by Donald M. Wilson

Groups of nerve cells controlling such activities as locomotion are
regulated not only by simple reflex mechanisms but also by behavior
patterns apparently coded genetically in the central nervous system.

Physicists can properly be concerned with atoms and subatomic particles as being important in themselves, but biologists often study simple or primitive structures with the long-range hope of understanding the workings of the most complex organisms, including man. Studies of viruses and bacteria made it possible to understand the basic molecular mechanisms that we believe control the heredity of all living things. Adopting a similar approach, investigators concerned with the mechanisms of behavior have turned their attention to the nervous systems of lower animals, and to isolated parts of such systems, in the hope of discovering the physiological mechanisms by which behavior is controlled.

One way to approach the study of behavior mechanisms is to ask: Where does the information come from that is needed to coordinate the observable activities of the nervous system? We know that certain behavior patterns are inherited.

This means that some of the informational input must be directly coded in the genetic material and therefore has an origin that is remote in time. Nonetheless, probably all behavior patterns depend to some degree on information supplied directly by the environment by way of the sense organs. Behavior that is largely triggered and coordinated by the nervous input of the moment is commonly called reflex behavior. Much of neurophysiological research has been directed at the analysis of reflex behavior mechanisms. Recent work makes it clear, however, that whole programs for the control of patterns of animal activity can be stored within the central nervous system [see "Small Systems of Nerve Cells," by Donald Kennedy; SCIENTIFIC AMERICAN Offprint 1073]. Apparently these inherited nervous programs do not require much special input information for their expression.

I should point out here that whereas there is now general agreement among

biologists that many aspects of animal behavior are under genetic control, it is not easy to show in particular cases that a kind of behavior is inherited and not learned. I believe, however, that this is a reasonable assumption for the cases to be discussed in this article, namely flight and walking by arthropods (insects, crustaceans and other animals with an external skeleton).

The studies I shall describe were begun as part of an effort to demonstrate how several reflexes could be coordinated into an entire behavior pattern. Until recently it was thought by most students of simple behavior such as locomotion that much of the patterning of the nervous command that sets the muscles into rhythmic movement flowed rather directly from information in the immediately preceding sensory input. Each phase of movement was assumed to be triggered by a particular pattern of input from various receptors. According to this hypothesis, known as the peripheral-

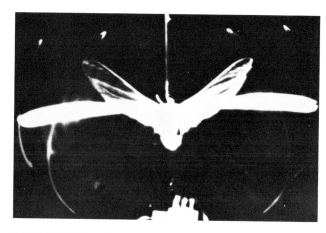

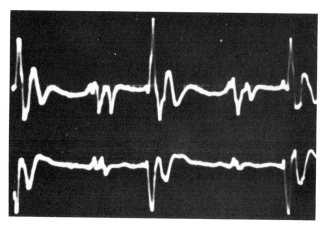

LOCUST WING position and wing-muscle action potentials were recorded in synchronous photographs. The flash that illuminates the locust (*left*) is triggered by the first muscle potential (*at left on oscilloscope trace*). The wing motion is traced by spots of white paint on each wing tip that reflect room light through the open shutter. The trace at the top shows three "doublet" firings of downstroke muscles controlling the forewing; the bottom trace shows similar firings for the hindwing. The smaller potentials visible between the large doublets are from elevator muscles more remote from the electrodes. The oscilloscope traces span 100 milliseconds.

control hypothesis, locomotion might begin because of a signal from external sense organs such as the eye or from brain centers, but thereafter a cyclic reflex process kept it properly timed.

This cyclic reflex could be imagined to operate as follows. An initiating input causes motor nerve impulses to travel to certain muscles, and the muscles cause a movement. The movement is sensed by position or movement receptors within the body (proprioceptors), which send impulses back to the central nervous system. This proprioceptive feedback initiates activity in another set of muscles, perhaps muscles that are antagonists of the first set. The sequence of motor outputs and feedbacks is connected so that it is closed on itself and cyclic activity results. Clearly such a system depends on a well-planned (probably inherited) set of connections among the many parts involved; thus both the central nervous system and its peripheral extensions (the

NERVE AND MUSCLE impulses were recorded during flight with this equipment. The locust is flying, suspended at the end of a pen-dulum, at the mouth of a wind tunnel. The scale at the right registers the insect's angle of pitch. The wires lead to amplifiers.

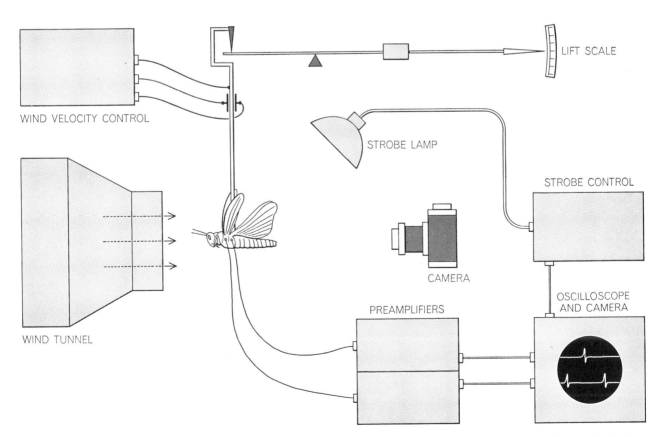

EXPERIMENTAL SETUP is diagrammed. The motion of the pen-dulum controls the wind-tunnel blower, so that the insect can fly at its desired wind speed. Muscle or nerve impulses are displayed on the oscilloscope, which in turn controls the stroboscopic flash lamp.

muscular and sensory structures) are crucial to the basic operation of the system.

An alternative hypothesis, known as the central-control hypothesis, suggests that the output pattern of motor nerve impulses controlling locomotion can be generated by the central nervous system alone, without proprioceptive feedback. This hypothesis has received much support from studies of embryological development. Only a few zoophysiologists have favored it, however, because the existence of proprioceptive reflexes had been clearly demonstrated. It seemed that if such reflexes exist, they must operate.

Proprioceptive reflexes certainly play an important role in the maintenance of posture. I suspect that this may be their basic and primitive function. In many animals—insects and man included—proprioceptive reflexes help to maintain a given body position against the force of gravity. A simple example was described by Gernot Wendler of the Max Planck Institute for the Physiology of Behavior in Germany. The stick insect, named for its appearance, stands so that its opposed legs form a flattened "M" [see illustration at right]. Sensory hairs are bent in proportion to the angle of the leg joints. The hairs send messages to the central nervous ganglia, concentrated groups of nerve cells and their fibrous branches that act as relay and coordinating centers. If too many impulses from the hairs are received, motor nerve cells are excited that cause muscles to contract, thereby moving the joint in the direction that decreases the sensory discharge. Thus the feedback is negative, and it results in the equilibration of a certain position.

If a weight is placed on the back of the insect, one would expect the greater force to bend the leg joints. Instead the proprioceptive feedback loop adjusts muscle tension to compensate for the extra load. The body position remains approximately constant, unless the weight is more than the muscles can bear. If the hair organs are destroyed, the feedback loop is opened and the body sags in relation to the weight, as one would expect in an uncontrolled system.

If the leg reflexes of arthropods are studied under dynamic rather than static conditions, one finds also that they are similar to the reflexes of vertebrates. When a leg of an animal is pushed and pulled rhythmically, the muscles respond reflexively with an output at the same frequency. At high frequencies of movement the reflex system cannot keep up and the output force developed by the muscles lags behind the input move-

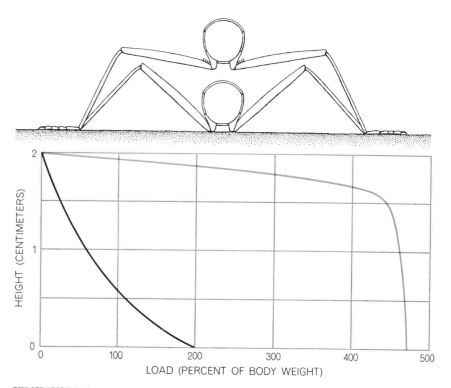

STICK INSECT keeps its body height nearly constant as the load on its leg muscles changes. Hairs at the first leg joint sense the angle of the joint and a reflex loop maintains the angle until the animal is overloaded (*colored curve*). If the sensory hairs are damaged, the reflex loop is opened and the body sags quickly as more weight is added to it (*black curve*).

ment. The peripheral-control hypothesis postulates an interaction of similar reflexes in each of the animal's legs. If one leg is commanded to lift, the others must bear more weight and postural reflexes presumably produce the increased muscle force that is needed. If any part oscillates, other parts oscillate too, perhaps in other phase relations. Although no such total system has been analyzed, one can imagine a sequence of reflex relationships that could coordinate all the legs into a smooth gait.

Against this background I shall describe the work on the nervous control of flight in locusts I began in the laboratory of Torkel Weis-Fogh at the University of Copenhagen in 1959. Weis-Fogh and his associates had already investigated many aspects of the mechanisms of insect flight, including the sense organs and their role in the initiation and maintenance of flight [see "The Flight of Locusts," by Torkel Weis-Fogh; SCIENTIFIC AMERICAN, March, 1956]. These studies, and the general climate of opinion among physiologists, tended to support a peripheral-control hypothesis based on reflexes. To test this hypothesis I set out to analyze the details of the reflex mechanisms.

An important consideration in the early phases of the work was how to study nervous activities in a small, rap-

idly moving animal. This was accomplished by having locusts fly in front of a wind tunnel while they were suspended on a pendulum that served as the arm of an extremely sensitive double-throw switch. The switch operated relays that controlled the blower of the tunnel, so that whenever the insect flew forward, the wind velocity increased and vice versa. Thus the insect chose its preferred wind speed, but it stood approximately still in space. Other devices measured aerodynamic lift and body and wing positions; wires that terminated in the muscles or on nerves conducted electrical impulses to amplifying and recording apparatus.

Early in the program of research we found that fewer than 20 motor nerve cells control the muscles of each wing, and that we could record from any of the motor units controlled by these cells during normal flight. We drew up a table showing when each motor unit was activated for various sets of aerodynamic conditions. The results of this rather tedious work were not very exciting but did provide a necessary base for further investigation. Moreover, I think we can say that these results constitute one of the first and most complete descriptions of the activity of a whole animal analyzed in terms of the activities of single motor nerve cells. In brief, we found that the output pattern consists of nearly synchronous impulses in two small populations of cooperating motor units, with activity alternating between antagonistic sets of muscles, the muscles that elevate the wings and the muscles that depress them. Each muscle unit normally receives one or two impulses per wingbeat or no impulse at all. The variation in the number of excitatory impulses sent to the different muscles serves to control flight power and direction.

We also found it possible to record from the sensory nerves that innervate, or carry signals to, the wings. These nerves conduct proprioceptive signals from receptors in the wing veins and in the wing hinge. The receptors in the

wing veins register the upward force, or lift, on the wing; the receptors in the wing hinge indicate wing position and movement in relation to the body. These sensory inputs occur at particular phases of the wing stroke. The lift receptors usually discharge during the middle of the downstroke; each wing-hinge proprioceptor is a stretch receptor that discharges one, two or several impulses toward the end of the upstroke [*see illustration on page 2727*].

Everything I have described so far about the motor output and sensory input of an insect in flight is consistent with the peripheral-control hypothesis. Motor impulses cause the movements

the receptors register. According to the hypothesis the sensory feedback should trigger a new round of output. Does this actually happen?

A useful test of feedback-loop function is to open the loop. This we did simply by cutting or damaging the sense organs or sensory nerves that provide the feedback. Cutting the sensory nerve carrying the information about lift forces caused little change in the basic pattern of motor output, although it did affect the insect's ability to make certain maneuvers. On the other hand, burning the stretch receptors that measure wing position and angular velocity always resulted in a drastic reduction in wingbeat frequency. These proprioceptors provide the only input we could discover that had such an effect. Most important of all, we found that, even when we eliminated all sources of sensory feedback, the wings could be kept beating in a normal phase pattern, although at a somewhat reduced frequency, simply by stimulating the central nervous system with random electrical impulses.

From these studies we must conclude that the flight-control system of the locust is not adequately explained by the peripheral-control hypothesis and patterned feedback. Instead we find that the coordinated action of locust flight muscles depends on a pattern-generating system that is built into the central nervous system and can be turned on by an unpatterned input. This is a significant finding because it suggests that the networks within the nerve ganglia are endowed through genetic and developmental processes with the information needed to produce an important pattern of behavior and that proprioceptive reflexes are not major contributors of coordinating information.

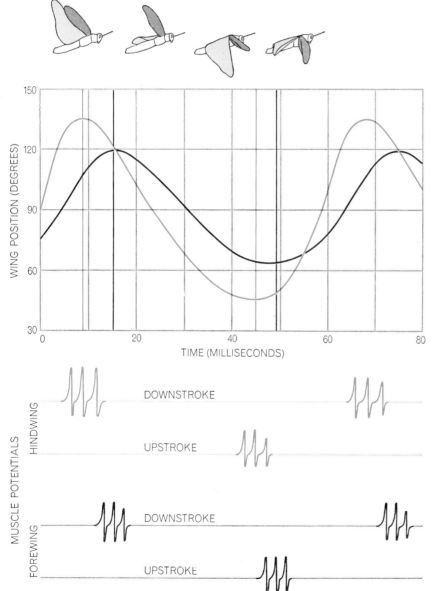

MUSCLE-POTENTIAL RECORDS are summarized in relation to wing positions in a flying locust. The curves (*top*) show the angular position (90 degrees is horizontal) of the hindwings (*color*) and forewings (*black*). The four simulated traces at the bottom show how the downstroke and upstroke muscles respectively fire at the high and low point for each wing.

Erik Gettrup and I were particularly curious to learn how the wing-hinge proprioceptor, a stretch receptor, helped to control the frequency of wingbeat. When Gettrup analyzed the response of this receptor to various wing movements, he found that to some degree it signaled to the central nervous system information on wing position, wingbeat amplitude and wingbeat frequency. We then cut out the four stretch receptors so that the wingbeat frequency was reduced to about half the normal frequency and artificially stimulated the stumps of the stretch receptors in an attempt to restore normal function. Under these conditions we found that electrical stimulation of the stumps could raise the frequency of wingbeat no matter what input pattern we used. Although the normal input

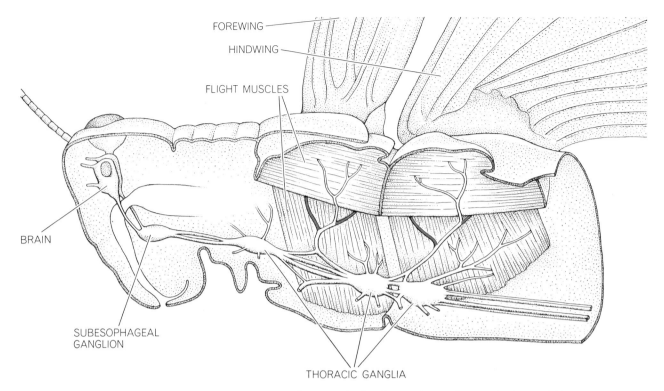

FOREWING
HINDWING
FLIGHT MUSCLES
BRAIN
SUBESOPHAGEAL
GANGLION
THORACIC GANGLIA

NERVES AND MUSCLES controlling flight in the locust are shown in simplified form. The central nervous system includes the brain and the various ganglia. From the thoracic ganglia, motor nerves lead to the wing's upstroke muscles (*vertical fibers*) and the downstroke muscles (*horizontal fibers*) above them. There are also sensory nerves (*color*) that sense wing position and aerodynamic forces.

from the stretch receptor arrives at a definite, regular time with respect to the wingbeat cycle, in our artificially stimulated preparations the effect was the same no matter what the phase of the input was.

We also found that the response to the input took quite a long time to develop. When the stimulator was turned on, the motor output frequency would increase gradually over about 20 to 40 wingbeat cycles. Hence it appears that the ganglion averages the input from the four stretch receptors (and other inputs too) over a rather long time interval compared with the wingbeat period, and that this averaged level of excitation controls wingbeat frequency. In establishing this average of the input most of the detailed information about wing position, frequency and amplitude is lost or discarded, with the result that no reflection of the detailed input pattern is found in the motor output pattern. We must therefore conclude that the input turns on a central pattern generator and regulates its average level of activity, but that it does not determine the main features of the pattern it produces. These features are apparently genetically programmed into the central network.

If that is so, why does the locust even have a stretch reflex to control wingbeat frequency? If the entire ordered pattern needed to activate flight muscles can be coded within the ganglion, why not also include the code for wingbeat frequency? The answer to this dual question can probably be found in mechanical considerations. The wings, muscles and skeleton of the flight system of the locust form a mechanically resonant system—a system with a preferred frequency at which conversion of muscular work to aerodynamic power is most efficient. This frequency is a function of the insect's size. It seems likely that even insects with the same genetic makeup may reach different sizes because of different environmental conditions during egg production and development. Hence each adult insect must be able to measure its own size, as it were, to find the best wingbeat frequency. This measurement may be provided by the stretch reflex, automatically regulating the wingbeat frequency to the mechanically resonant one.

What kind of pattern-generating nerve network is contained in the ganglia? We do not know as yet. Nonetheless, a plausible model can be suggested. The arguments leading to this model are not rigorous and the evidence in its favor is not overwhelming, but it is always useful to have a working hypothesis as a guide in planning future experiments. Also, it seems worthwhile to present a hypothesis of how a simply structured network might produce a special temporally patterned output when it is excited by an unpatterned input.

When neurophysiologists find a system in which there is alternating action between two sets of antagonistic muscles, they tend to visualize a controlling nerve network in which there is reciprocal inhibition between the two sets of nerve cells [*see top illustration on page 2728*]. Such a network can turn on one or both sets at first, but one soon dominates and the other is silenced. When the dominant set finally slows down from fatigue, the inhibiting signal it sends to the silent set also decreases, with the result that the silent set turns on. It then inhibits the first set. This reciprocating action is analogous to the action of an electronic flip-flop circuit; timing cues are not needed in the input. The information required for the generation of the output pattern is contained largely in the structure of the network and not in the input, which only sets the average level of activity. A nerve network that acts in this way can consist of as few as two cells or be made up of two populations of cells in which there is some mechanism to keep the cooperating units working together.

In the locust flight-control system several tens of motor nerve cells work together in each of the two main sets. The individual nerve cells within each set

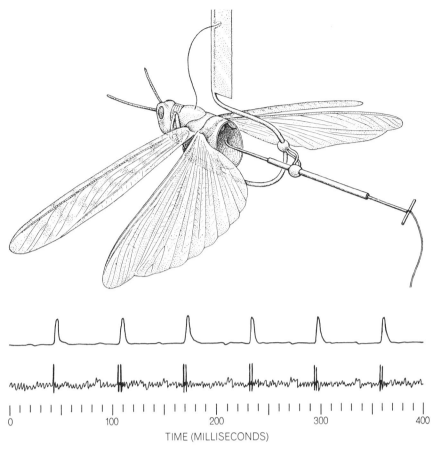

SENSORY DISCHARGES in nerves from the wing and wing hinge are recorded with wires manipulated into the largely eviscerated thoracic cavity of a locust. The top record is of downstroke muscle potentials, which are repeating at the wingbeat frequency. The bottom record is of a sensory (stretch) receptor from one wing, firing one or two times per wingbeat.

seem to share some excitatory interconnections. These not only provide the coupling that keeps the set working efficiently but also have a further effect of some importance. Strong positive coupling between nerve cells can result in positive feedback "runaway"; the network, once it is activated, produces a heavy burst of near-maximum activity until it is fatigued [see *middle illustration on following page*] and then turns off altogether until it recovers. A network of this kind can also produce sustained oscillations consisting of successive bursts of activity alternating with periods of silence without any patterned input. Thus either reciprocal inhibition or mutual excitation can give rise to the general type of burst pattern seen in locust motor units. Both mechanisms have been demonstrated in various behavior-control systems. It is likely that both are working in the locust and that these two mechanisms, as well as others, converge to produce a pattern of greater stability than might otherwise be achieved.

In summary, the model suggests that each group of cooperating nerve cells is mutually excitatory, so that the units of each group tend to fire together and produce bursts of activity even when their input is steady. In addition the two sets are connected to each other by inhibitory linkages that set the two populations into alternation. Notice that in these hypothetical networks the temporal pattern of activity is due to the network structure, not to the pattern of the input. Even the silent network stores most of the information needed to produce the output pattern.

The locust flight-control system consists in large part of a particular kind of circuit built into the central ganglia. Some other locomotory systems seem much more influenced by reflex inputs. As an example of such a system I shall describe briefly the walking pattern of the tarantula spider. There is much variation in the relative timing of the eight legs of this animal, but on the average the legs exhibit what is called a diagonal rhythm. Opposite legs of one segment alternate and adjacent legs on one side alternate so that diagonal pairs of legs are in step [see *illustration on page 2729*].

The tarantula can lose several of its legs and still walk. Suppose the first and third pairs of legs are amputated. If the spider's legs were coordinated by means of a simple preprogrammed circuit like the one controlling the locust's flight muscles, one would expect the spider to move the remaining two legs on one side in step with each other and out of step with the legs on the other side. A four-legged spider that did this would fall over. In actuality the spider adjusts relations between the remaining legs to achieve the diagonal rhythm. Other combinations of amputations give rise to other adaptations that also maintain the mechanically more stable diagonal rhythm.

Thus it appears that the pattern of coordination does depend on input from the legs. One can advance a possible explanation. Each leg is either driven by a purely central nervous oscillator or each leg and its portion of ganglion forms an oscillating reflex feedback loop. Suppose the several oscillators are negatively coupled. A pair of matched negatively coupled oscillators will operate out of phase. If the nearest leg oscillators are negatively coupled more strongly than the ones farther apart, the normal diagonal rhythm will result. For example, if left leg 1 has a strong tendency to alternate with right leg 1 and right leg 1 alternates with right leg 2, then left leg 1 must operate synchronously with right leg 2, to which it is more weakly connected. Now if some of the oscillators are turned off by amputating legs, so that either the postulated oscillatory feedback loop is broken or the postulated central oscillator receives insufficient excitatory input, new patterns of leg movements will appear that will always exhibit a diagonal rhythm.

The real nature of the oscillators involved in the leg rhythms is not known. These results and the postulated model nonetheless illustrate how sensory feedback could be used by the nervous system in such a way that the animal could adjust to genetically unpredictable conditions of the body or environment without recourse to learning mechanisms. Could this be the role of reflexes in general? We have seen that in the locust flight system much information for pattern generation is centrally stored—presumably having been provided genetically—and that the reflexes do seem to supply only information that could not have been known genetically.

A way to describe the two general models of muscle-control systems has been suggested by Graham Hoyle of the University of Oregon. He calls the cen-

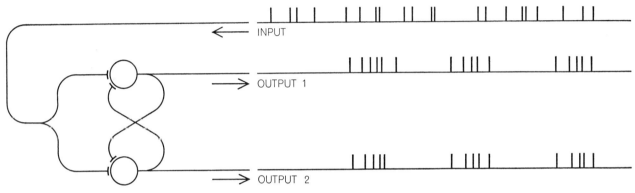

CROSS INHIBITION is one kind of interaction between nerve cells. The cells are connected in such a way that impulses from one inhibit the other (*color*). This can cause a pattern of alternating bursts, each cell firing (inhibiting the other) until fatigued. The hypothetical network shows how an unpatterned input can be transformed into a patterned output by structurally coded information.

IN CROSS EXCITATION the output from each cell excites the other. This makes for approximate synchrony. There may also be a positive feedback "runaway" until fatigue causes deceleration or a pause; once rested, the network begins another accelerating burst.

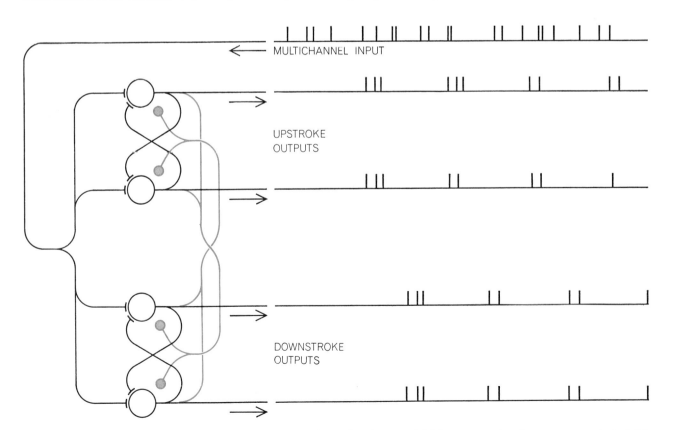

HYPOTHETICAL NETWORK of nerve cells in the locust might involve two cell populations, an upstroke group (*top*) and a downstroke group (*bottom*). Cells *within* a group excite one another but there are inhibitory connections *between* groups (*color*). The inhibition keeps the activity of one group out of phase with that of the other, so that upstroke and downstroke muscles alternate.

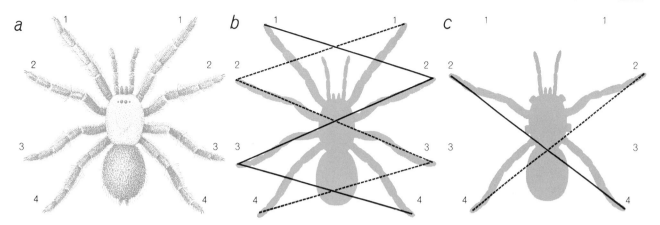

LEGS OF TARANTULA (*a*) move in diagonally arranged groups of four (*b*). The diagonal pattern persists if some legs are removed (*c*), suggesting that proprioceptive reflexes set the pattern or that changed input alters inherent central nervous system activity.

trally stored system of pattern generation a motor-tape system. In such a system a preprogrammed "motor score" plays in a stereotyped manner whenever it is excited. For a system in which reflexes significantly modify the behavioral sequences, Hoyle suggests the term "sensory tape" system. I prefer the term "sensory template." Such systems have a preprogrammed input requirement that can be achieved by various outputs. If the output at any moment results in an input that does not match the template, the mismatch results in a changed output pattern until the difference disappears. Such a goal-oriented feedback system can adapt to unexpected environments or to bodily damage. Spider and insect walking patterns show a kind of plasticity in which new movement patterns can compensate for the loss of limbs. In the past this kind of plasticity has often been interpreted as evidence for the reflex control of locomotory pattern. The locust flight-control system, on the other hand, certainly has a motor score that is not organized as a set of reflexes. Can a motor-score system show the plastic behavior usually associated with reflex, or sensory-template, systems?

For several years we thought that the locust flight-control system was relatively unplastic. We knew about the control of wingbeat frequency by the stretch reflex and about other reflexes, for example a reflex that tends to keep the body's angle of pitch constant. When I had made recordings of nerve impulses to some important control muscles before and after cutting out other muscles or whole wings, however, I had not found differences in the motor output. I therefore concluded that the flight-control system was not capable of a wide range of adaptive behavior. On this basis one

would predict that a damaged locust could not fly, or could fly only in circles.

The locust has four wings. Recently I cut whole wings from several locusts, threw the locusts into the air and found to my surprise that they flew quite well. The flying locust shows just as much ability to adapt to the loss of a limb as a walking insect or spider does. For the crippled locust to fly it must significantly change its motor output pattern. Why did the locust not show this change in the experiments in which I made recordings of nerve impulses to its muscles?

In all the laboratory experiments the locust was approximately fixed in space. If the insect made a motor error, it might sense the error proprioceptively, but it could not receive a feedback signal from the environment around it indicating that it was off course. The free-flying locust has at least two extraproprioceptive sources of feedback about its locomotory progress: signals from its visual system and signals from directionally sensitive hairs on its head that respond to the flow of air. Either or both of these extraproprioceptive sources can tell the locust that it is turning in flight. In the free-flying locust the signals are involved in a negative feedback control that tends to keep the animal flying straight in spite of functional or anatomical errors in the insect's basic motor system. When animals are studied in the laboratory under conditions that do allow motor output errors or anatomical damage to produce turning motions, then compensatory changes in the motor output pattern are observed provided only that the appropriate sensory structures are intact.

The locust flight-control system consists in part of a built-in motor score, but it also shows the adaptability expected of a reflex, or sensory-template, type of control. From these observations on plasticity in the locust flight system one

can see that there may be no such thing as a pure motor-tape or a pure sensory-template system. Many behavior systems probably have some features of each.

What we are striving for in studies such as the ones I have reported here is a way of understanding the functioning of networks of nerve cells that control animal behavior. Neurophysiologists have already acquired wide knowledge about single nerve cells—how their impulses code messages and how the synapses transmit and integrate the messages. Much is also known about the electrical behavior and chemistry of large masses of nerve cells in the brains of animals. The intermediate level, involving networks of tens or hundreds of nerve cells, remains little explored. This is an area in which many neurobiologists will probably be working in the next few years. I suspect that it is an area in which important problems are ripe for solution.

I shall conclude with a few remarks on the unraveling of the mechanisms of genetically coded behavior. As I see it, there are two major stages in the readout of genetically coded behavioral information. The first stage is the general process of development of bodily form, including the detailed form of the networks of the central nervous system and the form of peripheral body parts, such as muscles and sense organs that are involved in the reflexes. This stage of the genetic readout is not limited to neurobiology, of course. It is a stage that will probably be analyzed largely at the molecular level. The second stage involves a problem that is primarily neurobiological: How can information that is coded in the grosser level of nervous-system structure, in the shapes of whole nerve cells and networks of nerve cells, be translated into temporal sequences of behavior?

The Author

DONALD M. WILSON is professor of biology at Stanford University. He did his undergraduate work at the University of Southern California and obtained a doctorate in zoology from the University of California at Los Angeles in 1959. Thereafter he held postdoctoral research positions at the zoophysiological laboratory of the University of Copenhagen and in the department of molecular biology at the University of California at Berkeley. Wilson taught at Yale University and at Berkeley before going to Stanford last year. He writes: "My first training in biology was in marine zoology. Later, under the guidance of T. H. Bullock, I became interested in animal behavior mechanisms and neurophysiology. Currently I am especially interested in problems of genetic control and evolution of behavior mechanisms."

Bibliography

THE CENTRAL NERVOUS CONTROL OF FLIGHT IN A LOCUST. Donald M. Wilson in *The Journal of Experimental Biology,* Vol. 38, No. 2, pages 471–490; June, 1961.

EXPLORATION OF NEURONAL MECHANISMS UNDERLYING BEHAVIOR IN INSECTS. Graham Hoyle in *Neural Theory and Modeling: Proceedings of the 1962 Ojai Symposium,* edited by Richard F. Reiss. Stanford University Press, 1964.

SCIENTIFIC
AMERICAN May 1968, Vol. 218, No. 5, pp. 116-126

OFFPRINT 1108

TERRITORIAL MARKING BY RABBITS

by Roman Mykytowycz

The wild rabbit of Australia lives in colonies from which the rabbits of other colonies are excluded. The rabbits mark their home territory by means of odorous substances that are secreted by specialized glands.

Over the past few years it has become increasingly apparent that one of the principal ways animals communicate is by means of odorous glandular secretions that have been given the name pheromones. The best examples of such chemical communication are found among the ants, some species of which use pheromones to lead their fellows to food, to warn them of danger and to organize social behavior in the nest [see "Pheromones," by Edward O. Wilson; SCIENTIFIC AMERICAN Offprint 157]. It has also been shown that pheromones play an important role in the social behavior of mammals (perhaps including man), although exactly how they do so has been somewhat obscure. My colleagues and I have been investigating such mechanisms in the wild rabbit of Australia (*Oryctolagus cuniculus*), and we have been able to demonstrate a clear-cut relation between the animal's social behavior and pheromone-secreting glands.

As is well known, the rabbit was introduced into Australia about 100 years ago, and since that time it has become a serious pest. Over the past two decades the Division of Wildlife Research of the Australian Commonwealth Scientific and Industrial Research Organization has conducted a broad ecological study of the wild rabbit in order to gain a better understanding of the life of the species and to provide a basis for the development of more effective methods of controlling the rabbit population. One result of this study is that we have been able to detect a distinct social organization within the rabbit colony. The role individual animals play in such a society is reflected both in the size of certain glands and in the extent to which the glands secrete pheromones. These substances probably serve a number of functions. They may communicate in-

formation about age, sex, reproductive stage and group membership. They may also warn of danger, and they definitely serve to define the "territory" of the rabbit.

The area within which an animal confines its activities is not necessarily the same as its territory; in a strict sense the term territory refers to that part of an animal's home range which it protects, sometimes by fighting. The size, form and character of a territory of course vary according to the animal's way of life. Some animals occupy a territory permanently; others hold it only during a particular season when, for example, breeding or nesting takes place. The possession of a territory, and therefore available shelter and food, makes breeding possible for some individuals of a species and prevents breeding by other individuals. Hence territoriality is an important factor regulating the density of a population.

Many mammals are known to mark out a territory with substances manufactured by specialized glands. Indeed, nearly all species of mammals possess scent-producing glands, often in several places on the body. In the dromedary the glands are at the back of the head, in the elephant at the temples, in a number of deer species below the eyes (the American deer and some other species have such glands near the hoof and on the leg). The chamois's glands are around the horns, the golden hamster's between the ribs, the peccary's in the loin. The pika (mouse hare) and marmot (woodchuck) employ glands behind and under the eyes and on the cheek for territorial marking. We have identified odor-producing glands on the chest of the kangaroo, and other workers in Australia have found similar glands on the chest of the koala and the brush-tailed

possum. The rabbit has scent glands in the anal region, in the groin and under the chin.

In order to study the territorial behavior of the rabbit it was necessary to follow the activities of individual free-living animals. As in the study of other wild animals, there were certain difficulties. It is not easy to identify individual rabbits, even in daylight, and the animal is largely nocturnal. Live rabbits are hard to catch, and it was necessary to examine them regularly. To minimize these problems we confined our rabbit colonies within fenced areas. Each enclosure was small enough to allow close observation (from an elevated hiding place nearby) but not so small as to interfere with the rabbit's normal behavior. To facilitate the recognition of individual animals each one was marked with black dye in a distinctive pattern; tags of different shapes were also affixed to the rabbits' ears. We used a colored light-reflecting tape to mark the tags in a variety of patterns; at night under spotlights the rabbits could easily be identified. To keep track of the animals' movements in their burrows we drilled a few holes down to the burrow tunnels. When we were not using the holes to check the location of rabbits underground or to catch them, they were plugged with earth enclosed in wire netting.

When a rabbit was born in an enclosure, its ears were permanently marked by tattooing; in this way the behavior of the animal could be followed for its lifetime. The rabbits were observed every day and on certain occasions continuously for a 24-hour period. Some studies were continued for three years, which is twice the average lifespan of a free-living wild rabbit in Australia.

It was established by these observa-

RABBIT EXPERIMENT followed the behavior of individual wild rabbits (*Oryctolagus cuniculus*) aboveground and underground.   Holes dug down to the burrow tunnel gave access to the animals; when not in use, the holes were plugged with earth enclosed in wire

tions that the rabbit lives in small groups, each consisting of eight or 10 animals, depending on the density of the overall rabbit population. A group of rabbits occupies a distinct territory, with a warren—a burrow with a number of entrances—as its center. At the onset of the breeding season the males establish a hierarchy of descending dominance by fighting. The same is done by the females. The most dominant male and female rule over the group and its territory, which is defended by its occupants and recognized by rabbits of other groups. The rabbit hierarchy is constantly reinforced by chasing and submission.

The advantages of having a territory were evident in the different breeding patterns of female rabbits. High-ranking females give birth frequently and regularly, housing the young in a special breeding chamber dug as an extension to the burrow. Their young have a much higher survival rate and growth rate than the young of low-ranking females. Some of the subordinate females are chased away from a warren and forced to drop their litters in isolated breeding "stops"—short, shallow burrows dug at a distance of 10 to 50 yards from the warren. Both in the stop and later, when they come out of it, the young of such litters are vulnerable to predators, particularly foxes and crows. When they try to enter the central warren, they are attacked by the resident females and their young. Some subordinate females, although they are physi-

ologically fit, never give birth; the embryos they conceive do not develop to full term but are absorbed in the uterus.

It could also be seen that not all rabbit territories are of equal quality: some offer more protection and a better food supply than others. The most dominant animals tend to occupy the better territories. The breeding stops of subordinate females are often in places that are susceptible to flooding, another factor that contributes to the lower survival rate of their young.

A rabbit that crosses the border between its own territory and a neighboring one is immediately aware that it has done so. No matter how frequent and regular the visits, the animal's behavior changes. In its own territory it moves freely; it feeds and examines objects confidently. In a foreign territory the animal seems always on the alert. Its neck is stretched, the movement of its nostrils indicates that it is sniffing continuously and it does not feed. Although the interloper may be dominant in its own territory, when it is challenged outside by a rabbit permanently attached to the foreign territory, it will offer no resistance. It will not resist even if the challenger is half-grown.

It is not difficult to believe that a rabbit territory becomes saturated with the characteristic smell of the group and that within the area in which this group smell prevails the animals feel at home, much as a man may be able to perceive the odor of another man's home as be-

ing strange in contrast to the familiar smell of his own. Specific odors associated with an individual group or colony have been demonstrated in many social insects and also in some fishes. Apart from the effect on the odor of the food eaten by the animals, genetic factors are undoubtedly involved as well.

One component of a rabbit territory's odor is the smell of urine. During amatory behavior the males can be seen urinating on the females, and the animals also urinate on each other during aggressive displays. Young rabbits so marked by adult members of their group are identified with it. The smell of foreign rabbit urine releases aggression; females may even attack their own young when they have been smeared with foreign urine.

Another component of the territory odor comes from feces. The marking out of territory with feces seems to be common among animals, as has been pointed out by Heini Hediger of the University of Zurich. Among certain animals the odor of the feces comes not from the excrement itself but from the pheromone of anal glands; such glands have been identified in more than 100 species of mammals. Some animals whose anal glands are highly developed—for example the gray squirrel, the marten, the dormouse and the hyena—use the anal pheromone alone, rather than in combination with feces, for territorial marking.

The anal glands of the rabbit are well developed. They consist of two clusters

netting. Each adult animal was marked with dye in a distinctive pattern and wore an identification tag in its ear. The ears of rabbits born during the experiment were tattooed. To house their young, rabbits dig special breeding chambers as extensions to the burrow.

of brownish tissue forming a saddle-like mass around the end of the rectum. The secretions of the glands flow through a few ducts into the rectum, where they coat the pellets of feces passing out of the anus. Until recently it was generally accepted that the secretion facilitated the passage of the hard fecal pellets. We have found, however, that when the rabbit's anal glands are removed, defecation is not affected.

Our studies indicate that the fecal pellets serve to distribute the rabbit's anal pheromone.

When one examines a rabbit territory, one can see a number of places in which the animals have repeatedly deposited feces. Around a typical warren it is usual to find about 30 of these small dunghills. They are interconnected by paths that all the rabbits use in moving around the area. When a strange rabbit enters the territory, it inevitably encounters a dunghill—a warning signal that the area is occupied. It then displays the wary behavior I have described.

A similar warning is deposited at the entrances to breeding stops. When a stop contains young, it is sealed by the female with soil. Once a day the female reopens the burrow, enters it and suckles the young. When the entrance

CHINNING rabbit enclosed in an experimental pen was photographed at night, when the animal is active. As the rabbit presses against a wooden peg, droplets from its submandibular (underchin) glands are forced through pores of the skin, "marking" the object.

5

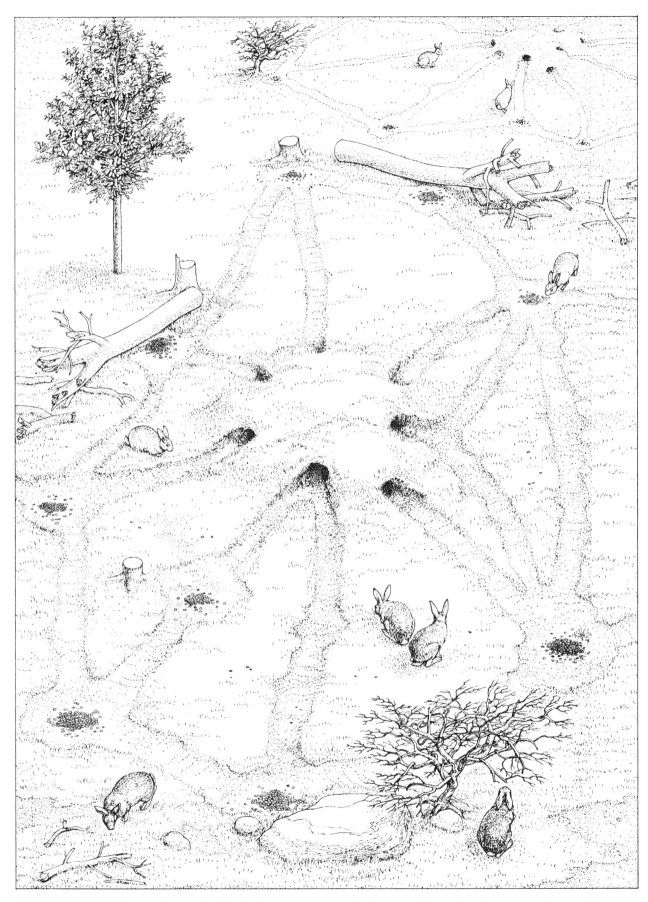

RABBIT TERRITORY, somewhat idealized in this drawing, lies around a central warren. Radiating from the entrances are intersecting runs. On the runs are small mounds of fecal pellets deposited by the rabbits of the colony. At top right a rabbit from another colony sniffs at a mound of pellets. It is warned by odor of an anal-gland secretion on the pellets that the territory is occupied.

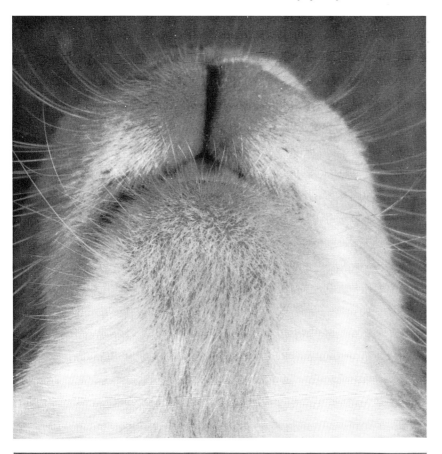

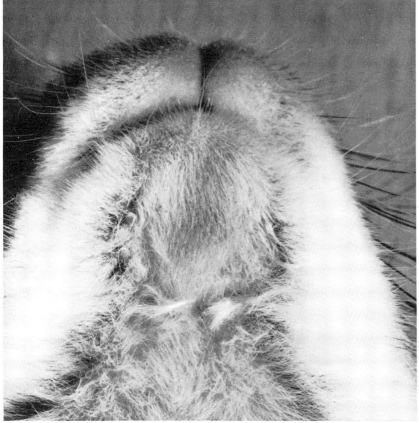

UNDERCHIN of the rabbit is the site of the subcutaneous glands that, together with the anal glands, function in territorial marking. The fur under the chin of the female rabbit (*top*) displays no trace of a glandular secretion. Under the male rabbit's chin (*bottom*) the fur is matted from the secretion, which the male produces more copiously than the female.

is resealed, the female is likely to deposit a few fecal pellets and some urine on top of the seal. It is remarkable that these earth seals are almost never disturbed by other rabbits. (The exceptions occur under abnormal circumstances, such as when the density of a confined population becomes exceptionally high and competition for breeding sites is intensive.)

Certain of our experiments support the idea that the rabbit's feces serve a communicative function, and indicate that the pellets of a dunghill differ from those that are randomly deposited around a territory. We presented a group of rabbits with artificial dunghills consisting of turf sprinkled with some of the pellets found at random locations in the territory of other rabbits. Around these artificial dunghills (and also on pieces of turf used as a control) the animals engaged mainly in digging and eating. When they were confronted with an actual dunghill from foreign territory, however, they did not eat, they sniffed intensely and they produced marking feces of their own. These pellets were similar to dunghill pellets in that their "rabbity" odor was (according to 30 human judges) decidedly stronger than the odor of the randomly distributed pellets. Presumably the difference between the two odors is perceived even more sharply by rabbits.

There is a highly significant relation between the size of a rabbit's anal gland and the place of the animal in the social hierarchy of a warren. The largest glands belong to dominant individuals, the smallest to subordinate animals. (Body weight is not the main factor determining glandular size.) The secretory activity of the anal gland also is higher in the dominant animals. Indeed, we found that the social rank of an individual rabbit could be guessed with a fair degree of accuracy merely from the appearance of the gland in section. It is also significant that the size and secretory activity of the gland were greatest during the breeding season— the period when territorial activity is most intense.

Male rabbits are mainly responsible for establishing dunghills (in Australia about 80 percent of the animals caught in traps set on dunghills are males), and the anal gland of the male rabbit is larger than the anal gland of the female. Our experiments indicate that the activity of the gland is under the control of sex hormones. When male rabbits were castrated before puberty, the growth and activity of the anal gland were inhibited; when the ovaries were removed from female rabbits, the size and activity of the gland were somewhat enhanced. When male sex hormones were given to both male and female rabbits, the gland grew larger and produced more secretion.

The rabbit's anal pheromone and its urine thus serve to establish its overall territory. For marking localized features such as logs, branches and blades of grass the animal employs a pheromone secreted by its chin gland. If one looks under the chin of a female rabbit, one cannot find any conspicuous marks, although the fur may be slightly moist or matted. Under the male rabbit's chin, however, there is a distinct yellowish encrustation and matted fur. One can feel the glands under the skin, and if pressure is applied, droplets of secretion can be forced out through a semicircular row of external pores. With this secretion (which is odorless to humans) the male rabbit marks not only objects that would be difficult to mark with feces or urine but also the entrances to its burrows, the fecal pellets of other rabbits, its own weathered pellets and its females and young. To describe marking with this gland we use the term "chinning."

Within its own territory a rabbit chins freely and frequently, particularly if it is a male. When it is in a foreign territory, it does not chin; when it is confronted on its own ground with foreign feces, it chins intensely. We have found that the individuals in a rabbit hierarchy that are most dominant chin more often than the subordinate animals. In fact, the frequency of chinning can

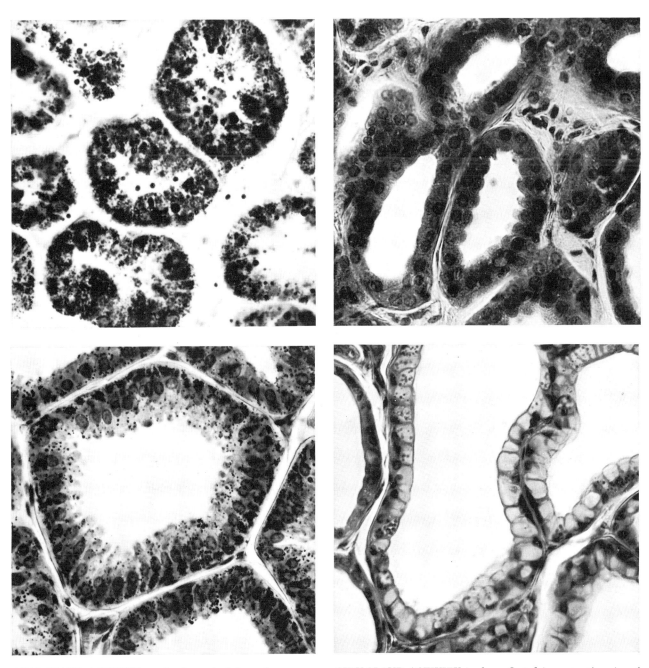

ANAL-GLAND ACTIVITY can be determined from the appearance of the gland in section. In the less actively secreting gland shown at top the tubules (*dark areas with light centers*) are smaller than those in the intensely secreting gland that appears below it.

CHIN-GLAND ACTIVITY is also reflected in contrasting size of tubules in less active gland (*top*) and more active one (*bottom*). These four photomicrographs were made by E. C. Slater of the Commonwealth Scientific and Industrial Research Organization.

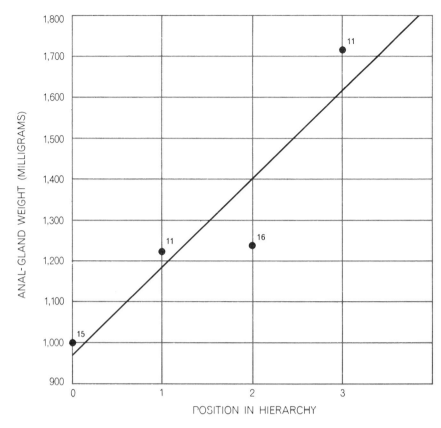

RELATION BETWEEN ANAL GLAND AND SOCIAL BEHAVIOR was established from observations of the activities of rabbits. The most dominant animals of a warren hierarchy (*here given the rating "3"*) were found to possess the heaviest anal glands. The number above a dot indicates the number of rabbits in the sample; the trend is indicated by a line.

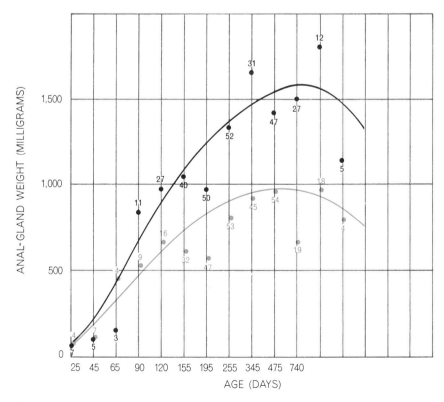

SEXUAL DIFFERENTIATION appears when the weight of the anal gland of male rabbits (*black*) is compared with that of female rabbits (*color*). The number of animals sampled is shown adjacent to a dot; the curves indicate the trend in the measurements. Age of the rabbits was estimated from the weight of the eye lens and is uncertain beyond 740 days.

be used as an indication of potential dominance: when two males completely strange to each other were brought together, the male with the highest chinning score always established his dominance over the other animal.

The rabbit's chin gland, like its anal gland, is larger and secretes more abundantly in dominant animals than in subordinate ones. This gland too, being larger in males than in females, appears to be under the control of sex hormones. Moreover, its activity, like that of the anal gland, fluctuates with the seasons; it secretes most freely at the time when the rabbit's territorial activity is at its most intense. Thus the animals that are dominant in the hierarchy are the ones most concerned with the demarcation and defense of territory. Indeed, their territorial behavior is rooted in their physiology.

Additional support for the territorial function of the anal and chin glands has come from the study of an animal that is closely related to the rabbit but that behaves differently with respect to territory. This animal is the European hare (*Lepus europaeus*), which like the rabbit has become widely distributed throughout Australia since being introduced there. Although a fully grown hare is four times as heavy as a rabbit and most of its glands (such as the thyroid and lachrymal glands) are larger, the anal and chin glands of the hare are only a tenth the size of those in rabbits.

This difference between the two animals reflects a difference in their territorial behavior. Unlike the rabbit, the hare does not live in a social group; it is a solitary animal and its home area is large. The hare protects only that part of its home range which is in its immediate vicinity, and it does not retain such a territory permanently. Although the animal makes dunghills (apparently at the sites it visits most frequently), the number of them is small compared with the number in a rabbit territory. There is no difference between the anal gland of the male hare and the gland of the female. In short, it appears to be unnecessary for the hare to mark its territory by odor.

Another difference in the glandular makeup of the hare and the rabbit also appears to have behavioral significance. The animals have similar glands in the groin, but in the hare these glands are larger than they are in the rabbit (and they are somewhat larger in the female hare than they are in the male). Observations of the hare's mating behavior suggest that the glands in the groin are the source of a powerful sexual attractant: when the solitary-living female hare is in estrus, she attracts males from some distance. The female of the gregarious rabbit species obviously has no need of a long-range attractant.

What we have learned about the wild rabbit in Australia is supported by studies of American species. The swamp rabbit (Sylvilagus aquaticus), a strongly territorial species, chins more frequently than the weakly territorial cottontail rabbit (Sylvilagus floridanus). Our examination of gland tissue from these species indicates that the anal and chin glands are larger in the swamp rabbit than in the cottontail. The social behavior of the two species was studied by Halsey M. Marsden and Nicholas R. Holler of the University of Missouri.

The pheromones of rabbits and other mammals have not yet been chemically separated and identified, as has been done with certain insect pheromones. With knowledge of the composition of specific odors it will be possible to establish their role in the life of a mammalian species. This will undoubtedly lead to a better understanding of the species' behavior. Some insect pheromones have even been synthesized; incorporated in traps, they have been used successfully in pest control. The synthesis of mammalian pheromones would no doubt be helpful in developing effective controls for mammals that are economically undesirable.

The existence of glands that function specifically and solely for territorial marking emphasizes the importance of territory and social organization in the life of an animal. In today's crowded world the question of space and territory is of interest not only to students of animal biology but also to all who are concerned with man's problems of overpopulation. The theory that in animals population control is achieved through social behavior of which territoriality is an integral part has been advanced by V. C. Wynne-Edwards of the University of Edinburgh [see "Population Control in Animals," by V. C. Wynne-Edwards; SCIENTIFIC AMERICAN Offprint 192]. More and more we are coming to realize that it is not only the availability of food that determines the size of a human population but also our own behavior and spacing, and that these factors must be considered in speculating on the fate of the human species. Every animal has, in addition to minimum requirements of things such as food, minimum requirements of living space and distance from others of its own species.

The Author

ROMAN MYKYTOWYCZ is principal research scientist with the Division of Wildlife Research of the Commonwealth Scientific and Industrial Research Organization of Australia. Born in the Ukraine, he studied veterinary science at the Zootechnical and Veterinary Institute of Lvov and obtained a doctorate in veterinary medicine from the University of Munich in 1948. He writes: "I then went to Australia, and after joining the CSIRO in 1950 I became engaged in investigations of the epidemiology of diseases of wildlife species, including the rabbit, the kangaroo and the Tasmanian shearwater or muttonbird. I was associated with experiments to control plague populations of rabbits by biological means, using the virus disease myxomatosis. While studying the fluctuations of numbers of animals in free-living populations, particularly of rabbits, I came to recognize the importance of social and territorial behavior of the animal as a population-regulating factor and became interested in ethology."

Bibliography

ANIMAL DISPERSION IN RELATION TO SOCIAL BEHAVIOR. V. C. Wynne-Edwards. Hafner Publishing Company, 1962.

FURTHER OBSERVATIONS ON THE TERRITORIAL FUNCTION AND HISTOLOGY OF THE SUBMANDIBULAR CUTANEOUS (CHIN) GLANDS IN THE RABBIT, ORYCTOLAGUS CUNICULUS (L.). R. Mykytowycz in Animal Behavior, Vol. 13, No. 4, pages 400–412; October, 1965.

OBSERVATIONS ON ODORIFEROUS AND OTHER GLANDS IN THE AUSTRALIAN WILD RABBIT, ORYCTOLAGUS CUNICULUS (L.), AND THE HARE, LEPUS EUROPAEUS P, I: THE ANAL GLAND. R. Mykytowycz in CSIRO Wildlife Research, Vol. 11, No. 1, pages 11–29; October, 1966.

PHEROMONES. Edward O. Wilson in Scientific American, Vol. 208, No. 5, pages 100–114; May, 1963.

SCIENTIFIC
AMERICAN   June 1968, Vol. 218, No. 6, pp. 78-88   OFFPRINT 1109

THE DISCOVERY OF DNA

by Alfred E. Mirsky

In 1869 Friedrich Miescher found a substance in white blood cells
that he called "nuclein." Cell biologists saw that it was a constituent
of chromosomes and hence must play a major role in heredity.

Deoxyribonucleic acid was discovered in 1869 by Friedrich Miescher. The reader may find it surprising that DNA, which has been the focus of so much recent work in biology, was isolated almost a century ago. If so, he will find it even more surprising that the function of DNA as the substance in chromosomes that transmits hereditary characteristics was recognized only a few years later by a number of biologists (not, however, by Miescher). Here I shall recall how it happened that Miescher discovered "nuclein" (as he called it) and what it meant to him and his contemporaries, and tell something of the investigative history of DNA until, some 25 years ago, it was conclusively shown to be the genetic material.

Miescher was the son of a Swiss physician who practiced in Basel and taught at the university there, and he followed his father into medicine. As a medical student he came under the influence of his uncle Wilhelm His, professor of anatomy and one of the outstanding investigators and teachers of his time. That influence was profound and lifelong. Miescher's later views on biology are often best understood in the light of His's attitudes and ideas; Miescher's writings are available today largely because His gathered and published the younger man's letters and papers after Miescher's death in 1895 at the age of 51.

His urged Miescher to go into histochemistry, the study of the chemical composition of tissues, because in "my own histological investigations I was constantly reminded that the ultimate problems of tissue development would be solved on the basis of chemistry." Miescher took the advice, and after receiving his degree in 1868 he went to the University of Tübingen first to learn or-

ganic chemistry and then to work in the laboratory of the biochemist F. Hoppe-Seyler. (The first laboratory devoted entirely to biochemistry, it was located incongruously in an ancient castle overlooking the Neckar River. In later years Miescher liked to tell his students how the narrow, deep-set windows and the dark vault of his room reminded him of a medieval alchemist's laboratory.) Within a few months, in the course of experiments on cells in pus, he discovered DNA.

Pus may seem an unlikely material for a study of cell composition, but Miescher considered the white blood cells present in pus to be among the simplest of animal cells, and there was a fresh supply of pus every day in the Tübingen surgical clinic. He was given the bandages removed from postoperative wounds, and he washed the white cells from the bandages for experiments. If the bandages were washed with ordinary saline solution, the cells swelled to form a gelatinous mass. In a dilute sodium sulfate solution, on the other hand, the cells were preserved and sedimented rapidly, making it easy to separate them from the blood serum and other material in pus.

Miescher undertook a general study of the chemical composition of the white cells. First he extracted them in various ways—with salt solutions, acid, alkali and alcohol. Earlier workers, including Hoppe-Seyler, had extracted pus cells with concentrated salt solutions and had obtained a gelatinous material that reminded them of myosin, the protein of muscle. Miescher got the same result. What is of interest to us is that the gelatinous substance consists largely of DNA! This did not become known until 1942, when it was demonstrated that

concentrated neutral salt solutions are an exceedingly useful medium for the extraction of polymerized (and therefore gelatinous) DNA. To extract DNA under these conditions, however, one needs a centrifuge. If Miescher had had a centrifuge, it is quite possible that he would have obtained DNA in its natural form instead of in the depolymerized form he eventually discovered.

Miescher's route to DNA was necessarily more indirect, as is frequently the case for the pathfinder. When he extracted pus cells with dilute alkali, he obtained a substance that precipitated on the addition of acid and redissolved when a trace of alkali was added. At this point Miescher noted: "According to recognized histochemical data I had to ascribe such material to the nuclei... and I therefore tried to isolate the nuclei." The isolation of the nucleus—or of any other cell organelle—had not been attempted before, as far as we know, although the nucleus had been identified in 1831. Miescher's primary observation, on which the isolation of nuclei depended, was that dilute hydrochloric acid dissolves most of the materials of a cell, leaving the nuclei behind. (This observation is the basis of what is still a valuable procedure for isolating nuclei.)

When Miescher examined the isolated nuclei under his microscope, he could still see contamination, which he suspected was cell protein. To obtain clean nuclei he therefore added the protein-digesting enzyme pepsin to the dilute hydrochloric acid he was using. More precisely (since this was in 1868), he made a hydrochloric acid extract of pig's stomach and applied it to pus cells. The isolated nuclei so prepared were somewhat shrunken but were clean enough for chemical study. The next step was to extract the isolated nuclei (rather than the whole cells) with dilute alkali. The extracted material precipitated on the addition of acid and redissolved readily in alkali. Miescher analyzed this material into its elements (finding, for example, 14 percent nitrogen and 2.5 percent phosphorus) and studied some of its other properties. He came to the conclusion that it did not fit into any known group of substances: it was a substance *"sui generis"* and he called it nuclein. The analytical data indicate that somewhat less than 30 percent of Miescher's first nuclein preparation consisted of DNA.

The year with Hoppe-Seyler was now ended. Hoppe-Seyler, who agreed that a new substance had been discovered, was sufficiently interested to repeat the preparation of nuclein after Miescher's departure, and sufficiently cautious to delay the publication of Miescher's paper until he had satisfied himself that the work was sound. The paper was published in 1871.

Miescher left Tübingen in the fall of 1869 to spend his vacation at home in Basel. He decided that during the two-month vacation he would broaden his study of nuclein by looking for it in various cells. The first material he chose was the hen's egg, because His had recently claimed that the microscopic particles called yolk platelets were genuine cells of connective-tissue origin. This was a controversial claim, and Miescher believed a biochemical approach—the search for nuclein in the platelets—would help to establish its validity. Following the procedure he had worked out for pus cells, he soon isolated from yolk platelets what he took to be nuclein. The phosphorus content and some other properties of platelet nuclein did differ somewhat from those of pus nuclein, but Miescher (like many another investigator!) was determined to find what he was looking for, and he was convinced that this was nuclein. In spite of the curious appearance of the platelets, he wrote, "nobody would any longer deny that they have genuine nuclei, because it is not in the optical properties but in the chemical nature of a structure that its role in the molecular events of a cell's life is rooted."

Miescher wrote up his new work and sent it off to Hoppe-Seyler; it was published along with the Tübingen research. The paper on pus cells remains a classic but the one on yolk platelets is forgotten. It was wrong. The microscopists, considering what Miescher called the "optical properties" of platelets, never did accept the idea that they were cells, and in time detailed chemical analysis showed (and Miescher had to agree) that what he had taken for platelet nuclein in fact had a very different composition. It is only recently that careful microscopic observation has demonstrated that yolk platelets are derived from the subcellular particles called mitochondria; they contain only a trace of DNA, whereas Miescher thought they contained a large amount of nuclein.

Having learned biochemistry at Tübingen, Miescher next spent a year at the University of Leipzig in Carl Lud-

CASTLE ON THE NECKAR RIVER in Germany was where nuclein, or deoxyribonucleic acid (DNA), was discovered. The castle housed F. Hoppe-Seyler's biochemical laboratory at the University of Tübingen, where Friedrich Miescher was a postdoctoral student.

wig's laboratory, a world center of physiology. Here he became convinced of the central role of the physical sciences in biology and in particular learned the importance of developing new instrumentation for research.

Miescher returned to Basel in 1870 and soon began the investigation on which he did his finest work: an analysis of the nuclein and other components of salmon spermatozoa. Wilhelm His introduced him to the salmon fishery then flourishing along the banks of the Rhine at Basel. The salmon, having swum all the way up from the North Sea to spawn, were sexually mature, so that huge quantities of ripe sperm were available. The nucleus is extremely large in any sperm cell, and among spermatozoa the salmon's is remarkable: its nucleus accounts for more than 90 percent of its mass. For a young investigator who had recently discovered nuclein in pus cells washed out of bandages, the sperm of the Rhine salmon must have seemed to present a God-given opportunity, and Miescher seized it. He was an intense and rapid worker. Within a year and a half he had completed most of his investigation—in spite of the fact that his laboratory was so crowded with medical students that he could only do his chemical analyses, an essential part of the job, at night and on Sundays. His classic paper on the sperm cell of the salmon was published in 1874.

By acidifying a suspension of sperm cells, Miescher first caused the cells to aggregate and settle. Then he treated them with hydrochloric acid (omitting the pepsin he had used with pus cells) and extracted from the sperm an organic base with a high content of nitrogen. It accounted for 27 percent of the mass of the sperm, and he called it "protamine." Then he treated the residue of the sperm with dilute alkali. This extracted the nuclein, which accounted for 49 percent of the sperm's mass.

Miescher saw that nuclein was an acid containing a number of acid groups (a "polybasic" acid) and that it combined with protamine, a base, to form an insoluble salt in the nucleus. He experimented with modifying the chemical equilibrium of the nuclein-protamine combination, a problem that still interests biochemists today. He found, for example, that if sperm washed with acetic acid and alcohol were treated with a sodium chloride solution, much of the protamine was released from combination and passed into solution. If *fresh* sperm were treated with the same salt solution, however, the material became a lumpy gel that could almost be cut with a pair of scissors, as in the case of the pus cells treated with salt solution. The reason was that the DNA in the preparation was present as an extremely long linear polymer. Miescher, as I have mentioned, never did isolate DNA in its natural state and learn the importance of its fibrous structure. He did, however, get some idea that nuclein consisted of large molecules because he found that it would not pass through a parchment filter, whereas protamine would do so readily.

Although most of Miescher's work was done on salmon sperm, he also investigated the sperm of other species, notably the carp and the bull. He was disappointed not to find protamine in either, or even in unripe salmon sperm, and in a letter to His he referred to its presence in ripe salmon sperm as a "miserable special case." Ten years later, however, the biochemist Albrecht Kossel discovered histone, a base analo-

FRIEDRICH MIESCHER, the discoverer of DNA, was born in 1844 and died in 1895. This portrait is the frontispiece of a collection of his letters and papers published posthumously.

gous to protamine, in the nucleus of red blood cells, and soon it was found also in the nucleus of lymphocytes, white cells from the thymus gland. Today it is known that either histones or protamines are present in saltlike combination with nuclein in the nuclei of all plant and animal cells.

Miescher's analyses of protamine and nuclein into their constituent elements were done with great care, and his results for nuclein are very close to those of later workers for what came to be called nucleic acid. (That term was introduced by the biochemist Richard Altmann in 1889. Altmann's method of preparation was different, but Miescher recognized that the substance was the same, and he did not object to the new name because he had been aware that nuclein was an acid.) To obtain good analytical results Miescher considered it essential to isolate his nuclein at low temperatures. He analyzed nuclein in the fall and winter, sometimes working from five in the morning until late at night in unheated rooms. (Later in his life, when he again took up his study of salmon sperm, he went back to spending long hours in the cold. His health, which had always been delicate, gave way and he died of a chest ailment.)

The process of elementary analysis was particularly important for Miescher because it was his main guide to the composition and identification of a substance. He could not know, for example, that protamine is a protein, and that it is basic and has a nitrogen content because it contains large amounts of the basic amino acid arginine. He knew that phosphoric acid was responsible for the acidity of sperm nuclein, and that nuclein was a substance *sui generis* quite distinct from proteins, but he did not know that nucleic acids had two kinds of subunit: purines and pyrimidines. The detailed chemistry of nucleic acids and the proteins associated with them in the cell nucleus was worked out by later biochemists, beginning with Kossel in 1884. It took longer still, as we shall see, to understand the biological function of these substances.

Miescher's work on nuclein led him to a deep interest in the life of the Rhine salmon. In conducting what he called his "sperm campaign" almost every year he was witness to one of the most impressive events of animal life: the migration of the salmon 500 miles from the sea on their way to spawn in the headwaters of the Rhine. Between 1874 and 1880 Miescher devoted himself to the physiology of the salmon, examining

more than 2,000 fish. He learned that the salmon spent eight to 10 months in fresh water, and that during this period, although they were extremely active, they did not feed; they did not, indeed, secrete any digestive juices. And yet during this time an extraordinary metabolic rearrangement occurred: a massive increase in the size of the sex organs. Miescher found that when female salmon left the sea, their ovaries accounted for .5 percent of their body weight; by the time the fish reached Basel the ovaries represented 26 percent of their weight. The growth, he discovered, was at the expense of the large "lateral trunk" muscle; the loss of protein and phosphate by this muscle was sufficient to account for the growth of the ovaries. From physiology Miescher moved on to ecology, making a number of suggestions for the conservation of the Rhine salmon.

In 1872, at the age of 28, Miescher had been appointed professor of physiology at Basel, succeeding His, who had moved to Leipzig. Miescher became a leading figure at the university. In time he built a new physiological institute (which he named the Vesalianum in honor of Andreas Vesalius, the 16th-century Belgian anatomist who had spent six months in Basel while his *De humani corporis fabrica* was being printed there), stocked it with new precision instruments as they became available and guided it in researches on such classical problems of physiology as respiration, the circulation and the effects of high altitude. From time to time he undertook public service projects. The canton of Basel asked his advice on the nutrition of inmates of prisons and other institutions, and he also gave a series of public lectures on nutrition and home economics. Living in a country that produced milk but did not consume much of it, he was aware of milk's nutritional value, and in one lecture he heaped scorn on the "sordid avarice" of peasants who withheld milk from their children in order to make "the last drop" into salable cheese. "If you ask these pale and feeble people what they eat," he said, "they reply: potatoes, coffee, more potatoes and schnapps to keep down hunger." Miescher's social concern led a colleague to comment that he would surely have been active in politics had he not been hard of hearing.

All his life Miescher's primary and absorbing interest was nuclein. From time to time he went back to investigating its chemistry; he was also preoccupied with the question of its biological

function. During his work on the white cells of pus he had made no mention, either in his letters or in the paper published in 1871, of the possible function of the nucleus. This is hardly surprising, since at that time and for some years to come the role of the nucleus in the life of the cell was simply not understood. As late as 1882 the leading cell biologist Walther Flemming, describing the latest work on the nucleus, conceded that "concerning the biological significance of the nucleus we remain completely in the dark."

When Miescher came to the investigation of sperm, however, he began to ask questions about the role of nuclein in fertilization and about the nature of fertilization. Does the sperm contain certain special substances that are effective in fertilization? Willy Kühne, the Heidelberg biochemist who had just introduced the word "enzyme," suggested that there might be enzymes in sperm. Since salmon sperm seemed to be a clean and uncontaminated material, Miescher looked in it for enzymes, but he failed to find anything that appeared to be promising. Then he went on to say: "If one wants to assume that a single substance ... is the specific cause of fertilization, then one should undoubtedly first of all think of nuclein." Coming on this passage written in 1874, the reader holds his breath for a moment—but only for a moment, because Miescher turned away from the idea. He supposed, one must remember, that the egg contained a rich supply of nuclein in its yolk granules. And in his opinion there was no special characteristic that distinguished sperm nuclein from the great mass of egg nuclein. Indeed, he believed "the riddle of fertilization is not hidden in a particular substance"; the sperm is acting as a whole through the cooperation of all its parts, and if one considers the magnitude of the sperm's contribution to heredity, it must work in a most complex way. This line of thought led Miescher to think of fertilization as a physical procedure in which a certain movement (*Bewegung*) of the sperm is transmitted to the egg. In casting about for the nature of the movement, Miescher pointed to the "molecular process" that occurs when a nerve stimulates a muscle as perhaps being analogous to the effect of the sperm on the egg.

When Miescher invoked physical motion to explain fertilization, it is clear that he was thinking of kinetic theory, which at that time was in its formative period. By 1892 another physicochemical approach to fertilization appealed to

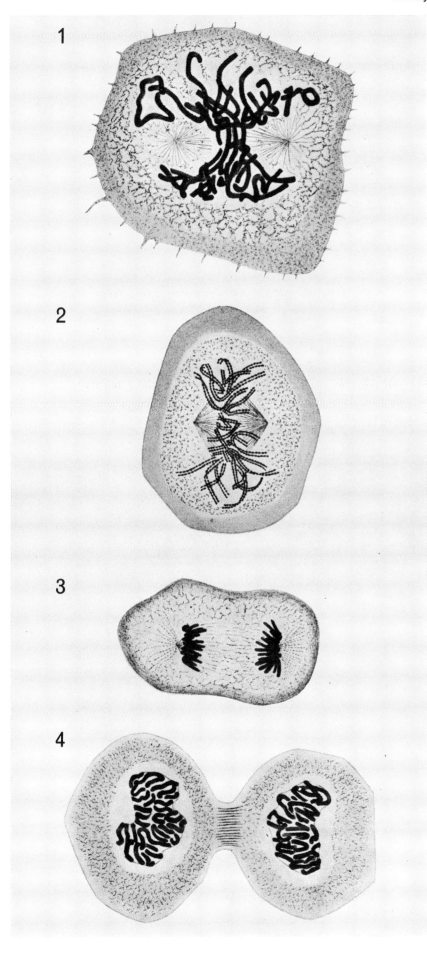

him and he wrote: "The key to sexuality lies for me in stereochemistry," that is, in the varying positions of the asymmetric carbon atoms in molecules. Miescher's desire to find explanations in physical science did not lead him to an understanding of the biological role of nuclein. At about the same time, however, another group of biologists—who were not interested in kinetic theory or stereochemistry—succeeded in laying some of the foundations of biology and at the same time recognizing the fundamental role of nuclein.

Miescher was nurtured in the great 19th-century school of "molecular biology." Both His and Carl Ludwig (in whose laboratory Miescher had spent a year) were leaders of the movement to analyze vital phenomena on the basis of physical science. Other outstanding figures in the movement were Claude Bernard and the neurophysiologists Emil du Bois-Reymond and Hermann von Helmholtz. The contributions of these men taken together constitute the foundation of modern physiology. Ludwig's work on kidney function, to take an example, is the basis of modern kidney physiology. In 1885 Michael Foster, professor of physiology at the University of Cambridge, spoke of "a new molecular physiology," maintaining that "the more these molecular problems of physiology ... are studied, the stronger becomes the conviction that the consideration of what we call 'structure' and 'composition' must, in harmony with the modern teachings of physics, be approached under the dominant conception of modes of motion.... The phenomena in question are the result not of properties of kinds of matter, in the vulgar sense of these words, but of kinds of motion." It was from this point of view that His considered fertilization to be a process in which the sperm communicates a mode

IN CELL DIVISION it is the chromosomes that provide continuity, as is shown in these drawings of salamander epithelial cells published in 1882 by Walther Flemming. In an early stage of cell division the chromosomes (which were known to contain nuclein) are ribbon-like and a "spindle" (two rayed structures) is beginning to form (1). The chromosomes seem to split (2); actually each has replicated lengthwise and the members of each resulting pair are separating. Each member goes to a different pole of the cell, so that two equal complements of chromosomes are established (3). Nuclear membranes surround each complement and the cell divides into two identical daughter cells (4).

of motion to the egg—a view that, as we have seen, Miescher accepted.

The biologists who at this time succeeded in discovering what actually happens in fertilization, and thereupon recognized the role of Miescher's nuclein, were (with several notable exceptions) not interested in physical science and in modes of motion. They were the founders of cell biology. The essential step taken by these biologists was the minute observation of what actually happens when a sperm fertilizes an egg. Although salmon sperm was an ideal material for Miescher's experiments on the chemistry of the sperm nucleus, it was of little value for observing the act of fertilization. For this purpose Oskar Hertwig of the University of Berlin and the Swiss biologist Hermann Fol chose the sea urchin and the starfish. In their classic experiments in the late 1870's

they observed that the sperm cell penetrates the egg and that the sperm nucleus then fuses with the egg nucleus. At the same time these experiments were being done Flemming, in another series of observations, described the changes that occur in the nucleus during cell division [*see illustration on page 2743*]. He was able to show chromosomes provide the continuing elements from one generation of cells to the next and that they do so by replicating during cell division.

The observations on fertilization by Hertwig and Fol and those on cell division by Flemming were brought together by Edouard van Beneden of the University of Liège in a wonderful series of observations on fertilization in the threadworm *Ascaris*, a parasite of horses. *Ascaris*, unlike most other animals, has nuclei that break down before the egg and sperm nuclei fuse, so that its chro-

mosomes can be seen with unusual clarity. Chromosome behavior can be followed readily because there are only two chromosomes in each nucleus. The male and female chromosomes, indistinguishable from each other, are brought together but do not fuse. Each chromosome replicates and soon there are two cells, formed in the way Flemming described for cell division. Van Beneden saw that in fertilization, as in cell division, continuity depends on chromosomes: the sperm's contribution to fertilization is ·a set of chromosomes homologous with those present in the egg [*see illustration on these two pages*]. (In the course of this work, which was published in 1883, van Beneden discovered meiosis, the halving of chromosome number that precedes the bringing together of egg and sperm chromosomes in fertilization.)

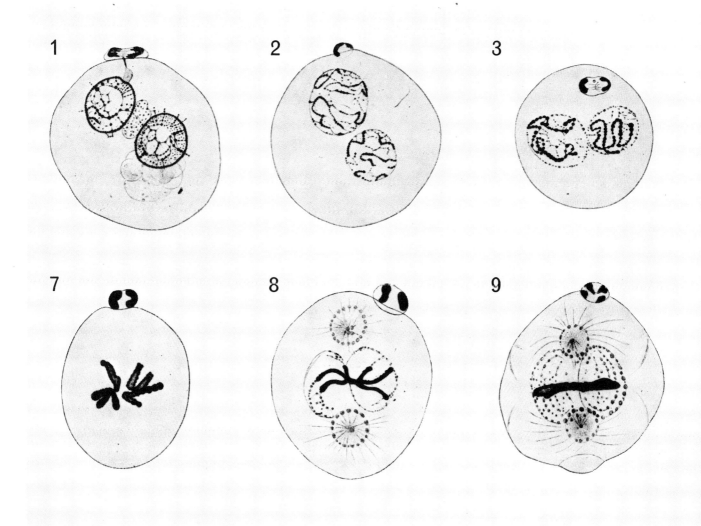

IN FERTILIZATION TOO chromosomes provide the continuity, as shown in drawings of the fertilized egg of *Ascaris*, a parasitic worm, published in 1883 by Edouard van Beneden. Sperm and egg nuclei approach each other (*1*). The nuclear membrane breaks down and the chromosomes become clearly visible; there are two chromosomes in each nucleus, half the normal "diploid" number in

Two years earlier Miescher's nuclein had been brought into the picture, not by Miescher but by an obscure young botanist named E. Zacharias. He showed that the characteristic material of chromosomes either was nuclein or was intimately associated with it. In his work on cell division Flemming had relied extensively on stains to make the nucleus and chromosomes readily visible. In the nucleus of cells that were not in the process of division there was a somewhat formless structure that took up certain stains; the same stains were taken up by the rod-shaped chromosomes as they emerged from the nucleus during mitosis. The material that took the stain was called chromatin, and Zacharias identified it as nuclein by following the same procedure that had led Miescher to the discovery of nuclein. He found that when a cell was digested with pepsin–hydro-

chloric acid its nucleus remained and retained its ability to be stained. If, however, the digested cell was extracted with dilute alkali (which, as Miescher had shown, removes nuclein), then no stainable material remained. Zacharias carried out these tests on an exceedingly wide range of cells, both plant and animal, with essentially uniform results. He tested cells in the process of division and found that chromatin could be stained after pepsin-digestion but not after extraction in alkali. Moreover, the spindle that forms during mitosis did not stain, and it was removed by pepsin–hydrochloric acid digestion. All of this pointed to the coupling of nuclein and chromatin. Zacharias' conclusions were quickly accepted by Flemming and many others.

In 1884 and 1885 four biologists published papers that summarized and

interpreted the work of the preceding decade, which had been so crowded with discoveries that those who participated felt the swift movement of events in much the same way that biologists do nowadays. Of the four summarizing accounts, three were by zoologists (Hertwig, Albrecht Kölliker and August Weismann) and one by a botanist (Eduard Strasburger). It was clear that an understanding of fertilization was at the same time an understanding of heredity. Continuity from one generation of an organism to the next was accomplished by the chromosomes in the nuclei of egg and sperm; continuity from one cell generation to the next was also accomplished by chromosomes in mitosis. At this point (in 1884) we suddenly come very close to what is one of the cornerstones of current biology: "I believe that I have at least made it highly prob-

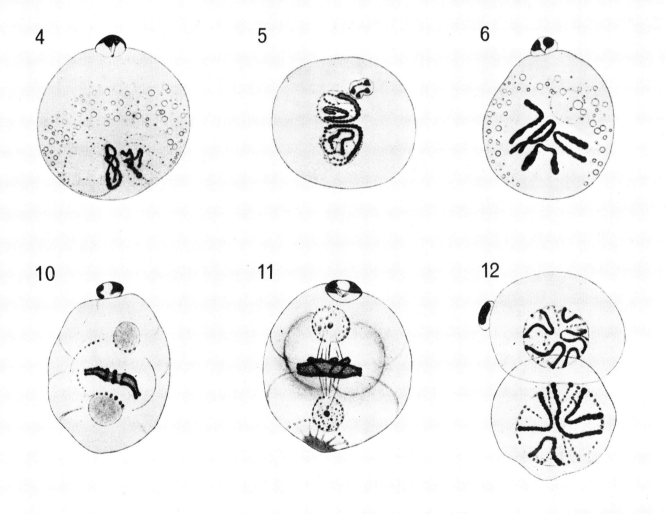

an *Ascaris* cell (2, 3, 4). The four chromosomes come together (5, 6). Each of the chromosomes has previously doubled; the replicated pairs separate slightly (7). Two "centrioles" appear and a spindle forms between them, and cell division begins (8, 9, 10, 11). Two cells are finally formed, each with a nucleus containing four chromosomes, two derived from the sperm and two from egg (12).

able," Hertwig said, "that nuclein is the substance that is responsible not only for fertilization but also for the transmission of hereditary characteristics.... Furthermore, nuclein is in an organized state before, during and after fertilization, so that fertilization is at the same time both a morphological and a physicochemical event."

Even before Hertwig recognized nuclein as the genetic material, the botanist Julius von Sachs had (in 1882) not only suggested this role for nuclein but also gone a step further in pointing out that the nucleins of egg and sperm could hardly be identical—that the nuclein brought into the egg by the sperm must be different from the nuclein already there. And so by 1885, only a decade after Miescher's paper on salmon sperm and 14 years after his first publication on nuclein, a number of biologists had reached a point of view that is at the heart of our present conception of DNA.

What was the attitude of the discoverer of nuclein? As far as we can tell from Miescher's letters (and those to His freely express his views on many problems of biology), he did not accept the new ideas. He doubted the association of nuclein with chromatin, which was an essential element in the ideas expressed by Hertwig and the others. Miescher had a rather low estimate of the value of staining. He wrote in a letter of 1890, "Here once again I must defend my skin against the guild of dyers who suppose there is nothing else [in the sperm head] but chromatin," and in a paper published posthumously he referred disparagingly to Zacharias' work on chromatin. Miescher's attitude was also conditioned by his belief (expressed in several letters in 1892 and 1893) that he had discovered a new substance in the sperm head, which he proposed calling "karyogen." It was this phosphorus-free substance (the nature of which is a mystery today), and not nuclein, that in his opinion was responsible for the special chromatin stain.

In a letter of 1893 Miescher wrote: "The speculations of Weismann and others are afflicted with half-chemical concepts, which are partly unclear and partly derived from an outmoded kind of chemistry. When, as is quite possible, a protein molecule has 40 asymmetric carbon atoms so that there can be a billion isomers,... my [stereochemical] theory is better suited than any other to account for the unimaginable diversity required by our knowledge of heredity." Weismann's speculations were based on the most advanced biology of the time; Miescher preferred to base his specula-

tions on recent advances in chemistry and paid relatively little attention to advances in biology. Time—and the rediscovery of Mendel's "units" of heredity and their identification with genes by Thomas Hunt Morgan—vindicated the idea, forcefully expressed by Weismann, that chromosomes transmit heredity.

Time did not, however, deal so consistently with the idea that nuclein is the material in chromosomes that transmits heredity. That idea appeared in the 1880's, as we have seen, and was widely held in the 1890's. As late as 1895 the American cytologist E. B. Wilson wrote: "Now, chromatin is known to be closely similar to, if not identical with, a substance known as nuclein.... And thus we reach the remarkable conclusion that inheritance may, perhaps, be effected by the physical transmission of a particular chemical compound from parent to offspring." In the next few years doubts arose concerning this conclusion because the amount of chromatin in the nucleus seemed too unstable to provide continuity; the amount varied considerably with the cycle of cell division and with changes in the physiological state of the cell. In the second edition of his book *The Cell in Development and Heredity,* published in 1900, Wilson described this fluctuation in staining but was able to explain it: "We may infer that the original chromosomes contain a high percentage of nucleinic acid; that their growth and loss of staining power is due to a combination with a large amount of albuminous substance...; that their final diminution in size and resumption of staining power is caused by a giving up of the albuminous constituent."

This analysis corresponds exactly with our present understanding, but it was soon to succumb to the apparent evidence of staining. When the large "lampbrush" chromosomes present in certain egg-cell precursors came under intensive study in the 1890's, there seemed to be no chromatin in them at all, and surely if chromatin (or nuclein) was the hereditary material, it must be present in an unbroken line from one cell generation to the next. In 1909 Strasburger wrote: "Chromatin cannot itself be the hereditary substance [because] the amount of it is subject to considerable variation in the nucleus, according to its stage of development." Strasburger was an eminent authority. So was Wilson, and the third edition of his book, which was published in 1925 and influenced a generation of biologists, took the Strasburger view: that the loss of staining in the enlarged

chromosomes indicates "a progressive accumulation of protein components and a giving up, or even a complete loss, of nuclein." Wilson emphasized, using italics, "These facts afford conclusive proof that *the individuality and genetic continuity of chromosomes does not depend upon a persistence of chromatin.*" Biologists maintained this position for a generation.

Then, in 1948 and 1949, groups at the Rockefeller Institute and in France independently measured the quantity of DNA in cell nuclei. They found that the amount of DNA per set of chromosomes is in general constant in the different cell types of any organism, even when there are striking differences in the intensity of staining, and that the amount of protein associated with a fixed quantity of DNA may vary considerably and so account for variations in staining capacity. Moreover, the DNA content per chromosome set is a characteristic of each particular species. The doubts and questions concerning staining that had been raised by Miescher and many others were essentially resolved by these measurements. Now the biological role of DNA, proposed in 1884 and 1885 when the chromosome theory of heredity was formulated, received solid support: It was demonstrated that the chromosome complements of egg and sperm carry identical amounts of DNA, which are combined at fertilization and then carried by successive replications to all cells of the organism. The continuity of chromosomes at fertilization and cell division has always been an essential element in the chromosome theory of heredity; the associated continuity of DNA (Miescher's nuclein) points to it as providing the molecular basis for heredity in the chromosomes.

Several years before this point was finally established investigators at the Rockefeller Institute—following a line of investigation with a different historical background—found that hereditary traits could be transmitted from one strain of bacteria to another by the transfer of DNA. Nucleic acid was thus shown to be the genetic material. It should be pointed out that if this substance had been discovered in bacteria, it would never have been called nucleic acid, because a bacterial cell does not have a formed nucleus. The use of such words as "nucleic acid" and "chromosome" in work on bacteria is a constant reminder that the contributions of Friedrich Miescher and his contemporaries form the background for the study of heredity in all living cells.

The Author

ALFRED E. MIRSKY is professor of general physiology at Rockefeller University, where he has worked since 1927. Since 1965 he has also been its librarian. Mirsky is one of the modern pioneers in the study of DNA. He has worked on many chemical aspects of the cell and the nucleus: chromosomes, nucleoprotein, hemoglobin and other proteins. Mirsky was graduated from Harvard College in 1922 and obtained a Ph.D. in physiology from the University of Cambridge in 1925. He was the author of "The Chemistry of Heredity" in the February 1953 issue of SCIENTIFIC AMERICAN and co-author with Vincent G. Allfrey of "How Cells Make Molecules" in the issue of September, 1961.

Bibliography

THE CELL IN DEVELOPMENT AND HEREDITY. Edmund B. Wilson. The Macmillan Company, 1928.

CELL, NUCLEUS, AND INHERITANCE: AN HISTORICAL STUDY. William Coleman in *Proceedings of the American Philosophical Society,* Vol. 109, No. 3, pages 124–158; June, 1965.

DIE HISTOCHEMISCHEN UND PHYSIOLOGISCHEN ARBEITEN. Friedrich Miescher. Verlag von F. C. W. Vogel, 1897.

ZELLSUBSTANZ, KERN UND ZELLTHEILUNG. Walther Flemming. Vogel Verlag Leipzig, 1882.

SCIENTIFIC
AMERICAN   June 1968, Vol. 218, No. 6, pp. 102-108   OFFPRINT **1110**

PLANTS WITHOUT CELLULOSE

by R. D. Preston

It was once believed that cellulose was the structural material in the cell walls of all green plants. Now it appears that other substances are employed for this purpose in certain marine algae.

Until quite recently the statement that all green plants contain cellulose seemed about as correct as the statement that all vertebrates contain bone. Cellulose is the structural constituent of the walls of the cells of green plants; it is a basically fibrous substance that gives the cell walls form and rigidity. Wood is about 40 percent cellulose. Cotton, linen and the fibrous products made from wood pulp—paper, rayon and so on—consist of little else.

It has therefore been startling to discover that there are two groups of green plants in which the cell walls are based not on cellulose but on two quite different substances. In one of these substances the molecular subunits are linked together in a way that is different from the linkages in cellulose; hence the molecules pack together differently in the fibrils, the structural units of the cell wall. In the other substance the linkages are the same as those in cellulose but the substance seems to form fibrils only occasionally.

The first indication that green plants lacking cellulose might exist came some time ago, when it proved impossible to demonstrate the presence of cellulose in certain plants. These demonstrations relied, however, on staining methods, which are notoriously unreliable. Consequently the cell walls of the anomalous plants did not attract much attention. Then about five years ago, by one of those coincidences that come about in science presumably because the atmosphere is right, several laboratories in different parts of the world independently began to study these odd plants. One of the laboratories was our own at the University of Leeds. Using approaches that were not available to the earlier workers, we have shown that such plants do indeed lack cellulose, and we have gone on to explore their molecular architecture.

The problem of how the cell wall of plants is built is an important one even apart from the fact that such essential materials as wood, cotton and paper are virtually cell walls in their natural state. The problem is important because the cell wall forms a rigid or semirigid envelope around each living unit in all plants. A plant can grow only at such a rate and in such a way as its constituent cells will allow; their growth is in turn restricted by the physical properties of their walls, and these properties of course depend on the walls' detailed molecular structure. Accordingly the shape of a plant and the rate at which it reaches its final form can in principle be traced back to the structure of the cell wall.

All the plants that lack cellulose are algae, and most of them are inhabitants of the warmer seas. They are interesting in their own right. Some of them, in spite of the fact that their cellular construction is comparatively simple, have achieved spectacular forms not surpassed by higher plants [see illustration on page 2750]. One of them, *Acetabularia,* is widely used as a model cell in studies of the relation between the nucleus and the surrounding cytoplasm.

As a group these seaweeds have considerably broadened our understanding of the structure of plant cell walls. Since their cell walls are different from those of all other green plants, they broaden the range of possible structures affecting plant growth, offering a basis against which we can assess knowledge obtained from cellulosic plants and thus define the factors involved in plant growth more sharply.

To understand how differently the walls of the noncellulosic plants are constructed, let us first consider the cell walls that contain cellulose. It is convenient, if somewhat oversimplified, to regard the cell wall as having two components. One of these components consists of partially crystalline microfibrils, that is, thin threads partly made up of long-chain molecules packed together in a regular parallel array. The microfibrils are between 100 and 200 angstrom units wide, about half as thick and so long as to be virtually endless. They are embedded in the second component of the cell wall, a less crystalline cementing matrix. The regular parallel arrays within the microfibrils are cellulose; the cementing matrix consists of substances that, like cellulose, are polysaccharides but are made up of different subunits.

The subunits of cellulose are the simple ring-shaped structures of the sugar glucose. The ring includes five carbon atoms, each of which is identified by number [see illustration on page 2751]. When one glucose molecule is linked to another, the carbon atom designated 1 is joined through an oxygen atom to carbon atom 4 of the other molecule. At either end of the dimer thus formed another glucose molecule can be attached in the

DIVERGENT STRUCTURE of the cell wall of plants is revealed in electron micrographs on opposite page. At top the tiny threads, or microfibrils, of the marine alga *Chaetomorpha melagonium* show distinctly. The microfibrils are composed of cellulose, as are those of all known higher plants. In the middle is a comparably magnified sample of the green alga *Penicillus dumetosus.* Short crossbars appear to connect the vertically running microfibrils. Although in other respects the fibrils resemble those in the micrograph at top, they are not made of cellulose. At bottom is the cell wall of the alga *Batophora,* which also lacks cellulose. No microfibrils are visible; instead the wall appears to consist mainly of granules. Magnification of the micrographs (from top to bottom) is 65,000, 75,000 and 100,000 diameters.

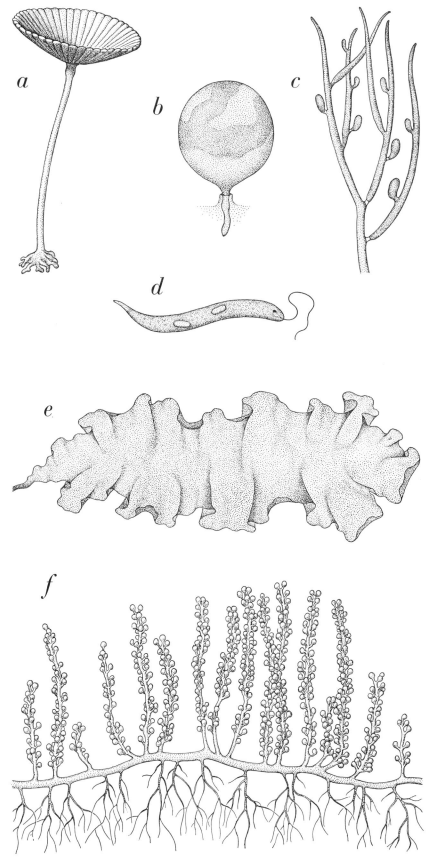

EXCEPTIONAL MARINE PLANTS differ from other green plants in cell structure. The single-celled *Acetabularia* (*a*) has its nucleus in the "stem," making it a useful model in cell studies. *Halicystis* (*b*) and *Derbesia* (*c*) are two generations of the same plant. The divergent feature in the structure of *Euglena* (*d*) is the paramylum, or food reserve, forming two lozenge-shaped bodies. *Porphyra* (*e*) is a papery red seaweed. *Caulerpa* (*f*) has structures like roots and branches but is a single cell. The plants shown here are not drawn to scale.

same way. If the process is continued, it ultimately gives rise to cellulose: a chain of perhaps 10,000 or 15,000 glucose units. The chain appears to be straight rather than coiled or folded (folded structures have been suggested but the evidence is not convincing), probably because of certain hydrogen bonds between its subunits. These straight chains, lying parallel to one another for at least part of their length, form a crystalline core in the microfibril.

The cementing matrix consists of such polysaccharides as xylan, mannan, araban, galactan, glucomannan and polygalacturonic acid. The subunits of these polysaccharides are sugars other than glucose. The molecular chains of matrix polysaccharides lie parallel to the microfibrils but are irregularly arranged and therefore only slightly crystalline. The cell wall owes its coherence to bonding between these amorphous substances and the microfibrils: hydrogen bonds, bridges through calcium and magnesium ions and, it is now believed, linkages to a specific protein found in cell walls.

This, then, is the general structure of the cell wall of the plants that contain cellulose. Let us now examine our anomalous algae. They are divided into two groups. Both were at one time classified in a single order, the Siphonales (a term that has since been dropped because it was not based on a legitimate family name). In one group cellulose is replaced by xylan, a polymer, or repeating chain, of the pentose sugar xylose; in the other group it is replaced by mannan, a polymer of the hexose sugar mannose. It will be convenient to examine the xylan seaweeds first.

All these plants are of filamentous construction. They range from a single-branched filament in *Bryopsis*, through an intricately formed filament simulating the habit of higher plants in *Caulerpa*, to a compact body of associated filaments in *Halimeda* or *Penicillus*. When subjected to X-ray analysis, the xylan plants all display the same diffraction pattern; it is markedly unlike the pattern obtained from cellulose. When the crystalline wall material of such plants is decomposed, it yields only xylose, indicating that the material is certainly a xylan.

The structure of the xylose molecule is much like that of the glucose molecule except that the sixth carbon atom, with its associated groups, is replaced by hydrogen. In the xylans found in the wall matrix of the cellulose-containing higher plants the xyloses are joined carbon 1 to 4, like the glucoses in cellulose. On the other hand, the X-ray pattern of the seaweed xylans is different from the pattern of higher-plant xylans; hence the seaweed xylan must be built up some other way. The only link possible is between carbon atoms 1 and 3.

This was the conclusion to which our studies led. It would perhaps have been the first demonstration of a link between the sugars of a polysaccharide by X-ray analysis alone. Before Eva Frei and I published our findings, however, 1,3-linked xylan had been detected in some of the seaweeds by chemical methods, first by I. M. Mackie and Elizabeth Percival of the University of Edinburgh and then by Y. Iriki and T. Miwa in Japan.

When the subunits of xylan are linked carbon 1 to carbon 4, they form a straight chain (although, as has been shown by Robert H. Marchessault of the State University of New York College of Forestry at Syracuse University, the chain is probably twisted around its long axis). With a 1,3 linkage the chain can only be curved. Nevertheless, in the electron microscope the fibrils of seaweed xylan are reasonably straight. In what way can curved chains pack together to form straight fibrils? One possibility is that the chains are coiled into straight helices. The helices would need to be fairly flat, because observations with the polarizing microscope show that the individual chains lie more nearly at right angles to the microfibrils than parallel to them. Evidence provided by X-ray analysis led us to conclude in 1964 that within a crystalline xylan fibril the chains are coiled in double helices. On the basis of these and other findings we worked out at that time a model of the xylan fibril.

Recently two of my colleagues, K. D. Parker and E. D. T. Atkins, have refined this model by applying helical diffraction theory to the X-ray diagram and by the use of polarized infrared spectrophotometry. They have demonstrated that three chains, not two, are coiled together, stabilized by interchain hydrogen bonding [see bottom illustration on next page]. At the moment it seems likely that the chains are coiled in a right-handed sense. We believe this represents an exceptionally precise determination of structure for a polysaccharide.

It is interesting that a specimen of xylan is more crystalline when it is wet than when it is dry; water is required (to the extent of 30 percent of the weight) if the chains are to lie in perfect order. This is quite different from the behavior of cellulose, which is equally crystalline dry or wet, water being unable to penetrate the crystal structure. It appears that insofar as crystallinity is indispensable to a structural polysaccharide, 1,3-linked xy-

lan is better suited to water plants than to plants on dry land. This may be why land plants have not, as far as we know, used the substance as a supporting framework for their cells.

Although the xylan seaweeds use xylose in preference to glucose as a structural subunit, glucose does appear among the polysaccharides that form the matrix substance of their walls. Thus the roles of the two sugars in the xylan seaweeds and higher plants are reversed. It has been reported that the glucose units in the walls of the seaweeds are 1,3-linked. It is not clear to what extent the reversed roles of glucose and xylose are related to the 1,3 linkage as against the 1,4 one.

A number of other polysaccharides spread through the plant kingdom are known to be 1,3-linked. We and others have examined some of them and we have found that their X-ray patterns are much alike and similar to that of the xylan. We are now looking into the possibility that most polysaccharides so linked—for example the laminarin of brown seaweeds, the paramylum of Euglena and the callose of higher plants—also form helices. It is quite likely that the most abundant polysaccharides on the earth are not, as we once thought, the straight-chain 1,4-linked polysaccharides cellulose and starch but rather the polysaccharides that are 1,3-linked and helical. These substances can be expected to have mechanical properties quite different from those of the straight-chain ones.

The second group of plants I wish to examine here is unusual in quite a different way. These plants, which include Acetabularia, are again basically filamentous. The only crystalline polysaccharide present in their walls is mannan. It is identical with the mannan found in higher plants. Mannose, the sugar of which mannan is composed, closely resembles glucose; the molecular structure of the two sugars differs only in the reversed positions of a hydroxyl group (OH) and a hydrogen on one of the carbons in the ring. Like cellulose, the units of mannan are linked through carbon 1 to carbon 4.

Although mannan is common among higher plants, it is detectable as a crystalline polysaccharide only in the seeds of one group of them. Even there it is not the only structural polysaccharide, the cell walls of the plants being basically cellulosic. This group is the palms, among them the date and the ivory-nut palm (the toughness of mannan is indicated by the name and the fact that laboratory-coat buttons are sometimes still

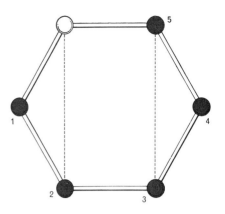

GLUCOSE

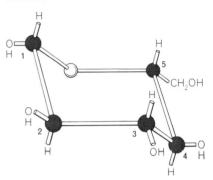

XYLOSE

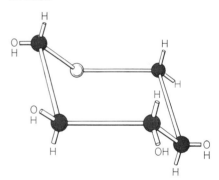

MANNOSE

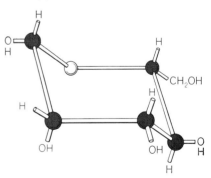

MOLECULAR DIAGRAMS indicate (color) where xylose and mannose differ from glucose. All these sugars have a hexagonal ring (top) composed of carbon atoms (black balls) and an oxygen atom (white ball). In the diagrams this hexagon is seen from the edge with two corners bent. Cellulose is formed from glucose; xylose and mannose form the substances that replace cellulose.

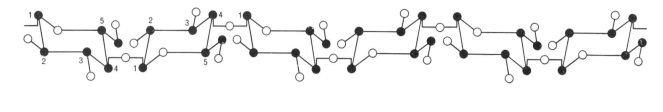

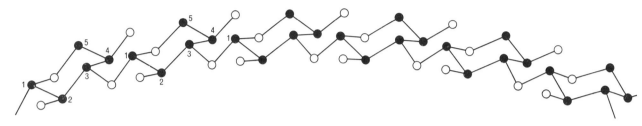

MOLECULAR CHAINS are linked through oxygen atoms. A short segment of the cellulose chain is shown at top with carbon atoms in the hexagonal ring of some of the units numbered. The linking in the chain of carbon atom 1 and carbon atom 4 gives rise to a regular alternation in the orientation of the glucose units and a straight chain. Below the cellulose chain appears the segment of a xylan chain that is composed of xylose units. The units are joined at carbon atoms 1 and 3, and this linkage produces a curved chain.

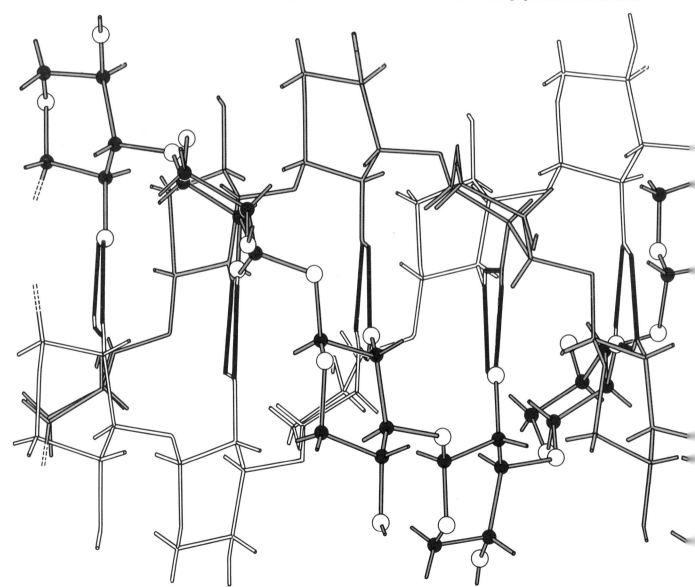

TWISTING OF XYLAN CHAINS gives rise to a straight micro-fibril. Each straight section of the three chains forming the helix represents a bond between atoms; the black linkages are hydrogen bonds that stabilize the chains. The chains are shown as being coiled in a right-handed sense; in one chain oxygen and carbon atoms are depicted. The drawing is based on a model worked out

made of "vegetable ivory" in spite of the availability of plastics).

Through X-ray analysis of the cell walls of mannan seaweeds it has been possible for the first time to provisionally elucidate the crystal structure of mannan. It appears to resemble the crystal structure of cellulose in that the chains of mannose units lie parallel to one another; it is dissimilar in that the chains pack together somewhat differently. Neither fact is surprising in view of the specific difference between the molecules of mannose and glucose. The molecular chains of mannan are commonly reported to be much shorter than those of cellulose, ranging up to 80 mannose units. My colleagues W. Mackie and D. B. Sellen have given reason to believe, however, that the chains of seaweed mannan approach the length of cellulose chains.

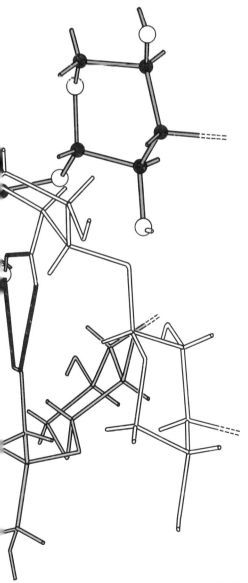

by the author and his colleagues at the University of Leeds. Hydrogen atoms are omitted from this illustration and the one at top.

The surprise comes when one examines under the electron microscope a sample of the cell-wall material that attests (on the basis of X-ray analysis) to mannan's crystal structure. One would expect to find much the same kind of well-oriented fibril that one sees in a comparable micrograph of cellulose or xylan. In spite of the most determined efforts such fibrils have not been detected by the methods that succeed with cellulose and xylan seaweeds. These methods involve the removal of the noncellulosic polysaccharides by chemical means. The polysaccharides that surround the crystalline regions of mannan, however, are themselves mostly mannan; they cannot be removed to reveal a possible underlying fibril structure without the risk of destroying the structure. Although the microfibrils of cellulose and xylan are often visible without the removal of the surrounding matrix, it is still not safe to conclude that in mannan walls microfibrils are completely absent. In fact, Mackie has shown recently that exceptionally mild treatment of mannan walls reveals occasional microfibrils. Nevertheless, in electron micrographs seaweed mannan appears most often to take the form of granules or at most very short rods (no longer than 2,000 angstroms). The granules must be mutually oriented in an orderly fashion to provide the observed X-ray pattern.

This presents problems even though microfibrils are occasionally present. If one assumes that the mechanical properties of the mannan walls are not much different from those of cellulose walls (this assumption is supported by manipulations of the walls), how does it come about that an array of granules behaves like an array of parallel threads? What mechanism specifies the particular order the granules display? It is possible to conceive of a mechanism by which threads can be arranged parallel to one another, and I have already proposed such a mechanism in general terms. It is much more difficult to imagine a mutual orientation of granules under the conditions that are known to exist at the cell surface.

In searching for an explanation we must bear in mind that the cell-wall layers examined in the electron microscope have been subjected to drastic drying in the process. Furthermore, under these conditions a structural element with one dimension of less than eight angstroms is not visible at all in the kind of material involved. One could therefore imagine the wall to be a parallel array of short molecular chains that overlap one another and that in small granule-shaped

regions, and in these regions only, are regularly spaced and therefore crystalline. Such a structure could give rise to both our X-ray patterns and our electron micrographs, and would be in agreement with the chain lengths found by Mackie and Sellen.

The mannan cell walls appear to have significance for certain theories of plant growth. The growth rate of plant cells is known to be under the control of auxins, or plant hormones. The view most widely held today is that the auxins achieve their effect by changing the coherence of the cell wall. The cell-wall components believed until recently to be the most intimately involved are the pectic compounds, derivatives of polygalacturonic acid that are found in the matrix substances of all higher plants.

Now, the auxins are also said to affect the development of the seaweed Acetabularia. The wall of this plant is lacking not only in cellulose but also—according to our analyses and those of other laboratories—in pectic compounds. We must therefore assume either that this plant is peculiar in still another way or that the wall substance affected by the auxins is not polygalacturonic acid. The latter seems to me the more likely, particularly in view of work that points in quite another direction. One of my students, E. W. Thompson, has detected in the walls of a noncellulosic plant (Codium tomentosum, which contains mannan) an abundance of a specific protein that D. A. Lamport of Michigan State University has shown to be present in the walls of higher plants. Lamport gives reason to believe the protein is a wall component closely associated with cell-wall extensibility, and we have reached the same conclusion both for mannan-containing and cellulose-containing seaweeds. This may prove not to be the whole story, but our work implies that it is at least an important part of it.

When we first began our study of these odd plants, we suspected that (to argue teleologically) plants in the sea might have experimented with a variety of polysaccharides as cell-wall skeletal materials. As it has turned out this suspicion was justified. It remains to be seen what other surprises may await us among the lower plants. We know already of two other groups of plants, also seaweeds, that combine in themselves the structures found separately in the mannan and xylan algae. In one of these groups the prominent member is Porphyra, familiar as the papery red seaweed that in season clothes many of the rocks along coasts in the Northern Hemi-

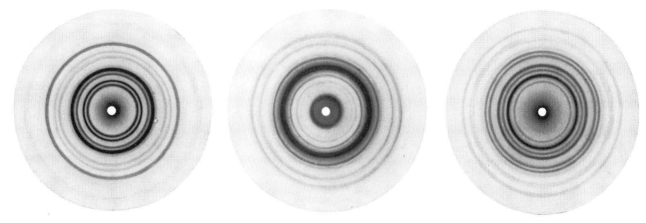

X-RAY DIFFRACTION PATTERNS indicate differences in the cell structure of seaweeds. At left is the pattern made by *Chaetomorpha*, whose cell walls contain cellulose. The middle pattern is of *Penicillus*, in which xylan replaces cellulose. At right is the pattern of *Batophora*, whose cell walls contain mannan, which is composed of mannose units and also substitutes for cellulose.

sphere. In this plant the cell walls contain crystalline 1,3-linked xylan, and the cuticle covering the walls contains crystalline 1,4-linked mannan.

The other example of combined structure, involving the plant *Halicystis* (commonly known as sea bottle and consisting of single "cells" or vesicles as large as a pea), is somewhat more complicated. *Halicystis* is a haplont, which is to say that each of its nuclei has only a single set of chromosomes. During reproduction the nuclei fuse in pairs, and the resulting cell ultimately develops to form a plant body called the diplont generation. Eventually some or all of the nuclei of the diplont plant undergo the division known as meiosis, by which the chromosome number in each daughter nucleus is halved. The cells then divide by normal mitosis to reproduce the haplont plant body, and the cycle is repeated.

Now, although *Halicystis* is composed of roughly globular large cells, its diplont phase is finely filamentous; the diplont is known by the name *Derbesia*. *Halicystis* and *Derbesia* are two generations of one and the same plant. Nevertheless, the walls of *Halicystis* contain crystalline 1,3-linked xylan together with some cellulose but no mannan, and the walls of *Derbesia* contain crystalline mannan but neither cellulose nor 1,3-linked xylan. This is a clear case of a correlation between the chromosome complement of a cell and the constitution of its wall.

It is a rather odd circumstance that although water plants such as those discussed here must have lived for many millions of years with their peculiar walls, as far as we know none of these walls is represented among land plants. It would be too much to say that the structure of all land plants is known—there may be some surprises waiting in this realm as well. Nonetheless, the consensus is that all green land plants will prove to have cellulose walls. It appears that several polysaccharides and polysaccharide combinations can make a serviceable cell wall for water plants but only one kind of wall serves those plants that have managed to colonize dry land. All we can conclude is that plants on dry land need in their cell walls a balance between the rigidity necessary to hold the plant upright and an extensibility—perhaps even a controlled breakdown—that will allow the plant to grow. Perhaps this balance can be achieved only in a wall based on cellulose.

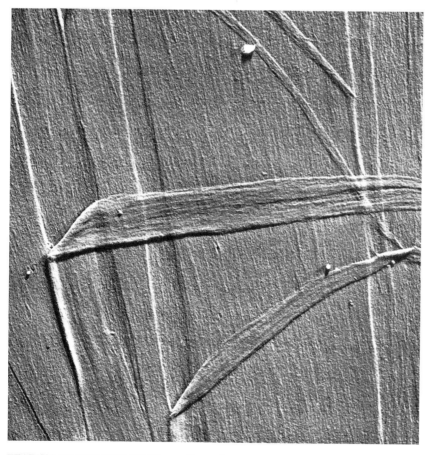

TEARING OF THE CELL WALL provides evidence of its underlying structure in the mannan seaweed *Dasycladus*. Like the cell wall of other mannan seaweeds, the wall of *Dasycladus* appears to be predominantly composed of granules. The fact that the specimen tears cleanly in a direction parallel to the grain of the wall indicates that the granules are in an ordered array. In this electron micrograph the specimen is enlarged about 6,500 diameters.

The Author

R. D. PRESTON is professor of plant biophysics and head of the Astbury Department of Biophysics at the University of Leeds, where he received his undergraduate and graduate training. After obtaining a Ph.D. in botany in 1931 he held the 1851 Exhibition Fellowship at the university for three years and then spent two years at Cornell University as a Rockefeller Foundation Fellow. In 1936 he returned to Leeds and took the position of lecturer in botany, becoming in turn senior lecturer, reader and professor. He has published some 150 articles and a book on the molecular architecture of the walls of plant cells. In 1954 he was elected a Fellow of the Royal Society.

Bibliography

CELL WALLS. D. R. Kreger in *Physiology and Biochemistry of Algae*, edited by Ralph A. Lewin. Academic Press, 1962.

CELLULOSE. R. D. Preston in *Scientific American*, Vol. 197, No. 3, pages 157–168; September, 1957.

THE MOLECULAR ARCHITECTURE OF PLANT CELL WALLS. R. D. Preston. John Wiley & Sons, Inc., 1952.

NON-CELLULOSIC STRUCTURAL POLYSACCHARIDES IN ALGAL CELL WALLS, I: XYLAN IN SIPHONEOUS GREEN ALGAE. Eva Frei and R. D. Preston in *Proceedings of the Royal Society*, Vol. 160, Series B, No. 980, pages 293–313; September 29, 1964.

NON-CELLULOSIC STRUCTURAL POLYSACCHARIDES IN ALGAL CELL WALLS, II: ASSOCIATION OF XYLAN AND MANNAN IN PORPHYRA UMBILICALIS. Eva Frei and R. D. Preston in *Proceedings of the Royal Society*, Vol. 160, Series B, No. 980, pages 314–327; September 29, 1964.

NON-CELLULOSIC STRUCTURAL POLYSACCHARIDES IN ALGAL CELL WALLS, III: MANNAN IN SIPHONEOUS GREEN ALGAE. Eva Frei and R. D. Preston in *Proceedings of the Royal Society*, Vol. 169, Series B, No. 1015, pages 127–145; January 16, 1968.

SCIENTIFIC
AMERICAN   July 1968, Vol. 219, No. 1, pp. 75-81   OFFPRINT **1111**

THE CONTROL OF PLANT GROWTH

by Johannes van Overbeek

The demonstration that the growth of plants can be turned on and off
at will by treating them with the proper combination of promotive and
inhibitory hormones suggests that this process also occurs in nature.

The growth of a plant basically calls for light and water. Simple though these requirements may sound, plant growth itself is of course quite complex. In the green plant cell sunlight splits the molecules of water; this is part of the process of photosynthesis. The

O of the H_2O is released into the air to provide the oxygen that all living things, including man, need for the process of respiration. The H_2 reacts with carbon dioxide and with nitrate to produce the major constituents of plant cells: sugar, starch, cellulose, protein and nucleic

acid. When plant tissues are consumed by humans, these substances supply the building blocks of human cells and the energy needed for human life processes.

Normal plant growth and development will not take place without exceedingly small quantities of specific, in-

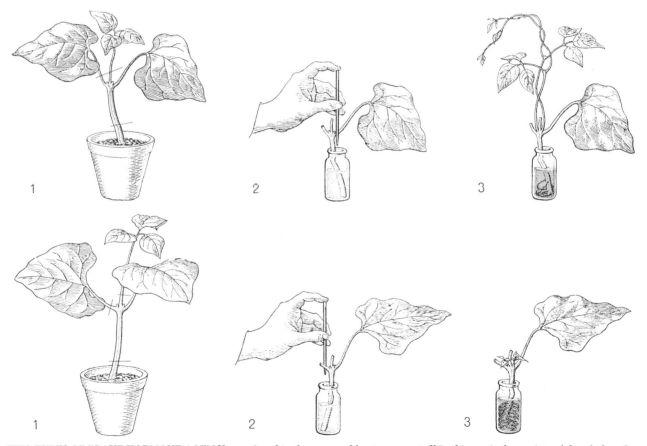

TWO TYPES OF PLANT-HORMONE ACTION were found in the author's early tests with dwarf bean plants. In both cases the original bean-plant cutting consisted of a piece of stem with one leaf and two nearly invisible buds (*step 1, top and bottom*); the fresh cuttings were then placed in small bottles of water. In one set of bottles a small amount of gibberellin, a naturally occurring plant-growth hormone, was dissolved in the water (*step 2, top*). Within a week the buds on these cuttings grew into long vinelike branches characteristic of pole beans (*step 3, top*); normally such branches

would not appear at all in this particular variety of dwarf plant. In addition the normal amount of root growth in water was observed at the base of the stem. In the other set of bottles a small amount of indolebutyric acid, a synthetic auxin, was added (*step 2, bottom*). After a week bud growth was not promoted abnormally, but instead root development was greatly augmented, so that the base of the stem had the appearance of a bottle brush (*step 3, bottom*). A definite growth response was observed using as little as one part per billion of gibberellin or one part per million of indolebutyric acid.

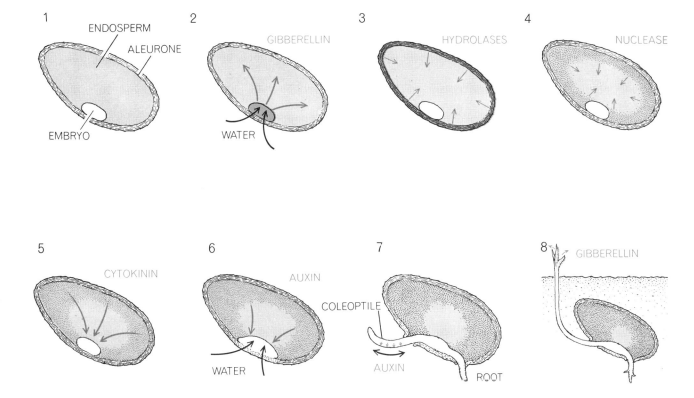

GERMINATION OF A CEREAL SEED below the surface of the soil (*1*) is regulated by a number of hormones working in sequence. First the absorption of water from the soil causes the embryo to produce a small amount of gibberellin (*2*). The gibberellin then diffuses into a layer of aleurone cells that surrounds the endosperm's food-storage cells, causing them to form enzymes (*3*) that in turn lead the endosperm cells to disintegrate and liquefy (*4*).

Cytokinins and auxins formed in this process (*5, 6*) then promote the growth of the embryo by making its cells divide and enlarge. If the shoot is pointing down into the soil, the auxins tend to migrate to the lower side of the seedling, causing it to grow faster and hence turning the growing point of the shoot upward toward the surface of the soil (*7*). Once the shoot has broken into the sunlight the plant begins to produce its own food by photosynthesis (*8*).

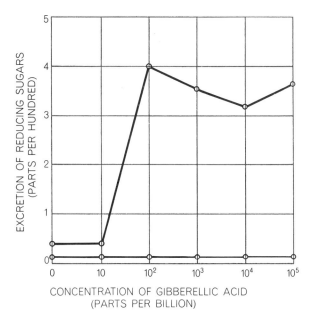

CONCENTRATION OF GIBBERELLIC ACID
(PARTS PER BILLION)

EVIDENCE OF GIBBERELLIN ACTION is contained in this graph, which relates the excretion of reducing sugars from barley endosperm tissue to the amount of gibberellin applied. The top curve shows the results when the endosperm cells are accompanied by aleurone cells; the bottom curve shows the results without the aleurone cells. Evidently the gibberellin acts on the aleurone cells to secrete the enzymes that hydrolyze the starch to form the sugars.

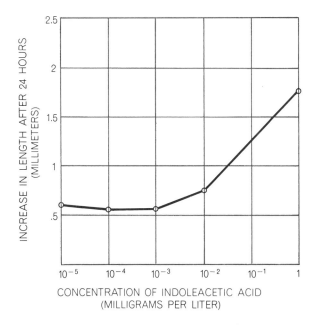

CONCENTRATION OF INDOLEACETIC ACID
(MILLIGRAMS PER LITER)

EVIDENCE OF AUXIN ACTION is contained in this graph, which shows the growth response of oat coleoptiles to various concentrations of the auxin indoleacetic acid (IAA). To obtain the results three-millimeter-long sections of the coleoptiles were floated on the surface of a shallow layer of hormone solution consisting of varying amounts of indoleacetic acid in distilled water. The measurements of length were made after a growing period of 24 hours.

ternally produced substances: the plant hormones. This basic rule was discovered in the 1920's by Frits W. Went, who was then working in the Netherlands. It turns out that these hormones must not only be present; they must be available in the proper balance and at the right time. Although the chemical study of the plant hormones began 40 years ago, only recently has enough knowledge been acquired to allow experiments that demonstrate the real power of the hormones in a dramatic fashion.

The basic discoveries concerning the powers of plant hormones came about in various ways. One of them grew out of an investigation of dwarf varieties ·of corn. In the mid-1930's at the California Institute of Technology I tried to find a physiological link between the gene deficiencies of dwarf plants and the inhibition of their growth. Could their failure to grow to normal size be traced to a hormone influence or deficiency? The only plant hormones known at the time were the auxins, typified by indole-

acetic acid. I experimented with these and found little evidence that auxins were critically involved in the dwarf plants' growth rate.

Twenty years later Bernard O. Phinney of the University of California at Los Angeles tested a new hormone on the same genetic dwarfs, and this time the outcome was different. The hormone was gibberellin, which had been discovered in Japan many years earlier but had just become available in the U.S. Phinney found that he could cause some of the corn dwarfs to grow into corn of normal size simply by placing a few drops of gibberellin solution in the center of the growing seedlings. Suddenly the old dream of being able to speed up the growth of plants at will had become an experimental reality.

The curious thing about gibberellin was that it had been discovered as the product of a fungus—a pathogenic fungus that made rice plants grow abnormally tall [see "Plant Growth Substances," by Frank B. Salisbury; SCIENTIFIC AMERICAN Offprint 110]. Strange-

ly enough, although the fungus apparently does not require gibberellin for its own growth, its cells produce huge quantities of the hormone. (In fact, the commercial production of gibberellin today depends on the fermentation of this fungus in vats, similar to the way penicillin is produced in vats by the cultivation of a mold.) By an odd quirk of nature the gibberellin fungus produces a chemical for which it has no use itself but which serves a vital function in higher plants. It has since been learned that all the higher plants produce gibberellin and require it for normal growth and development. We shall look into gibberellin's roles later in the article.

Another potent group of plant hormones came to light in the mid-1950's. As it happens, I was involved in the early stages of this exploration also. About 1940 the geneticist Albert F. Blakeslee asked my help on a physiological problem. In the course of solving the problem I found, with my research associate Marie E. Conklin, that coco-

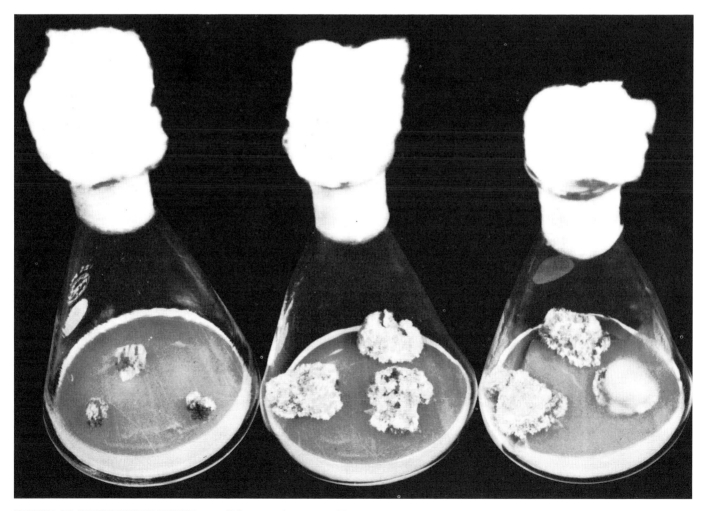

EFFECT OF CYTOKININ HORMONE on undifferentiated tissue from tobacco stems was studied by a group of investigators at the   University of Wisconsin headed by Folke Skoog. By manipulating the concentration of the cytokinin in the growth medium they were

nut milk contains a new growth factor, which later turned out to be cytokinin. With the help of graduate students and chemists at Cal Tech we tried to isolate it, but after a year we had to give up. The active ingredient was always hidden in the dirtiest, stickiest residue.

In the 1950's the University of Wisconsin biologist Folke Skoog and his associates took up the pursuit of the active factor. They had been using coconut milk to grow pieces of tobacco tissue in bottles, and in order to run down the active substance they turned to other possible sources. They found that it was present in a yeast extract in a soluble form. Absorption spectra and other markers of the active fraction suggested it was a purine. Recalling that nucleic acids contain purines, one of Skoog's associates, Carlos O. Miller, searched the laboratory shelves for bottles with a nucleic acid label. He found one marked "Herring Sperm DNA," and sure enough, this material proved to be capable of causing tobacco cells to grow and divide.

I am told that when this bottle was used up, a large supply of freshly prepared DNA was ordered, but to everyone's consternation it failed to show any biological activity. The Wisconsin biological laboratories were then ransacked for nonfresh samples of DNA, and all of these proved to be active. Miller therefore returned to the freshly prepared DNA and "aged" it rapidly in an autoclave, whereupon it became active. The logical conclusion was that the growth-promoting factor must be a breakdown product of nucleic acid!

By beautiful teamwork the Wisconsin biologists led by Skoog and biochemists led by F. M. Strong then succeeded in isolating the factor. It turned out to be indeed a nucleic acid component—a derivative of adenine, one of the purine bases that make up nucleic acid. The Wisconsin group named the new hormone kinetin. They went on to synthesize several similarly active compounds, and collectively these hormones are now called the cytokinins.

Skoog and his students proceeded to

experiment with combinations of cytokinins and auxins in the culture of tobacco tissues. Starting with stem tissue, they found that simply by manipulating the relative concentrations of cytokinins and auxins in the growth medium they were able to grow roots, shoots and even flowers from the original colony of stem cells. Their results overturned the old idea that there were specific hormones for the formation of roots, leaves and stems; instead it became evident that growth and differentiation in plants are determined by the interplay of at least two growth factors.

Further information on this subject has come to light within the past two years. The breakthrough was provided by a newly discovered hormone that was first identified three years ago by Frederick T. Addicott and a team of co-workers at the University of California at Davis. They found it in extracts from cotton bolls, and they named it abscisin II, because it was believed to be responsible for the premature drop (abscission) of bolls from the plant. Meanwhile the

able to grow normal plants from the original undifferentiated cells. In the flasks shown the concentration was (*left to right*) 0, .04, .2, 1, 5 and 25 micromoles per liter. The cytokinin used was 6-(γ,γ-dimethyallylamino)purine. The growth period was six weeks.

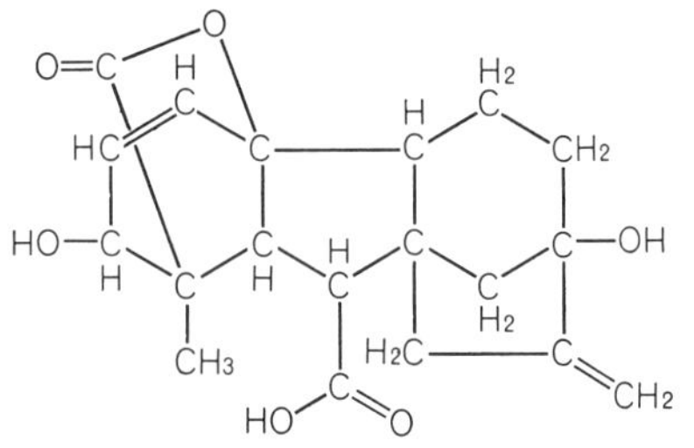

GIBBERELLIN

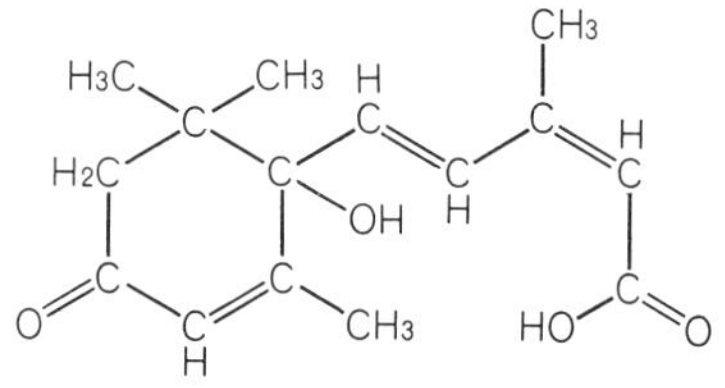

DORMIN

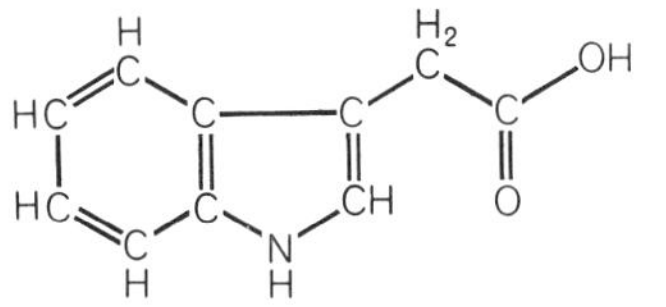

INDOLEACETIC ACID

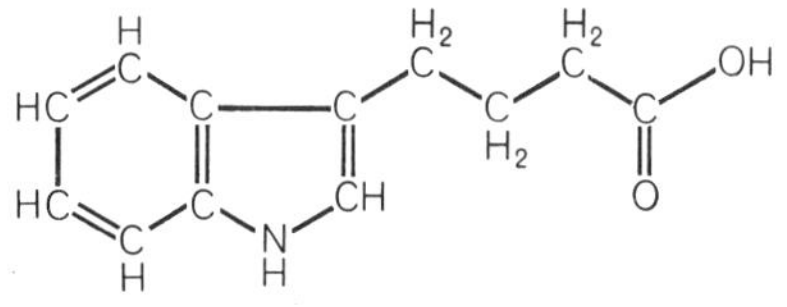

INDOLEBUTYRIC ACID

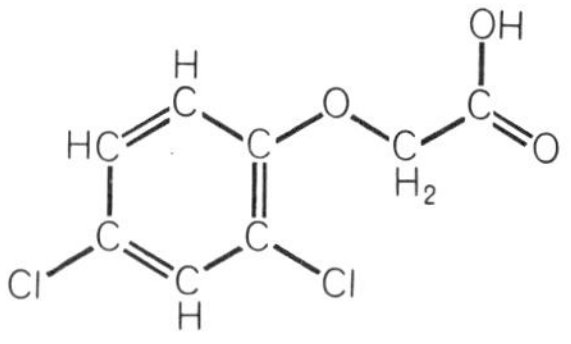

2,4-DICHLOROPHENOXYACETIC ACID

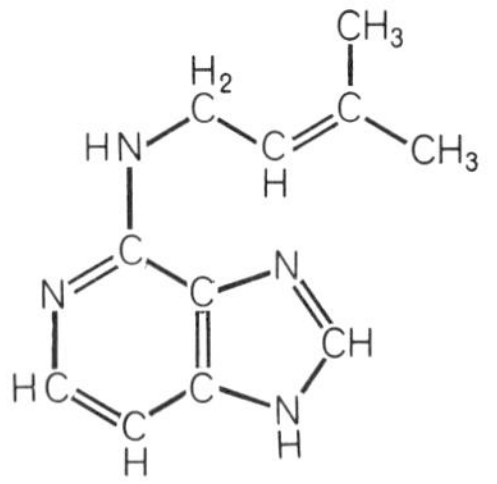

6-(γ,γ-DIMETHYLALLYLAMINO) PURINE

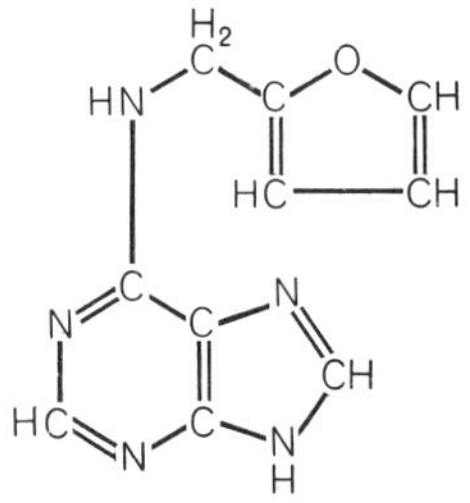

KINETIN

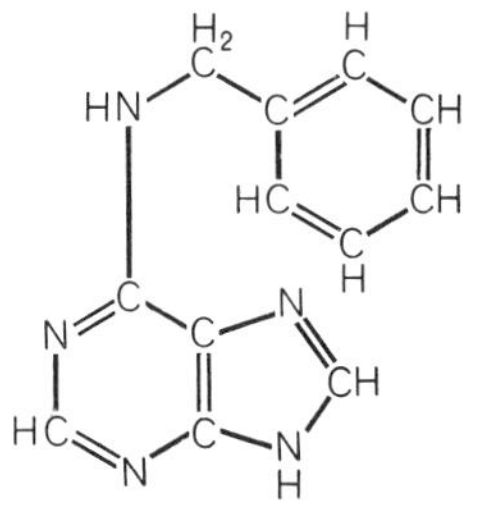

BENZYLADENINE

CHEMICAL STRUCTURES of eight of the plant-growth substances mentioned in the text are illustrated on this page. All are growth-promoting substances except for dormin, which is a growth-inhibiting hormone. Both gibberellin and dormin occur naturally; dormin has also been synthesized. Indoleacetic acid is a naturally occurring auxin. Indolebutyric acid and 2,4-dichlorophenoxyacetic acid are artificial auxins. 6-(γ,γ-dimethylallylamino)purine is a natural cytokinin. Kinetin and benzyladenine are artificial cytokinins.

same substance, found in maple leaves, was being investigated by the British chemist John W. Cornforth. It had been discovered by P. F. Wareing, an investigator of tree physiology; observing that it apparently prepared tree buds for their winter dormancy, Wareing named it dormin. Cornforth and his associates at the Milstead laboratory of Shell Research Ltd. soon succeeded in synthesizing the hormone and describing its stereochemical structure. Its full chemical name is 2-cyclohexene-1-penta-2,4-dienoic acid, 1-hydroxy-β,2,6,6-tetramethyl-4-oxo,cis-2-trans-4(d). Faced with the problem of selecting a short name for the hormone, the International Conference on Plant Growth Substances held in Ottawa in July, 1967, chose "abscisic acid." Unfortunately this name does not describe either the chemical structure or the phys-

iological activity of the substance. I shall refer to it here as dormin, because that term is descriptive of its physiological effects.

Dormin is an inhibitor of plant growth. For that reason some biologists object to calling it a hormone—a term that literally means "arousing to activity." Physiologists are now inclined, however, to classify both the promotive and the inhibitory growth regulators as hormones, because they operate in the same way (as chemical messengers) and are complementary in their actions.

In 1966, after dormin had been synthesized, Josef E. Loeffler, Iona Mason and I began detailed studies to work out its mode of action. We found that ordinary duckweed was extremely sensitive to the hormone. As little as one part per

billion of this substance in a culture solution would reduce the weed's growth rate, and one part per million was sufficient to keep this floating weed in a dormant state indefinitely. A striking feature of this action was its reversibility: as soon as the inhibiting hormone was removed, the plant tissues resumed their growth. Even after more than six weeks in the state of suspended growth, the culture could be revived simply by transferring it to a fresh medium in which the inhibiting hormone was absent.

In contrast to dormin, the cytokinins (but not auxins or gibberellin) strongly promoted the growth of duckweed cultures. We proceeded to study the mutual effects of dormin and a synthetic cytokinin (benzyladenine). Some 60 cultures were started in tubes under fluorescent light in a room at a constant temperature and were allowed to grow for a week, by which time their fresh weight had increased tenfold. Dormin, in the amount of one part per million, was then injected into some of the tubes. Within three days the growth of the cultures in these tubes slowed almost to a standstill. We now injected a very small amount of benzyladenine—one part per 10 million—in some of the tubes where growth had stopped. The cytokinin caused the cultures to resume their normal growth rate, although the dormin was still present. In short, the cytokinin overcame the dormin's inhibitory effect. We found that simply by supplying the growth medium with suitable concentrations of the opposing hormones we could apply stop-or-go control to the plant cells' growth.

It seems reasonable to conclude that in nature the growth of plants is similarly regulated by a combination of promoting and inhibiting hormones. Thus growth is controlled by an interplay of counteracting mechanisms much like that involved in driving a car. Just as we would not dream of operating a car with an accelerator and no brakes, so a plant apparently needs brakes for proper control. Dormin furnishes the brake. The accelerator, according to the particular plant and the conditions, may be cytokinin, gibberellin and/or auxin. Dormin has been detected in many plants in nature. We can speculate that the bursting forth of buds in the spring may be due in part to the decline of dormin and in part to the production of accelerating hormones such as the cytokinins.

How do the accelerating and the braking mechanisms work in the plant? By means of tracer studies with radioactive phosphate we discovered that dormin inhibits the synthesis of nucleic acids by the plant cells, and this is followed by a slowdown of growth. Conversely, the injection of cytokinin greatly accelerates the synthesis of nucleic acids. We therefore conclude that cytokinin speeds up the growth of the duckweed by increasing nucleic acid production and dormin retards it by reducing this production.

Even before our experiments Skoog's group at Wisconsin had demonstrated in the early 1950's that auxins increase the rate of nucleic acid synthesis. Our finding that dormin slows this rate has since been confirmed by Wareing, who presented the confirmation at a symposium on plant-growth regulators in London last winter. Obviously, since cell division and the growth of tissues require the production of nucleic acids (the cells' genetic material), this evidence of the involvement of the promoting and inhib-

itory hormones goes far toward explaining their effects on growth. It appears that the plant hormones control the fundamental biochemistry—the nucleic acid chemistry—of plant life.

We now have enough information in hand to put together a coherent picture of how the hormone system may start the growth of important crop plants such as wheat, oats, barley and rice. What happens when we sow a dry cereal seed in moist soil? The seed consists of two parts: the germ, or embryo, from which the plant will develop, and a store of food, called the endosperm, that will

nourish the developing seedling until it puts out green leaves that will enable it to produce its own food by photosynthesis. The stored food, initially in solid, undissolved form, is locked up in the endosperm cells. As the Austrian botanist Gottlieb Haberlandt showed many years ago, some action by the embryo is needed to release the food and allow it to become liquefied. What, then, is the key in the embryo that unlocks the food cellar? In 1960 Haraguro Yomo in Japan and L. G. Paleg in Australia independently discovered that the key is gibberellin.

Apparently the absorption of water

ROLE OF GIBBERELLIN in the germination process of a cereal seed is elucidated by this photograph, made by Joseph E. Varner of Michigan State University. The three barley seeds in the photograph have been cut in half and their embryos have been removed. Normally it is the embryo that produces gibberellin, which regulates the hydrolysis, or digestion, of the food-storage cells of the endosperm. The open surfaces of the three seeds have been treated with plain water (*bottom*), a solution of gibberellin in water at a concentration of one part per billion (*center*) and a solution of gibberellin in water at a concentration of 100 parts per billion (*top*). The photograph, taken 48 hours later, shows that in the gibberellin-treated seeds the digestion of the starch-filled storage tissue is already taking place.

from the soil causes the embryo to produce a small amount of gibberellin (of the order of a fraction of a part per million), which then diffuses into a layer of aleurone cells that surrounds the endosperm's food-storage cells. The gibberellin initiates a series of events that has been investigated in detail by the biochemist Joseph E. Varner at the Plant Research Laboratory of Michigan State University. Under the influence of the hormone the aleurone cells soon begin to synthesize enzymes. One (amylase) hydrolyzes the starch in the food cells into sugar; others break down those cells, disintegrating their nucleic acids and proteins. In brief, this is what happens: (1) in response to the uptake of water the embryo secretes gibberellin, (2) the gibberellin in turn causes the aleurone cells to form enzymes, (3) the enzymes go to work on the food-storage cells and cause them to disintegrate and liquefy.

Now we can reason that in the course of these events cytokinins and auxins are formed. The splitting of the nucleic acids by the newly formed enzymes can be expected to generate cytokinins, which as we have seen are derived from the breakdown of nucleic acid. Similarly, the breakdown of proteins can give rise to auxin; Skoog showed many years ago that the amino acid tryptophan is converted to indoleacetic acid in the cells of the coleoptile (the protective sheath around the seedling).

The newly generated hormones now proceed to promote the growth of the embryo. The cytokinins make its cells divide. The auxin assists by facilitating enlargement of the cells—the other requisite for cell growth. It does so by weakening the cell walls so that the cells take up water by osmosis and thus expand. The process by which auxin softens the cell walls is complex; Joe L. Key and his associates at Purdue University have found that it involves the synthesis of nucleic acids.

So finally we have our seedling growing. The shoot may, however, be pointing down into the soil. How does it make its way up to the surface and the sun? Again auxin is involved. By some geotropic process not yet understood auxin tends to migrate to the lower side of a seedling that is lying on its side. This causes the lower side to grow more rapidly than the upper, and hence the growing point of the shoot turns upward toward the soil surface. Once the shoot has broken through into the sunlight the leaves unfold, photosynthesis takes over

and the self-sustaining life of the plant begins.

As biologists we study plant hormones primarily because they can teach us a great deal about the process of growth. A fuller understanding of such life processes is the basic motivation, of course, for the work of any biologist. At the same time, plant scientists are always mindful of the practical implications of their studies. The green plant is, after all, the essential link to the sun's energy that sustains all life on the earth. It is the ultimate source of all man's food, and in itself it could supply our every food requirement. The Dutch horticulturist G. J. A. Terra has shown, for example, that calorie for calorie green leaves are as rich in essential proteins as the best of meats. A full appreciation of that fact in tropical countries could rescue those peoples from the ravages of the common protein-deficiency disease kwashiorkor. With the world population now growing very rapidly, plants have become more important than ever to man. There is no doubt that in order to cope with the increasing need for food we shall have to improve the efficiency of our agriculture. In several ways this is already being accomplished by the use of plant hormones, natural and synthetic.

One of the most useful of these, of course, is the synthetic weed killer known as 2,4-D. This substance, an auxin that retains its activity in plants much more persistently than the natural indoleacetic acid, can upset a plant's hormone balance so that growth occurs at places in the plant where it should not. It can cause roots to form on the stems in the air and at the same time slow down the normal root development underground. These abnormalities eventually lead to the death of the plant. Fortunately 2,4-D is selective in its action; it particularly attacks a number of useless weeds. As little as 500 grams per hectare (about half a pound per acre) can produce abnormal growth in a susceptible weed. Cereal plants, on the other hand, are relatively insensitive to 2,4-D; they inactivate the hormone, possibly by tying it to their proteins. Consequently the use of 2,4-D as a weed killer in grain fields has increased crop yields significantly. Synthetic auxins of the 2,4-D type have also been applied to other uses—for example to stop the premature drop of apples and pears. One might list a number of other applications of hormonal aids, for instance the use of auxins such as naphthaleneacetic acid and indolebutyric acid to propagate plants from cuttings and the use of gib-

berellins to speed up the malting process and to increase the size of grapes.

Studies of the potential uses of the newer hormones—the cytokinins and inhibitors such as dormin—are just beginning. I have obtained a patent for the utilization of cytokinins for preserving fresh vegetables. A cytokinin produced in the Shell laboratories has been found to be capable of generating viable seeds from Persian-grape plants that normally produce only male flowers; the hormone changes the developmental pattern of the flower from male to perfect hermaphrodite.

What uses dormin will have remains to be seen. Artificial growth inhibitors have been employed for some time and have produced interesting results. Materials such as CCC, B-Nine and so on have shown that flowering and fruiting can be promoted by slowing down vegetative growth. Azalea growers in the Eastern U.S. produce neat little plants that look like veritable balls of flowers by treating the plants with dwarfing agents. In the Netherlands I have seen young apple trees made to bear fruit two years before they would normally do so, simply through the use of growth inhibitors. The synthetic inhibiting chemical TIBA is widely used in the U.S. to shorten soybean plants and make them produce more branches so that they will bear more seeds.

In the long run, however, probably the most effective results will be obtained not by spraying chemicals on plants but by breeding them to produce a suitable balance of hormones of their own making. A good example of such a plant is the marvelous new variety of rice called IR-8, which has a short stem, is open in structure and gives a high yield of grain. This semidwarf variety, developed at the International Rice Research Institute near Manila, may have a genetic constitution that tips the hormone balance in favor of a vegetative-growth inhibitor. One can guess that its growth promoter may be a gibberellin and its inhibitor may be a dormin; that question remains to be determined by further research.

In the future we may measure the quality of a plant in terms of its hormone balance, just as we now define the nutritional value of a protein by its amino acid composition. Plant breeders will then be able to produce plants with specified properties by deliberately selecting genes to provide a particular ratio of growth promoters such as gibberellin or cytokinin and growth inhibitors such as dormin.

The Author

JOHANNES VAN OVERBEEK is director of the Institute of Life Science and of the department of biology at Texas A&M University. Born in Holland, he was graduated from the University of Leiden in 1928 and obtained a Ph.D. from the State University of Utrecht in 1933. He then spent nine years in research and teaching at the California Institute of Technology. From 1943 to 1947 he headed the work in plant physiology at the Institute of Tropical Agriculture in Puerto Rico. For the next 20 years, until he took up his present work, he was chief plant physiologist at the agricultural research laboratory of the Shell Development Company at Modesto, Calif. Van Overbeek is also a farmer; he owns a commercial vineyard in California and in that capacity he is a member of the Farm Bureau, the Allied Grape Growers and the United Vintners.

Bibliography

BIOCHEMISTRY AND PHYSIOLOGY OF PLANT GROWTH SUBSTANCES. Edited by F. Wightman and G. Setterfield. The Runge Press (in press).

THE LORE OF LIVING PLANTS. Johannes van Overbeek and Harry K. Wong. Scholastic Book Services, 1964.

PAPERS ON PLANT GROWTH AND DEVELOPMENT. Watson M. Laetsch and Robert Cleland. Little, Brown and Company, 1967.

PLANT HORMONES AND REGULATORS. J. van Overbeek in *Science*, Vol. 152, No. 3723, pages 721–731; May 6, 1966.

SCIENTIFIC
AMERICAN   July 1968, Vol. 219, No. 1, pp. 108-114

OFFPRINT 1112

HIDDEN LIVES

by Theodore H. Savory

Concealed at the surface of the ground, dwelling in conditions of maximum security, are a multitude of small invertebrate animals. Some may well retain the form in which life on land first appeared.

A few inches above and below the surface of the ground lie the boundaries of a space wherein countless multitudes of small animals live the whole of their lives "in the dark and the damp." Lift, if you will, a rotting log and you will find sheltering under it a group of woodlice; raise a half-buried piece of rock and you will disturb a resting centipede; stir a heap of fallen leaves and a dozen of the insects that are aptly called springtails leap to sudden activity.

One does not remember having seen woodlice or centipedes running in the fields like spiders or basking in the sunshine like butterflies. They are examples of the many small animals that lead hidden lives; they are some of the cryptozoa, the animals in hiding. The name cryptozoa was suggested for them in 1895 by Arthur Dendy, an Australian naturalist.

The cryptozoa form an assemblage of invertebrates unrelated in terms of conventional systematics but united ecologically by their occupation of a particular environment. This environment is the immediate neighborhood of the earth's surface, where there are leaves and humus, mosses and stones, fungi and soil and salts. It may be called the cryptosphere. It is a region as well defined as any other ecological zone; it is universal and because it is largely responsible for the existence of the cryptozoa its physical nature may be described first.

The most significant components are illumination, humidity and temperature, the first two being represented in the phrase from J. L. Cloudsley-Thompson quoted in my opening sentence: "the dark and the damp." Light is less vital to animals than it is to plants; indeed, most small animals tend to shun the light rather than to welcome it. The bacteria and fungi, on whose cooperation the cryptozoa largely depend, are organisms of the darkness rather than of the sunshine.

But darkness is not so essential as dampness. Almost every class of terrestrial animals can be traced back to an aquatic ancestor, and all have been compelled from the first to face and to overcome the risk of desiccation. The problem of avoiding or limiting water loss is one that is encountered again and again in the study of the cryptozoa.

Temperature is clearly a condition that affects this matter. The relative constancy of temperature in the cryptosphere is one of its favorable features, since less elaborate adaptations are necessary to secure control of escaping water. Incidentally, the region challenges experimental biologists to exercise ingenuity in the design of apparatus for recording humidity and temperature in inaccessible spots.

The most significant fact about the physical conditions of the cryptosphere is their approach to constancy; there is in general only slight change in illumination, temperature and humidity in this obscure and sheltered world. Perhaps only in the remotest parts of large caves and in the greatest depths of the oceans are the physical conditions of the environment less subject to variation, either daily or seasonal.

The advantage to the cryptozoa themselves is obvious. Few stimuli fall on their sense organs, which in many instances are simple in form and function. Thus the cryptozoa are not obliged to move frequently in response to external impulses, which is only another way of saying that in the cryptosphere there is protection against all those unpleasant climatic changes that can collectively be summarized by the word "exposure."

There is freedom in a life that is predictable to the point of monotony.

The stage having been set, let us turn to the players. They are readily discernible if one shakes fallen leaves in a wide-mesh strainer while watching a white sheet spread below it. One is likely to be struck, first, by the number of creatures that are thus revealed; second, by their variety, and third, by their size.

Firstly, as to numbers. Simple sifting does not provide a reliable record of the population. More sophisticated methods, carried out with the help of devices such as the Berlese funnel (which enables one to extract all the fauna from a ground sample) and backed by statistics of sampling that have no place here, have yielded results that are frankly bewildering. A few examples can be quoted in illustration.

In 1936 W. R. S. Ladell estimated the insect population near Cambridge to be nearly 48 million per acre on fallow land and 96 million per acre on grass. At the Rothamsted Experimental Station four years later K. D. Baweja gave a figure of more than 100 million for the number of insects per acre, and the amazing figure of 1,400 million insects per acre was computed at Cambridge in 1948.

A little rumination on these population densities leads to a balanced view of the animal population of the world. Estimates are often made of the numbers of various interesting animals the world contains—animals such as elephants, ostriches, giraffes and gorillas—and the figures that are given generally run from some tens of thousands to hundreds of thousands. Comparing such estimates with the totals of the various cryptozoa in one's own garden leads to the conclusion that most of the animals in the world are securely hidden in obscurity.

This planet is undoubtedly and overwhelmingly adapted to the support, protection and survival of its cryptozoa.

Secondly, as to diversity. The class or order to which each specimen on the collecting sheet belongs may not always be obvious, but almost every naturalist will at once perceive the dominance of the phylum Arthropoda (which includes the arachnids, the crustaceans and the many-legged "myriapods," as well as the insects) and the presence of a few representatives of other groups.

R. F. Lawrence of the Natal Museum in Pietermaritzburg, who has studied the cryptozoic fauna of South Africa, emphasizes the surprising fact that (excluding the Protozoa) only five phyla are represented: these are, in addition to the Arthropoda, the Platyhelminthes, Nematoda, Annelida and Mollusca. More interesting than this are the proportions in which they are found. The worms (flatworms, roundworms and earthworms: Platyhelminthes, Nematoda and Annelida respectively) do not exceed 3 percent of the total. The snails (Mollusca) and woodlice (Crustacea) account for about 4 percent each. Of the remaining nine-tenths about half are Arachnida, a third are Insecta and the rest are myriapods of different kinds. Evidently the cryptosphere is predominantly the domain of the Arachnida.

To these generalizations let us add some analysis of the chief classes of the five phyla that are represented. Most of the insects are wingless and accordingly belong to the subclass Apterygota, the present-day representatives of the first insects that came into existence. In

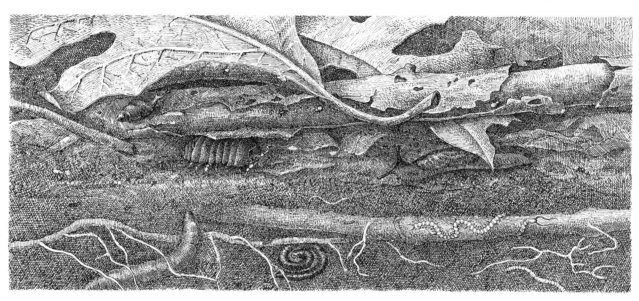

THE CRYPTOSPHERE, a distinct ecological region where diverse invertebrate animals dwell in seclusion, extends a few inches above and below the surface of the ground. Depicted here is a typical section of the cryptosphere in a wooded area in the Temperate Zone. The relative proportions of the various fauna in such an area are not indicated; the fauna are shown about twice actual size.

INHABITANTS OF THE CRYPTOSPHERE appear apart from their natural environment. Often found in woodland debris and just below the ground surface are the land snail (a), the woodlouse (b), also known as sow bug and pill bug, and the slug (c). Usually inhabiting lower levels of the cryptosphere are the earthworm (d), the millipede (e) and the centipede (f). The smallest cryptozoic animals (not labeled in the drawing) include mites, roundworms, false scorpions and wingless insects, such as the springtail.

PHYLUM	CLASS	ORDER
ARTHROPODA	ARACHNIDA	ACARI (MITES AND TICKS)
		ARANEIDA (SPIDERS)
		PALPIGRADI
		PSEUDOSCORPIONES (FALSE SCORPIONS)
		RICINULEI
		SCHIZOMIDA
	CRUSTACEA	ISOPODA (WOODLICE)
	INSECTA SUBCLASS: APTERYGOTA (WINGLESS)	COLLEMBOLA (SPRINGTAILS)
		DIPLURA
		PROTURA
		THYSANURA (BRISTLETAILS)
	MYRIAPODA: CHILOPODA (CENTIPEDES) DIPLOPODA (MILLIPEDES)	
	PAUROPODA	
	SYMPHYLA	

PRIMITIVE ORDERS predominate among the arthropods of the cryptosphere. For example, arachnids of the orders Palpigradi, Pseudoscorpiones (false scorpions) and Schizomida have a fully segmented abdomen. This is a characteristic of primitive arthropods.

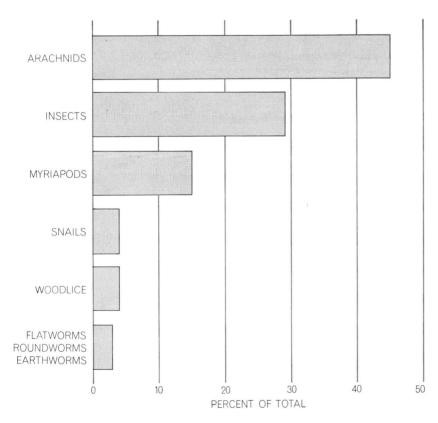

DISTRIBUTION OF THE CRYPTOZOA is based on the studies in South Africa of R. F. Lawrence of the Natal Museum in Pietermaritzburg. The dominant phylum is clearly the Arthropoda, the Arachnida accounting for approximately half of the total number of species.

other words, the cryptosphere houses the primitive orders rather than the specialized forms with wings and no distaste for daylight.

Among the myriapods there is a large proportion of the simpler centipedes and millipedes, but one's attention is distracted from such familiar creatures by the knowledge that certain others may be present. Here, and here only, are to be found the two rather unfamiliar classes: Symphyla and Pauropoda. A few lines on each will be justified, to underline the lesson they teach.

A symphylid has 12 pairs of legs and looks like a tiny centipede; like a centipede, it has a pair of long antennae. Since their discovery in 1762 they have been found all over the world wherever the minimum temperature is above 15 degrees Fahrenheit. These myriapods are generously supplied with four pairs of mouthparts and, on their last segment, a pair of spinning tubules. They are vegetarians and surprisingly are often inclined to feed on the surface of the ground, even in strong sunshine.

The Pauropoda are not so well known, perhaps because there are only two million to the acre, less than one-tenth of the number of the Symphyla. Also vegetarians, they resemble millipedes, having the sex orifice on the third segment of the body. They have nine pairs of legs and their antennae fork at the tip.

It is difficult to resist the temptation to see in the Symphyla the representatives of the ancestors of the centipedes and in the Pauropoda the representatives of the ancestors of the millipedes. Whether this is wishful thinking or not, these two small classes direct us to the fundamental characteristic that unites such a large proportion of the cryptozoa, namely their primitive nature.

For instance, the class of Arachnida is chiefly represented by mites of the family Oribatidae and, among other orders, by the Pseudoscorpiones, Palpigradi and Schizomida. The false scorpions, which look like tiny tailless scorpions, are the most familiar of these. The fully segmented abdomen is the most obvious evidence of their primitive organization [see "False Scorpions," by Theodore H. Savory; SCIENTIFIC AMERICAN Offprint 1039]. In the Palpigradi and Schizomida segmentation also survives in the cephalothorax (the joined head and thorax that constitutes the forward portion of the body). There the carapace is divided into three parts; the abdomen has 11 visible segments in the Palpigradi and 12 in the Schizomida. The bodies of all these cryptozoa, whether they are insects, myriapods or arachnids, are indisputably of the type known as primitive.

Thirdly, as to size. Manifestly one cannot expect the animals of the cryptosphere to be other than small. And, in fact, Lawrence finds that 95 percent of the cryptozoa are less than half an inch long and that some 30 percent are less than a millimeter long.

The small size of all cryptozoic animals is virtually the governing factor in their behavior and their general biology. In the first place a thin, partially sclerotized exoskeleton (a partially hard outer cuticle) usually serves to protect their small bodies, affording them sufficient support without undue increase in weight. These animals thereby avoid one of the first problems that confront larger

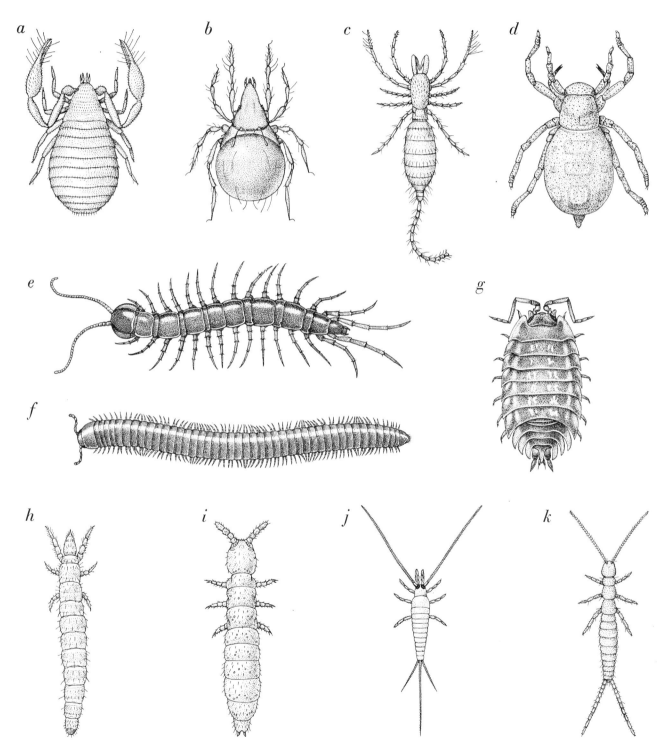

THE CRYPTOZOA depicted here are grouped by taxonomic class. The arachnids are represented by the false scorpion *Chelifer* (*a*), the mite *Belba* of the family Oribatidae (*b*), the palpigrade *Koenenia* (*c*) and *Ricinoides* (*d*) of the order Ricinulei. Below them appear the centipede *Lithobius* (*e*) and the millipede *Arctobolus* (*f*). The woodlouse *Oniscus* (*g*) is a crustacean. At bottom are representatives of the four orders of wingless insects: *Acerentomon* (*h*) of the Protura; *Achorutes* (*i*) of the Collembola, or springtails; *Petrobius* (*j*) of the Thysanura, or bristletails, and *Campodea* (*k*) of the Diplura. The organisms are not drawn to actual scale.

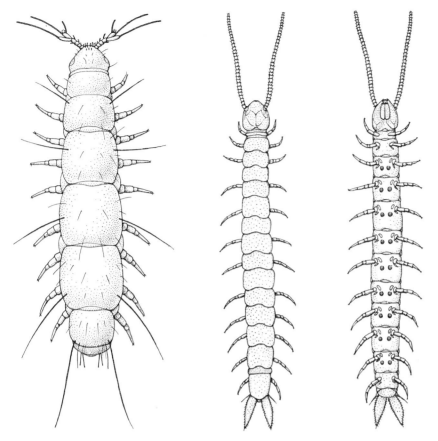

EARLY FORM of the millipede and centipede may be preserved in the pauropod (*left*) and the symphylid (*center*). In the ventral view of the symphylid (*right*) the eversible sacs (*color*) can be seen. The sacs' function is unknown; they distinguish many cryptozoic animals.

land animals, many of which, arriving on dry land from an aquatic ancestry, encounter a set of problems in vital mechanics when they are no longer surrounded and supported by water.

The thin cuticle is also permeable to gases; it allows the diffusion of oxygen directly into the body tissues and the escape of carbon dioxide in the same way. Thus in many cryptozoa respiratory organs are either absent or simple and rather ineffective; for this reason the animals have little or no control over the rate of respiration in response to changes in external circumstances. But since the chief characteristic of the cryptosphere is the constancy of its physical conditions, its inhabitants can be said to be slightly restricted by their inefficient respiration rather than seriously handicapped.

The loss of carbon dioxide through the cuticle, however, argues a simultaneous escape of water vapor, a much more important matter. The area of surface per unit of volume is greater for a smaller animal than it is for a larger one, and this in turn bears on the problems of the loss of heat by radiation and the loss of water by evaporation. It is scarcely an

exaggeration to say that these problems are governing considerations in the lives and in the evolution of every group in the cryptozoic fauna.

The loss of water is at once the simplest and the commonest way in which the internal conditions of the body are upset. The different aqueous solutions contained within a body must of necessity remain in osmotic balance. The preservation of this state of equilibrium, homeostasis, is secured by various devices or organs described as osmoregulatory. These embrace a wide range of mechanisms: simple contractile vacuoles, or cavities, in amoeba; somewhat more complex "flame" cells (named for their beating cilia) and solenocytes in flatworms; nephridia in annelids and, finally, kidneys in vertebrates.

For invertebrate animals excessive loss of water is nearly always fatal, and this unwelcome desiccation is hastened by exposure to wind, by remaining in the direct rays of the sun and by an increased metabolic rate within the animal itself. Wind is rare in the region we have called the cryptosphere. Among the cryptozoa few are more vulnerable to wind exposure than slugs. They are

seldom to be seen abroad by daylight, but people who have the enthusiasm to search for slugs after sunset find them in incredible numbers. The searchers maintain emphatically that a windy night, more than anything else, keeps the slug community at home.

Here, then, is the simplest way in which animals that are sensitive to desiccation have come to avoid its effects: they have perforce assumed a habit of life in which conditions or circumstances favoring water loss are shunned. Since the heat and the light of the sun are inseparable, most members of the cryptozoa have acquired a rhythmic mode of living; they limit their activities to the hours after sunset, thereby acquiring the superficial description "animals of nocturnal habits."

The changes that follow the setting of the sun are obviously important to the cryptozoa. The larger predators (such as birds) are more easily avoided in the dark; on the other hand, the predatory cryptozoa find their own tiny prey easier to secure in the busy after-dark traffic. The fall in temperature brings about a rise in the relative humidity and as a result a decreased evaporating power of the atmosphere. Here are ideal conditions for the ordinary activities of the ill-protected bodies of the cryptozoa.

Ulterior consequences are well known. Rhythmic behavior, which originated as a direct response to change in physical conditions, eventually becomes ingrained, or, as it is called, endogenous. Endogenous rhythms are found to persist when these animals are kept in artificially constant surroundings. Even without complex laboratory equipment an example can be seen in the nightly web-spinning of many spiders, which are, in fact, less subject to the risk of desiccation than other cryptozoa.

An alternative and very different method of water conservation is possible. This is the production of a layer of wax in the epicuticle, or outermost layer of the exoskeleton, where it acts like a raincoat in reverse, serving to keep water not out but in. A wax layer is found on many flies and spiders, which, because of its presence, are able to live in the open and often show no great tendency to avoid the sunshine. Other members of the cryptozoic population are not so lucky; they show a limited tolerance to unfavorable conditions, indicating a less efficient protection. They can perhaps charitably be described as "potentially diurnal."

By means of these adaptations the small animals of the cryptosphere avoid the direct action of the sun's rays. Let us compare their situation to that of birds and mammals, creatures that are normally several degrees warmer than the air surrounding them. They are constantly losing heat, but through various devices, such as fur, feathers, blubber or fat, they tend to prevent or control it. This is harder for the small mammal or bird with its relatively large surface; every student of physiology is told that a ton of mice eat 10 times as much food as a ton of horses. But for the cryptozoa the case is reversed: their need is to keep heat out because a rise in temperature hastens evaporation. It also produces an expansion of the liquid contents of the body, causing pressure against the exoskeleton and a possible occlusion of the blood vessels or other conduits. Thus it is that many of these creatures die quickly if they are forced to remain in the sun.

There is another source of heat that inevitably promotes the loss of water: the production of heat internally by sustained activity. The muscular actions needed to escape from any source of danger are clearly limited at least in part by the size of the running creature. Hence a large spider compelled to run by gentle prodding is exhausted sooner than a small spider, whose greater ratio of surface to volume enables it to cool more quickly.

Evidently the conditions under which the cryptozoa live are in their way as definite and as exacting as the apparently more rigorous conditions of the littoral zone, the arctic or the desert. Is there perhaps a general cryptozoic pattern or form—a consequence of the factors described above—by which a zoologist could say with confidence, "This is one of the cryptozoa"?

The answer to this question is more suggestive than definitive. A cryptozoan is an animal that is not more than five millimeters long, is fully segmented, weakly sclerotized, devoid of color or pattern, with small eyes or none but with sensitively tactile antennae, and with no conspicuous respiratory organs. Such an animal possesses all the characteristics that are most widely distributed among the cryptozoa.

In addition to these distinguishing marks I should mention the remarkable and mysterious organs usually vaguely termed "eversible sacs." They are found below certain segments of the body or legs in Symphyla, Palpigradi, wingless insects and some other groups. Their function is uncertain but if any one organ can be said to indicate a cryptozoic life, this is probably it.

By these steps we are led to another conclusion. The cryptozoa not only include the greater number of the world's animals, they also preserve for us the form in which life on land, as distinct from life in water, made its first appearance. Early marine arthropods must have crept ashore from seas that were overcrowded and were becoming intolerable. They must luckily have found it possible to feed on the plants they met, and undoubtedly they must have survived only if they constantly hid themselves from the scorching sun. Today

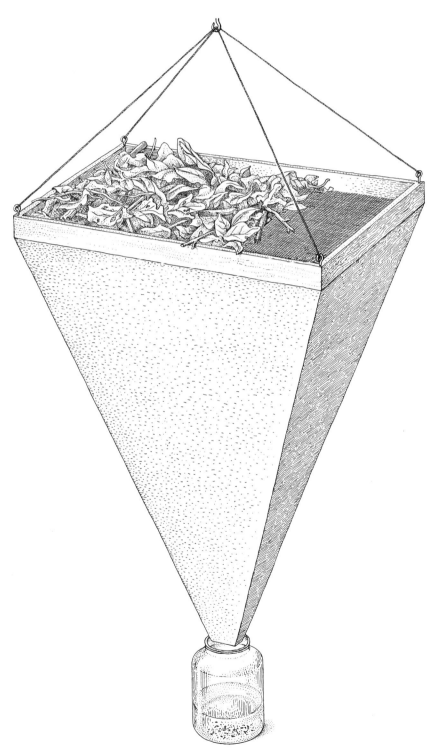

BERLESE FUNNEL, named after its inventor, the naturalist Antonio Berlese, separates fauna from a sample of organic debris. Heaped on wire mesh fitted across the funnel, the debris gradually dries out; the cryptozoic fauna in the debris, seeking dark, moist conditions, move downward; ultimately they fall into the bottle of killing liquid below the funnel.

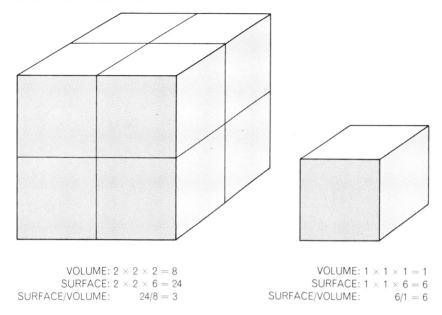

VOLUME: 2 × 2 × 2 = 8
SURFACE: 2 × 2 × 6 = 24
SURFACE/VOLUME: 24/8 = 3

VOLUME: 1 × 1 × 1 = 1
SURFACE: 1 × 1 × 6 = 6
SURFACE/VOLUME: 6/1 = 6

SURFACE AREA per unit of volume is greater for the smaller entity than for the larger one. Thus the small animals of the cryptosphere are particularly vulnerable to the loss of water by evaporation. To avoid desiccation they have been forced to develop nocturnal habits.

their descendants occupy the same stable environment; they have been described as prisoners in a cul-de-sac, unable to evolve in their conditions of protective security.

On this opinion two comments may be made. The first is that the optimum conditions of the cryptosphere are not likely to have urged its inhabitants to seek a change; hence the primitive groups are still represented among them. Not every order of insects or arachnids can, however, be traced back to cryptozoic ancestors; therefore one must suppose that on occasion there appeared mutants that found an open-air life tolerable. One imagines that such must be called, in contrast, the phanerozoa—the animals open to view.

Second, some of the present-day cryptozoa show definite specializations that can best be regarded as modifications of or additions to the primitive forms, but which have neither enabled them nor driven them to seek the world above. The false scorpions, for example, are an order of the Arachnida, with the fully segmented abdomen that is a sign of a primitive arthropod. They have evolved silk-secreting glands in the small organs near the mouth known as the chelicerae; their pedipalps, or pincers, have grown into large and formidable weapons; their life history is unusually complex. But essentially they remain hidden. The same is true of the much rarer arachnid

Ricinulei, one of whose secondary characteristics is the great thickness of the exoskeleton, another the elaboration of male organs on the tips of the third pair of legs.

The Symphyla too show signs of specialization: they have developed spinning organs on the last segment and have also acquired an unusual escape reaction that enables them to reverse direction rapidly by rotating their bodies with the last segment as a pivot. Is it more than a coincidence that false scorpions are also addicted to suddenly rushing backward?

Clearly the cryptozoa can be described as a self-sufficient society, on the whole decidedly primitive in form, that continues to exploit successfully the favorable circumstances in which it has established itself. What is true of myriapods and arachnids may also be true of other groups. There are phyla and classes in the animal kingdom in which the course of evolution is not easy to detect. The progress from a primitive to a specialized type is obscure, often made more conjectural by uncertainty as to the start. All species of animals alive today have their own specializations accrued during past centuries; it has been wisely said that there are now no primitive organisms but only some primitive organs. Therefore when the question arises, "What is the nearest living representative of the primitive arachnid, of the primitive insect, of the primitive myriapod?" a helpful suggestion can be offered. These genera should be sought in the cryptosphere; if they have left recognizable descendants, they are likely to be found among the cryptozoa.

The Author

THEODORE H. SAVORY is vice-principal of Stafford House, a tutorial college at Kensington in England. Although he describes himself as having spent most of his adult years leading "the uneventful life of a British public school master," he is an authority on arachnids, the group of organisms that includes spiders, daddy longlegs and scorpions. "Years of grubbing about in woods for the spiders that were the first objects of my search," he writes, "caused me to realize how many and various are the small animals that share the safety of the same surroundings; it is from this experience that my interest in these so-called cryptozoa has grown." After being graduated from the University of Cambridge in 1918, Savory spent 31 years teaching science at Malvern College and seven years as senior biology master at the Haberdashers' School in Hampstead. His article in his fourth in SCIENTIFIC AMERICAN.

Bibliography

ARACHNIDA. Theodore Savory. Academic Press, 1964.

THE BIOLOGY OF THE CRYPTIC FAUNA OF FORESTS WITH SPECIAL REFERENCE TO THE INDIGENOUS FORESTS OF SOUTH AFRICA. R. F. Lawrence. Balkema, 1953.

RHYTHMIC ACTIVITY IN ANIMAL PHYSIOLOGY AND BEHAVIOR. J. L. Cloudsley-Thompson. Academic Press, 1961.

SCIENTIFIC
AMERICAN   August 1957, Vol. 197, No. 2, pp. 48-54

OFFPRINT **1113**

HOW FISHES SWIM

by Sir James Gray

In which the speed of small fishes is measured in the laboratory
and their power calculated. Similar observations in nature suggest
that water may flow over a dolphin completely without turbulence.

The submarine and the airplane obviously owe their existence in part to the inspiration of Nature's smaller but not less attractive prototypes —the fish and the bird. It cannot be said that study of the living models has contributed much to the actual design of the machines; indeed, the boot is on the other foot, for it is rather the machines that have helped us to understand how birds fly and fish swim [see "Bird Aerodynamics," by John H. Storer; SCIENTIFIC AMERICAN Offprint 1115]. But engineers may nevertheless have something to learn from intensive study of the locomotion of these animals. Some of their performances are spectacular almost beyond belief, and raise remarkably interesting questions for both the biologist and the engineer. In this article we shall consider the swimming achievements of fishes and whales.

Looking at the propulsive mechanism of a fish, or any other animal, we must note at once a basic difference in mechanical principle between animals and inanimate machines. Nearly all machines apply power by means of wheels or shafts rotating about a fixed axis, normally at a constant speed of rotation. This plan is ruled out for animals because all parts of the body must be connected by blood vessels and nerves: there is no part which can rotate freely about a fixed axis. Debarred from the use of the wheel and axle, animals must employ levers, whipping back and forth, to produce motion. The levers are the bones of its skeleton, hinged together by smooth joints, and the source of power is the muscles, which pull and push the levers by contraction.

The chain of levers comprising a vertebrate's propulsive machine appears in its simplest form in aquatic animals. Each vertebra (lever) is so hinged to its neighbors that it can turn in a single plane. In fishes the backbone whips from side to side (like a snake slithering along the ground), in whales the backbone undulates up and down. A swimming fish drives itself forward by sweeping its tail sidewise; as the tail and caudal fin are bent by the resistance of the water, the forward component of the resultant force propels it [see drawing on page 2774]. As the tail sweeps in one direction, the front end of the body must tend to swing in the opposite direction, since it is on the opposite side of the hinge, but this movement is usually small—partly because of the high moment of inertia of the front end of the body and partly on account of the resistance which the body offers to the flow of water at right angles to its surface. Thus the head end of the fish acts as a fulcrum for the tail, operating as a flexible lever.

At the moment when the tail fin sweeps across the axis of propulsion, it is traveling rapidly but at a constant speed. During other phases of its motion the speed changes, accelerating as the tail approaches the axis and decelerating after it passes the axis. The whole cycle can be regarded as comparable to that of a variable propeller blade which periodically reverses its direction of rotation and changes pitch as it does so.

How efficient is this propulsion system? Can the oscillating tail of a fish approach in efficiency the steadily running screw propellers that drive a submarine, in terms of the ratio of speed to applied power? To attempt an answer to this question we must first know how fast a fish can swim. Here the biologist finds himself in an embarrassing position, for our information on the subject is far from precise.

As in the case of the flight of birds, the speed of fish is a good deal slower than most people think. When a stationary trout is startled, it appears to move off at an extremely high speed. But the human eye is a very unreliable instrument for judging the rate of this sudden movement. There are, in fact, very few reliable observations concerning the maximum speed of fish of known size and weight. Almost all the data we have are derived from studies of fish under laboratory conditions. These are only small fish, and in addition there is always some question whether the animals are in as good athletic condition as fish in their native environment.

With the assistance of a camera, a number of such measurements have been made by Richard Bainbridge and others in our zoological laboratory at the University of Cambridge. They indicate that ordinarily the maximum speed of a small fresh-water fish is about 10 times the length of the fish's body per second; these speeds are attained only briefly at moments of great stress, when the fish is frightened by a sudden stimulus. A trout eight inches long had a maximum speed of about four miles per hour. The larger the fish, the greater the speed: Bainbridge found, for instance, that a trout one foot long maintained a speed of 6.5 miles per hour for a considerable period [see table on page 2776].

It is by no means easy to establish a fair basis of comparison between fish of different species or between different-sized members of the same species. Individual fish—like individual human beings—probably vary in their degree of athletic fitness. Only very extensive observations could distinguish between average and "record-breaking" performances. On general grounds one would expect the speed of a fish to increase

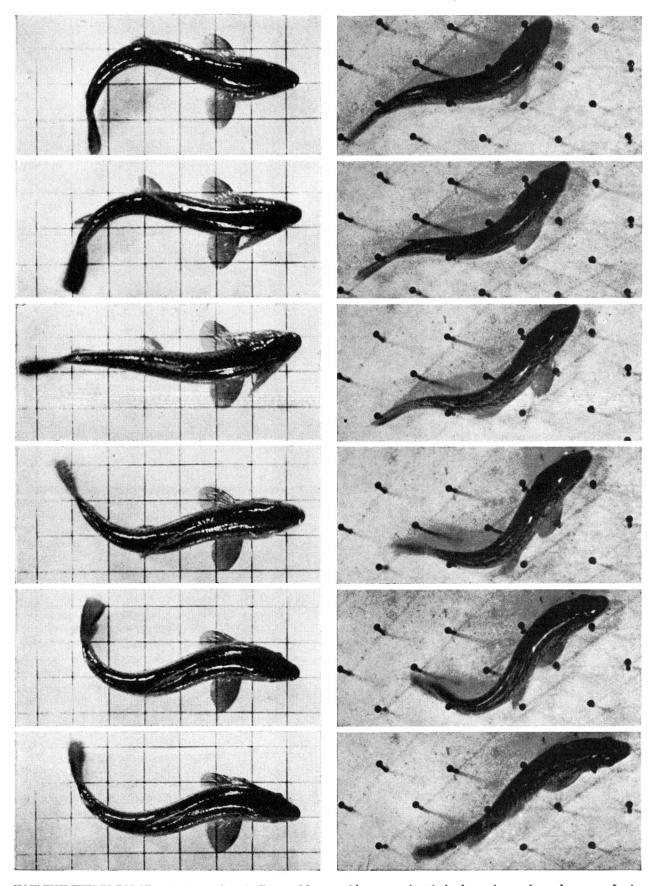

HOW FISH EXERTS FORCE against its medium is illustrated by these two sequences of photographs showing trout out of the water. In the sequence at left the fish has been placed on a board marked with squares; it wriggles but makes no forward progress. In the sequence at right the fish has been placed on a board covered with pegs; its tail pushes against the pegs and moves it across the board.

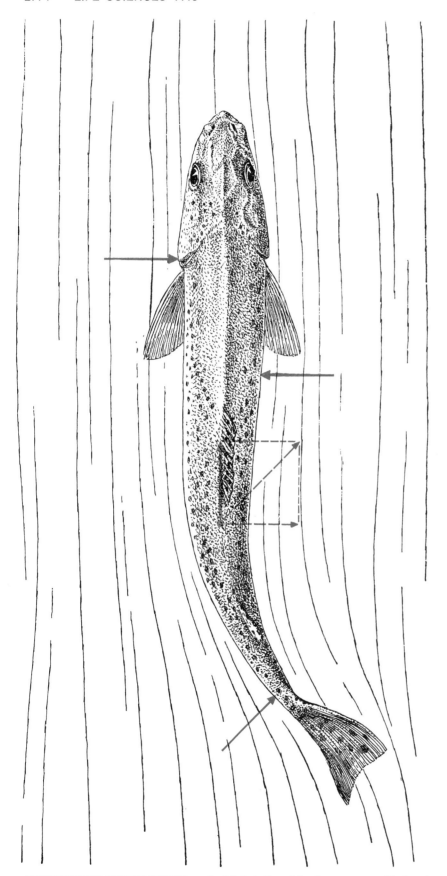

HOW MEDIUM EXERTS FORCE on the fish is indicated by the arrows on this drawing of a trout. As the tail of the fish moves from right to left the water exerts a force upon it (*two diagonal arrows*). The forward component of this force (*heavy vertical arrow*) drives the fish forward. The lateral component (*broken horizontal arrow*) tends to turn the fish to the side. This motion is opposed by the force exerted by the water on sides of the fish.

with size and with the rapidity of the tail beat. Bainbridge's data suggest that there may be a fairly simple relationship between these values: the speed of various sizes of fish belonging to the same species seems to be directly proportional to the length of the body and to the frequency of beat of the tail—so long as the frequency of beat is not too low. In all the species examined the maximum frequency at which the fish can move its tail decreases with increase of length of the body. In the trout the maximum observed frequencies were 24 per second for a 1.5-inch fish and 16 per second for an 11-inch fish—giving maximum speeds of 1.5 and 6.5 miles per hour respectively.

The data collected in Cambridge indicate a very striking feature of fish movement. Evidently the power to execute a sudden spurt is more important to a fish (for escape or for capturing prey) than the maintenance of high speed. Some of these small fish reached their maximum speed within one twentieth of a second from a "standing" start. To accomplish this they must have developed an initial thrust of about four times their own weight.

This brings us to the question of the muscle power a fish must put forth to reach or maintain a given speed. We can calculate the power from the resistance the fish has to overcome as it moves through the water at the speed in question, and the resistance in turn can be estimated by observing how rapidly the fish slows down when it stops its thrust and coasts passively through the water. It was found that for a trout weighing 84 grams the resistance at three miles per hour was approximately 24 grams—roughly one quarter of the weight of the fish. From these figures it was calculated that the fish put out a maximum of about .002 horsepower per pound of body weight in swimming at three miles per hour. This agrees with estimates of the muscle power of fishes which were arrived at in other ways. It seems reasonable to conclude that a small fish can maintain an effective thrust of about one half to one quarter of its body weight for a short time.

As we have noted, in a sudden start the fish may exert a thrust several times greater than this—some four times its body weight. The power required for its "take-off" may be as much as .014 horsepower per pound of total body weight, or .03 per pound of muscle. The fish achieves this extra force by a much more violent maneuver than in ordinary swimming. It turns its head end sharply to one side and with its markedly flexed tail executes a wide and powerful sweep against the water—in short, the fish takes off by arching its back.

This sort of study of fishes' swimming performances may seem at first sight to be of little more than academic inter-est. But in fact it has considerable prac-tical importance. The problem of the salmon industry is a case in point. The seagoing salmon will lay eggs only if it can get back to its native stream. To reach its spawning bed it must journey upriver in the face of swift currents and sometimes hydroelectric dams. In de-signing fish-passes to get them past these obstacles it is important to know pre-cisely what the salmon's swimming ca-pacities are.

Contrary to popular belief, there is little evidence that salmon generally sur-mount falls by leaping over them. Most of the fish almost certainly climb the falls by swimming up a continuous sheet of water. Very likely the objective of their

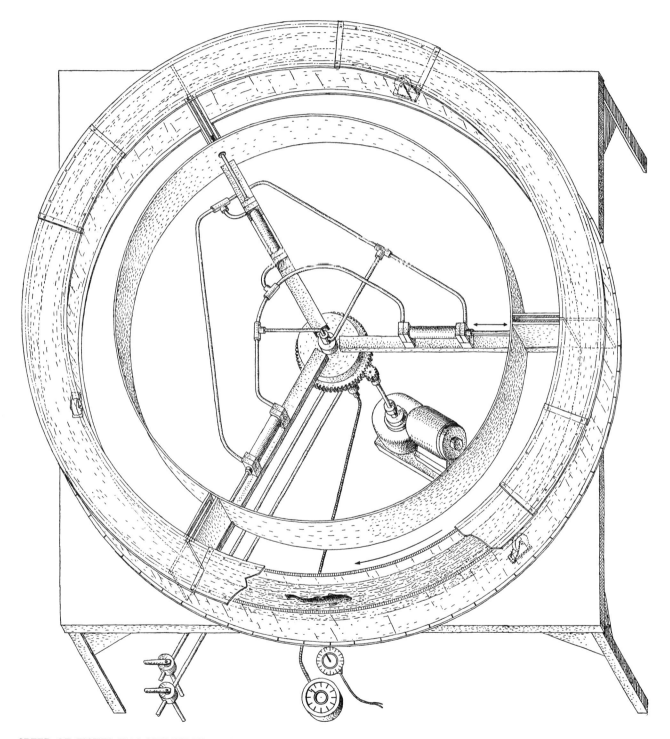

SPEED OF FISHES WAS MEASURED in this apparatus at the University of Cambridge. The fish swims in a circular trough which is rotated by the motor at right center. The speed of the trough is adjusted so that the swimming fish is stationary with respect to the observer. The speed of the fish is then indicated on the speedometer at bottom. Above the speedometer is a clock. When the apparatus is started up, the water is made to move at the same speed as the trough by doors which open to let the fish pass.

SPECIES	LENGTH (FEET)	MAXIMUM OBSERVED SPEED		RATIO OF MAXIMUM SPEED TO LENGTH
		(FEET PER SECOND)	(MILES PER HOUR)	
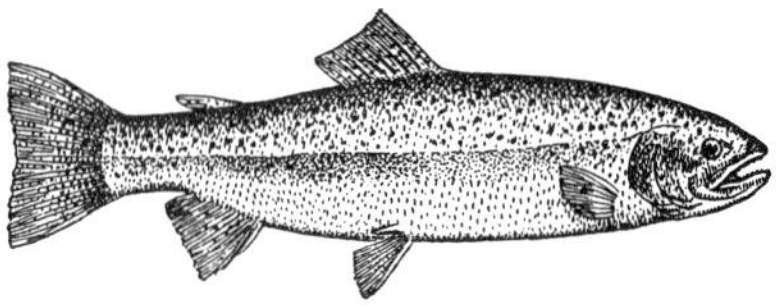 TROUT	.656 .957	5.552 10.427	3.8 6.5	8.5 11
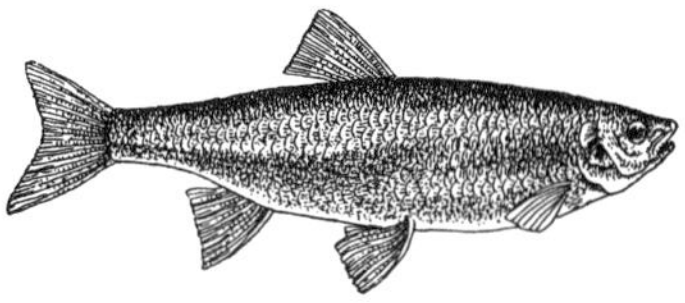 DACE	.301 .594 .656	5.229 5.552 8.812	3.6 3.8 5.5	17.8 9 13.5
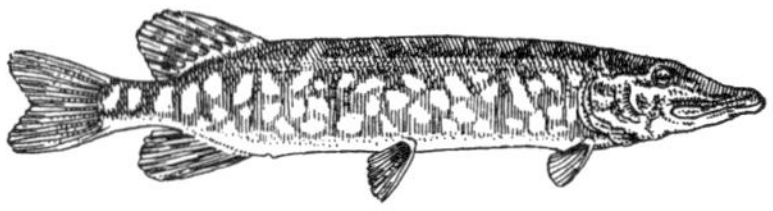 PIKE	.529 .656	6.850 4.896	4.7 3.3	13 7.5
 GOLDFISH	.229 .427	2.301 5.552	1.5 3.8	10.3 13
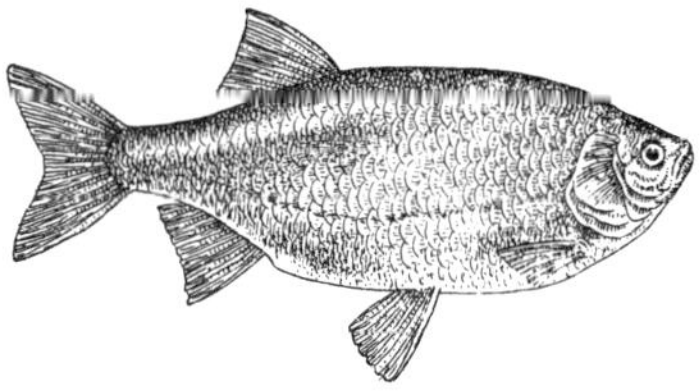 RUDD	.730	4.240	2.9	6
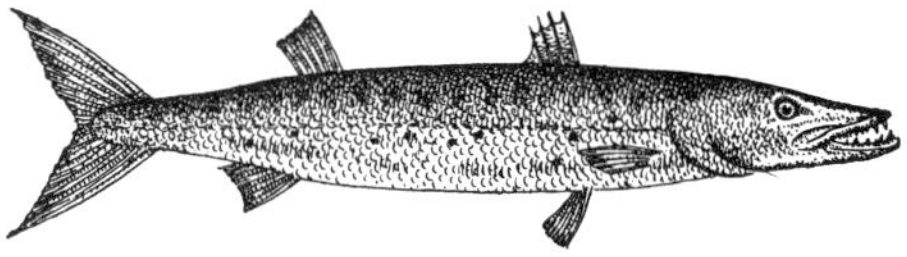 BARRACUDA	3.937	39.125	27.3	10
 DOLPHIN	6.529	32.604	22.4	5
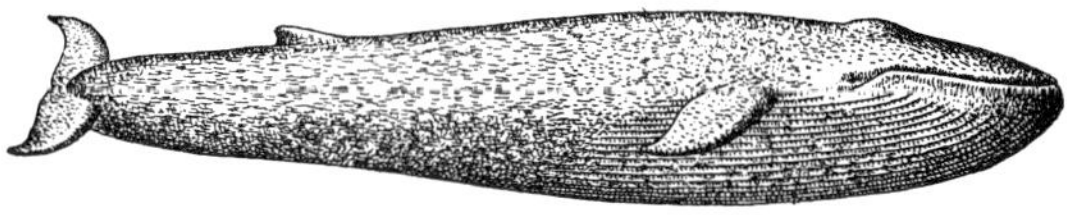 WHALE	90	33	20	.33

SPEED OF FISHES IS LISTED in this table. The speed of the first five fishes from the top was measured in the laboratory; that of the barracuda, dolphin and whale in nature. The barracuda is the fastest known swimmer. Whale in the table is the blue whale.

leap at the bottom of the fall is to pass through the fast-running water on the surface of the torrent and reach a region of the fall where the velocity of flow can be negotiated without undue difficulty. The brave and prodigious leaps into the air at which spectators marvel may well be badly aimed attempts of the salmon to get into the "solid" water!

A salmon is capable of leaping about six feet up and 12 feet forward in the air; to accomplish this it must leave the water with a velocity of about 14 miles per hour. The swimming speed it can maintain for any appreciable time is probably no more than about eight miles per hour. Accurate measurements of the swimming behavior of salmon in the neighborhood of falls are badly needed—and should be possible to obtain with electronic equipment.

At this point it may be useful to summarize the three main conclusions that have been reached from our study of the small fish. Firstly, a typical fish can exert a very powerful initial thrust when starting from rest, producing an acceleration about four times greater than gravity. Secondly, at times of stress it can exert for a limited period a sustained propulsive thrust equal to about one quarter or one half the weight of its body. Thirdly, the resistance exerted by the water against the surface of the moving fish (*i.e.*, the drag) appears to be of the same order as that exerted upon a flat, rigid plate of similar area and speed. Fourthly, the maximum effective power of a fish's muscles is equivalent to about .002 horsepower per pound of body weight.

Such is the picture drawn from studies of small fishes in tanks. It has its points of interest, and its possible applications to the design of fish-passes, but it poses no particularly intriguing or baffling hydrodynamic problems. Recently, however, the whole matter of the swimming performance of fishes was given a fresh slant by a discovery which led to some very puzzling questions indeed. D. R. Gero, a U. S. aircraft engineer, announced some startling figures for the speed of the barracuda. He found that a four-foot, 20-pound barracuda was capable of a maximum speed of 27 miles per hour! This figure not only established the barracuda's claim to be the world's fastest swimmer but also prompted a new look into the horsepower of aquatic animals.

A more convenient subject for such an examination is the dolphin, whose attributes are somewhat better known than those of the barracuda. (The only essential difference between the propulsive machinery of a fish and that of a dolphin, small relative of the whale, is that the dolphin's tail flaps up and down instead of from side to side.) The dolphin is, of course, a proverbially fast swimmer. More than 20 years ago a dolphin swimming close to the side of a ship was timed at better than 22 miles per hour, and this speed has been confirmed in more recent observations. Now assuming that the drag of the animal's body in the water is comparable to that of a flat plate of comparable area and speed, a six-foot dolphin traveling at 22 miles per hour would require 2.6 horsepower, and its work output would be equivalent to a man—of the same weight as the dolphin—climbing 28,600 feet in one hour! This conclusion is so clearly fantastic that we are forced to look for some error in our assumptions.

Bearing in mind the limitations of animal muscle, it is difficult to endow the dolphin with much more than three tenths of one horsepower of effective output. If this figure is correct, there must be something wrong with the assumption about the drag of the animal's surface in the water: it cannot be more than about one tenth of the assumed value. Yet the resistance could have this low value only if the flow of water were laminar (smooth) over practically the whole of the animal's surface—which an aerodynamic or hydrodynamic engineer must consider altogether unlikely.

The situation is further complicated when we consider the dolphin's larger relatives. The blue whale, largest of all the whales, may weigh some 100 tons. If we suppose that the muscles of a whale are similar to those of a dolphin, a 100-ton whale would develop 448 horsepower. This increase in power over the dolphin is far greater than the increase in surface area (*i.e.*, drag). We should therefore expect the whale to be much faster than the dolphin, yet its top speed appears to be no more than that of the dolphin—about 22 miles per hour. There is another reason to doubt that the whale can put forth anything like 448 horsepower. Physiologists estimate that an exertion beyond about 60 or 70 horsepower would put an intolerable strain on the whale's heart. Now 60 horsepower would not suffice to drive a whale through the water at 20 miles per hour if the flow over its body were turbulent, but it would be sufficient if the flow were laminar.

Thus we reach an impasse. Biologists are extremely unwilling to believe that

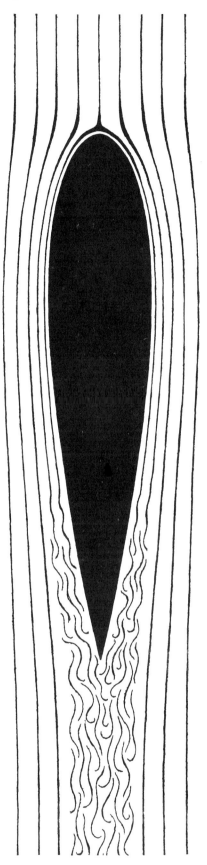

LAMINAR AND TURBULENT FLOW are depicted in this diagram of a streamlined body passing through the water. The smooth lines passing around the body indicate laminar flow; the wavy lines, turbulent flow.

fishes or whales can exert enough power to drive themselves through the water at the recorded speeds against the resistance that would be produced by turbulent flow over their bodies, while engineers are probably equally loath to believe that laminar flow can be maintained over a huge body, even a streamlined body, traveling through the water at 20 miles per hour.

Lacking direct data on these questions, we can only speculate on possible explanations which might resolve the contradiction. One point that seems well worth re-examining is our assumption about the hydrodynamic form of the swimming animal. We assumed that the resistance which the animal (say a dolphin) has to overcome is the same as

that of a rigid body of the same size and shape moving forward under a steady propulsive force. But the fact of the matter is that the swimming dolphin is not a rigid body: its tail and flukes are continually moving and bending during each propulsive stroke. It seems reasonable to assume, therefore, that the flow of water over the hind end of the dolphin is not the same as it would be over a rigid structure. In the case of a rigid model towed through the water, much of the resistance is due to slowing down of the water as it flows past the rear end of the model. But the oscillating movement of a swimming animal's tail accelerates water in contact with the tail; this may well reduce or prevent turbulence of flow. There is also another possibility which might be worth investiga-

tion. When a rigid body starts from rest, it takes a little time for turbulence to develop. It is conceivable that in the case of a swimming animal the turbulence never materializes, because the flukes reverse their direction of motion before it has an opportunity to do so.

It would be foolish to urge these speculative suggestions as serious contributions to the problem: they can only be justified insofar as they stimulate engineers to examine the hydrodynamic properties of oscillating bodies. Few, if any, biologists have either the knowledge or the facilities for handling such problems. The questions need to be studied by biologists and engineers working together. Such a cooperative effort could not fail to produce facts of great intrinsic, and possibly of great applied, interest.

DOLPHINS (called porpoises by seamen) were photographed by Jan Hahn as they swam beside the bow of the *Atlantis*, research vessel of the Woods Hole Oceanographic Institution, in the Gulf of Mexico. The speed of these dolphins was about 11 miles per hour.

The Author

SIR JAMES GRAY has been head of the department of zoology of the University of Cambridge for 20 years. In addition to his academic duties, he is the British Government's Development Commissioner for Fishery Research and a trustee of the British Museum. Born in 1891, he was elected a fellow of King's College, Cambridge, in 1914. During World War I he was a captain in the Queen's Royal West Surrey Regiment and won the Military Cross and Croix de Guerre with Palm. A long-time student of animal movement, Sir James writes that he first studied fish as a result of trying to explain the movement of single cells. The lash of a spermatozoon's tail led him to observe eels, and eels led him to fishes and dolphins. He is the author of the well-known book *How Animals Move*.

Bibliography

ASPECTS OF THE LOCOMOTION OF WHALES. R. W. L. Gawn in *Nature*, Vol. 161, No. 4,080, pages 44-46; January 10, 1948.

THE PROPULSIVE POWERS OF BLUE AND FIN WHALES. K. A. Kermack in *The Journal of Experimental Biology*, Vol. 25, No. 3, pages 237-240; September, 1948.

STUDIES IN ANIMAL LOCOMOTION. VI: THE PROPULSIVE POWERS OF THE DOLPHIN. J. Gray in *The Journal of Experimental Biology*, Vol. 13, No. 2, pages 192-199; April, 1936.

WHAT PRICE SPEED? G. Gabrielli and Th. von Kármán in *Mechanical Engineering*, Vol. 72, No. 10, pages 775-781; October, 1950.

SCIENTIFIC
AMERICAN   May 1960, Vol. 202, No. 5, pp. 148-157   OFFPRINT **1114**

HOW ANIMALS RUN

by Milton Hildebrand

Many animals, both predators and prey, have evolved the ability to run two or three times faster than a man can. What are the adaptations that make these impressive performances possible?

A man (but not necessarily you or I!) can run 220 yards at the rate of 22.3 miles per hour, and a mile at 15.1 miles per hour. The cheetah, however, can sprint at an estimated 70 miles per hour. And the horse has been known to maintain a speed of 15 miles per hour not just for one mile but for 35 miles.

Other animals are capable of spectacular demonstrations of speed and endurance. Jack rabbits have been clocked at 40 miles per hour. The Mongolian ass is reported to have run 16 miles at the impressive rate of 30 miles per hour. Antelopes apparently enjoy running beside a moving vehicle; they have been reliably timed at 60 miles per hour. The camel has been known to travel 115 miles in 12 hours. Nearly all carnivorous mammals are good runners: the whippet can run 34 miles per hour; the coyote, 43 miles per hour; the red fox, 45 miles per hour. One red fox, running before hounds, covered 150 miles in a day and a half. A fox terrier rewarded with candy turned a treadmill at the rate of 5,000 feet per hour for 17 hours.

I have been attracted by such performances as these to undertake an investigation of how the living running-machine works. The subject has not been thoroughly explored. One study was undertaken by the American photographer Eadweard Muybridge in 1872. Working before the motion-picture camera was invented, Muybridge set up a battery of still cameras to make photographs in rapid sequence. His pictures are still standard references. A. Brazier Howell's work on speed in mammals and Sir James Gray's studies on posture and movement are well known to zoologists. Many investigators have added to our knowledge of the anatomy of running vertebrates, but the analysis of function has for the most part been limited to deductions

from skeletons and muscles. The movements of the running animal are so fast and so complex that they cannot be analyzed by the unaided eye.

In my study I have related comparative anatomy to the analysis of motion pictures of animals in action. The method is simple: Successive frames of the motion picture are projected onto tracing paper, where the movements of the parts of the body with respect to one another and to the ground can be analyzed. The main problem is to get pictures from the side of animals running at top speed over open ground. With an electric camera that exposes 200 frames per second I have succeeded in photographing the movements of a cheetah that had been trained by John Hamlet of Ocala, Fla., to chase a paper bag in an enclosure 65 yards long. However, the animal never demonstrated its top speed, but merely loped along at about 35 miles per hour. I have used the same

STRIDE OF A CANTERING HORSE is shown in these photographs from Eadweard Muybridge's *The Horse in Motion*, published in 1878. The sequence runs right to left across the

camera to make pictures of horses running on race tracks, and I am presently collecting motion-picture sequences of other running animals from commercial and private sources.

All cursorial animals (those that can run far, fast and easily) have evolved from good walkers, and in doing so have gained important selective advantages. They are able to forage over wide areas. A pack of African hunting dogs, for example, can range over 1,500 square miles; the American mountain lion works a circuit some 100 miles long; individual arctic foxes have on occasion wandered 800 miles. Cursorial animals can seek new sources of food and water when their usual supplies fail. The camel moves from oasis to oasis, and in years of drought the big-game animals of Africa travel impressive distances. The mobility of cursorial animals enables them to overcome seasonal variations in climate or in food supply. Some herds of caribou migrate 1,600 miles each year. According to their habit, the predators among the cursorial animals exploit superior speed, relay tactics, relentless endurance or surprise to overtake their prey. The prey species are commonly as

swift as their pursuers, but sometimes they have superior endurance or agility.

Speed and endurance are the capacities that characterize all cursorial vertebrates. But one could not make a definitive list of the cursorial species without deciding quite arbitrarily how fast is fast and how far is far. Even then the list would be incomplete, because there are reliable data on speed for only a few animals; in most cases authors quote authors who cite the guesses of laymen. Many cursors are extinct. On the basis of fossils, however, we can surmise that many dinosaurs were excellent runners; that some extinct rhinoceroses, having had long and slender legs, were very fast; and that certain extinct South American grazing animals, having evolved a horselike form, probably had horselike speed.

In order to run, an animal must overcome the inertia of its body and set it into motion; it must overcome the inertia of its legs with every reversal in the direction of their travel; it must compensate for forces of deceleration, including the action of the ground against its descending feet. A full cycle of motion is called a stride. Speed is the product of length of stride times rate of stride. The giraffe achieves a moderate speed with

a long stride and a slow rate of stride; the wart hog matches this speed with a short stride and a rapid rate. High speed requires that long strides be taken at a rapid rate, and endurance requires that speed be sustained with economy of effort.

Although longer legs take longer strides, speed is not increased simply by the enlargement of the animal. A larger animal is likely to have a lower rate of stride. Natural selection produced fast runners by making their legs long in relation to other parts of the body. In cursorial animals the effective length of the leg—the part that contributes to length of stride—is especially enhanced. The segments of the leg that are away from the body (the foot, shank and forearm) are elongated with respect to the segments close to the body (the thigh and upper arm). In this evolutionary lengthening process the bones equivalent to the human palm and instep have become the most elongated.

Man's foot does not contribute to the length of his leg, except when he rises on his toes. The boar, the opossum, the raccoon and most other vertebrates that walk but seldom run have similar plantigrade ("sole-walking") feet. Carnivo-

top row and continues across the bottom row. With these and similar photographs Muybridge settled the controversy of whether or not a horse "even at the height of his speed [has] all four of his feet . . . simultaneously free from contact with the ground."

rous mammals, birds, running dinosaurs and some extinct hoofed mammals, on the other hand, stand on what corresponds to the ball of the human foot; these animals have digitigrade ("finger-walking") feet. Other hoofed mammals owe an even further increase in the effective length of their legs to their unguligrade ("hoof-walking") posture, resembling that of a ballet dancer standing on the tips of her toes. Where foot posture and limb proportions have been modified for the cursorial habit, the increased length and slenderness of the leg is striking [*see illustration on page 2786*].

The effective length of the front limb of many runners is also increased by the modification of the structure and function of the shoulder. The shoulder joint of amphibians, reptiles and birds is virtually immobilized by the collarbone, which runs from the breast bone to each shoulder blade, and by a second bone,

the coracoid bone. Because mammals do not have a coracoid bone their shoulder blade has some freedom of movement. In the carnivores this freedom is increased by the reduction of the collarbone to a vestige; in the ungulates the collarbone is eliminated. In both carnivores and ungulates the shoulder blade is oriented so that it lies against the side of a narrow but deep chest rather than against the back of a broad but shallow chest, as it does in man. Thus mounted, the shoulder blade pivots from a point about midway in its length, and the shoulder joint at its lower end is free to move forward and backward with the swing of the leg. The exact motion is exceedingly difficult to ascertain in a running animal, but I have found that it adds about 4.5 inches to the stride of the walking cheetah.

The supple spine of the cat and the dog increases the length of stride of these animals still further. The body of such an animal is several inches longer

when the back is extended than when it is flexed. By extending and flexing its back as its legs swing back and forth the animal adds the increase in its body length to its stride. Timing is important in this maneuver. If the animal were to extend its back while its body was in mid-air, its hindquarters would move backward as its forequarters moved forward, with no net addition to the forward motion of the center of mass of its body. In actuality the running animal extends its back only when its hind feet are pushing against the ground. The cheetah executes this maneuver so adeptly that it could run about six miles per hour without any legs.

With the extra rotation of its hip and shoulder girdles and the measuring-worm action of its back, the legs of the running cursor swing through longer arcs, reaching out farther forward and backward and striking and leaving the ground at a more acute angle than they would if the back were rigid. This clear-

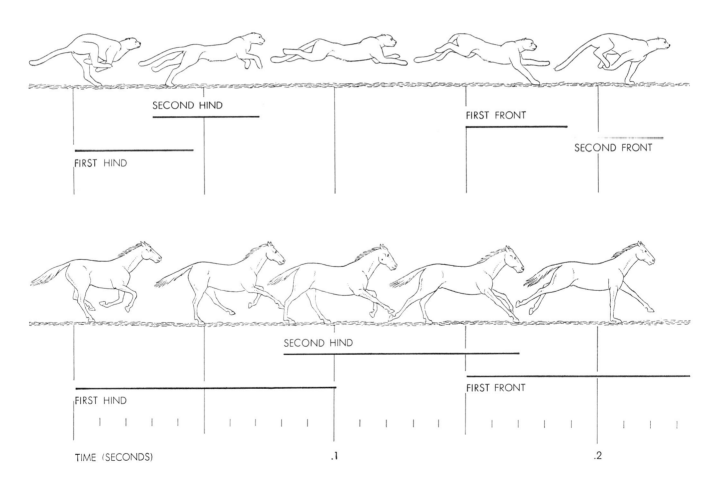

STRIDES OF THE CHEETAH AND THE HORSE in full gallop are contrasted in these illustrations. The sequence and duration of their footfalls, indicated by the horizontal lines under each animal, relate to the time-scale at bottom, which is calibrated in 10ths of a second. The cheetah has two unsupported periods, which account for about half its stride; the horse has one unsupported period,

ly increases stride length, but it also aggravates a problem. The body of the animal tends to rise when its shoulders and hips pass over its feet, and tends to fall when its feet extend to the front or rear. Carnivores offset this bobbing motion by flexing their ankles and wrists, thus shortening their legs. Ungulates do the same by sharply flexing the fetlock joint at the moment that the body passes over the vertical leg. The cheetah, a long-striding back-flexer, supplements its wrist-flexing by slipping its shoulder blade up its ribs about an inch, and thus achieves a smooth forward motion.

Since running is in actuality a series of jumps, the length of the jump must be reckoned as another important increment in the length of the stride. Hoofed runners have one major unsupported period, or jump, in each stride: when the legs are gathered beneath the body. The galloping carnivore has two major unsupported periods: when the back is flexed, and again when it is extended. In the horse all of these anatomical and functional adaptations combine to produce a 23-foot stride. The cheetah, although smaller, has a stride of the same length.

Fast runners must take their long strides rapidly. The race horse completes about 2.5 strides per second and the cheetah at least 3.5. It is plain that the higher the rate of stride, the faster the runner must contract its muscles. One might infer that cursorial animals as a group would have evolved the ability to contract their muscles faster than other animals. Within limits that is true, but there is a general principle limiting the rate at which a muscle can contract. Assuming a constant load on the muscle fibers, the rate of contraction varies inversely with any of the muscle's linear dimensions; the larger muscle therefore contracts more slowly. That is why an animal with a larger body has a slower rate of stride and so loses the advantage of its longer length of stride.

The familiar mechanical principle of gear ratio underlies the fast runner's more effective use of its trim musculature. In the linkage of muscle and bone the gear ratio is equal to the distance between the pivot of the motion (the shoulder joint, for example) and the point at which the motion is applied (the foot) divided by the perpendicular distance between the pivot and the point at which the muscle is attached to the bone. Cursorial animals not only have longer legs; their actuating muscles are also attached to the bone closer to the pivot of motion. Their high-gear muscles, in other words, have short lever-arms, and this increases the gear ratio still further. In comparison, the anatomy of walking animals gives them considerably lower gear-ratios; digging and swimming animals have still lower gear ratios.

But while high gears enable an automobile to reach higher speed, they do

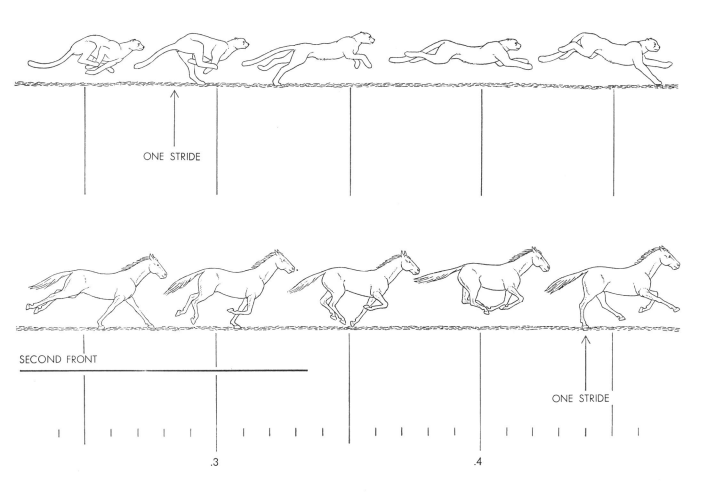

ONE STRIDE

SECOND FRONT

ONE STRIDE

.3   .4

which accounts for about a quarter of its stride. Although both the cheetah and the horse cover about 23 feet per stride, the cheetah attains speeds on the order of 70 miles per hour, to the horse's 43, because it takes about 3.5 strides to the horse's 2.5. The size of the horse has been reduced disproportionately in these drawings for the sake of uniformity in the stride-lines and time-scale.

SWIVELING SHOULDER BLADES of the horse and the cheetah add several inches to their stride length. The faster cheetah gains a further advantage from the flexibility of the spine, which in addition to adding the length of its extension to the animal's stride, adds the speed of its extension to the velocity of its travel. Horse's relatively longer leg partially compensates for its rigid spine.

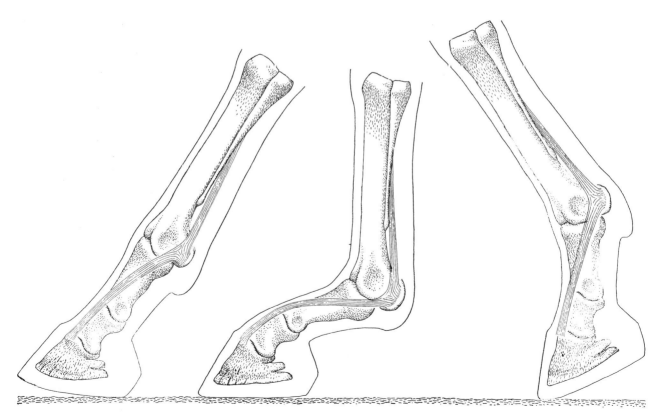

SPRINGING LIGAMENTS in the legs of horses, shown here, and other hoofed runners reduce the need for heavy muscles. Impact of the foot against the ground (*left*) bends the fetlock joint (*middle*) and stretches an elastic ligament (*shown in color*) that snaps back when the foot leaves the ground (*right*). The springing action at once straightens the foot and gives the leg an upward impetus.

so at the expense of power. The cursorial animal pays a similar price, but the exchange is a good one for several reasons. Running animals do not need great power: air does not offer much resistance even when they are moving at top speed. Moreover, as the English investigators J. M. Smith and R. J. G. Savage have noted, the animal retains some relatively low-gear muscles. Probably the runner uses its low-gear muscles for slow motions, and then shifts to its high-gear muscles to increase speed.

Since the speed at which a muscle can contract is limited, the velocity of the action it controls must be correspondingly limited, even though the muscle speed is amplified by an optimum gear-ratio. A larger muscle, or additional muscles, applied to action around the same joint can produce increased power but not greater speed. Several men together can lift a greater weight than one can lift alone, but several equally skilled sprinters cannot run faster together than one of them alone. The speed of a leg can be increased, however, if different muscles simultaneously move different joints of the leg in the same direction. The total motion they produce, which is represented by the motion of the foot, will then be greater than the motion produced by any one muscle working alone. Just as the total speed of a man walking up an escalator is the sum of his own speed plus that of the escalator, so the independent velocities of each segment of the leg combine additively to produce a higher total velocity.

The trick is to move as many joints as possible in the same direction at the same time. The evolution of the cursorial body has produced just this effect. By abandoning the flat-footed plantigrade posture in favor of a digitigrade or unguligrade one, the cursorial leg acquired an extra limb-joint. In effect it gained still another through the altered functioning of the shoulder blade. The flexible back of the cursorial carnivore adds yet another motion to the compound motion of its legs; the back flexes in such a way that the chest and pelvis are always rotating in the direction of the swinging limbs.

The supple spine of the carnivore contributes to stride rate by speeding up the motion of its body as well as of its legs. The spine is flexed when the runner's first hind foot strikes the ground, and by the time its second hind foot leaves the ground the animal has extended its spine and thus lengthened its body. In the brief interval when its hind

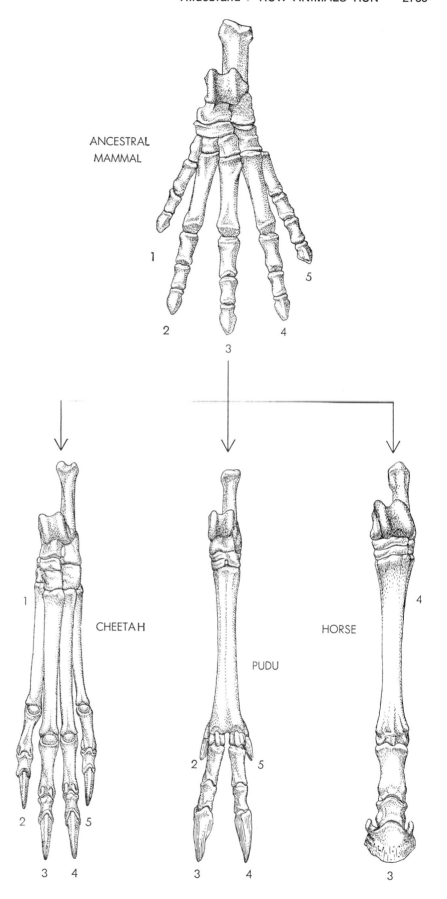

MODERN CURSORIAL FOOT EVOLVED from the broad, five-digited foot of an ancestral mammal (*top*). Lateral digits were lost and metatarsal bones, the longest in the foot, were further elongated. Resultant foot is lighter and longer. Pudu is a deer of the Andes.

feet are planted, the forequarters, riding on the extending spine, move farther and faster than the hindquarters. Similarly when the front feet are on the ground, the hindquarters move faster than the forequarters. So although the speed that the driving legs can impart to the forequarters or hindquarters is limited by their rate of oscillation, the body as a whole is able to exceed that limit. In a sense the animal moves faster than it runs. For the cheetah the advantage amounts to about two miles per hour—enough to add the margin of success in a close chase.

In addition to the obvious tasks of propelling the animal's body and supporting its weight, the locomotor mus-

cles must raise the body to compensate for the falling that occurs during the unsupported phases of the stride. The load they must raise is proportional to the mass of the body, which is in turn proportional to the cube of any of its linear dimensions. A twofold increase in body length thus increases weight eightfold. The force that a muscle can exert, on the other hand, increases only as the square of its cross section. Thus against an eightfold increase of load, bigger muscles can bring only a fourfold increase of force. As body size increases, the capacity of the muscles to put the body in forward motion and to cause its legs to oscillate cannot quite keep up with the demands placed upon them. These fac-

tors in the nature of muscle explain why the largest animals can neither gallop nor jump, why small runners such as rabbits and foxes can travel as fast as race horses without having marked structural adaptations for speed and why the larger cursorial animals must be highly adapted in order to run at all.

If the bigger runners are to have endurance as well as speed, they must have not only those adaptations that increase the length and rate of their stride, but also adaptations that reduce the load on their locomotor structures and economize the effort of motion. In satisfying this requirement natural selection produced a number of large and fast runners that are able to travel for long distances at somewhat less than their maximum speed. In these animals the mass of the limbs is minimized. The muscles that in other animals draw the limbs toward or away from the midline of the body (the "hand-clapping" muscles in man) are smaller or adapted to moving the legs in the direction of travel, and the muscles that manipulate the digits or rotate the forearm have disappeared. The ulna in the forearm and the fibula in the shank —bones involved in these former motions—are reduced in size. The ulna is retained at the point where it completes the elbow joint, but elsewhere becomes a sliver fused to its neighbor; the fibula is sometimes represented only by a nubbin of bone at the ankle.

The shape of the cursorial limb embodies another load-reducing principle. Since the kinetic energy that must be alternately developed and overcome in oscillating the limb is equal to half the mass times the square of its velocity, the load on muscles causing such motions can be reduced not only by reducing the mass of the faster-moving parts of the limb but also by reducing the velocity of the more massive parts. Accordingly the fleshy parts of the limb are those close to the body, where they do not move so far, and hence not so fast, as the more distant segments. The lower segments, having lost the muscles and bones involved in rotation and in digit manipulation, are relatively light.

The rigor of design imposed by natural selection is especially evident in the feet of cursorial animals. The feet of other animals tend to be broad and pliable; the bones corresponding to those of the human palm and instep are rounded in cross section and well separated. In the foot of the cursorial carnivore, on the other hand, these bones are

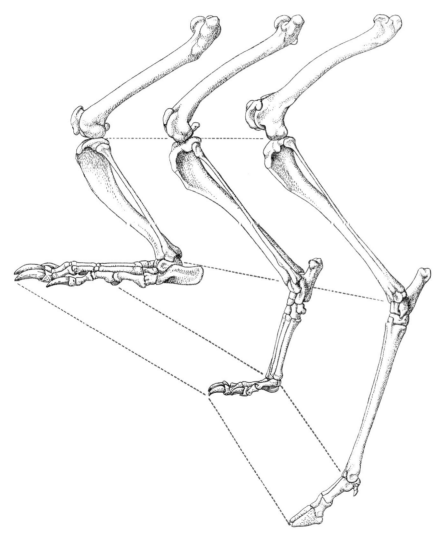

ADAPTATION OF THE LEG FOR SPEED is illustrated by the hind-leg bone of the slow badger (*left*), moderately fast dog (*middle*) and highly adapted deer (*right*). The lengthened metatarsus of the latter two has yielded a longer foot and an altered ankle posture that is better suited to running. The thigh bones of all three animals have been drawn to the same scale to show that the leg segments farthest from the body have elongated the most.

crowded into a compact unit, each bone having a somewhat square cross section. In the ungulates the ratio of strength to weight has been improved still further by reduction of the number of bones in the foot. The ungulates have tended to lose their lateral toes; sometimes the basal elements of the other toes are fused into a single bone. This process gave rise to the cannon bone: the shank of the hoofed mammals [*see illustration on page 2785*]. In compensation for the bracing lost as the bones and muscles of

their lower limbs were reduced or eliminated, these animals evolved joints that are modified to function as hinges and allow motion only in the line of travel.

The burden on the muscles of hoofed animals is relieved by an especially elegant mechanism built into the foot. When the hoof of the running animal strikes the ground, the impact bends the fetlock joint and stretches certain long ligaments called the suspensory or springing ligaments [*see bottom illustration on page 2784*]. Because ligaments

are elastic, they snap back as the foot leaves the ground, thereby straightening the joint and giving the leg an upward push. Charles L. Camp of the University of California has found that these built-in pogo-sticks evolved from foot muscles at the time that the animals forsook river valleys for the open plains. The exchange was advantageous, for by means of this and the other adaptations, nature has reconciled the limitations of muscle mechanics with the exacting requirements of speed.

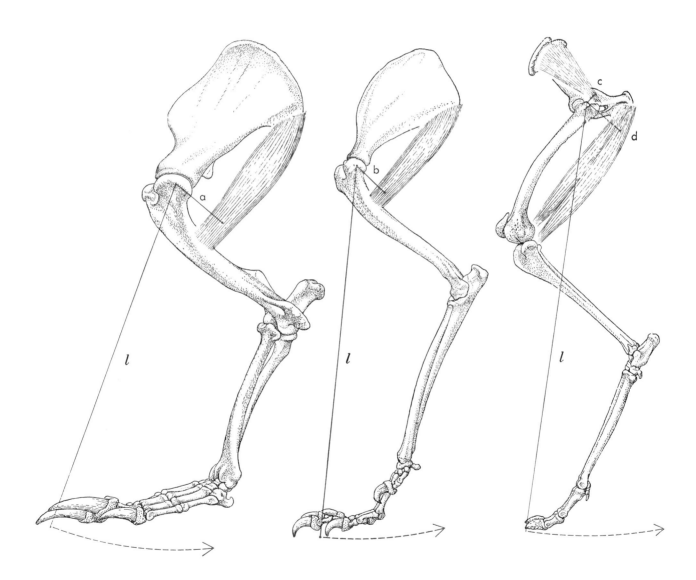

POWER AND SPEED are alternatively achieved in the badger (*left*) and the cheetah (*middle*) by placement of the teres major muscle. In the cheetah the small distance (b) between the muscle insertion and the joint it moves yields a higher rate of oscillation than in the badger, in which the distance (a) is greater. The higher oscillation rate, coupled with a longer leg (*l*), yields a faster stride. In the vicuña (*right*) the gluteus muscle (c) develops about five times the velocity but only a fifth the force of the larger semimembranosus muscle (d). The animal may use the latter to overcome inertia; the former, for high speed. Legs are not in same scale.

The Author

MILTON HILDEBRAND is associate professor of zoology at the University of California campus in Davis, Calif. He was born in Philadelphia in 1918, and decided at the age of six that he wanted to be a zoologist. "As a boy," he reports, "I collected skeletons. My private teaching collection now includes about 1,000 specimens from all over the world." He took his B.A., M.A. and Ph.D. degrees at the University of California; "this scholastic inbreeding," he says, "is largely responsible for my present study in England." During World War II he was an officer in the ski troops, and taught skiing, winter warfare, rock-climbing and mountain combat. During the Italian campaign he served as a battalion pack officer in charge of mule transport in the Apennines.

Bibliography

How Animals Move. James Gray. Cambridge University Press, 1953.

Motions of the Running Cheetah and Horse. Milton Hildebrand in *Journal of Mammalogy*, Vol. 40, No. 4, pages 481-495; November, 1959.

Quadrupedal and Bipedal Locomotion of Lizards. Richard C. Snyder in *Copeia*, No. 2, pages 64-70; June, 1952.

Some Locomotory Adaptations in Mammals. J. Maynard Smith and R. J. G. Savage in *Journal of the Linnean Society—Zoology*, Vol. 42, No. 288, pages 603-622; February, 1956.

Speed in Animals. A. Brazier Howell. University of Chicago Press, 1944.

SCIENTIFIC
AMERICAN   April 1952, Vol. 186, No. 4, pp. 24–29   OFFPRINT **1115**

BIRD AERODYNAMICS

by John H. Storer

It is even more like that of the airplane than is generally
assumed. A bird does not fly through the air as a man swims
through the water; it employs the airfoil and the propeller.

THE flight of birds has always excited man's envy and wonder. At first sight the process looks simple enough: a bird seems to lift and drive itself forward by beating its wings against the air in much the same way as a swimmer propels himself through water by flapping his arms. When men first tried to fly, they built their flying machines ("ornithopters") on this principle, with mechanical wings that flapped. But the machines never got off the ground.

For this is not at all the way birds fly. Paradoxically it was the development of the modern propeller plane that finally taught us how birds fly—not the other way around. A bird is actually a living airplane. It flies by the same aerodynamical principles as a plane and uses much of the same mechanical equipment—wings, propellers, steering gear, even slots and flaps for help in taking off and landing.

The slow-motion camera shows that a bird does not push itself along by beating its wings back against the air. On the downstroke the wings move forward, not backward. And when the bird lifts its wings for the next stroke, it does not lose altitude, as might be expected, but sails on steadily on a level course. The easiest way to understand its flight is to consider first how an airplane flies.

The air, like any fluid, has weight, and it presses against every surface of anything submerged in it—downward from above, upward from below and inward from all sides. At sea level the air presses on all surfaces with a force of 14.7 pounds per square inch. The air therefore will supply the force to support flight, provided the flying object can somehow reduce the pressure on its upper surface to less than the lifting pressure, and decrease the pressure against its front surface or increase that from behind. Birds and airplanes do this by means of properly shaped wings and propellers which they manipulate to drive themselves forward at a certain required angle and speed.

We can study the aerodynamical problems involved by blowing a stream of smoke, which makes the air currents visible, against an obstruction in a wind tunnel. When the smoke stream hits the obstruction, it does not flow smoothly around the surface and close up again immediately behind it. Instead, it breaks up and is deflected away from the obstruction so that the air no longer presses against the object's sides with the same force. Moreover, the air stream does not close up again until it has moved some distance past the obstruction, so the pressure on the rear surface of the obstacle also is reduced. There remains a disproportionate pressure on the front surface of the obstacle: what would be known as "drag" if the object instead of the air stream were moving.

Now suppose we place in the air stream an object so shaped that it fills in the spaces that were left vacant when the air was deflected by the first obstruction. The air flows smoothly around this new object, and the pressure is more nearly even on all sides. Drag is reduced. We have "streamlined" the obstacle. By altering this shape just a little, we can change the relative pressures on its different surfaces. Let us flatten the bottom surface slightly, reducing the downward deflection of the air stream. Now the upward pressure of air against the bottom surface is more nearly normal, while the downward pressure on the top surface remains subnormal as before. Presto! We have more pressure from below than from above. If the streamlined model is light enough, the moving air will lift it. We have the beginning of a wing.

If the front edge of this embryo wing is tilted upward just a little so that the air strikes the bottom surface more directly, the lifting force on the wing is increased. The more the wing is tilted, the more lift it will give—up to a certain point. As the angle of tilt approaches the vertical, the pressure against the bottom surface begins to push the wing backward rather than upward. Eventually,

if a plane's wing is tilted too much, the lifting force vanishes, the drag is so great that it stops the plane, and we have what is known as a "stall." The plane must regain the proper angle and speed or it will crash.

In the air a pilot controls the lifting power of his plane both by tilt and by speed: the more speed, the more lift. In taking off or landing, however, he must rely mainly on tilt: to get enough lifting force he must hold his wing at the greatest possible angle against the air, up to the point of stalling. The angle to which he can tilt the wing without stalling can be increased by placing a very small auxiliary wing in front of or behind the main wing. The "slot" formed between the main wing and the small auxiliary airfoil increases the speed of the air flow over the wing and so maintains its lifting power, even after it has passed the normal stalling point.

Once we have a streamlined wing, the next step necessary for flying is to move it through the air fast enough to generate lift. This we accomplish by equipping the machine with propellers, which are actually another set of wings, whose "lift" is exerted forward rather than upward. For mechanical reasons the blades of a propeller function better with a shape and angle slightly different from those of the wings, but the principle on which they work is just the same.

So we have, basically, a single mechanism which, placed in one position, holds an airplane up, and in another, drives it forward. Now we can look at a bird's anatomy and find exactly the same mechanism used in just the same two ways.

THE WING of a bird consists of two parts, which have two very different functions. It is divided into an inner half, operated from the shoulder joint, and an outer half, which is moved separately by a "wrist" midway along the wing. The inner half of the wing is devoted almost exclusively to giving lift. It is held rather rigidly at a slight angle, sloping like the wing of a plane. It also

RING-BILLED GULL has a long, narrow wing characteristic of birds that do much gliding over water. The wing's anatomy is shown at the top of page 2792. Birds that do little gliding have shorter and broader wings.

BROWN PELICAN is another gliding bird. The feathers at the tip of its left wing are turned up by a current of air. When the wing beats down during active flight, this same effect occurs and pushes the bird forward.

has the streamlined shape of a plane's wing: its upper feathers are arched to make a curved surface.

At the front edge of the wrist, where the inner and outer wings join, is a small group of feathers called the alula. This is the bird's auxiliary airfoil for help in taking off and landing. The bird can raise the alula to form a slot between that structure and the main wing. Without the alula a bird cannot take off or land successfully.

But where is the propeller? Astonishing as it may seem, every bird has a pair of them, though where they might be is certainly far from obvious. They can be seen in action best in a slow-motion picture of a bird in flight. During the downward beat of the wings the primary feathers at the wing tips stand out almost at right angles to the rest of the wing and to the line of flight. These feathers are the propellers. They take on this twisted form for only a split second during each wing beat. But this ability to change their shape and position is the key to bird flight. Throughout the entire wing beat they are constantly changing their shape, adjusting automatically to air pressure and the changing requirements of the wing as it moves up and down.

This automatic adjustment is made possible by special features of the feather design. The front vane of a wing-tip feather (on the forward side of the quill) is much narrower than the rear vane. Out of this difference comes the force that twists the feather into the shape of a propeller. As the wing beats downward against the air, the greater pressure against the wide rear vane of each of these feathers twists that vane upward until the feather takes on the proper shape and angle to function as a propeller. The degree and shape of its twist is controlled largely by the design of the quill, which is rigid at its base but flattened and flexible toward the end.

With their specialized design the primary feathers are beautifully adapted to meet the varied demands of bird flight. An airplane's propeller rotates around a pivot in one direction; the bird's propeller, in contrast, oscillates rapidly down and up, and it must automatically adapt its shape, position, angle and speed to the changing requirements of the moment. The feathers are not fastened immovably to the bone of the wing but are held by a broad flexible membrane, which allows considerable freedom of movement to each feather. While the bird is flying easily, only the tips of the feathers twist to become propellers. But if the bird is in a hurry and beats its wings more strongly against the air, the whole outer section of the wing, from the wrist out, may be twisted by the greater pressure into one big propeller.

The path of the propeller on the

FLYING egret shows the different functions of the inner and outer halves of a bird wing. In the first two pictures the wings are moving downward. In the third the inner half rises ahead of the outer to maintain lift.

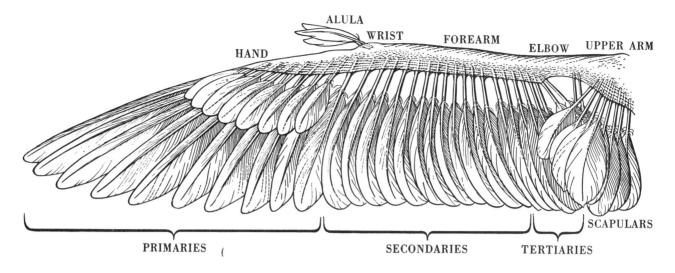

ALULA   WRIST   FOREARM   ELBOW   UPPER ARM

HAND

PRIMARIES   SECONDARIES   TERTIARIES   SCAPULARS

ANATOMY of a herring gull's wing reflects the different functions of its inner and outer halves. The inner half of the wing, from the shoulder to the wrist, is adapted primarily to lift; the outer half, from the wrist to the wing tip, to control and propulsion. Only the principal feathers of the wing are shown in this illustration.

downstroke is downward and forward; on the upstroke, upward and backward. The amount of forward and backward motion varies with the bird's wingbeat. When the bird beats its wings fast, as in taking off, the increased pressure drives the wing tips forward on a more nearly horizontal path; in leisurely flight the movement is more nearly vertical. The inner wing, by maintaining the proper angle, supports the bird's weight through the entire wing beat. This angle is constantly adjusted to maintain a steady lifting force.

In free flight the bird's powerful breast muscles sweep the whole wing up and down from the shoulder. The inner wing does not actually need to move, but it acts as a handle to move the propeller and gives the latter greater speed and power. I have a slow-motion movie of a low-flying white heron skimming some bushes so closely that it did not have room to make a full downward wing beat. The bird held the inner half of each wing extended horizontally and beat the outer half up and down from the wrist. To move the propeller fast enough without the help of the breast muscles must have required great effort. But this flight demonstrated perfectly the true function of each half of the wing. The inner half was the wing of a living airplane, lifting the bird. The outer half was the propeller, driving it forward.

Like an airplane, a bird has special equipment for steering and balancing. It steers by turning its tail, up, down or sidewise. (It can also spread the tail wide to give added lifting surface when needed.) The bird balances itself by means of its wings; if it tips to one side, it can restore itself to an even keel by increasing the lift of that wing, either by beating more strongly with it or by changing its angle.

OF ALL the powers of birds in the air probably none has caused more wonder than their soaring ability. To see a bird rise in the air and sail on motionless wings into the distance until at last it disappears from sight gives one a sense of magic. We now know how it is done, but it is still difficult to realize what is happening as we watch it.

Actually the bird is coasting downhill in relation to the flow of air. It rises because it is riding a rising current of air which is ascending faster than the bird is sinking in the current.

Ascending air currents on which birds can soar or glide arise from two different kinds of situations. One is an obstruction, such as an ocean wave, a shore line or a hillside, which deflects the wind upward. It is common to see a pelican or albatross sailing over the water on motionless wings just above the crest of a moving wave, or a gull hanging motionless against a wind current that rises over a headland, or a hawk soaring on the air current that sweeps up a mountainside.

The second type of rising current is heated air, known as a thermal. A field warmed by the sun heats the air above it, causing it to expand and rise. If the field is surrounded by a cooler forest, the heated pocket of air may rise in the form of a great bubble or of a column. Everyone has seen birds soaring in wide circles over land; usually they are coasting around the periphery of a rising air column. Over the ocean, when the water warms colder air above it, the air rises in a whole group of columns, packed together like the cells of a honeycomb. If the wind then freshens, it may blow the columns over until they lie horizontally on the water. The flat-lying columns of air may rotate around their axes, each in the opposite direction from its neighbor. This has been demonstrated in the laboratory by blowing smoke-filled air over a warmed surface at increasing speed, corresponding to an increase in the wind over the ocean. If you put your two fists together and rotate them, the right clockwise and the left counterclockwise, you will see that the two

	MILES PER HOUR
Great Blue Heron [cruising]	18-29
Great Blue Heron [pressed]	36
Canada Goose [easy flight]	20
Canada Goose [pressed by plane]	45-60
Mallard [pressed by plane]	55-60
Duck Hawk [pressed by plane]	175-180
Pheasant [average top speed]	60
Woodcock	5-13
Ruby-throated Hummingbird [easy flight]	45-55
Barn Swallow	20-46
Crow	25-60
Sharp-shinned Hawk	16-60
Osprey	20-80

FLYING SPEEDS of 13 species of birds are listed in this table. The conditions under which the observations were made are in parentheses.

LANDING American egret demonstrates the use of feathers in its alulas and wing tips to maintain balance and control. The alula is a small bunch of feathers on the leading edge of the wing. By opening slots with these feathers and those of the wing tips, the bird can control its lift while losing flying speed.

inner faces of the fists rise together. Just so two adjoining air cells rotating in opposite directions will push up between them a ridge of rising air. Birds can glide in a straight line along such a ridge.

At the Woods Hole Oceanographic Institution Alfred H. Woodcock studied the soaring of sea gulls at different seasons. During the summer, when the air is warmer than the water, gulls seldom do any soaring. But they do a great deal of it in the fall, when the water is warmer than the air and produces many updrafts. The gull's movements may clearly mark the outlines of the rising air columns. When the wind is relatively light, under 16 miles an hour, the gulls soar in spirals, showing that the columns are standing upright. But as the wind freshens and tilts over the columns, the birds' soaring patterns begin to change; when the wind speed reaches 24 miles per hour, all the gulls soar in straight lines. The spectacle is all but incredible, with the birds sailing into the strong wind on motionless wings and gaining altitude as they go, until they disappear in the distance. I watched it once, and it is a never-to-be-forgotten sight.

How fast do birds fly? A great deal of nonsense has been uttered on this subject. The measurement of a bird's speed capabilities is a very uncertain and tricky thing. The wind, the angle of the bird's flight, whether it is being pressed—these factors and many others affect its speed.

The cruising and top speeds of some common birds are listed in the table at the bottom of the opposite page. Birds vary greatly, of course, in their speed requirements and possibilities. The pheasant and grouse, which have short wings adapted to maneuvering in underbrush, must fly with a rapid wing beat and considerable speed to stay in the air. The same is true of ducks, which do not need large wings because they have an easy landing field on the water. Herons, on the other hand, must be able to land slowly to protect their long, slender legs, which they use for wading to find food. Their big, cumbersome wings are suited for slow landing, but they produce so much friction and drag in the air that herons cannot fly very fast.

As the table shows, 60 miles an hour is fast for a bird, and the fastest known species, the duck hawk, does not exceed 175 to 180 miles per hour. These speeds, of course, are far slower than the speeds of modern planes. They involve very different problems in aerodynamics, and different streamline designs. But some of them do approach the speed of the early planes, and it is interesting to see how closely the designs produced by nature approach the best results of the human engineer.

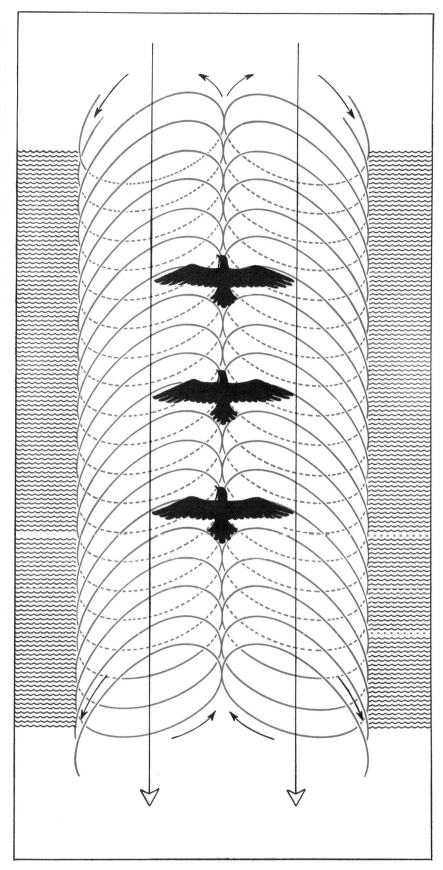

SEA BIRDS sometimes soar against a prevailing wind (*vertical arrows*) by riding an updraft between two counter-rotating cylinders of air. These cylinders result from the fact that when the sea is warmer than the air, many vertical columns of warm air rise from the surface of the water. If the wind freshens, the columns lie parallel to the surface. Birds are not drawn to scale.

The Author

JOHN H. STORER is a student of the aerodynamics of birds. He is author of the book *The Flight of Birds*.

Bibliography

THE FLIGHT OF BIRDS: ANALYZED THROUGH SLOW-MOTION PHOTOGRAPHY. John H. Storer. Cranbrook Institute of Science, Bulletin No. 28; 1948.

SPECIAL NOTE TO TEACHERS: Each article in this volume, plus more than 660 others, is available as a separate, self-bound SCIENTIFIC AMERICAN Offprint. Offprints may be ordered in any combination and in any quantity. Teachers who want to adopt articles for their courses, therefore, can ensure that each student has his own set. Students' sets are collated by the publisher before shipment.

SCIENTIFIC AMERICAN   March 1955, Vol. 192, No. 3, pp. 88-96

OFFPRINT 1116

BIRDS AS FLYING MACHINES

by Carl Welty

A sequel to Offprint 1115 on the aerodynamics of birds. Among the remarkable adaptations birds have made to life in the air are high power and light weight.

The great struggle in most animals' lives is to avoid change. A chickadee clinging to a piece of suet on a bitter winter day is doing its unconscious best to maintain its internal status quo. Physiological constancy is the first biological commandment. An animal must eternally strive to keep itself warm, moist and supplied with oxygen, sugar, protein, salts, vitamins and the like, often within precise limits. Too great a change in its internal economy means death.

The spectacular flying performances of birds—spanning oceans, deserts and whole continents—tend to obscure the more important fact that the ability to fly confers on them a remarkably useful mechanism to preserve their internal stability, or homeostasis. Through flight birds can search out the external conditions and substances they need to keep their internal fires burning clean and steady. A bird's wide search for specific foods and habitats makes sense only when considered in the light of this persistent, urgent need for constancy.

The power of flight opens up to birds an enormous gaseous ocean, the atmosphere, and a means of quick, direct access to almost any spot on earth. They can eat in almost any "restaurant"; they have an almost infinite choice of sites to build their homes. As a result birds are, numerically at least, the most successful vertebrates on earth. They number roughly 25,000 species and subspecies, as compared with 15,000 mammals and 15,000 fishes.

At first glance birds appear to be quite variable. They differ considerably in size, body proportions, color, song and ability to fly. But a deeper look shows that they are far more uniform than, say, mammals. The largest living bird, a 125-pound ostrich, is about 20,000 times heavier than the smallest bird, a hummingbird weighing only one tenth of an ounce. However, the largest mammal, a 200,000-pound blue whale, weighs some 22 million times as much as the smallest mammal, the one-seventh-ounce masked shrew. Mammals, in other words, vary in mass more than a thousand times as much as birds. In body architecture, the comparative uniformity of birds is even more striking. Mammals may be as fat as a walrus or as slim as a weasel, furry as a musk ox or hairless as a desert rat, long as a whale or short as a mole. They may be built to swim, crawl, burrow, run or climb. But the design of nearly all species of birds is tailored to and dictated by one pre-eminent activity—flying. Their structure, outside and inside, constitutes a solution to the problems imposed by flight. Their uniformity has been thrust on them by the drastic demands that determine the design of any flying machine. Birds simply dare not deviate widely from sound aerodynamic design. Nature liquidates deviationists much more consistently and drastically than does any totalitarian dictator.

Birds were able to become flying machines largely through the evolutionary gifts of feathers, wings, hollow bones, warm-bloodedness, a remarkable system of respiration, a strong, large heart and powerful breast muscles. These adaptations all boil down to the two prime requirements for any flying machine: high power and low weight. Birds have thrown all excess baggage overboard. To keep their weight low and feathers dry they forego the luxury of sweat glands. They have even reduced

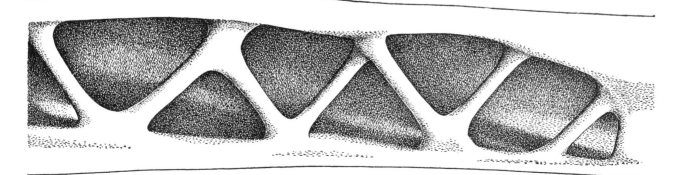

INTERNAL STRUCTURE of the metacarpal bone of a vulture's wing is shown in this drawing of a longitudinal section. The braces within the bone are almost identical in geometry with those of the Warren truss commonly used as a steel structural member.

their reproductive organs to a minimum. The female has only one ovary, and during the nonbreeding season the sex organs of both males and females atrophy. T. H. Bissonette, the well-known investigator of birds and photoperiodicity, found that in starlings the organs weigh 1,500 times as much during the breeding season as during the rest of the year.

As early as 1679 the Italian physicist Giovanni Borelli, in his *De motu animalium*, noted some of the weight-saving features of bird anatomy: ". . . the body of a Bird is disproportionately lighter than that of man or of any quadruped . . . since the bones of birds are porous, hollowed out to extreme thinness like the roots of the feathers, and the shoulder bones, ribs and wing bones are of little substance; the breast and abdomen contain large cavities filled with air; while the feathers and the down are of exceeding lightness."

The skeleton of a pigeon accounts for only 4.4 per cent of its total body weight, whereas in a comparable mammal such as a white rat it amounts to 5.6 per cent. This in spite of the fact that the bird must have larger and stronger breast bones for the muscles powering its wings and larger pelvic bones to support its locomotion on two legs. The ornithologist Robert Cushman Murphy has reported that the skeleton of a frigate bird with a seven-foot wingspread weighed only four ounces, which was less than the weight of its feathers!

Although a bird's skeleton is extremely light, it is also very strong and elastic— necessary characteristics in an air frame subjected to the great and sudden stresses of aerial acrobatics. This combination of lightness and strength depends mainly on the evolution of hollow, thin bones coupled with a considerable fusion of bones which ordinarily are separate in other vertebrates. The bones of a bird's sacrum and hip girdle, for example, are molded together into a thin, tube-like structure—strong but phenomenally light. Its hollow finger bones are fused together, and in large soaring birds some of these bones have internal trusslike supports, very like the struts inside airplane wings. Similar struts sometimes are seen in the hollow larger bones of the wings and legs.

To "trim ship" further, birds have evolved heads which are very light in proportion to the rest of the body. This has been accomplished through the simple device of eliminating teeth and the accompanying heavy jaws and jaw muscles. A pigeon's skull weighs about

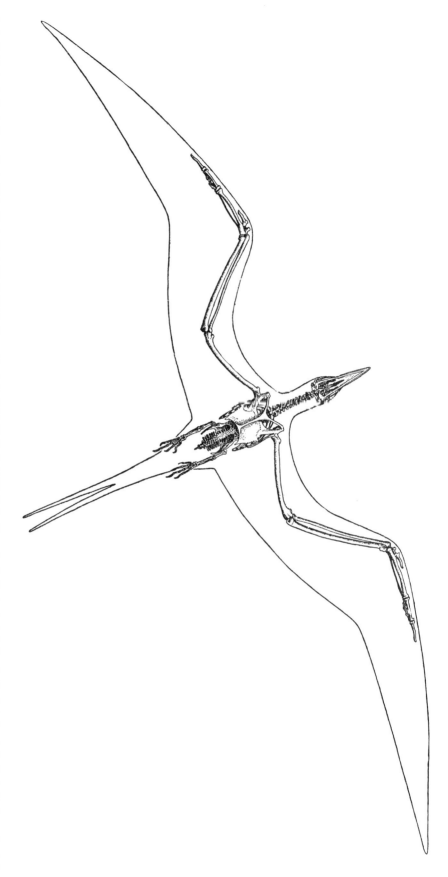

FRIGATE BIRD has a seven-foot wing span, but its skeleton weighs only four ounces. This is less than the weight of its feathers. The skeleton is shown against the outline of the bird.

one sixth as much, proportionally, as that of a rat; its skull represents only one fifth of 1 per cent of its total body weight. In birds the function of the teeth has been taken over largely by the gizzard, located near the bird's center of gravity. The thin, hollow bones of a bird's skull have a remarkably strong re-

inforced construction [see *photograph on page 2799*]. Elliott Coues, the 19th-century U. S. ornithologist, referred to the beautifully adapted avian skull as a "poem in bone."

The long, lizard-like tail that birds inherited from their reptilian ancestors has been reduced to a small plate of bone

at the end of the vertebrae. The ribs of a bird are elegantly long, flat, thin and jointed; they allow extensive movement for breathing and flying, yet are light and strong. Each rib overlaps its neighbor—an arrangement which gives the kind of resilient strength achieved by a woven splint basket.

Feathers, the bird's most distinctive and remarkable acquisition, are magnificently adapted for fanning the air, for insulation against the weather and for reduction of weight. It has been claimed that for their weight they are stronger than any wing structure devised by man. Their flexibility allows the broad trailing edge of each large wing-feather to bend upward with each downstroke of the wing. This produces the equivalent of pitch in a propeller blade, so that each wingbeat provides both lift and forward propulsion. When a bird is landing or taking off, its strong wingbeats separate the large primary wing feathers at their tips, thus forming wing-slots which help prevent stalling. It seems remarkable that man took so long to learn some of the fundamentals of airplane design which even the lowliest English sparrow demonstrates to perfection [see "Bird Aerodynamics," by John H. Storer; SCIENTIFIC AMERICAN Offprint 1115].

Besides all this, feathers cloak birds with an extraordinarily effective insulation—so effective that they can live in parts of the Antarctic too cold for any other animal.

The streamlining of birds of course is the envy of all aircraft designers. The bird's awkwardly angular body is trimmed with a set of large quill, or contour, feathers which shape it to the utmost in sleekness. A bird has no ear lobes sticking out of its head. It commonly retracts its "landing gear" (legs) while in flight. As a result birds are far and away the fastest creatures on our planet. The smoothly streamlined peregrine falcon is reputed to dive on its prey at speeds up to 180 miles per hour. (Some rapid fliers have baffles in their nostrils to protect their lungs and air sacs from excessive air pressures.) Even in the water, birds are among the swiftest of animals: Murphy once timed an Antarctic penguin swimming under water at an estimated speed of about 22 miles per hour.

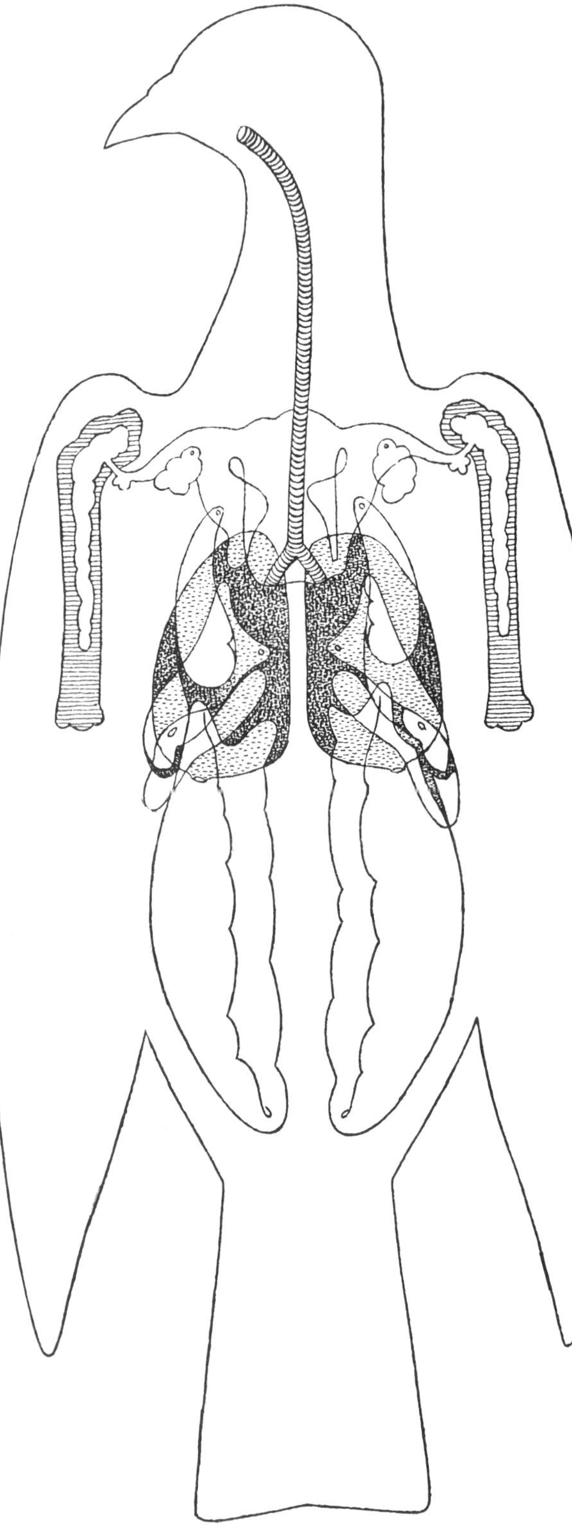

A basic law of chemistry holds that the velocity of any chemical reaction roughly doubles with each rise of 10 degrees centigrade in temperature. In nature the race often goes to the metabolically swift. And birds have evolved the highest operating temperatures of all animals. Man, with his conservative 98.6

AIR SACS connected to the lungs of a pigeon not only lighten the bird but also add to the efficiency of its respiration and cooling. The lungs are indicated by the two dark areas in the center. Two of the air sacs are within the large bones of the bird's upper "arm."

degrees Fahrenheit, is a metabolic slow-poke compared with sparrows (107 degrees) or some thrushes (113 degrees). Birds burn their metabolic candles at both ends, and as a result live short but intense lives. The average wild songbird survives less than two years.

Behind this high temperature in birds lie some interesting circulatory and respiratory refinements. Birds, like mammals, have a four-chambered heart which allows a double circulation, that is, the blood makes a side trip through the lungs for purification before it is circulated through the body again. A bird's heart is large, powerful and rapid-beating [*see table of comparisons on page 2800*]. In mammals and birds the heart rate, and the size of the heart in proportion to the total body, increases as the animals get smaller. But the increases seem significantly greater in birds than in mammals. Any man with a weak heart knows that climbing stairs puts a heavy strain on his pumping system. Birds do a lot of "climbing," and their circulatory systems are built for it.

The blood of birds is not significantly richer in hemoglobin than that of mammals. The pigeon and the mallard have about 15 grams of hemoglobin per 100 cubic centimeters of blood—the same as man. However, the concentration of sugar in their blood averages about twice as high as in mammals. And their blood pressure, as one would expect, also is somewhat higher: in the pigeon it averages 145 millimeters of mercury; in the chicken, 180 millimeters; in the rat, 106 millimeters; in man, 120 millimeters.

In addition to conventional lungs, birds possess an accessory system of five or more pairs of air sacs, connected with the lungs, that ramify widely throughout the body. Branches of these sacs extend into the hollow bones, sometimes even into the small toe bones. The air-sac system not only contributes to the birds' lightness of weight but also supplements the lungs as a supercharger (adding to the efficiency of respiration) and serves as a cooling system for the birds' speedy, hot metabolism. It has been estimated that a flying pigeon uses one fourth of its air intake for breathing and three fourths for cooling.

The lungs of man constitute about 5 per cent of his body volume; the respiratory system of a duck, in contrast, makes up 20 per cent of the body volume (2 per cent lungs and 18 per cent air sacs). The anatomical connections of the lungs and air sacs in birds seem to provide a one-way traffic of air through most of the system, bringing in a constant stream of unmixed fresh air, whereas in the lungs

of mammals stale air is mixed inefficiently with the fresh. It seems odd that natural selection has never produced a stale air outlet for animals. The air sacs of birds apparently approach this ideal more closely than any other vertebrate adaptation.

Even in the foods they select to feed their engines birds conserve weight. They burn "high-octane gasoline." Their foods are rich in caloric energy—seeds, fruits, worms, insects, rodents, fish and so on. They eat no low-calorie foods such as leaves or grass; a wood-burning engine has no place in a flying machine. Furthermore, the food birds eat is burned quickly and efficiently. Fruit fed to a young cedar waxwing passes through its digestive tract in an average time of 27 minutes. A thrush that is fed blackberries will excrete the seeds 45

minutes later. Young bluejays take between 55 and 105 minutes to pass food through their bodies. Moreover, birds utilize a greater portion of the food they eat than do mammals. A three-weeks-old stork, eating a pound of food (fish, frogs and other animals), gains about a third of a pound in weight. This 33 per cent utilization of food compares roughly with an average figure of about 10 per cent in a growing mammal.

The breast muscles of a bird are the engine that drives its propellers or wings. In a strong flier, such as the pigeon, these muscles may account for as much as one half the total body weight. On the other hand, some species—*e.g.*, the albatross—fly largely on updrafts of air, as a glider does. In such birds the breast muscles are greatly re-

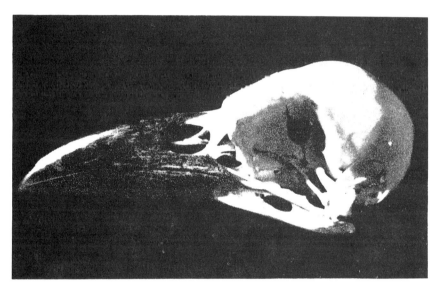

SKULL OF A CROW achieves the desirable aerodynamic result of making the bird light in the head. Heavy jaws are sacrificed. Their work is largely taken over by the gizzard.

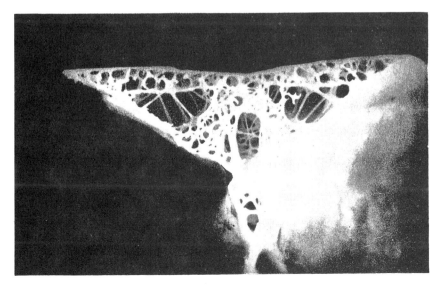

FRONTAL BONE in the skull of a crow is cut through to show its hollow and braced internal construction. The skull of the bird accounts for less than 1 per cent of its total weight.

HEART	PERCENT OF BODY WEIGHT	HEART BEATS PER MINUTE
FROG	.57	22
MAN	.42	72
PIGEON	1.71	135
CANARY	1.68	514
HUMMINGBIRD	2.37	615

HEART WEIGHT and pulse rate are compared for a number of animals. The hearts of birds are relatively large for body size.

duced, and there are well-developed wing tendons and ligaments which enable the bird to hold its wings in the soaring position with little or no effort.

A bird may have strong breast muscles and still be incapable of sustained flight because of an inadequate blood supply to these muscles. This condition is shown in the color of the muscles; that is the explanation of the "white meat" of the chicken and the turkey—their breast muscles have so few blood vessels that they cannot get far off the ground. The dark meat of their legs, on the other hand, indicates a good blood supply and an ability to run a considerable distance without tiring.

After a ruffed grouse has been flushed four times in rapid succession, its breast muscles become so fatigued that it can be picked up by hand. The blood supply is simply inadequate to bring in fuel and carry away waste products fast enough. Xenophon's *Anabasis* relates the capture of bustards in exactly this manner: "But as for the Bustards, anyone can catch them by starting them up quickly; for they fly only a short distance like the partridge and soon tire. And their flesh was very sweet."

In birds the active phase of the breathing cycle is not in inhaling but exhaling. Their wing strokes compress the rib case to expel the air. Thus instead of "running out of breath" birds "fly into breath."

Probably the fastest metabolizing vertebrate on earth is the tiny Allen's hummingbird [see "The Metabolism of Hummingbirds," by Oliver P. Pearson; SCIENTIFIC AMERICAN, January, 1953]. While hovering it consumes about 80 cubic centimeters of oxygen per gram of body weight per hour. Even at rest its metabolic rate is more than 50 times as fast as man's. Interestingly enough, the hovering hummingbird uses energy at about the same proportionate rate as a hovering helicopter. This does not mean that man has equalled nature in the efficiency of energy yield from fuel. To hover the hummingbird requires a great deal more energy, because of the aerodynamic inefficiency of its small wings and its very high loss of energy as dissipated heat. The tiny wings of a hummingbird impose on the bird an almost incredible expenditure of effort. Its breast muscles are estimated to be approximately four times as large, proportionately, as those of a pigeon. This great muscle burden is one price a hummingbird pays for being small.

A more obvious index of the efficiency of bird's fuel consumption is the high mileage of the golden plover. In the fall the plover fattens itself on bayberries in Labrador and then strikes off across the open ocean on a nonstop flight of 2,400 miles to South America. It arrives there weighing some two ounces less than it did on its departure. This is the equivalent of flying a 1,000-pound airplane 20 miles on a pint of gasoline rather than the usual gallon. Man still has far to go to approach such efficiency.

The Author

CARL WELTY is professor of zoology and chairman of the department of biology at Beloit College in Wisconsin. He was brought up in Fort Wayne, Ind., except for summers spent on a run-down fruit farm in Michigan, which his mother, a widow, had bought to "turn her four boys out to pasture" each summer. These summers left him with a taste for nature, and when he later went to Earlham College, a Quaker school in Indiana, he majored in biology. He took a master's degree at Haverford College in 1925 and spent the following winter in a logging camp in northern Ontario. He remembers spending "11 hours per day logging in the bush and two hours teaching arithmetic and reading to French-Canadians under the auspices of the Frontier College of Toronto. Pay: board, bunk and $25 a month! Hardest work I ever did." From there he went to Parsons College, Iowa, where he taught biology and met his wife, a fellow teacher. She is a freelance writer. Welty has been at Beloit College since 1934. In 1946 he took a year's leave to head a team of American Quakers engaged in postwar relief work in Coblenz, Germany.

Bibliography

THE BIOLOGY OF BIRDS. J. A. Thomson. The Macmillan Company, 1923.

THE BIRD: ITS FORM AND FUNCTION. C. W. Beebe. Henry Holt and Company, 1906.

FUNCTIONAL ANATOMY OF THE VERTEBRATES. D. P. Quiring. McGraw-Hill Book Company, Inc., 1950.

A HISTORY OF BIRDS. W. P. Pycraft. Methuen & Co., 1910.

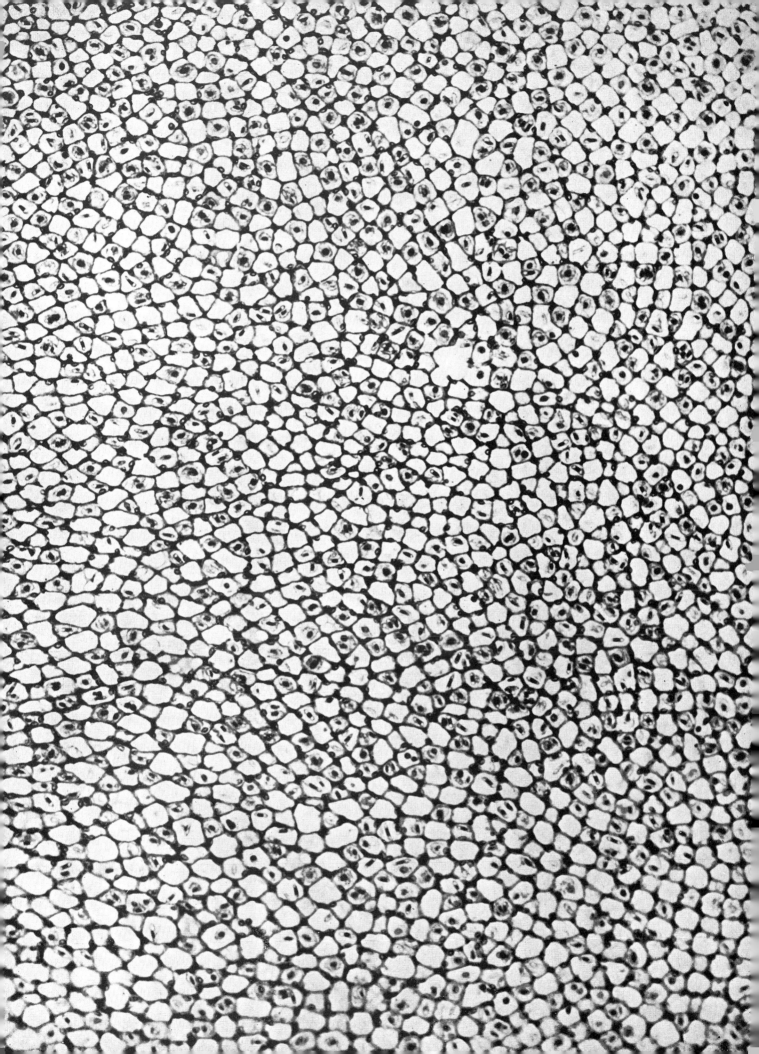

SCIENTIFIC
AMERICAN April 1957, Vol. 196, No. 4, pp. 96-107

OFFPRINT **1117**

"THE WONDERFUL NET"

by P. F. Scholander

These words are a translation of *rete mirabile,* an arrangement
of blood vessels in which animals can conserve heat and oxygen
pressure by applying the principles of counter-current exchange.

A man standing barefoot in a tub of ice water would not survive very long. But a wading bird may stand about in cold water all day, and the whale and the seal swim in the arctic with naked fins and flippers continually bathed in freezing water. These are warm blooded animals, like man, and have to maintain a steady body temperature. How do they avoid losing their body heat through their thinly insulated extremities? The question brings to light a truly remarkable piece of biological engineering. It seems that such animals block the loss of heat by means of an elementary physical mechanism, familiar enough to engineers, which nature puts to use in a most effective way. In fact, the same principle is employed for several very different purposes by many members of the animal kingdom from fishes to man.

The principle is known as counter-current exchange. Consider two pipes lying side by side so that heat is easily transmitted from one to the other. Suppose that fluids at different temperatures start flowing in opposite directions in the two pipes: that is, a cold stream flowing counter to a warm [*see diagram on next page*]. The warmer stream will lose heat to the colder one, and if the transfer is efficient and the pipes long enough, the warm stream will have passed most of its heat to the counter current by the time it leaves the system. In other words, the counter current acts as a barrier to escape of heat in the direction of the warm current's flow. This method of heat exchange is, of course, a common practice in industry: the counter-flow system is used, for example, to tap the heat of exhaust gases from a furnace for preheating the air flowing into the furnace. And the same method apparently serves to conserve body heat for whales, seals, cranes, herons and other animals with chilly extremities.

Claude Bernard, the great 19th-century physiologist, suggested many years ago that veins lying next to arteries in the limbs must take up heat from the arteries, thus intercepting some of the body heat before it reaches the extremities. Recent measurements have proved that there is in fact some artery-to-vein transfer of heat in the human body. But this heat exchange in man is minor compared to that in animals adapted to severe exposure of the extremities. In those animals we find special networks of blood vessels which act as heat traps. This type of network, called *rete mirabile* (wonderful net), is a bundle of small arteries and veins, all mixed together, with the counter-flowing arteries and veins lying next to each other. The retes are generally situated at the places where the trunk of the animal deploys into extremities—limbs, fins, tail and so on. There the retes trap most of the blood heat and return it to the trunk. The blood circulating through the extremities is therefore considerably cooler than in the trunk, but the limbs can function perfectly well at the lower temperature. It has been found that many arctic animals have a leg temperature as low as 50 degrees Fahrenheit or even less.

Anatomists have confirmed that the whale, the seal and the long-legged wading birds possess such retes. However, these networks have also been discovered in the extremities of many tropical animals. It is not surprising to find heat-trapping retes in a water-dwelling animal such as the Florida manatee, for even tropical waters are chilling to a constantly immersed body, but why should the retes appear in tropical land animals like the sloth, the anteater and the armadillo? The answer may be that these animals are hypersensitive to cool air. The sloth, for instance, begins to shiver when the air temperature drops below 80 degrees F. It has to adjust to this situation almost every night, and the retes may well be the means by which it makes the adjustment: that is, the sloth may let its long arms and legs cool to the temperature of the night air, as a reptile does, to preserve its body heat. Recent measurements have shown that it takes a sloth two hours to rewarm a chilled arm from 59 to 77 degrees, whereas an animal without retes, such as a monkey, accomplishes this in 10 minutes.

There is another finding, however, which at first sight is more puzzling. Many animals that spend a great deal of time in cold water or live in the arctic snow seem to lack retes to sidetrack body heat from their poorly insulated legs or feet. Among them are ducks, geese, sea gulls, the fox and the husky (the Eskimo dog). The absence of retes in these animals is not difficult to explain, however, when we consider that all of them are heavily insulated over most of their bodies. Their principal problem lies in getting rid of body heat rather than in conserving it. Consider, for instance, the situation of a husky. It is so well insulated that it can sleep on the snow at 40

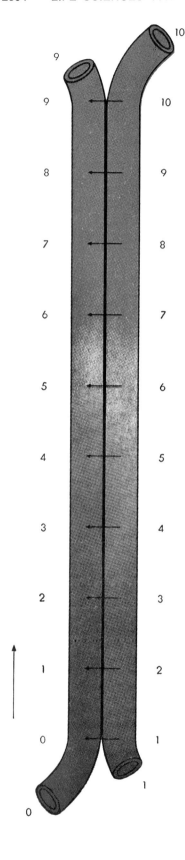

THE PRINCIPLE of counter-current exchange is demonstrated in two pipes lying side by side. Hot water enters one pipe (*top right*); cold water enters the other (*bottom left*). The fluids flow in opposite directions. Under ideal conditions heat will diffuse almost completely from one to the other.

degrees below zero without raising its normal rate of metabolism at rest. When this animal gets up after a cold night and begins to run in the warm sun, increasing its metabolic rate 10- or 20-fold because of the exercise, it is immediately faced with the problem of dissipating a good deal of excess heat. Because of its thick fur covering, it can lose heat only through exposed surfaces such as its tongue, face and legs. An arteriovenous network impeding the transport of heat to its legs would be a severe handicap. The same is true of the duck and other extremely well-insulated birds, which probably depend upon their webbed feet for heat dissipation.

As I mentioned at the beginning, heat conservation is only one of the functions performed by counter-current networks such as I have described. Indeed, there are more dramatic manifestations of this sort of system in the animal world. Nowhere in nature is counter-current exchange more strikingly developed nor more clearly illustrated than in the swim-bladder wall of deep-sea fishes. Here the function of the "wonderful network" is to prevent the loss of oxygen from the fish's air bladder.

A deep-sea fish keeps its swim bladder filled with gas which is more than 90 per cent oxygen. At depths of 9,000 feet or so it must maintain an oxygen pressure amounting to 200 to 300 atmospheres—nearly double the pressure in a fully charged steel oxygen cylinder. On the other hand, the oxygen pressure in the bladder's surroundings—in the fish's bloodstream and in the sea water outside—is no more than a fifth of an atmosphere. So the oxygen pressure difference across the thin swim-bladder wall is some 200 atmospheres. What is more, blood is constantly streaming along this wall through myriads of blood vessels embedded in it. Oxygen from the bladder, under the enormous pressure of 200 atmospheres, must diffuse into these blood vessels. How is it, then, that the streaming blood does not quickly drain the oxygen from the bladder? The answer, of course, is a counter-current exchange system. Very little oxygen escapes from the swim-bladder wall to the rest of the fish's body, because the outgoing veins, highly charged with oxygen, give it up to adjacent incoming arteries. There is a network of thousands of looping capillaries, so closely intermingled that diffusion of oxygen from veins to arteries goes on at a high rate.

What would be the most efficient arrangement of veins and arteries to give the maximum surface for transfer from

one type of vessel to the other? We can treat this as a problem in topology and ask: How can we arrange black and white polygons (representing the cross sections of the blood vessels) so that black always borders white? If we allow only four polygons to meet at each corner, there are two different possible solutions: a checkerboard of squares or a pattern of hexagons with triangles filling the open corners. Under the microscope we observe that evolution has produced precisely these two patterns in the swim-bladder retes of deep-sea fishes [*see photographs and diagrams on opposite page*].

From the number and dimensions of the capillaries, the speed of the blood flow and other information we can calculate the amount of the oxygen-pressure drop across the rete, or, in other words, how effectively the rete traps oxygen. The calculation indicates that across a rete only one centimeter long, the oxygen pressure is reduced by a factor of more than 3,000. That is to say, the leak of oxygen through the rete is insignificant: translating the situation into terms of heat, if boiling water were to enter such a rete from one end and ice water from the other, the exchange of heat would be all but complete—to within one 10,000th of a degree!

To put it another way, the counter-current exchange in the swim-bladder rete is so efficient that in a single pass a rete one centimeter long is capable of raising a given concentration 3,000-fold, which leaves industrial engineering far behind. In speaking of concentration we include several different kinds: heat, gas pressure, the concentration of a solution and so on. A counter-current exchange system can establish a steep gradient in any of these quantities.

It has recently been suggested that counter-current exchange may be involved in the process whereby the kidneys filter the blood and produce urine. During the process of conversion of blood fluid to urine, the concentration of salts and urea in the fluid may be increased three- or four-fold. Just how is this concentration carried out?

The machines that perform the transformation are the units called nephrons, of which a kidney contains several millions. A nephron consists of a glomerulus capsule (a small ball of capillaries), a long, twisting tubule and a collecting duct [*see diagram on page 2808*]. From the blood in the glomerulus capillaries a filtered fluid passes through the capsule wall into the tubule. The fluid travels along the tortuous course of the tubule,

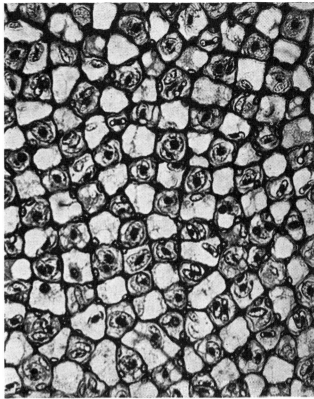

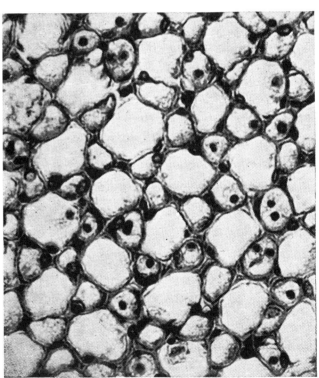

THE APPLICATION of counter-current exchange reaches the ultimate in the swim-bladder retes of deep-sea fishes. Drawings at left show two ways in which veins (*white*) and arteries (*black*) might be arranged to gain the greatest possible area of exchange. One pattern is non-staggered; it gives rise to a checkerboard (*top left*). The other is staggered; it gives rise to stars (*bottom left*). Photomicrographs show the checkerboard in the deep-sea eel (*top right*) and the star pattern in the rosefish (*bottom right*).

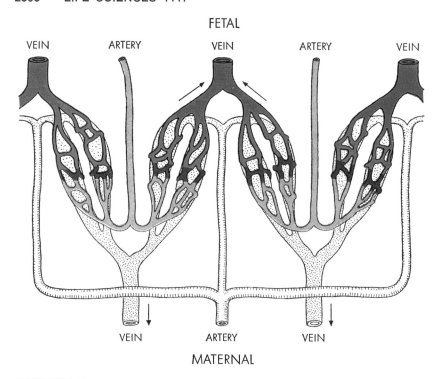

FETAL

VEIN   ARTERY   VEIN   ARTERY   VEIN

VEIN   ARTERY   VEIN

MATERNAL

OXYGEN IS EXCHANGED by counter-current flow in the placenta of the ground squirrel. Oxygen-rich maternal blood (*white*) enters the arteries (*bottom*) and flows counter to the oxygen-poor fetal blood (*gray*). The fetal blood picks up oxygen (*red*) and leaves through the fetal vein (*top*). Oxygen-poor maternal blood (*stippled*) returns to the maternal heart.

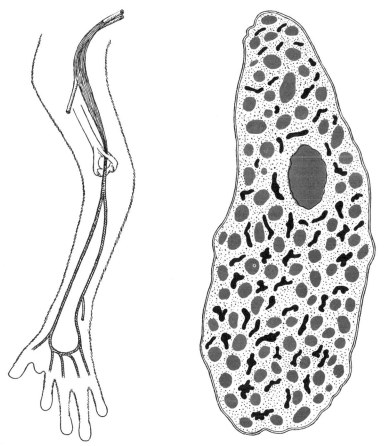

HEAT IS CONSERVED by rete bundles in the loris. They occur where the limbs join the body (*upper left*). Arterial blood enters the upper end of the rete and flows counter to the venous blood. Heat diffuses from the arteries to the veins and is returned to the body. A cross section of the rete (*right*) shows the arrangement of arteries (*red*) and veins (*black*).

doubles back at the loop of Henle [*see diagram*], and by the time it leaves the collecting duct it has become concentrated urine. The conversion does not take place in the glomerulus capsule; it has been established that the fluid emerging from the capsule has essentially the same salt concentration as the blood. The problem is to determine exactly where in the system the change in concentration is produced.

B. Hargitay and Werner Kuhn of the University of Basel recently proposed that it takes place primarily in the loop of Henle. They pointed out that only animals with a Henle loop in the kidney nephron—namely, mammals and birds—can produce a concentrated urine.

The loop of Henle, with its two arms running parallel and fairly close together, is a structure which reminds us of the arteriovenous capillary loops that make up the rete of a fish [*see diagram on page 2807*]. Hargitay and Kuhn reasoned that salts or water might migrate from one arm of Henle's loop to the other, and that as a result salts might be concentrated in the bend of the loop. This part of the loop is situated in an internal structure of the kidney called the papilla, which also contains the collecting ducts and adjacent blood capillaries. The investigators assumed that the fluid concentrated in the loop would be transmitted to the ducts and capillaries and become concentrated urine. To test their idea, they first froze rat kidneys and examined small sections of the tissue under a microscope. The sections of tissue in the papilla, around the loop of Henle, proved to have the same melting point as the rats' urine, while the tissue in the cortex (outer part) of the kidney had the same melting point as frozen blood. Since the melting point depends on the salt concentration this finding tended to confirm the idea that the primary site for the concentration of urine is located in the bend of Henle's loop. But it was possible that the brutal freezing process had damaged the tissues so that urine escaped from the collecting ducts and diffused to the area around the loop. To make a clearer test, H. Wirz, a colleague of Hargitay and Kuhn, developed a technique for drawing samples of blood from the capillaries around Henle's loop. This blood proved to have the same melting point as the urine, *i.e.*, its salt concentration was as much as three times higher than that of blood in other parts of the animal's body.

Thus Hargitay and Kuhn seem to have strong support for their thesis that a counter-current exchange in Henle's loop plays a part in the formation of urine by the kidney. But the question still needs further research.

Various other examples of counter-current exchange in animals have been discovered. One of them concerns the breathing of fishes. A fish requires a far more efficient and resourceful breathing apparatus than an animal that lives in the air. Each quart of air contains some 200 cubic centimeters of oxygen, but a quart of sea water has only about five cubic centimeters, and oxygen diffuses through water slowly. The fish therefore has to be remarkably efficient in extracting oxygen from the water that flows over its gills. It can, in fact, take up as much as 80 per cent of the oxygen in the water. Anatomical studies and experiments have proved that fishes employ a counter-current system in this process. The blood in the capillaries of the fish's gill plates flows in the direction counter to the flow of water over its gills. When experimenters reversed the direction of the water current over the gills, fishes extracted only one fifth as much oxygen as they did normally!

In many species of animals the mother and the fetus she is carrying share their blood substances by means of a counter-current exchange system. This apparently is not true of the human animal, for the fetus's capillaries are bathed directly by the mother's blood. But in the rabbit, the sheep, the squirrel, the cow, the cat, the dog and other animals, the mother's blood vessels are intermingled with the fetus's in the placenta, and by counter-current flow they exchange oxygen, nutrients, heat and wastes.

In sum, the principle of counter-current exchange is employed in many and various ways in the world of living

THE WRECKFISH swims at great depths and must therefore keep its swim bladder filled with oxygen at tremendous pressures. It does this by means of counter-current bundles (*solid red*) in the swim-bladder wall (*first and second drawings*). One of these bundles (*third drawing*) and a single counter-current capillary (*fourth drawing*) are schematically depicted. Very little oxygen (*red shading*) escapes beyond the swimbladder wall because it diffuses (*small arrows*) from the outgoing vein into the adjacent incoming artery. The diagram (*bottom*) represents build-up of pressure (*P*) by means of a pressure difference (*p*) between the counter-current artery and vein.

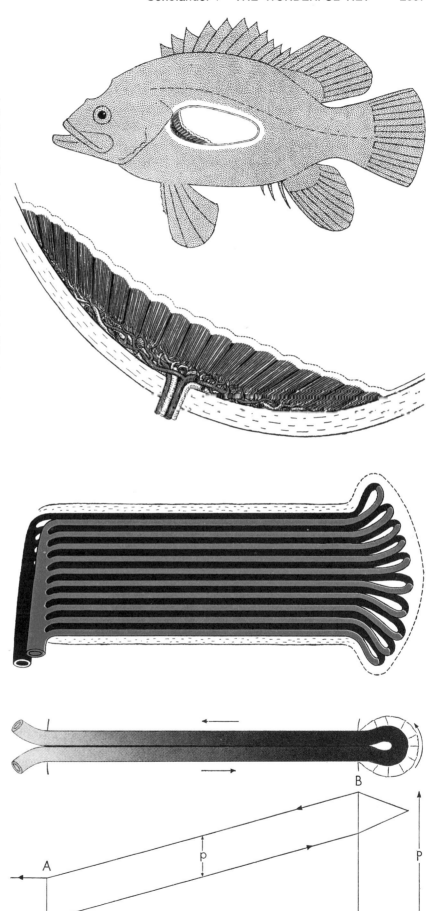

RETE GRADIENTS

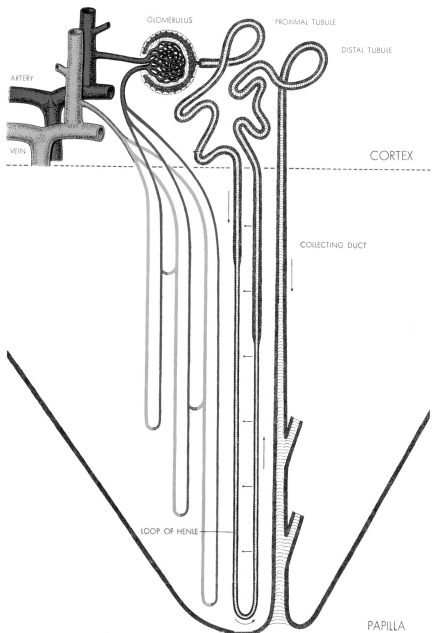

GLOMERULUS   PROXIMAL TUBULE

DISTAL TUBULE

ARTERY

VEIN

CORTEX

COLLECTING DUCT

LOOP OF HENLE

PAPILLA

things. We cannot fail to be impressed by the marvels of bio-engineering that nature has achieved in its development of "the wonderful net."

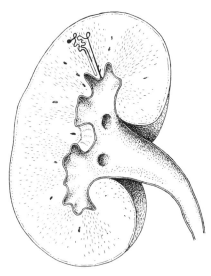

THE NEPHRON filters urea and other dissolved substances out of the blood and concentrates them in the urine. It may accomplish this with counter-current exchange. The drawing above is a cross section of the human kidney. At its upper left is a schematic representation of one of the kidney's several million nephrons. In the drawing at left the nephron is enlarged. Arterial blood (*red*) is depicted entering the capillaries of the glomerulus. The dissolved substances filter out of the glomerulus into the proximal tubule. The solution then travels down the tubule and doubles back at the loop of Henle. If the dissolved substances diffused from the ascending tubule into the descending tubule, the urine would be concentrated (*yellow shading*) near the bend of the loop. The concentration might then be transmitted to the portion of the collecting ducts located in the papilla.

The Author

P. F. SCHOLANDER is on the staff of the Scripps Institution of Oceanography. Scholander was born in Orebro, a town in Sweden, and later moved to Norway. As a student at the University of Oslo, he says, "I found an intellectual outlet from medical studies in music and lichenology, which took me on expeditions to Greenland and Spitsbergen. In 1932 I undeservedly received my degree in medicine, and in 1934 a Ph.D. in botany was bestowed upon me. I then turned to the study of diving in seals, whales and penguins, and have worked in comparative physiology ever since." Scholander came to the U.S. in 1939 as a Rockefeller Fellow to study diving with Laurence Irving of Swarthmore College. During World War II he was an aviation physiologist for the U.S. Air Force and became a naturalized U.S. citizen. He later studied climatic adaptation in the Arctic Research Laboratory at Point Barrow in Alaska, did research on cell respiration at the Harvard Medical School and worked at the Woods Hole Oceanographic Institution.

Bibliography

COUNTER-CURRENT VASCULAR HEAT EXCHANGE IN THE FINS OF WHALES. P. F. Scholander and William E. Schevill in *Journal of Applied Physiology*, Vol. 8, No. 3, pages 279-282; November, 1955.

EVOLUTION OF CLIMATIC ADAPTATION IN HOMEOTHERMS. P. F. Scholander in *Evolution*, Vol. 9, No. 1, pages 15-26; March, 1955.

THE RABBIT PLACENTA AND THE PROBLEM OF PLACENTAL TRANSMISSION. Harland W. Mossman in *The American Journal of Anatomy*, Vol. 37, No. 3, pages 433-497; July, 1926.

SECRETION OF GASES AGAINST HIGH PRESSURES IN THE SWIMBLADDER OF DEEP SEA FISHES. II: THE RETE MIRABILE. P. F. Scholander in *The Biological Bulletin*, Vol. 107, No. 2, pages 260-277; October, 1954.

TEMPERATURE OF SKIN IN THE ARCTIC AS A REGULATOR OF HEAT. Laurence Irving and John Krog in *Journal of Applied Physiology*, Vol. 7, No. 4, pages 355-364; January, 1955.

SCIENTIFIC
AMERICAN   January 1959, Vol. 200, No. 1, pp. 109-116   OFFPRINT **1118**

SALT GLANDS

by Knut Schmidt-Nielsen

A special organ which eliminates salt with great efficiency
enables marine birds to meet their fluid needs by drinking
sea water. Similar organs have been found in marine reptiles.

As the writers of stories about castaways are apt to point out, a man who drinks sea water will only intensify his thirst. He must excrete the salt contained in the water through his kidneys, and this process requires additional water which is taken from the fluids of his body. The dehydration is aggravated by the fact that sea water, in addition to common salt or sodium chloride, also contains magnesium sul-

fate, which causes diarrhea. Most air-breathing vertebrates are similarly unable to tolerate the drinking of sea water, but some are not so restricted. Many birds, mammals and reptiles whose ancestors dwelt on land now live on or in the sea, often hundreds of miles from any source of fresh water. Some, like the sea turtles, gulls and albatrosses, return to the land only to reproduce. Whales, sea cows and some sea snakes, which bear

living young in the water, have given up the land entirely.

Yet all these animals, like man, must limit the concentration of salt in their blood and body fluids to about 1 per cent—less than a third of the salt concentration in sea water. If they drink sea water, they must somehow get rid of this excess salt. Our castaway can do so only at the price of dehydrating his tissues. Since his kidneys can at best se-

PETREL EJECTS DROPLETS of solution produced by its salt gland through a pair of tubes atop its beak, as shown in this high-speed photograph. The salt-gland secretions of most birds drip from the tip of the beak. The petrel, however, remains in the air almost continuously and has apparently evolved this "water pistol" mechanism as a means of eliminating the fluid while in flight.

crete a 2-per-cent salt solution, he must eliminate up to a quart and a half of urine for every quart of sea water he drinks, with his body fluids making up the difference. If other animals drink sea water, how do they escape dehydration? If they do not drink sea water, where do they obtain the water which their bodies require?

The elimination of salt by sea birds and marine reptiles poses these questions in particularly troublesome form. Their kidneys are far less efficient than our own: a gull would have to produce more than two quarts of urine to dispose of the salt in a quart of sea water. Yet many observers have seen marine birds drinking from the ocean. Physiologists have held that the appearance of drinking is no proof that the birds actually swallow water, and that the low efficiency of their kidneys proves that they do not. Our experiments during the past two years have shown that while the physiologists are right about the kidneys, the observations of drinking are also correct. Marine birds do drink sea water. Their main salt-eliminating organ is not the kidney, however, but a special gland in the head which disposes of salt more rapidly than any kidney does. Our studies indicate that all marine birds and probably all marine reptiles possess this gland.

The obvious way to find out whether birds can tolerate sea water is to make them drink it. If gulls in captivity are given only sea water, they will drink it without ill effects. To measure the exact amount of sea water ingested we administered it through a stomach tube, and found that the birds could tolerate large quantities. Their output of urine increased sharply but accounted for only a small part of the salt they had ingested. Most of the salt showed up in a clear, colorless fluid which dripped from the tip of the beak. In seeking the source of this fluid our attention was drawn to the so-called nasal glands, paired structures of hitherto unknown function found in the heads of all birds. Anatomists described these organs more than a century ago, and noted that they are much larger in sea birds than in land birds. The difference in size suggested that the glands must perform some special function in marine species. Some investigators proposed that the organs produce a secretion akin to tears which serves to rinse sea water from the birds' sensitive nasal membranes.

We were able to collect samples of the secretion from the gland by inserting a thin tube into its duct. The fluid turned out to be an almost pure 5-per-cent solution of sodium chloride—many times saltier than tears and nearly twice as salty as sea water. The gland, it was plain, had nothing to do with rinsing the nasal membranes but a great deal to do with eliminating salt. By sampling the output of other glands in the bird's head, we established that the nasal gland was the only one that produces this concentrated solution.

The nasal glands can handle relatively enormous quantities of salt. In one experiment we gave a gull 134 cubic centimeters of sea water—equal to about a tenth of the gull's body weight. In man this would correspond to about two gallons. No man could tolerate this much sea water; he would sicken after drinking a small fraction of it. The gull, however, seemed unaffected; within three hours it had excreted nearly all the salt. Its salt glands had produced only about two thirds as much fluid as its kidneys, but had excreted more than 90 per cent of the salt.

The fluid produced by the salt gland is about five times as salty as the bird's blood and other body fluids. How does the organ manage to produce so concentrated a solution? Microscopic examination of the gland reveals that it consists of many parallel cylindrical lobes, each composed of several thousand branching tubules radiating from a central duct like bristles from a bottle brush. These tubules, about a thousandth of an inch in diameter, secrete the salty fluid.

A network of capillaries carries the blood parallel to the flow of salt solution in the tubules, but in the opposite direction [*see illustration on opposite page*]. This arrangement brings into play the principle of counter-current flow, which seems to amplify the transfer of salt from the blood in the capillaries to the fluid in the tubules. A similar arrangement in the kidneys of mammals appears to account for their efficiency in the concentration of urine [see "The Wonderful Net," by P. F. Scholander; SCIENTIFIC AMERICAN Offprint 1117]. No such provision for counter-current flow is found in the kidneys of reptiles, and it is only slightly developed in birds.

Counter-current flow, however, does not of itself account for the gland's capacity to concentrate salt. The secret of this process lies in the structure of the tubules and the cells that compose them.

The microscopic structure of a salt-gland tubule resembles a stack of pies with a small hole in the middle. Each "pie" consists of five to seven individual

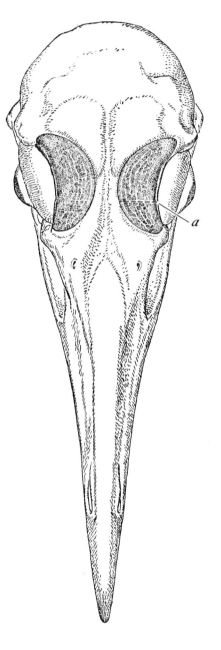

STRUCTURE of salt gland is essentially the same in all sea birds. In the gull the glands lie above the bird's eyes, as shown at left. Cross section of a gland (*a*) shows that it consists of many lobes (*b*). Each of these

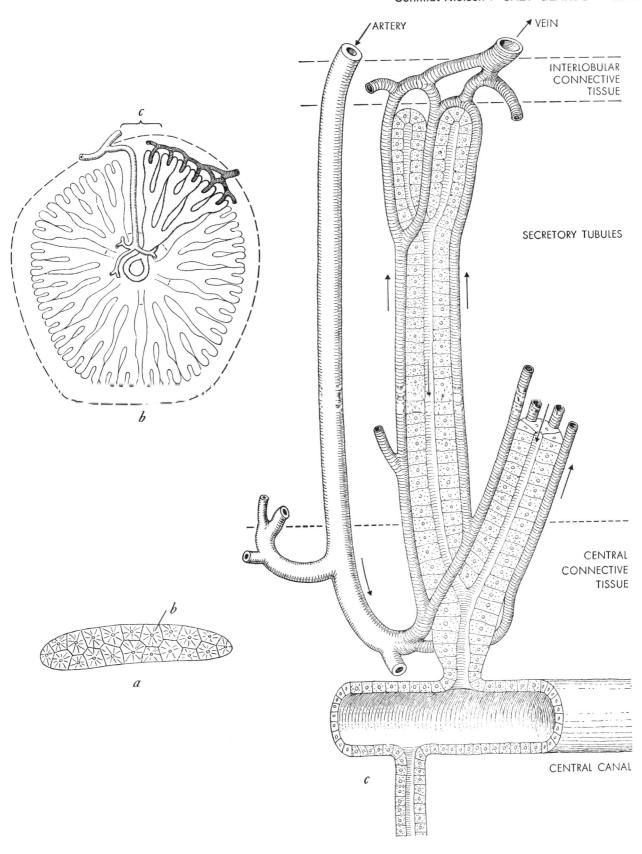

ARTERY

VEIN

INTERLOBULAR
CONNECTIVE
TISSUE

SECRETORY TUBULES

CENTRAL
CONNECTIVE
TISSUE

CENTRAL CANAL

c

b

b

a

c

lobes contains several thousand branching tubules which radiate from a central duct like the hairs of a bottle brush. Enlargement of a single tubule (*c*) reveals that it is surrounded by capillaries in which blood flows counter to the flow of salt secretion in the tubule. This counter-current flow, which also occurs in the kidneys of mammals, facilitates the transfer of salt from the blood to the tubule. The tubule wall, only one cell thick, consists of rings of five to seven wedge-shaped cells. These rings, stacked one on top of another, encircle a small hole, or lumen, through which the salty secretion flows from the tubule into the central canal of the lobe.

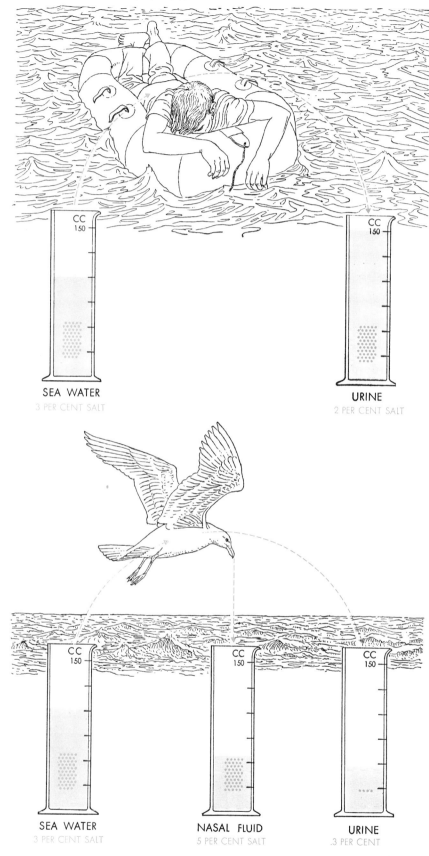

cells arranged like wedges. The hole, or lumen, funnels the secretion into the central duct. When we inject dye into the lumen, colored fluid seeps out into a system of irregular crevices in the walls of the tubule. More detailed examination with the electron microscope reveals a similar, interlocking system of deep folds which extend inward from the outer surface of the tubule. This structure may be important in that it greatly multiplies the surface area of the cell. It is worth noting that cells with similar, though shallower, folds are found in the tubules of the mammalian kidney.

Evidently some physiological mechanism in the cell "pumps" sodium and chloride ions against the osmotic gradient, from the dilute salt solution of the blood to the more concentrated solution in the lumen. Nerve cells similarly "pump" out the sodium which they absorb when stimulated [see "The Nerve Impulse and the Squid," by Richard D. Keynes; SCIENTIFIC AMERICAN Offprint 58]. Of course the mechanisms in the two processes may be quite different. In the tubule cells the transport of sodium and chloride ions seems to involve the mitochondria, the intracellular particles in which carbohydrates are oxidized to produce energy.

The similarities between the salt gland and the mammalian kidney should not obscure their important differences. For one thing, the salt gland is essentially a much simpler organ. The composition of its secretions, which apart from a trace of potassium contain only sodium chloride and water, indicates that its sole function is to eliminate salt. In contrast, the kidney performs a variety of regulatory and eliminative tasks and produces a fluid of complex and variable composition, depending on the animal's physiological needs at a particular time.

The salt gland's distinctive structure, elegantly specialized to a single end, enables it to perform an almost unbelievable amount of osmotic work in a short time. In one minute it can produce up to half its own weight of concentrated salt solution. The human kidney can produce at most about a twentieth of its weight in urine per minute, and its normal output is much less.

Another major difference between the two glands is that the salt gland functions only intermittently, in response to the need to eliminate salt. The kidney, on the other hand, secretes continuously, though at a varying rate. The salt gland's activity depends on the concentration of salt in the blood. The injection of salt solutions into a bird's bloodstream causes

SALT EXCRETION IN MEN AND BIRDS is compared in these drawings. Castaway at top cannot drink sea water because in eliminating the salt it contains (*colored dots*) he will lose more water than he has drunk. His kidney secretions have a salt content lower than that of sea water. Gull (*below*) can drink sea water even though its kidneys are far less efficient than a man's. It eliminates salt mainly through its salt, or "nasal," glands. These organs, more efficient than any kidney, secrete a fluid which is nearly twice as salty as sea water.

the gland to secrete, indicating that some center, probably in the brain, responds to the salt concentration. The gland responds to impulses in a branch of the facial nerve, for electric stimulation of this nerve causes the gland to secrete.

While the structure and function of the salt gland is essentially the same in all sea birds, its location varies. In the gull and many other birds the glands are located on top of the head above the eye sockets [*see illustrations on this page*]; in the cormorant and the gannet they lie between the eye and the nasal cavity. The duct of the gland in either case opens into the nasal cavity. The salty fluid flows out through the nostrils of most species and drips from the tip of the beak, but there are some interesting variations on this general scheme. The pelican, for example, has a pair of grooves in its long upper beak which lead the fluid down to the tip; the solution would otherwise trickle into the pouch of the lower beak and be reingested. In the cormorant and the gannet the nostrils are nonfunctional and covered with skin; the fluid makes its exit through the internal nostrils in the roof of the mouth and flows to the tip of the beak.

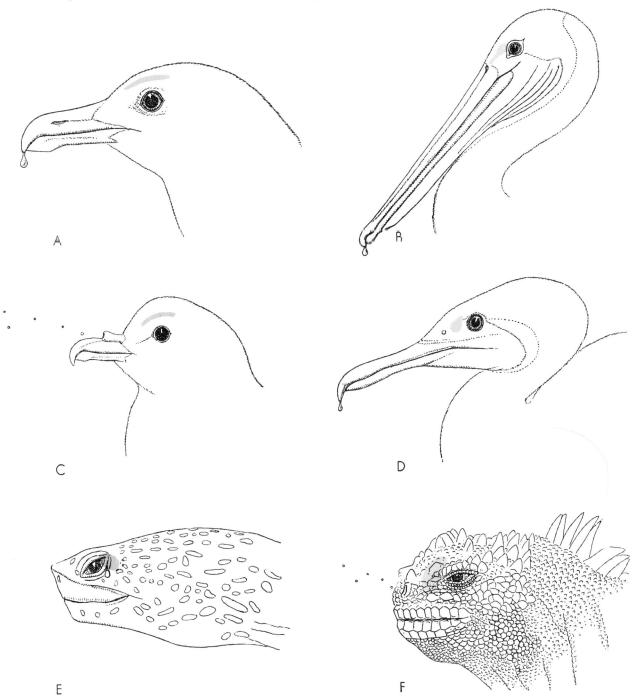

LOCATION OF SALT GLAND (*color*) varies in different species of marine birds and reptiles. In the gull (*A*) the gland's secretions emerge from the nostril and drip from the beak; in the cormorant (*D*) the fluid flows along the roof of the mouth. The pelican (*B*) has grooves along its upper beak which keep the fluid from dripping into its pouch; the petrel (*C*) ejects the fluid through tubular nostrils. In the turtle (*E*) the gland opens at the back corner of the eye; in the marine iguana (*F*) it opens into the nasal cavity.

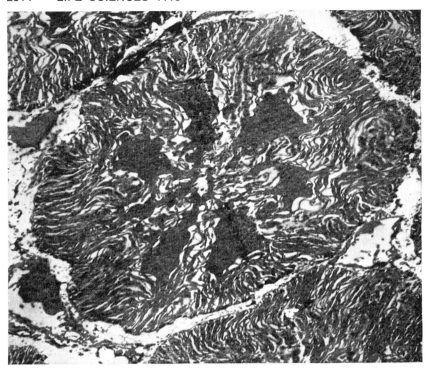

CROSS SECTION OF SALT-GLAND TUBULE is shown magnified about 5,700 diameters in this electron micrograph made by William L. Doyle of the University of Chicago. To emphasize the cell-structure the specimen was kept in a solution which shrank and distorted the cells and their nuclei. Most of the material of the cells lies in folded, leaflike layers; cells with a somewhat similar structure are found in the kidney tubules of mammals.

The petrel displays an especially interesting mechanism for getting rid of the fluid. Its nostrils are extended in two short tubes along the top of its beak. When its salt glands are working, the bird shoots droplets of the fluid outward through the tubes [see illustration on page 2809]. This curious design may reflect a special adaptation to the petrel's mode of life. Though the bird remains at sea for months at a time, it rarely settles down on the water to rest. Presumably the airstream from its almost continuous flight would hamper the elimination of fluid from the bird's nostrils, were it not for the water-pistol function of the tubes.

Our studies so far have demonstrated the existence of the salt gland in the herring gull, black-backed gull, common tern, black skimmer, guillemot, Louisiana heron, little blue heron, double-crested cormorant, brown pelican, gannet, petrel, albatross, eider duck and Humboldt penguin. These species, from a wide variety of geographical locations, represent all the major orders of marine birds. There is little doubt that this remarkable organ makes it possible for all sea birds to eliminate salt and live without fresh water.

The discovery of the salt gland in sea birds prompted us to look for a similar organ in other air-breathing sea animals.

In *Alice's Adventures in Wonderland* the Mock Turtle weeps perpetual tears because he is not a real turtle; real turtles, at least the marine species, also weep after a fashion. A. F. Carr, Jr., a distinguished specialist in marine turtles, gives us a vivid account of a Pacific Ridley turtle that came ashore to lay its eggs. The animal "began secreting copious tears shortly after she left the water, and these continued to flow after the nest was dug. By the time she had begun to lay, her eyes were closed and plastered over with tear-soaked sand and the effect was doleful in the extreme." Thus Carr makes it clear that the turtle's tears do not serve to wash its eyes free of sand, an explanation that otherwise might seem reasonable. The suggestion that the turtle weeps from the pangs of egg-laying is even wider of the mark.

With the loggerhead turtle as our subject, we have found that the sea turtle's tears come from a large gland behind its eyeball. The tears have much the same composition as that of the salt-gland secretions of the sea bird. Thus it would seem more than likely that the turtle's "weeping" serves to eliminate salt. The salt gland of the turtle has a structure similar to that of the gland in sea birds, with tubules radiating from a central duct, and it seems that this structure is essential for the elaboration of a fluid with a high salt concentration. The similarity is the more striking because the location of the gland in the turtle indicates that it has a different evolutionary origin. Still a third independent line of evolution may be represented by the salt gland in the Galápagos marine iguana, the only true marine lizard.

Anatomical studies of the other marine reptiles—the sea snakes and the marine crocodiles—have established that their heads contain large glands whose function may be similar to that of the salt gland. When we succeed in obtaining living specimens of these creatures, we expect to determine whether their glands have the same function.

Investigations of marine mammals thus far indicate that these animals handle the elimination of salt from their systems in a more conventional manner. The seal and some whales apparently satisfy their need for water with the fluids of the fish on which they feed. The elimination of such salt as these fluids contain requires kidneys of no more than human efficiency. But other whales, and walruses, whose diet of squid, plankton or shellfish is no less salty than sea water, must surely eliminate large quantities of excess salt even if they do not drink from the ocean itself. Our knowledge of their physiology suggests that their kidneys, which are more powerful than ours, can eliminate all the salts in their food. Some mammalian kidneys do function at this high level. The kangaroo rat, whose desert habitat compels it to conserve water to the utmost, can produce urine twice as salty as the ocean, and thrives in the laboratory on a diet of sea water and dried soybeans [see "The Desert Rat," by Knut and Bodil Schmidt-Nielsen; SCIENTIFIC AMERICAN Offprint 1050].

We should like to study salt excretion in whales, but these animals are obviously not easy to work with. We have undertaken, however, some pilot studies on seals. When we injected them with salt solutions that stimulate the salt glands of birds and reptiles, they merely increased their output of urine. Methacholine, a drug which also stimulates the salt gland, gave equally negative results. Whatever the seal's need to eliminate salt, its kidneys are evidently adequate to the task. We must therefore assume that the salt gland has evolved only in the birds and reptiles, animals whose kidneys cannot produce concentrated salt solutions.

The Author

KNUT SCHMIDT-NIELSEN has long studied the water metabolism of animals (see "The Desert Rat," by Knut and Bodil Schmidt-Nielsen; SCIENTIFIC AMERICAN, July, 1953). In addition to rats and sea-water-drinking birds, his recent subjects have included camels, whose physiology he studied in the Sahara Desert. Born in Norway, he studied at the University of Oslo, then went to Denmark to work at the University of Copenhagen under the well-known physiologist August Krogh (his wife and collaborator is Krogh's daughter). At Krogh's suggestion Schmidt-Nielsen augmented his training at the Carlsberg Laboratory, where he learned microchemical methods for the analysis of tiny samples of biological material. Since receiving the degree of Doctor of Philosophy at Copenhagen in 1946, he has worked at Swarthmore College, the University of Oslo, Stanford University, the University of Cincinnati and Duke University, where he is now professor in the department of zoology.

Bibliography

THE SALT GLAND OF THE HERRING GULL. Ragner Fänge, Knut Schmidt-Nielsen and Humio Osaki in *Biological Bulletin*, Vol. 115, pages 162-171; October, 1958.

SALT GLANDS IN MARINE REPTILES. Knut Schmidt-Nielsen and Ragnar Fänge in *Nature*, Vol. 182, No. 783-785; September 20, 1958.

SCIENTIFIC
AMERICAN April 1959, Vol. 200, No. 4, pp. 105-120

OFFPRINT **1119**

HOW REPTILES REGULATE THEIR BODY TEMPERATURE

by Charles B. Bogert

Although they lack internal controls, they can maintain a high temperature by their behavior. A lizard, for example, can raise its temperature by changing the position of its body in the sun.

More than 50 years ago Sir Charles Martin, a distinguished British physiologist, compared the regulation of body temperature in a number of mammals with that of a lizard. He showed that the mammals were able to maintain their temperatures within a fairly narrow range during wide variations in the temperature of the laboratory environment. The temperature of the lizard, on the other hand, rose and fell almost as rapidly as that of the environment. Observations of this sort long ago established the textbook aphorism to the effect that "reptiles have the temperature of the surrounding atmosphere."

It is true that reptiles are "cold-blooded" animals and have no mechanism of temperature regulation such as that of

mammals. The laboratory observations correctly reflect what happens to a lizard's body temperature when the laboratory temperature is changed. But the conclusion drawn from these observations holds true only for the lizard in the laboratory. In their natural habitats during the day, lizards forage, mate, defend territories and flee at body temperatures that may be even higher than our own, and they maintain their temperatures within narrow limits despite wide variation in air temperature. The greater earless lizard (*Holbrookia texana*), an inhabitant of the foothills of the U. S. Southwest, has a mean temperature of 101.3 degrees Fahrenheit (38.5 degrees centigrade), slightly above our own, and while the lizard is active its temperature is within 3.3 de-

grees of this level 75 per cent of the time. At 14,600 feet in the Peruvian Andes, with the temperature of the thin air at the freezing point, Oliver P. Pearson of the University of California found that the lizard *Liolaemus multiformis* had a body temperature of 87.8 degrees F. (31 degrees C.); at temperatures as much as eight degrees below freezing he found other lizards abroad, a trifle sluggish, with body temperatures of 58 degrees (14.4 degrees C.), or 34 degrees above the temperature of the air. In my own studies over the past 14 years I have measured the temperatures of lizards of many different North American species. I have found that members of the same or closely related species show the same high and constant temperature in widely different

CHUCKWALLA (*Sauromalus obesus*) is found in deserts of the southwestern states. When this lizard cannot take shelter from the sun, it pants to cool itself by the evaporation of moisture from its lungs and places itself parallel to the sun's rays in order to expose a minimum of body surface to the radiant heat. This animal was photographed at three o'clock in the afternoon on a clear day.

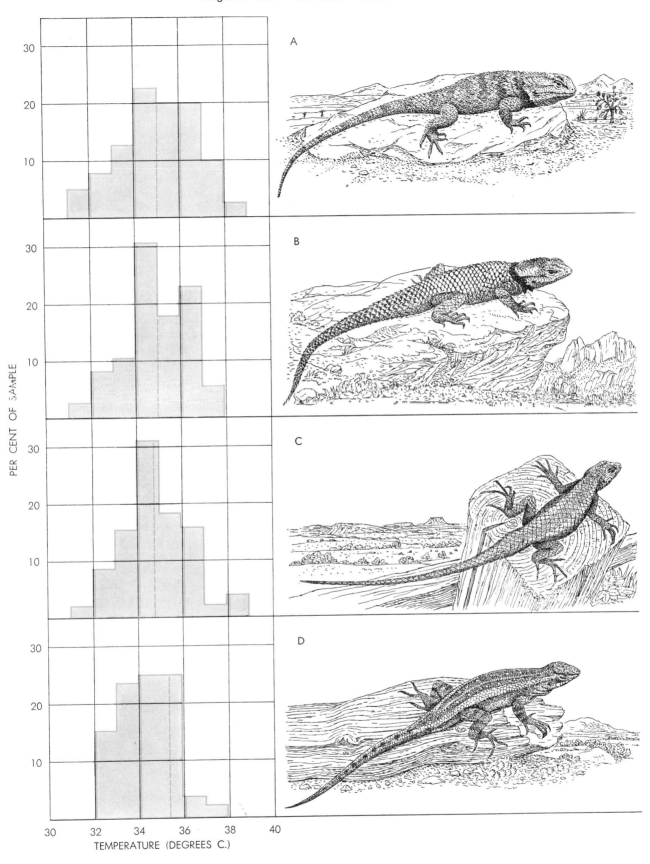

TEMPERATURES OF SPINY LIZARDS of four species were tested by the author, who sampled 39 or more active animals of each kind in their native habitats. In each species the average temperature (*broken line*) fell close to 35 degrees C. (95 degrees F.). The lizards tested were *Sceloporus magister* (A), *Sceloporus jarrovi* (B), *Sceloporus undulatus* (C), *Sceloporus variabilis* (D).

environments, although members of distantly related species maintain distinctly different mean temperatures in the same environment.

Why were the laboratory observations so misleading? For much the same reason that a man with a heavy iron ball chained to one leg cannot demonstrate how fast he can run! The analogy is to the point, because lizards regulate their temperatures to a large degree by their behavior. Many of these creatures are "heliotherms," deriving the heat they need to energize their body chemistry directly from the sun. In consequence of this dependence they have developed basking to a fine art. Lizards do not merely crawl out of their nocturnal shelter and rest in the sunlight. When their temperatures are below the threshold for normal activity, they orient their bodies at right angles to the sun's rays in order to maximize their exposure and even seek inclined surfaces to achieve such orientation with respect to the slanting rays of the morning sun. In the desert, where the ground becomes warmer than the air, lizards often press their bodies close to the surface, shifting slowly from side to side in the loose sand to secure better conduction of heat. On a rocky mountainside that warms up more slowly, they do their basking on mats of dead grass that insulate them from the cold ground. When a lizard's temperature approaches the upper limit of tolerance, on the other hand, it faces the sun, exposing the least possible surface, or returns to cooler temperatures in the shade or underground.

The size and shape of their bodies and the pigmentation of their skins play a part in determining and regulating the rate at which lizards absorb heat from their surroundings. But the decisive factor is behavior. In the artificial situation of the laboratory the lizard could not show what it can do.

When our department at the American Museum of Natural History and R. B. Cowles of the University of California at Los Angeles undertook to study heat regulation in reptiles, we had a general understanding of the factors involved. We expected to find, however, that temperature tolerances played an important part in determining the distribution of species and that the various species would show different optimal body temperatures in their various environments. In framing these assumptions, it turned out, we underestimated the efficiency of our subjects' heat-regulating behavior and equipment.

We chose the spiny lizards as our subjects, because their 50 species are abundant all over North America, from coast to coast and from southern Canada to Panama. Few groups of lizards have penetrated more environments. As many as five species may occur in a single locality, each in its own ecological niche. Some live on the ground, others on trees or shrubs; they variously frequent rocky hillsides, canyon walls, sand dunes, grassy plains and even human habitations. We sought them out in coastal areas, foothills, plateaus and mountains; in arid regions with little vegetation; in pine barrens, short-tree forests, pine forests and high-altitude cloud forests. We made our measurements in habitats ranging from sea level to near the timber line at 12,500 feet on the Nevado de Toluca in Mexico.

In all of this diversity of habitat, to our surprise, measurements indicate that spiny lizards go about their active lives at a mean body temperature of about 93 degrees (34 degrees C.). This approximates the average for all species, but does not imply that spiny lizards

have no leeway in temperature. They function at apparently full efficiency with body temperatures between 86 and 104 degrees (30 and 40 degrees C.). In their natural environments, however, these extremes are exceptional. Once their basking has brought them to the temperature threshold at which activity begins, spiny lizards maintain their body temperatures within 4.5 degrees of the 93-degree mean for about 80 per cent of the time and over the entire range of environmental temperature to which they voluntarily subject themselves during their daily routine.

These lizards regulate their heat intake largely by exposing themselves to direct sunlight, prolonging their forays by suitably orienting their bodies to the sun much as they do when basking, or by retreating from the sunlight when their temperatures run high. Comparisons of the temperature curves of various spiny lizards reveal only minor differences between species. Most of the peaks fall near the 93-degree average for the group. For species living in the tropical lowlands of Mexico and Honduras,

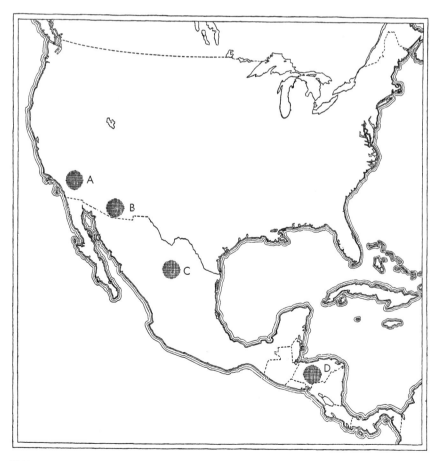

HABITATS of the lizards shown on the opposite page are marked with the corresponding letters on this map. They are (A) the Mojave Desert of California, elevation 3,000 to 4,000 feet; (B) the Chiricahua Mountains of Arizona, elevation 5,000 to 9,000 feet; (C) La Goma, Durango, Mexico, elevation 4,000 feet; (D) El Zamorano, Honduras, 2,600 to 3,000 feet.

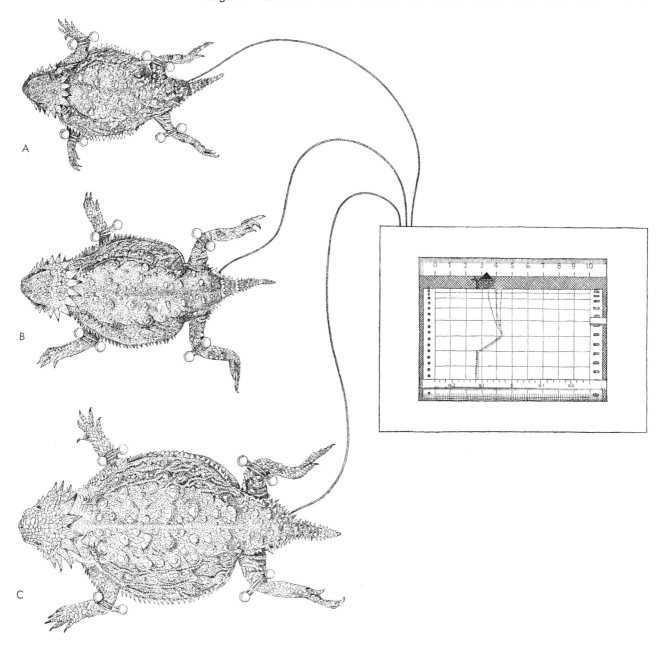

REGAL HORNED LIZARDS (*Phrynosoma solare*) weighing respectively 12.4 grams (A), 29.4 grams (B) and 85.5 grams (C) were exposed to the midday sun while their body temperatures were continuously measured by means of wires from the cloaca of each to a recording potentiometer. The colors of the lizards were nearly the same at start but then changed, smallest lizard becoming palest.

the curves shift scarcely three or four degrees toward the warmer end of the scale. For those in higher mountains, the shift is two or three degrees toward the lower end of the scale.

We do not know whether such slight differences result from variations in the physiological adaptation of the animals or from the limitations of behavioral regulation. We sometimes lose sight of the fact that temperature refers to the amount of heat per unit of mass, or the degree of heat concentration, as measured on one scale or another. Optimum heat concentrations for biological processes, however, lie not at points but within zones on the scale. The "normal temperature" of human beings is actually an average for a zone between limits set roughly at 98 to 99.5 degrees (37 to 37.5 degrees C.). Hence the differences observed between the mean temperatures of the spiny lizard in different environments may still lie within the zone of optimum temperature for their biological activity.

In the Arizona desert the air temperatures recorded at sites of capture averaged 90 degrees (32.2 degrees C.); the temperature of the spiny lizards averaged 95 degrees (35 degrees C.). In contrast, the average body temperature of cloud-forest lizards is 91 degrees (32.8 degrees C.) when the air around them averages 66 degrees. The greater differential between air and body temperature in the case of the cloud-forest lizard should accelerate the loss of heat to the air. Nevertheless the lizards absorb enough solar heat to compensate for these losses and thus keep their body temperature within the zone permitting them to be active.

So, we learned, spiny lizards do not have different body temperatures reflecting different physiological adaptation to different environmental temperatures. Instead these reptiles restrict the fluctuation of their body temperature to a relatively narrow zone suitable to the similar physiological needs of all species. Probably the optimum zone for activity in spiny lizards is an ancestral trait. If so, their physiological adaptation became stabilized for a narrow zone centering around 93 degrees early in the evolution of the stock. The subsequent diversification and dispersal of the stock is largely a history of changes in such characteristics as size, shape, pigmentation and basking behavior, as species after species became adapted for survival under various combinations of environmental conditions.

From one standpoint the evolution of the spiny lizard, as a minor current in the broad stream of evolution, followed the course of least resistance. Once the delicate equilibrium of its physiological processes had been established in a particular zone of temperature, any adaptation to lower internal temperatures would have entailed revision of the whole complex system. The lizard's behavior, on the other hand, provided the necessary leeway to permit it to invade different environments successfully. There, under the pressure of selection, the various species developed adaptations in pigmentation and other physical characteristics associated with the regulation of body heat.

Spiny lizards are not the only lizards whose normal physiological activity is restricted to a narrow zone of temperature characteristic of the group as a whole. Other groups exhibit a similar identity. The whip-tailed lizards (*Cnemidophorus*), with species widely distributed in North, Central and South America, are active at body temperatures in a zone higher on the scale. They maintain a mean temperature close to 104 or 106 degrees (40 or 41 degrees C.), with no significant difference between the temperatures of populations in climates and landscapes as widely different as Arizona, Florida and Honduras.

Whip-tailed and spiny lizards often occur side by side in the same habitat. Invariably they show an average difference of 10 or 12 degrees between the means of their respective ranges in temperature, even though the identical sources of external heat are available to both. This alone is evidence of the effectiveness of behavioral control of heat intake and dissipation, augmented by pigmentation and enhanced by adaptation in structure.

This interaction of structure and behavior makes it difficult to design experiments testing one or another of the reptile's temperature-regulating at-

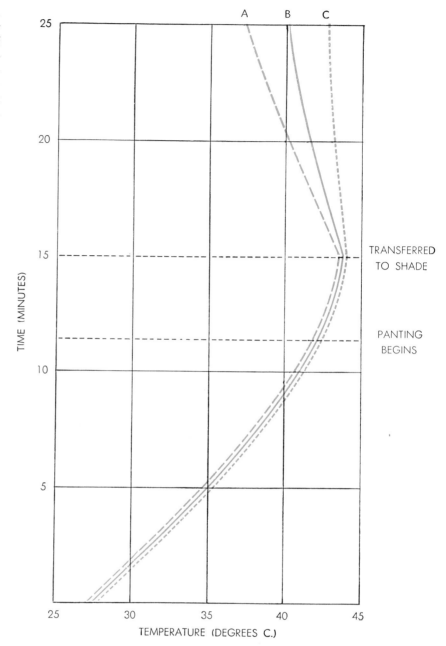

TEMPERATURE CURVES of the three lizards in the experiment depicted on page 2819 were virtually identical as the animals absorbed heat. The smaller animals lightened in color, which aided the absorption of heat and compensated for their smaller ratio of surface to mass. Actually the lines on the chart were nearly superimposed. They were separated by transferring the smallest lizard (A) to the shade, while the next largest lizard (B) had only its head in the sun and the largest lizard (C) had its head and half its body in the sun.

tributes. Unquestionably size is subject to adaptive changes. Whenever we compare adequate samples from various portions of the range of a widely distributed reptile, we nearly always find that the average size of adults in one area is larger than that of adults in another area. Often the trend in size parallels the trend in climate, with the larger animals restricted to warmer areas or to regions with longer growing-seasons. Such correlations are suggestive, but without experimental evidence we cannot be certain that heat and size are directly re-

lated. The length or bulk of a reptile may be governed by such other variables as the food supply or gene combinations.

The problem of evaluating the adaptive significance of size is not simplified by the existence of adults and young of the same species in the same environments at the same time. In the desert-dwelling whip-tailed lizard (*Cnemidophorus tigris*) we found that juveniles of the species, weighing two or three grams, keep their bodies at mean temperatures identical with those of

TEXAS HORNED LIZARD, like the regal horned lizard, has the ability to change the color of its skin. When its body temperature is low, the animal is quite dark (*left*), but the same animal becomes paler (*right*) after it has been exposed to higher temperatures.

adults weighing up to 16 grams. However, we found a difference between adults and juveniles when we recorded the daily variation of their temperatures. The smaller lizards restrict their temperature fluctuations to a narrow zone of five or six degrees, while the activity zones in adults range over 10 or 11 degrees, from 99 to 110 degrees (37.2 to 43.3 degrees C.). The juveniles may be more responsive to heat and so may adjust their exposure more sensitively, or their body temperature may adjust more quickly by virtue of their smaller bulk.

Other things being equal, the temperature of a smaller lizard should rise or fall faster because it has more surface in proportion to its mass. In the first rough experiment we performed to test the validity of this generalization from the physics of inanimate objects, however, we discovered how important "other things" can be if they are not equal. Our subjects were adult specimens of two species of spiny lizard. The larger of the two was a green spiny lizard (*Sceloporus formosus*), a species restricted to open areas in moist forests of broad-leaved trees at elevations above 5,000 feet in central Honduras; its green skin is marked with black pigment sparsely distributed in a reticulated pattern of lines and smaller blotches. The other was a much smaller, slate-colored spiny lizard (*Sceloporus variabilis*) that lives at elevations up to 3,000 feet in the arid valleys below the cloud forests on the mountain summits. In their very different environments the two species keep nearly the same average body temperature.

Though our two specimens had roughly the same bodily proportions, the greenish one, weighing 27.8 grams, had four times the bulk of the other, which weighed only 6.9 grams. The temperature of both lizards was 77 degrees (25 degrees C.) when they were placed in full sunlight with the air temperature at about 90 degrees. Temperatures were recorded at intervals of three minutes. During the first nine minutes the body temperature of the larger lizard lagged less than a degree behind that of the smaller lizard. But after 12 minutes the temperature of the larger lizard rose slightly above that of the smaller. At the conclusion of the experiment, after 18 minutes, the temperature of the larger lizard was 109 degrees (43 degrees C.), and that of the smaller was 108.7 degrees (42.6 degrees C.). If the two lizards had absorbed heat at rates predicated solely on their weight, the heavier should have required approximately 10 more minutes to reach the temperature attained by the smaller animal in 18 minutes. Though inexact, the results of our simple experiment suggest that the cloud-forest lizard is better equipped, figuratively, "to make hay while the sun shines." Because the pigments in its skin absorb heat so rapidly, it can attain its threshold temperature quickly enough and often enough during the year to permit it to forage and fuel itself.

We suspected that the outcome of this experiment may have been influenced by changes in the pigmentation of one or the other reptile in the course of the experiment. To find out how important such changes are in regulating the absorption of heat we performed an experiment with individuals of different weights but belonging to the same species. This time we used regal horned lizards (*Phrynosoma solare*) weighing respectively 12.4, 29.4 and 85.5 grams. The experiment was conducted in

August on a clear day, with no wind, in the foothills of the Chiricahua Mountains in Arizona. The body temperature of each lizard was 80 degrees (27 degrees C.) at the start of the experiment; within 15 minutes their temperatures simultaneously reached 109 degrees (43 degrees C.), with the curves on the recording instrument indicating that they had risen at a virtually identical rate. About halfway through the experiment the temperature of the smallest lizard ran a degree ahead of the others, but shifted back to the curves being plotted for the other two, as though some mechanism were regulating the rate of heat absorption. This proved to be the case. Although the three lizards were not conspicuously different in color at the beginning of the experiment, we could discern distinct differences at the conclusion. The largest lizard was the darkest of the three, and the smallest lizard the palest.

While their broadly flattened bodies are adapted for the rapid absorption of heat, it is apparent that horned lizards are equally well equipped by pigmenta-tion to regulate the rate at which they absorb heat. The black-pigmented cells, or melanophores, of their skin expand laterally when the animal is cold, thus darkening the body and increasing the rate at which it absorbs radiant energy. When the body is warm, the same cells contract, thereby exposing light pigments in adjacent cells that reflect infra-red radiation. To match such efficiency we would need a mechanism that automatically exchanged our dark winter clothing for white linens with the advent of hot weather.

There appears to be an upper limit of bulk beyond which behavioral and physiological adaptations can no longer secure adequate regulation of a reptile's internal body heat. At the Archbold Biological Station in Florida, E. H. Colbert, R. B. Cowles and I conducted a suggestive experiment with a five-foot, 30-pound alligator as the subject. We found that it took the summer sun 7.5 minutes to increase the animal's temperature by two degrees. To see what this signifies for large reptiles, let us con-sider the plight of a 10-ton dinosaur under the same circumstances. This creature would have to bask in a blazing sun for more than an hour to elevate its body temperature to the same extent. Suppose our hypothetical dinosaur were active in daytime, and subject to cooling at night, as it would be in any desert region today. If its temperature dropped even four or five degrees below its threshold for activity, the dinosaur would have to bask for a large part of the following day in order to regain the threshold temperature of activity. The odds favor the deduction that dinosaurs, at least the larger ones, lived under fairly constant environmental tempera-tures. This is an important piece of evi-dence favoring the conclusion that the earth's climate was once quite uniformly tropical, for the distribution of fossils shows that the large dinosaurs roamed the earth far beyond the borders of the modern tropics.

All the truly large reptiles still abroad in the world today—pythons in Asia as long as 33 feet, anacondas in South America as long as 28 feet, monitor liz-

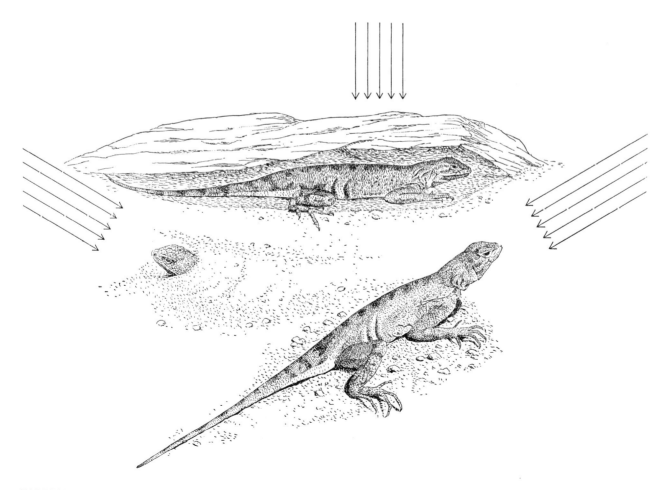

EARLESS LIZARD of the Southwest regulates its temperature within narrow limits by means of its behavior. The morning sun (*arrows at left*) warms the blood in the animal's head while the rest of it remains hidden in the sand until it is warm enough to be active. At noon (*top*) the lizard seeks shelter from the hot sun, but later it emerges and lies parallel to the sun's rays (*bottom*).

ards in Indonesia as long as 10 feet, crocodilians in the Americas as long as 23 feet and tortoises in the Galápagos Islands weighing more than 500 pounds—are residents of the tropics. The only exception is the enormous leatherback turtle *Dermochelys*, which is known to reach and possibly to exceed

1,500 pounds and is perhaps the largest living reptile. But this creature is protected by the constant warmth of its oceanic environment. The occasional specimen that turns up on the coast of Nova Scotia is probably carried there by the Gulf Stream.

In general the tropics afford stable

temperatures that fluctuate in the narrow and comfortable range from 68 to 86 degrees year in and year out. The hottest places on earth are found not in the tropics but in the desert regions of the zones so inappropriately called "temperate." In winter these same deserts may be bitterly cold, and the daily fluctuation in temperature throughout the year far exceeds any encountered in the tropics. The night temperature of a tropical forest may drop only a few degrees below that of the day. Thus none of the large reptiles is ever exposed to either the freezing or the high temperatures encountered by reptiles in the middle-latitude deserts.

The green iguana (*Iguana iguana*), the largest lizard found on the American continent, illustrates the dependence of large reptiles on the constancy of the tropical temperature-environment. This animal, widely distributed in the lowlands of Latin America, reaches a length of at least 5.5 feet and a weight of 13 pounds. In a population we studied on the west coast of Mexico we found that the body temperature of the green iguana fluctuates much more than smaller lizards living in the same area or in the deserts 700 miles to the north. Green iguanas spend virtually all of their time high up in the crowns of large trees, feeding on leaves, buds or fruits during the day, and sleeping at night with only branches or foliage to conceal them from predators and to protect them from heat loss. They evidently do some basking; we found that the temperatures of nearly 50 iguanas were 10 to 15 degrees above the level of the air. But it is doubtful that there has ever been sufficient stress from heat fluctuations in the environment of the green iguana to induce adaptive changes in its behavior.

We had an opportunity to test this supposition by exposing several green iguanas to summer conditions in the desert. Instead of fleeing from the heat of direct sunlight, as any reptile native to the desert would have done, they literally sat in the sun until they died. The increase in their body temperatures was slowed a little by respiratory cooling when their breathing turned to panting, but this merely prolonged their discomfiture before breathing stopped. Green iguanas are equally unprepared for exposure to low temperatures. One specimen at the San Diego Zoo had to be moved indoors overnight because it showed no disposition to seek shelter of its own accord. However, one evening preceding a cold snap the creature chanced to be overlooked. Next morning it was found in a state of cold narcosis,

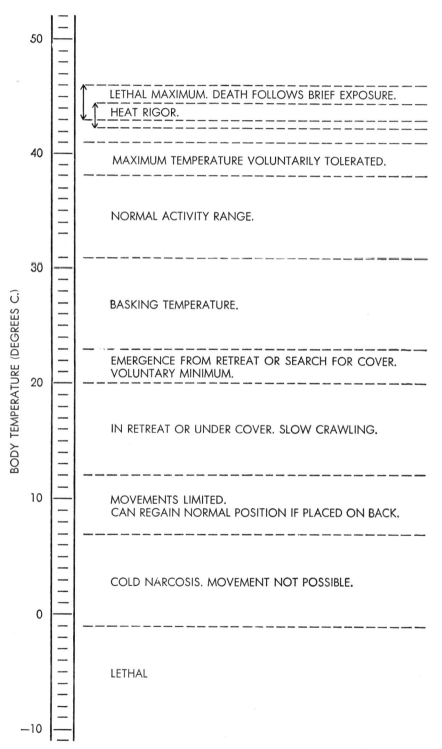

SIGNIFICANCE OF BODY TEMPERATURE to the behavior of reptiles is indicated by this chart giving the approximate temperatures for various activities of spiny lizards. The effects of exposure to heat levels near the extremes depend on duration. Even temperatures near the upper limit for the lizards' normal activity become lethal if exposure is prolonged.

suspended by a single claw accidentally hooked to a twig.

Considering that all reptiles would be just about equally vulnerable to cold and heat were it not for their behavioral adaptations, it is remarkable that so many lizard species are established in the temperate zone. It should be noted that snakes and turtles also have adapted themselves by much the same means to survive in rigorous climates. Snakes in particular have gotten around the surface-to-mass limitation by the lengthening of their bodies, which exposes more surface per unit of mass to absorb solar energy. In Canada snakes outnumber lizards and turtles combined, and in Europe one viper (*Vipera berus*) has penetrated to regions within the Arctic Circle. However, one European lizard (*Lacerta vivipara*) ranges even farther to the north, farther from the Equator than any other living reptile.

In common with the majority of reptiles near the northern limit of their distribution, both the snake and the lizard give birth to living young. Some snakes and lizards transfer nutrients and oxygen from maternal blood to the developing embryo by means of mammal-like placental structures. Most of these live-bearing reptiles, however, incubate the eggs within the body of the female, where behavioral regulation of temperatures keeps the eggs at heat levels within the optimum zone for development. Such modification of the reproductive pattern, peculiar to lizards and snakes, gives them a leeway in dispersal not open to turtles and crocodilians. Snakes and lizards are accordingly the most widely distributed reptiles, with lizards having a slight edge in number of species. Of the two groups it is the lizard that best exemplifies the complexity of adaptations involving the coordination of structure, physiology and behavior in response to the interplay of selective forces in the environment.

Perhaps the most amazing behavioral adaptation is that of the earless lizard, which is almost never found abroad with its body temperature below 96 degrees. We exposed its secret only by observing what it did in a laboratory cage provided with sources of radiant heat. From its overnight retreat, submerged in the sand, the earless lizard first thrusts its inconspicuous head above the surface; there it waits until the blood coursing through a large sinus in its head has absorbed enough heat from the sunlight to raise the temperature of its entire body. When its temperature is well above the threshold for efficient activity, this wary reptile emerges from the sand, preheated and ready to take off at top speed.

The Author

CHARLES M. BOGERT is chairman and curator of the Department of Amphibians and Reptiles at the American Museum of Natural History in New York. He was born on a ranch at Mesa, Colo. "The reptiles my elder brother—a trapper —brought in, particularly such rarities as a gaudily banded red, yellowish and black king snake (*Lampropeltis doliata*) left a lasting impression on me," he reports. "There were eight offspring in our family when the mortgage was foreclosed in 1913. Our departure from Mesa must have decreased the population by at least 8 per cent. We moved to Los Angeles and I took a few horned lizards with me on the train, hidden in my luggage. By the time I was in high school I had managed to trade a rawhide rope for a Model-T Ford. This enabled me to make trips to the Mojave desert, collecting tortoises, lizards and snakes, which I kept under observation in a large cage in the back yard." Bogert did undergraduate and graduate work at the University of California at Los Angeles, where he began a long collaboration with Raymond B. Cowles on thermoregulation in reptiles.

Bibliography

A PRELIMINARY STUDY OF THE THERMAL REQUIREMENTS OF DESERT REPTILES. Raymond Bridgman Cowles and Charles Mitchill Bogert in *Bulletin of the American Museum of Natural History*, Vol. 83, Article 5, pages 265-296; 1944.

RATTLESNAKES: THEIR HABITS, LIFE HISTORIES, AND INFLUENCE ON MANKIND. Laurence M. Klauber. University of California Press, 1956.

THERMOREGULATION IN REPTILES—A FACTOR IN EVOLUTION. Charles M. Bogert in *Evolution*, Vol. 3, No. 3, pages 195-211; September, 1949.

TEMPERATURE TOLERANCES IN THE AMERICAN ALLIGATOR AND THEIR BEARING ON THE HABITS, EVOLUTION AND EXTINCTION OF THE DINOSAURS. Edwin H. Colbert, Raymond B. Cowles and Charles M. Bogert in *Bulletin of the American Museum of Natural History*, Vol. 86, Article 7, pages 331-373; 1946.

SCIENTIFIC AMERICAN November 1961, Vol. 204, No. 5, pp. 107-116

OFFPRINT **1120**

DESERT GROUND SQUIRRELS

by George A. Bartholomew and Jack W. Hudson

Two little animals of the Mohave Desert have evolved remarkable
adaptations to heat and aridity. Each has adapted in its own way,
which apparently enables them to live together without competing.

Among the handful of animals that inhabit the hot, dry and sparsely vegetated Mohave Desert of California are two species of ground squirrel: the antelope ground squirrel and the mohave ground squirrel. Both species live in burrows, both are active and aboveground during the day and both feed on the small amount of plant life that is available.

This is an uncommon situation in nature. Species as closely related as these two, and as much alike in their food and habitat requirements, seldom live together even in more favorable environments. In his *Origin of Species* Charles Darwin suggested the reason. "As the species of the same genus usually have ...much similarity in habits and constitution, and always in structure," he wrote, "the struggle will generally be more severe between them, if they come into competition with each other, than between the species of distinct genera." Implicit in Darwin's statement is a concept now fundamental to biology. It is known as the principle of competitive

exclusion and it says, in brief, that two noninterbreeding populations that stand in precisely the same relationship to their environment cannot occupy the same territory indefinitely. They cannot, in other words, live in "sympatry" forever. Sooner or later one will displace the other.

Such a displacement could be under way in the Mohave Desert right now. The mohave ground squirrel may well be a species in the process of extinction. Not only does it have a smaller total population and a narrower geographical distribution than the antelope ground squirrel (which is one of the commonest ground squirrels of the southwestern U.S.); it also appears to be less numerous in the small section of the Mohave Desert to which it is restricted. But since no historical information is available on the population trends of the two animals, there is no way of knowing exactly what the present difference in their number portends.

In any case the mohave ground squirrel is not as yet extinct. This raises a

number of intriguing questions. Do the two species have the same way of life and the same relationship to their environment? If so, the competition between them must be severe. Or are there differences in their adaptation to their common environment? If there are, do these differences reduce competition between them sufficiently to permit them, at least temporarily, a period of peaceful coexistence?

These questions become more intriguing when one considers the nature of the desert environment. Aridity and heat make particularly severe demands on animals, and animals that live in deserts must be equipped with special physiological and behavioral adaptations to meet these demands. The camel, for example, withstands aridity because it can tolerate a high degree of dehydration, can restore its body fluids quickly and can travel long distances in search of water and succulent vegetation. It withstands heat through its tolerance of a wide range of body temperatures and

ANTELOPE GROUND SQUIRREL is found in the Mohave Desert and throughout the southwestern U.S. It is active during the day all year round, in spite of extremes of heat and aridity. Both of these photographs were made in the laboratory by Jack W. Hudson.

MOHAVE GROUND SQUIRREL is found only in one corner of the Mohave Desert. It is active during the day from March to August but remains in its burrow the rest of the year. Before retiring underground it becomes very fat, as this photograph shows.

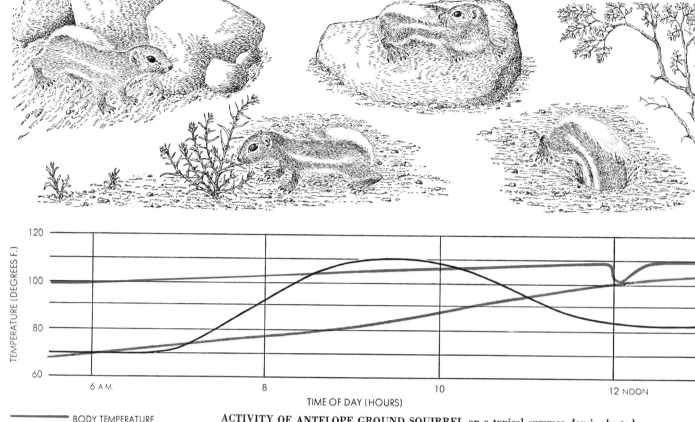

BODY TEMPERATURE
ENVIRONMENTAL TEMPERATURE
LEVEL OF ACTIVITY

ACTIVITY OF ANTELOPE GROUND SQUIRREL on a typical summer day is charted. On emerging from its burrow animal runs to feeding area. Then it suns and grooms itself. When its body temperature rises too high, it goes to a special retreat burrow to cool

through the insulating qualities of its coat [see "The Physiology of the Camel," by Knut Schmidt-Nielsen; SCIENTIFIC AMERICAN Offprint 1096]. The desert rat of the U.S. Southwest has adapted equally well but in quite different ways. To combat aridity it conserves its body water; the desert rat's kidney is so efficient that it uses only about a fourth of the amount of water that the human kidney requires to excrete the same amount of urea. This adaptation enables the animal to meet a substantial fraction of

its water needs by the oxidation of foodstuff, as opposed to drinking. The desert rat deals with heat by avoiding it: the animal remains in its burrow during the daylight hours, emerging only at night, when the air and soil are cool [see "The Desert Rat," by Knut and Bodil Schmidt-Nielsen; SCIENTIFIC AMERICAN Offprint 1050].

Like the desert rat, most small, burrowing desert rodents are nocturnal. But both the mohave ground squirrel and the antelope ground squirrel are diurnal.

They emerge from their burrows near sunrise and forage outdoors throughout the day. They do so even in summer, when the air temperature may reach 110 degrees Fahrenheit or higher, and when the surface temperature of the soil may rise above 150 degrees F. Since the desert is as arid as it is hot, they must sustain their exposure to heat with a minimum loss of water for evaporative cooling.

In appearance and temperament the antelope ground squirrel resembles the

RETREAT BURROW OF ANTELOPE GROUND SQUIRREL is shown at right in this drawing. It is usually dug in soft soil close

to desert vegetation and is about one foot deep and 12 to 15 feet long. The animal seems to use this burrow to unload body heat

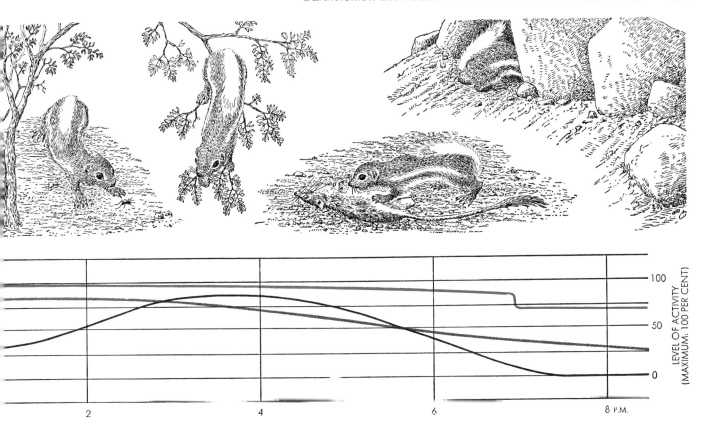

off. In early afternoon it stays in the shade. Before retiring to its home burrow it returns to feeding area, and at any time may catch insects or feed on dead animals. On the graph, dip in body tem-

perature is shown only at noon. But dips occur often, whenever animal goes underground to unload heat. At all other times its temperature is a few degrees above the environmental temperature.

chipmunk. Its body is about six inches long; its weight is about 90 grams. It has two white stripes down its grayish-brown back. It carries its tail high, exposing a white rump; this suggests the appearance of the pronghorn antelope, for which it is named. An extraordinarily active and high-strung animal, the antelope ground squirrel is constantly in motion, dashing from place to place, often traveling hundreds of feet from its home burrow. That it can maintain such hyperactivity even in soaring tempera-

tures is in itself evidence of unusual adaptive mechanisms.

For every animal the ability to adapt to external temperatures depends on two internal factors: the range of body temperatures in which it can function effectively and the rate at which it can produce body heat. Below a lower critical environmental temperature the body loses so much heat that internal temperature can be maintained only if the animal can step up its production of body heat sufficiently. Above an upper critical

environmental temperature the body retains so much metabolic heat that internal temperature can be held within the required range only if the animal can get rid of heat, in most cases by evaporative cooling; that is, by sweating or panting at a sufficient rate. Between the upper and lower critical temperatures—in the thermal neutral zone—an animal can maintain its optimum body temperature without having either to increase its metabolic rate or to lose body water. Such stratagems as contracting or dilat-

and to store food, but not as a living place; the dens in the burrow contain neither nests nor fecal matter. The antelope ground squir-

rel's living burrows have not been excavated. They are probably dug under rocky buttes, like that seen at left side of drawing.

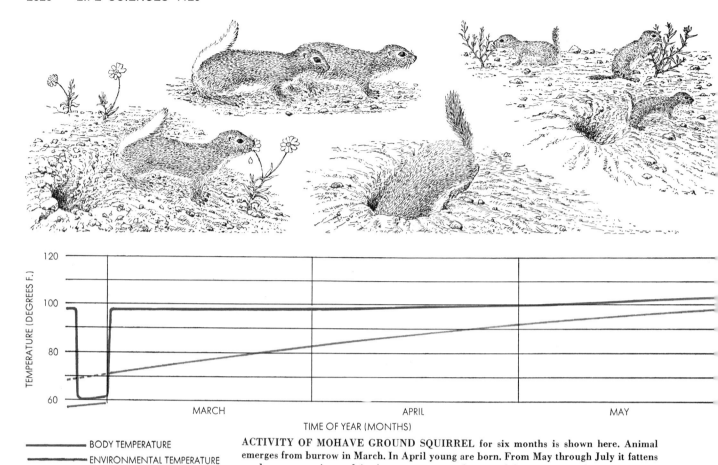

BODY TEMPERATURE
ENVIRONMENTAL TEMPERATURE

ACTIVITY OF MOHAVE GROUND SQUIRREL for six months is shown here. Animal emerges from burrow in March. In April young are born. From May through July it fattens on desert vegetation and in August returns underground for seven months. Broken line

ing cutaneous blood vessels and depressing or raising hair or feathers allow the animal to function at a minimum cost of energy for temperature maintenance.

In terms of this analysis of the adjustment of body temperature to environmental temperature, the adaptation of the antelope ground squirrel is admirable. It has a broad thermal neutral zone and one that accommodates to high environmental temperatures. Between environmental temperatures of 90 and

107 degrees F. its metabolic rate remains virtually constant. No other non-sweating mammal has a thermal neutral zone extending so high [*see illustration on page 2830*].

Unlike man, the antelope ground squirrel can tolerate a high body temperature; in other words, it can "run a fever" without debility. It can therefore permit its temperature to rise with the temperature of the environment. Like the camel, it can store heat, and it does not have to dispose of heat until its body

temperature reaches an extreme point. The antelope ground squirrel shows no serious discomfort even when its body temperature goes above 110 degrees. Throughout the thermal neutral zone it runs a temperature a few degrees above that of the environment [*see top illustration on pages 2826 and 2827*]. Instead of expending energy to cool itself and thereby adding to its heat load—as man must do—this animal actually disposes of a portion of its metabolic heat to the lower-temperature environment by

LIVING BURROW OF MOHAVE GROUND SQUIRREL is seen here at various stages and times of year. Burrow is dug in soft sand near the desert plants the animal eats. It is about 18 feet long and three feet deep. First panel shows burrow in early spring,

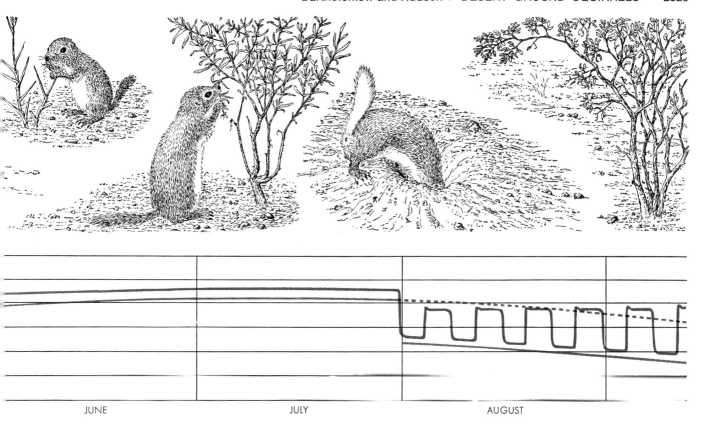

on graph is desert air temperature when animal is in burrow. Gray line under it is burrow temperature then. Rise and fall of animal's temperature in August corresponds to its periods of wakefulness and torpor. Body temperature is always higher than environmental temperature, although when animal is active its temperature fluctuates sharply. A mean body temperature is shown here.

conduction, convection and radiation.

But environmental temperatures in the desert are commonly far higher than tissues can tolerate, and small animals heat up rapidly. The antelope ground squirrel must therefore unload some of its accumulated body heat at intervals during the day. It does this either by flattening itself against the soil in a shaded area or by retreating underground to its burrow. When its body temperature gets dangerously high, it has only to return to the relative coolness of its burrow and remain quiet for a few minutes until its fever has subsided. In our laboratory at the University of California at Los Angeles, antelope ground squirrels have lowered their body temperature from above 107 degrees to about 100 degrees within three minutes after being transferred from an environmental temperature of 104 degrees to one of 77 degrees.

The antelope ground squirrel contends with heat in still another way. Under protracted heat stress it will begin to drool. The animal then systematically spreads the saliva over its cheeks and head with its forepaws as though it were grooming itself. On very hot days, when it has had to tolerate air temperatures of 104 degrees or more for several hours, the antelope ground squirrel may be soaking wet around the head.

Drooling, with its high cost in water losses, is a last resort. But even when the temperature is not extreme, the antelope ground squirrel loses a considerable amount of body water. At 100

when animal emerges. Second shows animal digging new burrow. Third shows it closing burrow in August before retiring underground. In last panel it is winter and animal is torpid. Periods of torpor probably last longer in winter months than in summer.

degrees, long before it has begun to drool, this hyperactive animal gives up water equal to 10 per cent of its body weight in respiration and evaporation through its skin in the course of a day. This is 15 per cent of its total body water. Fortunately the animal withstands dehydration well. Although it gives up three times more water every day than it can extract from its food by oxidation, it can survive from three to five weeks on a completely dry diet. If the antelope ground squirrel is to maintain itself in a healthy state, however, it must find sources of preformed water. It is therefore hardly surprising that the animal is omnivorous, eating insects as well as desert vegetation. When it is seen on the highways, as it often is, it is probably feeding on the corpse of some animal, perhaps another of its species that has been hit by a car.

The antelope ground squirrel is able to stretch its scanty water supply because, like the desert rat, it loses a minimal amount of water in the excretion of nitrogenous wastes. On a dry diet this animal can produce urine with a mean concentration of 3,700 milliosmols (the maximum concentration of human urine is about 1,300 milliosmols). The urine of the desert rat is somewhat more concentrated. But the antelope ground squirrel's urine is still 10 times more concentrated than its body fluids. Its ability to turn salty water to physiological use is even more impressive. The desert rat can maintain itself on sea water; the antelope ground squirrel can drink water approximately 1.4 times saltier than sea water and still remain in good health. No other mammal can process water of such high salinity. This capacity is important in the desert, where the little surface water that is available is usually highly mineralized.

The structure of the animal's kidney explains its efficient use of water. As in several other desert mammals, the renal papilla—that part of the kidney which contains the ascending and descending kidney tubules—of the antelope ground squirrel is extremely large, extending as far down as the ureter [see illustration on page 2831]. In urine formation the kidney first extracts a filtrate containing all the constituents of blood except proteins and blood cells. This filtrate is then converted to urine by the selective reabsorption of water and essential solutes in the kidney tubules. The longer the tubules, the greater the amount of water they can absorb and the greater the amount the body retains. As the antelope ground squirrel's tolerance for high body temperatures constitutes its major physiological adaptation to heat, so the efficiency of its kidney embodies its major adaptation to aridity.

Considering the success with which the antelope ground squirrel occupies its narrow desert niche, how does the mohave ground squirrel manage to find a place beside it? The question cannot be fully answered, because less is known about the life history of the mohave ground squirrel. This in itself is significant, because it appears that the mohave ground squirrel manages to persist largely by staying out of the way of the antelope ground squirrel.

Of the two animals the mohave ground squirrel is the bigger and fatter, and it has the temperament that goes with its more generous proportions. Its body length is about six and a half

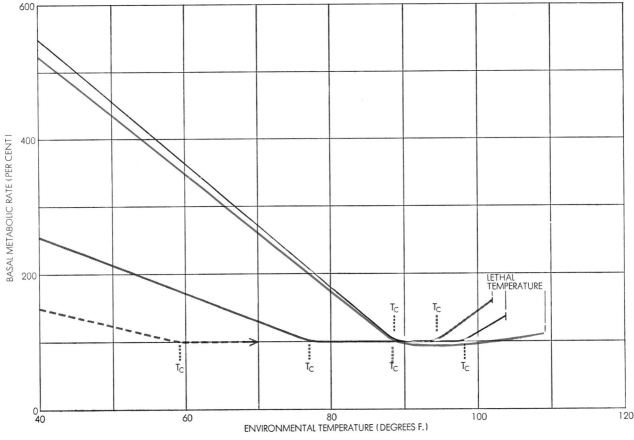

THERMAL NEUTRAL ZONE is range of external temperatures to which an animal is best adjusted. In it body temperature and metabolism can be held at optimum levels. At critical temperatures (T_C) metabolic rate begins to rise. Lemming is an arctic rodent. Kangaroo rat and ground squirrels are desert dwellers.

inches; its weight in its natural habitat, about 150 grams. In temperament it is placid, docile and sedentary. This little brown animal seldom wanders far from its home burrow, which it digs in loose sand, generally in the shade of the desert plants that provide it with its food. For more than half the year, from August to March, it remains in its burrow. During the spring months, when the desert vegetation is at its annual peak, the mohave ground squirrel emerges to reproduce and to fatten itself in preparation for its return underground.

The major proportion of the mohave ground squirrel's life is therefore normally concealed from observation. Fortunately it seems to show a comparable pattern of behavior in the laboratory. Here, as in its natural habitat, the animal is active throughout the day from March to August. During the remainder of the year, however, even at room temperature, and in spite of the continuous availability of food and water, it is intermittently torpid for periods lasting from several hours to several days. If food and water are at hand, it will eat and drink in its periods of wakefulness. If they are not, it does not seem to be disturbed. We do not know whether or not in its natural habitat it stores food in its burrow. We do know that it is usually thin in early spring, when it emerges from its burrow, and that it can add as much as 100 grams to its body weight in the period before its retirement underground. We also know that in the laboratory, where the animal becomes exceedingly fat, it loses an appreciable amount of weight during its period of dormancy only if no food and water have been made available to it.

Since this pattern of intermittent dormancy extends from late summer to early spring, it involves what would normally be considered two separate processes— hibernation and the summer dormancy called estivation. Our studies indicate, however, that in the mohave ground squirrel the two processes are merely aspects of the same physiological phenomenon. From early August to the end of February, whether the temperature in the laboratory is one that would normally be associated with estivation or whether it is one at which hibernation would be expected to occur, the same events take place. As the animal becomes torpid, its oxygen consumption and its body temperature drop sharply. Then both level off, and the body temperature stabilizes at the environmental temperature or very slightly above it. During the time the animal is dormant its torpor is more pronounced than

deep sleep, its breathing is suspended for long periods and its heart rate is profoundly reduced. On arousal it restores its body temperature to normal through increases in breathing movements, acceleration of heartbeat, shivering (which releases heat) and increased oxygen consumption. In the laboratory arousal may come about spontaneously or it may be induced by a touch or a sound. In either event it is extremely rapid. Although the animal can take as long as six hours to enter torpor, it can wake in less than one hour. Oxygen consumption can reach its peak in 15 to 20 minutes, and body temperature can rise from 68 to 86 degrees in 20 to 35 minutes.

Such rapid alterations in temperature do not occur during the five-month period in which the mohave ground squirrel is active. Even then, however, its body temperature is remarkably variable and fluctuates over a broad range. We have measured a deep-body temper-

ature as low as 88 degrees in individual animals engaged in normal activity, and yet the animal does not seem to suffer any ill effects from body temperatures as high as 107 degrees. Its thermal neutral zone does not, however, extend as high as that of the antelope ground squirrel. The metabolism of the mohave ground squirrel begins to rise at an environmental temperature of about 98 degrees. Its tolerance for high body temperatures is of major adaptive value in June and July, when the desert is particularly hot.

Obviously the mohave ground squirrel's dormancy serves the function traditionally associated with hibernation: it conserves energy. At an environmental temperature of 68 degrees the oxygen consumption of a dormant mohave ground squirrel is only about a tenth that of the same animal active at the same temperature. Fat is undoubtedly the major energy source. Since the oxidation

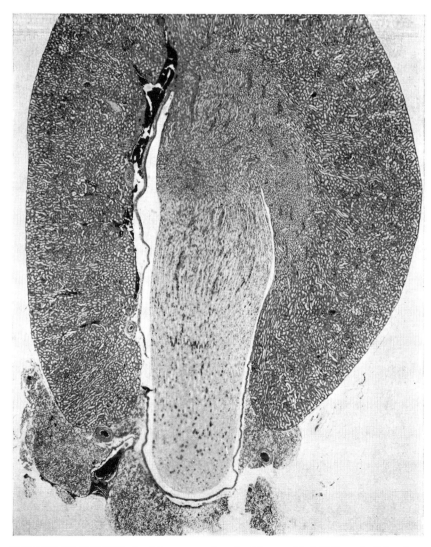

SECTION OF KIDNEY of antelope ground squirrel is magnified 20 times in this micrograph by Hudson. Mass outlined in white is papilla, containing renal tubules. In this species it is very large. This permits reabsorption of much water and production of concentrated urine.

of one gram of fat requires two liters of oxygen, a simple calculation shows that a torpid squirrel, weighing 300 grams and consuming oxygen at a rate of .08 cubic centimeter per gram per hour, will burn .29 gram of fat a day. Some 50 grams of fat would therefore supply it with all its energy requirements for 172 days; this is just half the fat supply it usually accumulates in the active months of the year. According to this calculation a mohave ground squirrel should be able to remain torpid for a whole year if it did not arouse at all. We do not know, of course, how much its energy requirements are increased by periods of arousal during the months of dormancy. Not being able to observe the animals in their burrows, we do not know how long these periods last. But from laboratory evidence we deduce that the cycles of torpor and wakefulness are repeated every week, with three to five days spent in torpor. The mohave ground squirrel should therefore be able to get along on its accumulated fat.

Energy conservation is not the only function that dormancy serves. Like the antelope ground squirrel, the mohave ground squirrel loses considerable body water in evaporative cooling and, like the antelope ground squirrel, it must have preformed water in its food. The sedentary mohave ground squirrel does not go in search of water; it gets its water almost entirely from the desert plants it eats. During the period the animal is underground this vegetation is in decline and the desert is at its driest. Thus dormancy is an important adaptation to seasonal aridity.

The mohave ground squirrel's seven months of estivation and hibernation serves still another function: it minimizes competition with the more active and abundant antelope ground squirrel. During the time the mohave ground squirrel is aboveground the food and water available are probably adequate to sustain both animals. But during its months of dormancy both food and water are in short supply.

The two animals have adapted well to desert life—the antelope ground squirrel by its tolerance for high body temperatures and the efficiency of its kidney, and the mohave ground squirrel by its avoidance underground of the most rigorous months of the year. One may conclude that these adaptations are sufficiently different to permit the two animals to live in sympatry in spite of Darwin's stern injunction.

The Authors

GEORGE A. BARTHOLOMEW and JACK W. HUDSON are colleagues in the department of zoology at the University of California at Los Angeles. Bartholomew received his A.B. and M.A. degrees from the University of California and, after wartime service as a scientist with the U.S. Navy, obtained his Ph.D. in 1947 at Harvard University. He then went to U.C.L.A., where he is professor of zoology. Bartholomew is now in Australia on a Guggenheim Fellowship and a National Science Foundation grant to study the physiology of marsupials and reptiles in Queensland. Hudson was graduated from Occidental College in 1948 and received his M.A. there the next year. While teaching biology at Occidental he did doctoral research under Bartholomew at U.C.L.A., receiving his Ph.D. in 1960. He is now a lecturer in zoology there.

Bibliography

THE COMPETITIVE EXCLUSION PRINCIPLE. Garrett Hardin in *Science*, Vol. 131, No. 3409, pages 1292-1297; April 29, 1960.

EFFECTS OF SODIUM CHLORIDE ON WEIGHT AND DRINKING IN THE ANTELOPE GROUND SQUIRREL. George A. Bartholomew and Jack W. Hudson in *Journal of Mammalogy*, Vol. 40, No. 3, pages 354–360; August 20, 1959.

HEAT REGULATION IN SOME ARCTIC AND TROPICAL MAMMALS AND BIRDS. P. F. Scholander, Raymond Hock, Vladimir Walters, Fred Johnson and Laurence Irving in *The Biological Bulletin*, Vol. 99, No. 2, pages 237–258; October, 1950.

HIBERNATION. Charles P. Lyman and Paul O. Chatfield in *Scientific American*, Vol. 183, No. 6, pages 18–21; December, 1950.

WATER METABOLISM OF DESERT MAMMALS. Knut Schmidt-Nielsen and Bodil Schmidt-Nielsen in *Physiological Reviews*, Vol. 32, No. 2, pages 135–166; April, 1952.